NONEQUILIBRIUM STATISTICAL THERMODYNAMICS

STUDIES IN SOVIET SCIENCE

PHYSICAL SCIENCES

1973

Densification of Metal Powders during Sintering
V. A. Ivensen

The Transuranium Elements
V. I. Goldanskii and S. M. Polikanov

Gas-Chromatographic Analysis of Trace Impurities
V. G. Berezkin and V. S. Tatarinskii

A Configurational Model of Matter
G. V. Samsonov, I. F. Pryadko, and L. F. Pryadko

Complex Thermodynamic Systems
V. V. Sychev

Crystallization Processes under Hydrothermal Conditions
A. N. Lobachev

Migration of Macroscopic Inclusions in Solids
Ya. E. Geguzin and M. A. Krivoglaz

1974

Theory of Plasma Instabilities
Volume 1: Instabilities of a Homogeneous Plasma
A. B. Mikhailovskii

Theory of Plasma Instabilities
Volume 2: Instabilities of an Inhomogeneous Plasma
A. B. Mikhailovskii

Nonequilibrium Statistical Thermodynamics
D. N. Zubarev

STUDIES IN SOVIET SCIENCE

NONEQUILIBRIUM STATISTICAL THERMODYNAMICS

D. N. Zubarev
V. A. Steklov Mathematical Institute
Academy of Sciences of the USSR
Moscow, USSR

Translated from Russian by

P. J. Shepherd
Department of Physics
University of Exeter
Exeter, England

Edited by

P. Gray
School of Mathematics and Physics
University of East Anglia
Norwich, England

and

P. J. Shepherd

CONSULTANTS BUREAU • NEW YORK-LONDON

Library of Congress Cataloging in Publication Data

Zubarev, Dmitriĭ Nikolaevich.
Nonequilibrium statistical thermodynamics.

(Studies in Soviet science)
Translation of Neravnovesnaía statisticheskaía termodinamika.
Bibliography: p.
1. Statistical thermodynamics. I. Title. II. Series.
QC311.5.Z813 536'.7'0182 73-83904
ISBN 0-306-10895-X

The original Russian text, published by Nauka Press in Moscow in 1971, has been corrected by the author for the present edition. This translation is published under an agreement with the Copyright Agency of the USSR (VAAP).

Д. Н. Зубарев

NERAVNOVESNAYA STATISTICHESKAYA TERMODINAMIKA
D. N. Zubarev

A Division of Plenum Publishing Corporation
227 West 17th Street, New York, N.Y. 10011

United Kingdom edition published by Consultants Bureau, London
A Division of Plenum Publishing Company, Ltd.
4a Lower John Street, London W1R 3PD, England

Printed in the United States of America

Preface

In this book an attempt is made to give a unified account of the present state of nonequilibrium statistical thermodynamics as a natural generalization of the equilibrium theory.

From a logical point of view, it would be desirable to develop the statistical theory of nonequilibrium processes first, and treat the theory of statistical equilibrium as its limiting case. Such an approach, however, is scarcely worthwhile at the present time, since nonequilibrium and equilibrium statistical thermodynamics are at very different stages of development. In Chapters I and II, therefore, we give a brief account of the basic ideas of the classical and quantum statistical mechanics of equilibrium systems, to the extent that this is necessary for the derivation of the basic thermodynamic relations for the case of statistical equilibrium.

The purpose of these introductory chapters is to recall the general method of Gibbsian statistical ensembles, since later, in Chapters III and IV, attempts are made to take over the ideas of statistical ensembles to nonequilibrium statistical thermodynamics.

A separate chapter is devoted to classical statistical mechanics, although it would be possible to develop it as the limiting case of quantum statistical mechanics when quantum corrections can be neglected. However, we shall not follow this route, since classical statistical mechanics is of interest in itself and is perfectly adequate for many problems. The methods of classical and quantum statistics have much in common as regards the fundamental formulation of the problem. They come up against very similar difficulties when attempts are made to justify them. The passage from quantum statistics to the classical limit is treated later, at the end of Chapter II.

In Chapter III we turn to the study of nonequilibrium processes and investigate the response of statistical ensembles to external perturbations of the mechanical type. By this, we mean perturbations arising as the result of switching on an external field, when the perturbation energy can be represented in the form of an additional term in the Hamiltonian. For the initial condition, we use the state of statistical equilibrium. An account is given of fluctuation-dissipation theorems, dispersion relations and sum rules, and of their applications, in particular to a system of charged particles.

Chapter IV is devoted to thermal perturbations which, generally speaking, cannot be represented uniquely by means of some perturbation energy; for example, we consider perturbations induced by spatial and temporal variation of temperature, pressure or concentration of particles. This case requires a more explicit construction of statistical ensembles than does the case of mechanical perturbations.

Using the idea of "quasi-integrals of motion" to simplify the description of a system, we construct a nonequilibrium statistical operator, which is then applied to different problems: to the derivation of a system of equations for transport of energy, momentum and number of particles in a many-component system, and to the derivation of relaxation equations, kinetic equations and equations of the Kramers–Fokker–Planck type. It is shown that this nonequilibrium statistical operator can be obtained from the extremum of the information entropy, when the fixed quantities defining the nonequilibrium state are given not only at a given moment of time but at all past times. This chapter is largely based on the researches of the author.

It is assumed that the reader is familiar with the fundamentals of quantum and classical equilibrium statistical mechanics, as given in the usual university courses.

The manuscript of the book was read by V. A. Moskalenko, Yu. L. Klimontovich, V. P. Kalashnikov, A. E. Marinchuk, L. A. Pokrovskii, A. G. Bashkirov, G. O. Balabanyan, M. V. Sergeev,

S. V. Tishchenko, and M. Yu. Novikov; the author is grateful to these for advice and comments.

The author is also deeply indebted to Academician N. N. Bogolyubov for fruitful discussion of various problems in the theory of nonequilibrium processes.

D. Zubarev

Introduction

Nonequilibrium statistical thermodynamics forms the theoretical basis for nonequilibrium thermodynamics [1], just as ordinary statistical thermodynamics forms the basis for equilibrium thermodynamics. It studies processes of transport of energy, momentum, and particles in different physical systems (gases, liquids, solids) on the basis of statistical mechanics. Its problem is to derive the equations of nonequilibrium thermodynamics by the methods of statistical mechanics "from first principles" (insofar as this is possible), i.e., from the equations of quantum or classical mechanics, to find expressions for the kinetic coefficients in terms of microscopic properties, to substantiate their symmetry properties and to prove the fluctuation-dissipation theorems.

The most highly developed method in the theory of irreversible processes is the method of the kinetic equation for the distribution function, proposed by Boltzmann and substantiated and further developed by Bogolyubov [2], Kirkwood [3], Born and Green [4], Van Hove [5], and others [6, 7]. This method enables us to derive the equations of nonequilibrium thermodynamics and to calculate the kinetic coefficients explicitly; in practice this method is very important, but it is applicable only to gases of sufficiently low density or with sufficiently weak interaction between the particles. Therefore, the problem of deriving the equations of irreversible thermodynamics from statistical mechanics for more general systems arises.

The usual linear phenomenological nonequilibrium thermodynamics is applicable to any system provided that the system is in a weakly nonequilibrium state, i.e., close to the state of complete statistical equilibrium. It should be noted that this is not a consistently macroscopic theory. Along with the axiomatic thermodynamic method, it essentially uses arguments on the microscopic level, namely the fact that the particles obey the equations of motion of mechanics. As an example, one can derive Onsager's reciprocity relations from the invariance of the equations of motion with respect to time reversal. Here, however, only the existence of the equations of motion is used, and not their concrete form, which is related to the form of the Hamiltonian. Nonequilibrium statistical thermodynamics, goes further in this direction, starting from the very beginning from the description of the system by a definite Hamiltonian and using the equations of motion explicitly.

Nonequilibrium statistical thermodynamics is a further development of the equilibrium theory; however, whereas the latter is a well-developed theory, of which the foundations had been laid as long ago as the beginning of the century by Gibbs [8], nonequilibrium statistical thermodynamics is still in the process of development and is far from being completed.

Until recently, there was a widely held opinion, still shared by many, that the theory of irreversible processes cannot have its own single universal method such as the Gibbsian method, applicable to any system, and that it permits an exact formulation of a problem only in the limiting cases of systems for which it is possible to construct a kinetic equation.

The development of the theory of irreversible processes in the last ten years (see the reviews [9-13]) demonstrate that considerable strides have been made toward constructing a statistical thermodynamics of irreversible processes for arbitrary systems, and the theory is already beginning to perfect its method. This gives support to the idea of Callen and Welton which they expressed in 1951 in a paper on the general theory of fluctuations and generalized noise [14]: "It is felt that the relationship between equilibrium fluctuations and irreversibility which is here developed provides a method for a general approach to a theory of irreversibility, using statistical ensemble methods." In this book we make an attempt to give a preliminary summing up of the results which have been achieved along this path.

We shall study nonequilibrium processes in thermodynamic macroscopic systems, e.g., in gases, liquids, and solids, by the methods of statistical mechanics. Therefore, we shall assume throughout that the system being studied consists of a large number of particles and obeys the laws of quantum (or classical) mechanics with a known Hamiltonian. (For an account of equilibrium statistical mechanics, see, e.g., [15-19].)

The method of the nonequilibrium statistical operator, on which Chapter IV of this book is based, was proposed in papers by the author [20]. This method was further developed jointly with V. P. Kalashnikov [21], L. L. Buishvili [22], and A. G. Bashkirov [23]. A number of important applications of the method have been made by L. A. Pokrovskii [24].

Contents

Chapter I

Equilibrium Statistical Thermodynamics of Classical Systems

The statistical thermodynamics of both equilibrium and nonequilibrium processes starts from the equations of mechanics (quantum or classical) for the aggregate of particles composing the system. Integration of this system of equations is practically impossible because of the extremely large number of variables, but even if this were possible, we should nevertheless be unable to establish the initial conditions for such a large number of equations, as this lies far beyond the bounds of experimental possibility. Therefore, to study such systems, we use the methods of statistical mechanics, based on the introduction of distribution functions in classical statistical mechanics or of statistical operators in quantum statistical mechanics.

In this chapter, we shall examine the basic concepts of the classical statistical mechanics of equilibrium systems, i.e., the method of Gibbsian statistical ensembles for systems of particles obeying classical mechanics.

§ 1. Distribution Functions

1.1. Distribution Functions of Systems of Interacting Particles

We shall consider a system of N identical interacting particles, enclosed in a finite but macroscopically large volume V. For simplicity we shall assume that the particles do not possess internal degrees of freedom.

In the case of classical mechanics, the dynamical state of each particle is defined by giving its coordinate $\mathbf{q}$ and momentum $\mathbf{p}$, where $\mathbf{q}$ and $\mathbf{p}$ denote the aggregate of the three Cartesian coordinates and momentum components, q^α, p^α $(\alpha = 1, 2, 3)$, while the state of the whole system is defined by giving the aggregate of the coordinates $\mathbf{q}_1, \ldots, \mathbf{q}_N$ and momenta $\mathbf{p}_1, \ldots, \mathbf{p}_N$ of all the particles, or by giving the point $(\mathbf{p}_1, \ldots, \mathbf{p}_N; \mathbf{q}_1, \ldots, \mathbf{q}_N)$ in the 6N-dimensional space.

The dynamical evolution of the system is determined by Hamilton's equations

$$\frac{d\boldsymbol{q}_k}{dt} = \frac{\partial H}{\partial \boldsymbol{p}_k}, \qquad \frac{d\boldsymbol{p}_k}{dt} = -\frac{\partial H}{\partial \boldsymbol{q}_k} \qquad (k = 1, 2, \ldots, N), \tag{1.1}$$

where

$$H = H\ (\boldsymbol{p}_1, \ldots, \boldsymbol{p}_N; \boldsymbol{q}_1, \ldots, \boldsymbol{q}_N; t)$$

is the total Hamiltonian (or Hamiltonian function) of the system, which we assume to be known. For example, for a system of N particles with centrally symmetric pair interaction described by a potential $\phi(|\mathbf{q}_i - \mathbf{q}_k|)$, the Hamiltonian has the form

$$H = \sum_k \frac{\boldsymbol{p}_k^2}{2m} + \frac{1}{2} \sum_{i \neq k} \phi(|\boldsymbol{q}_i - \boldsymbol{q}_k|). \tag{1.2}$$

The corresponding equations of motion are

$$\dot{\boldsymbol{q}}_k = \frac{\boldsymbol{p}_k}{m}, \quad \dot{\boldsymbol{p}}_k = -\sum_{i \neq k} \frac{\partial \phi(|\boldsymbol{q}_i - \boldsymbol{q}_k|)}{\partial \boldsymbol{q}_k} = \boldsymbol{F}_k \quad (k = 1, 2, \ldots, N), \tag{1.3}$$

where $\mathbf{F}_k$ is the force with which all the other particles act on the k-th particle. The fact that the volume V is finite can be taken into account by adding to (1.2) an additional potential function $U_V(\mathbf{q}_1, \ldots, \mathbf{q}_N)$ that depends on the particle coordinates, is constant in the volume V and rapidly increases to infinity as the coordinates of any of the particles approach the boundary.

In statistical mechanics, we employ a probabilistic interpretation of dynamic processes. Following Gibbs, we consider not the given system but an aggregate of a large number (in the limit, infinity) of copies of it, all in macroscopically identical conditions, i.e., we introduce a statistical ensemble "representing" the macroscopic state of the system.

By identical external conditions in the macroscopic sense we mean that all the members of the ensemble are characterized by the same values of the macroscopic parameters (apart from possible fluctuations) and by the same types of contact with the surrounding bodies: reservoirs of energy or of particles, or moving pistons. This imposes limitations on the coordinates and momenta of the particles, which are otherwise arbitrary.

To each system belonging to the ensemble, there corresponds a point $(p_1, \ldots, p_N; q_1, \ldots, q_N)$ in phase space, or, more briefly, (p, q). In the course of time, each phase point moves along its own trajectory in phase space in accordance with the equations (1.1) or (1.3).

The statistical ensemble is specified by the distribution function

$$f(p, q, t),$$

which has the meaning of the probability density of the distribution of systems in phase space. It is defined in such a way that

$$dw = f(p, q, t)\, dp\, dq \tag{1.4}$$

is the probability of finding the system at time t in the element of phase space $dp\,dq$ close to the point (p, q), i.e., $(p_1, \ldots, p_N; q_1, \ldots, q_N)$.

1.2. Normalization

The distribution function (1.4) must satisfy the normalization condition

$$\int f(p, q, t)\, dp\, dq = 1, \tag{1.4a}$$

since the sum of the probabilities of all possible states must be equal to unity.

However, such a normalization of the distribution function is inconvenient. Classical statistical mechanics is the limiting case of quantum statistics at sufficiently high temperatures, when the quantum effects can be neglected, and the normalization (1.4a) does not correspond to the passage from quantum statistics to the classical limit. Therefore, it is more convenient to use a distribution function with another normalization.

It is known from quantum mechanics that the classical mechanical concept of the coordinates and momentum of a particle can be introduced only in the framework of the semiclassical approximation, if we are not to arrive at a contradiction with quantum mechanics. The minimum dimension of a phase cell for one-dimensional motion of the i-th particle in the semiclassical approximation is equal to Planck's constant $h = 2\pi\hbar$,

$$\Delta q_i^x \Delta p_i^x \geqslant h.$$

Consequently, the minimum dimension of a cell in the phase space of one particle is equal to h^3, and in the phase space of N particles is equal to h^{3N}. The quantity h^{3N} is thus the natural unit of phase volume. It is therefore convenient to introduce a distribution function normalized to unity over the dimensionless phase volume $dpdq/h^{3N}$.

In addition, we must take into account that permutation of identical particles in quantum mechanics does not change the state, and this property must also be conserved in classical statistics, which is its limiting case. Since the number of permutations for N identical particles is N!, the element of phase volume must be reduced by a factor of N!, since we need only consider different states.

It follows from the above considerations that it is convenient to introduce a dimensionless distribution function referred to an element of phase volume expressed in units h^{3N} with allowance for the identity of the particles, i.e., $dpdq/N!h^{3N}$. Consequently, the distribution function $f(p, q, t)$ is more conveniently defined not by (1.4) but by the relation

$$dw = f(p, q, t) \frac{dp\, dq}{N!\, h^{3N}}. \tag{1.5}$$

In this case, the normalization condition for the distribution function has the form

$$\int f(p, q, t)\, d\Gamma = 1, \tag{1.5a}$$

where

$$d\Gamma = \frac{dp\, dq}{N!\, h^{3N}} \tag{1.5b}$$

is the dimensionless element of phase volume. The integration in (1.5a) now corresponds to summation over all the different states. It can be shown that if classical mechanics is regarded as the limiting case of quantum mechanics, precisely this normalization condition is obtained (cf. §14).

It should be noted that the factor $1/N!$ in the phase volume was introduced by Gibbs [1] before quantum mechanics was discovered, in order to avoid the paradox bearing his name, i.e., the increase in entropy on mixing identical gases at identical temperatures and identical pressures. He distinguished "specific" phases p, q and "generic" phases, for which the phase volume is less by a factor of N!, and normalized the distribution functions over the generic phases.

A knowledge of the distribution function $f(p, q, t)$ enables us to calculate the average value of any dynamic variables $A(p, q)$,

$$\langle A\rangle = \int A(p, q) f(p, q, t)\, d\Gamma, \tag{1.6}$$

where the normalization (1.5a) is assumed.

§ 2. Liouville's Equation

2.1. Liouville's Theorem on the Invariance of Extension in Phase

The possibility of introducing a distribution function for the probability density is based on Liouville's theorem, a purely mechanical theorem which does not invoke any probabilistic considerations.

According to Liouville's theorem, the extension in phase for systems obeying the equations of mechanics in the Hamiltonian form (1.1) remains constant during the motion of the systems. That is, if at the initial moment of time the phase points (p^0, q^0) occupied continuously a certain region G_0 of initial values in phase space, and at the moment of time t they occupy the region G_t, then the corresponding extensions in phase are equal:

$$\int_{G_0} dp^0\, dq^0 = \int_{G_t} dp\, dq, \tag{2.1}$$

or, for infinitesimally small elements of phase volume,

$$dp^0\,dq^0 = dp\,dq. \tag{2.2}$$

In other words, the motion of the phase points depicting the system in phase space is similar to the motion of an incompressible fluid.

In order to prove Liouville's theorem, we transform the integral in the right-hand side of (2.1) by replacing the integration variables p and q by p^0 and q^0. Then

$$\int_{G_t} dp\,dq = \int_{G_0} \frac{\partial(p, q)}{\partial(p^0, q^0)}\,dp^0\,dq^0,$$

where $\partial(p, q)/\partial(p^0, q^0)$ is the Jacobian of the transformation from the variables p, q to p^0, q^0. It has the form of a determinant with elements $\partial x_i/\partial x_k^0$, where x_i is the aggregate of momenta and coordinates p_i, q_i and x_k^0 is the aggregate of p_k^0, q_k^0.

We shall now show that, by virtue of Hamilton's equations, the Jacobian is equal to unity, i.e.,

$$\frac{\partial(p, q)}{\partial(p^0, q^0)} = 1. \tag{2.3}$$

The equality (2.3) can be proved directly by differentiating the Jacobian with respect to time (see [2, 20]). Following Gibbs [1], however, it is simpler to use from the beginning the property of functional determinants

$$\frac{\partial(p, q)}{\partial(p^0, q^0)} = \frac{\partial(p', q')}{\partial(p^0, q^0)}\,\frac{\partial(p, q)}{\partial(p', q')},$$

where p', q' are the values of the momenta and coordinates corresponding to the arbitrary time t'. We shall differentiate this equality with respect to t, putting t_0 and t' constant:

$$\frac{d}{dt}\,\frac{\partial(p, q)}{\partial(p^0, q^0)} = \frac{\partial(p', q')}{\partial(p^0, q^0)}\,\frac{d}{dt}\,\frac{\partial(p, q)}{\partial(p', q')}.$$

Since t' is arbitrary, after the differentiation we can put t' = t. Then only the terms in the Jacobian standing on the principal diagonal will give a nonzero result†

$$\frac{d}{dt}\frac{\partial(p, q)}{\partial(p^0, q^0)} = \frac{\partial(p, q)}{\partial(p^0, q^0)} \sum_i \left(\frac{\partial \dot{p}_i}{\partial p_i} + \frac{\partial \dot{q}_i}{\partial q_i}\right) \tag{2.4}$$

(cf. [2, 20]). But by virtue of the equations of motion (1.1)

$$\frac{\partial \dot{p}_i}{\partial p_i} + \frac{\partial \dot{q}_i}{\partial q_i} = 0, \tag{2.5}$$

and, consequently,

$$\frac{d}{dt}\frac{\partial(p, q)}{\partial(p^0, q^0)} = 0, \tag{2.6}$$

i.e., the Jacobian does not depend on the time.

By making use of the initial condition

$$\left.\frac{\partial(p, q)}{\partial(p^0, q^0)}\right|_{t=t_0} = 1,$$

we see that the Jacobian (2.3) is indeed equal to unity; Liouville's theorem is thereby proved.

---

†Introducing for the Jacobian the notation

$$D = \det \frac{\partial x_i}{\partial x_k^0} = \frac{\partial(p,q)}{\partial(p^0, q^0)},$$

where $x_i = (p_i, q_i)$, and differentiating it with respect to time, we obtain

$$\frac{d}{dt} D = \sum_{i,k} \frac{dD}{da_{ik}} \dot{a}_{ik} = \sum_{i,k} D_{ik} \dot{a}_{ik},$$

where $a_{ik} = \partial x_i / \partial x_k^0$ and D_{ik} is the minor of the element a_{ik}. We have

$$\dot{a}_{ik} = \frac{d}{dt}\frac{\partial x_i}{\partial x_k^0} = \frac{\partial \dot{x}_i}{\partial x_i^0} = \sum_l \frac{\partial \dot{x}_i}{\partial x_l}\frac{\partial x_l}{\partial x_k^0} = \sum_l a_{lk}\frac{\partial \dot{x}_i}{\partial x_l};$$

consequently,

$$\frac{d}{dt} D = \sum_{ikl} D_{ik} a_{lk} \frac{\partial \dot{x}_i}{\partial x_l} = D \sum_{il} \delta_{il} \frac{\partial \dot{x}_i}{\partial x_l} = D \sum_i \frac{\partial \dot{x}_i}{\partial x_i},$$

since

$$\sum_k D_{ik} a_{lk} = D \delta_{il}.$$

Thus, the relation (2.4) is proved.

2.2. Liouville's Equation

In deriving and formulating Liouville's theorem, we have, up to now, nowhere used the concept of the distribution function, as is natural for a theorem of mechanics. If we now introduce distribution functions, as was done in § 1, it is possible to give other formulations of Liouville's theorem.

From Liouville's theorem, it follows that the distribution function is constant along the phase trajectories, and this may be regarded as one of the formulations of this theorem.

Indeed, in the motion in phase space of the points describing the systems, the number of phase points does not change; all the phase points situated at time t in the volume element dpdq will go over at time t' into the element dp'dq'. Consequently,

$$f(p, q, t)\,dp\,dq = f(p', q', t')\,dp'\,dq',$$

and since, by virtue of Liouville's theorem, dpdq = dp'dq', we obtain

$$f(p, q, t) = f(p', q', t'), \tag{2.7}$$

i.e., f is constant along the phase trajectories, as we wished to prove.

We shall give one more very convenient formulation of Liouville's theorem, one which is most often applied in practical problems — we shall introduce Liouville's equation for the distribution function.

Assuming the time t to be infinitesimally close to t' = t + dt, from (2.7) we shall have

$$f(p, q, t) = f(p + \dot{p}\,dt,\ q + \dot{q}\,dt,\ t + dt).$$

Assuming further that the function f is differentiable, we obtain a differential equation for it:

$$\frac{df}{dt} = \frac{\partial f}{\partial t} + \sum_k \left(\frac{\partial f}{\partial \boldsymbol{p}_k} \cdot \dot{\boldsymbol{p}}_k + \frac{\partial f}{\partial \boldsymbol{q}_k} \cdot \dot{\boldsymbol{q}}_k \right) = 0. \tag{2.8}$$

Taking Hamilton's equations into account, Eq. (2.8) is Liouville's equation:

$$\frac{\partial f}{\partial t} = \sum_k \left(\frac{\partial H}{\partial \boldsymbol{q}_k} \cdot \frac{\partial f}{\partial \boldsymbol{p}_k} - \frac{\partial H}{\partial \boldsymbol{p}_k} \cdot \frac{\partial f}{\partial \boldsymbol{q}_k} \right). \tag{2.9}$$

The sum on the right-hand side of (2.9) is called the **Poisson bracket** for the functions H and f:

$$\{H, f\} = \sum_k \left(\frac{\partial H}{\partial q_k} \cdot \frac{\partial f}{\partial p_k} - \frac{\partial H}{\partial p_k} \cdot \frac{\partial f}{\partial q_k} \right) \equiv \frac{\partial H}{\partial q} \frac{\partial f}{\partial p} - \frac{\partial H}{\partial p} \frac{\partial f}{\partial q}; \tag{2.10}$$

consequently, Liouville's equation can be written in the form

$$\frac{\partial f}{\partial t} = \{H, f\}. \tag{2.11}$$

This is the basic equation for the construction of statistical ensembles for both equilibrium and nonequilibrium cases, and it enables us to calculate f at any moment of time t, if it is known at time $t = t_0$; it also enables us to study the response of statistical systems to external perturbations (cf. Chapter III).

Liouville's equation has the form of a continuity equation for the motion of the phase points in phase space. We can obtain a simple intuitive interpretation of it if we regard the motion of phase points in the 6N-dimensional phase space as the motion of a "fluid" with density f. The rate of flow is represented by a vector $\dot{p}_1, \ldots, \dot{p}_N; \dot{q}_1, \ldots, \dot{q}_N$ in this space. Consequently, the condition for conservation of the phase points, i.e., the continuity equation in phase space, has the form

$$\frac{\partial f}{\partial t} + \sum_k \left[\frac{\partial}{\partial p_k} \cdot (f\dot{p}_k) + \frac{\partial}{\partial q_k} \cdot (f\dot{q}_k) \right] = 0, \tag{2.12}$$

where the quantity in brackets is the 6N-dimensional divergence of the flux vector. Performing the differentiation in the summand and taking into account that, by virtue of Hamilton's equations, the relation (2.5) holds, we see that equation (2.12) takes the form (2.8), i.e., coincides with Liouville's equation. It follows from (2.5) that the motion of the "fluid" is incompressible.

For the case of statistical equilibrium, f and H do not depend explicitly on time, and Liouville's equation has the form

$$\{H, f\} = 0, \tag{2.13}$$

i.e., the distribution function in this case is an integral of motion.

In fact, Liouville's equation is a linear differential equation in partial derivatives, while Hamilton's equations are the corresponding characteristic system of ordinary differential equations. Therefore, the total integral of equation (2.9) is an arbitrary function of all the integrals of the system (1.1).

2.3. The Time Evolution of Distribution Functions

To study the evolution of distributions functions in time, it is convenient to write Liouville's equation (2.11) in the form

$$i\frac{\partial f}{\partial t} = -Lf, \tag{2.14}$$

where L is a linear operator defined by the relation

$$iLf = \{H, f\}; \tag{2.14a}$$

it is called **Liouville's operator**.†

It is convenient to represent Liouville's equation in the form (2.14) because the operator L is Hermitian, and one can use the known properties of Hermitian operators. It is easy to see that L is Hermitian. Indeed, for arbitrary functions $\varphi_m(p, q)$ and $\varphi_n(p, q)$ vanishing at the boundaries of the phase volume, we obtain by means of integration by parts of the Poisson bracket

$$\int \varphi_m^* (L\varphi_n)\, dp\, dq = \int \varphi_n (L^* \varphi_m^*)\, dp\, dq. \tag{2.15}$$

The relation (2.15) is the Hermiticity condition for the operator L.

There is a formal analogy between Liouville's equation (2.14) and the Schrödinger equation

$$i\hbar\frac{\partial \Psi}{\partial t} = H\Psi,$$

since L and H are linear Hermitian operators. This analogy has been widely used by Prigogine [3] to carry over the methods of quantum mechanics to classical statistical mechanics.

From the Hamiltonian (1.2), Liouville's operator L has the form

$$L = i\sum_k \left[\frac{\boldsymbol{p}_k}{m}\cdot\frac{\partial}{\partial \boldsymbol{q}_k} + \boldsymbol{F}_k\cdot\frac{\partial}{\partial \boldsymbol{p}_k}\right] \tag{2.16}$$

and does not depend explicitly on time.

†Liouville's operator is sometimes defined with the opposite sign [3].

Using the Liouville operator (2.14a), we can write a formal solution of Liouville's equation, if the initial value of the distribution function at t = 0 is known. This solution has the form

$$f(p, q, t) = e^{iLt} f(p, q, 0), \tag{2.17}$$

if L does not depend explicitly on time.

Differentiating (2.17) with respect to t, we see that this function does indeed satisfy Liouville's equation

$$\frac{\partial f(p, q, t)}{\partial t} = iL\, e^{iLt} f(p, q, 0) = iL\, f(p, q, t)$$

and the initial condition

$$f(p, q, t)\,|_{t=0} = f(p, q, 0).$$

We shall often apply this formal integration of Liouville's equation.

We shall obtain the equation of motion for the dynamic variable A(p, q, t), where the latter argument indicates the explicit time dependence. For this, we shall differentiate A(p, q, t) with respect to time, assuming that p and q depend on time in accordance with Hamilton's equations. We shall have

$$\frac{dA}{dt} = \frac{\partial A}{\partial t} + \{A, H\}. \tag{2.18}$$

We shall show that here the average value of the derivative of A is equal to the derivative of the average value. In fact, the average value of A(p, q, t) at time t is equal to

$$\langle A \rangle = \int A(p, q, t) f(p, q, t)\, d\Gamma, \tag{2.19}$$

where f(p, q, t) satisfies Liouville's equation (2.11). Differentiating (2.19) with respect to time, using Liouville's equation for f and integrating the Poisson bracket by parts, we obtain

$$\frac{d}{dt} \langle A \rangle = \int \left(\frac{\partial A}{\partial t} + \{A, H\} \right) f\, d\Gamma,$$

i.e.,

$$\frac{d}{dt}\langle A\rangle = \left\langle \frac{dA}{dt} \right\rangle, \tag{2.19a}$$

as we wished to prove.

If A does not depend explicitly on time, then

$$\frac{dA}{dt} = \{A,\ H\} = -iLA. \tag{2.20}$$

If the value A(0) of the dynamic variable at t = 0 is known, and if L does not depend explicitly on time, (2.20) can be solved formally

$$A(t) = e^{-iLt} A(0), \tag{2.21}$$

i.e., using the Liouville operator, we can express the evolution of dynamic variables in time.

The operator e^{-iLt} is called the **evolution operator**. It acts on an arbitrary function φ(p(0), q(0)) converting it to φ(p(t), q(t)), i.e.,

$$e^{-iLt}\varphi(p(0),\ q(0)) = \varphi(p(t),\ q(t)), \tag{2.21a}$$

where p(t) and q(t) are the solutions of Hamilton's equation with H not depending explicitly on time and with the initial conditions $p(t)|_{t=0} = p(0)$ and $q(t)|_{t=0} = q(0)$.

The relation (2.21a) follows from the properties of the Liouville operator, since

$$-iLq = \frac{dq}{dt}, \quad -iLp = \frac{dp}{dt}$$

and

$$(-iL)^n q = \left(\frac{d}{dt}\right)^n q, \quad (-iL)^n p = \left(\frac{d}{dt}\right)^n p,$$

whence we obtain

$$e^{-iLt} q(0) = \sum_n \frac{t^n}{n!}\left(\frac{d^n q}{dt^n}\right)_{t=0} = q(t), \qquad e^{-iLt} p(0) = p(t),$$

since this is a Taylor series with derivatives determined by the equations of motion. Similar properties are also valid for powers of the coordinates and momenta, e.g., $-iLq^2 = dq^2/dt$. The relation (2.21a) for the evolution operator can be proved using these properties, by expanding φ in a Taylor series in p and q.

2.4. Entropy

The logarithm of the distribution function with a minus sign

$$\eta = -\ln f(p, q, t) \tag{2.22}$$

plays a special role in statistical mechanics (Gibbs calls $-\eta$ the phase exponent). The quantity η is convenient because it is additive for multiplicative distribution functions and, as we shall see below, is related to the entropy of the system.

It is easy to see that η, like f, satisfies Liouville's equation:

$$\frac{\partial \eta}{\partial t} = \{H, \eta\}. \tag{2.23}$$

This equation, which was obtained by Gibbs, turns out to be very useful in the theory of irreversible processes, although much less attention is given to it than to Liouville's equation. The convenience of Eq. (2.23) is associated with the fact that the properties of H are much closer to those of η than to those of $f(p, q)$.

The average of the logarithm of the distribution function with a minus sign, i.e., the average value of η, is called the **Gibbs entropy**. With the dimensionless normalization (1.5a) of the distribution function, the Gibbsian entropy is equal to

$$S = \langle \eta \rangle = -\int f(p, q, t) \ln f(p, q, t) \frac{dp\, dq}{N!\, h^{3N}}. \tag{2.24}$$

For a dilute gas, the states of the different particles can be regarded as statistically independent, and therefore the complete distribution function is equal to a product of the distribution functions for the individual particles:

$$f(p, q, t) = \frac{N!}{N^N} \prod_{i=1}^{N} f_1(\mathbf{p}_i, \mathbf{q}_i, t) \qquad \left(\ln N! \cong N \ln \frac{N}{e}\right), \tag{2.25}$$

where the single-particle distribution functions $f_1(\mathbf{p}_i, \mathbf{q}_i, t)$ have the normalization

$$\int f_1(\mathbf{p}_1, \mathbf{q}_1, t) \frac{d\mathbf{p}_1\, d\mathbf{q}_1}{h^3} = N. \tag{2.26}$$

The factor $N!/N^N$ in (2.25) is introduced to make the normalizations (1.5a) and (2.26) consistent. In fact,

$$\int f(p, q, t)\, d\Gamma = \left\{ \frac{1}{N} \int f_1(\mathbf{p}_1, \mathbf{q}_1, t) \frac{d\mathbf{p}_1\, d\mathbf{q}_1}{h^3} \right\}^N = 1.$$

For the distribution function (2.25), the entropy (2.24) is

$$S = S_{\mathrm{B}}, \tag{2.27}$$

where

$$S_{\mathrm{B}} = -\int f_1(p_1, q_1, t) \ln \frac{f_1(p_1, q_1, t)}{e} \frac{dp_1\, dq_1}{h^3} \tag{2.28}$$

is the Boltzmann entropy.

In the general case, when the multiplicative property (2.25) does not hold, the Boltzmann entropy may also be defined by formula (2.28), where $f_1(p_1, q_1, t)$ is the single-particle distribution function, i.e., the function obtained from $f(p, q, t)$ by integration over all the coordinates and momenta except p_1 and q_1:

$$f_1(p_1, q_1, t) = \int f(p_1, q_1, \ldots, p_N, q_N; t) \frac{dp_2\, dq_2 \ldots dp_N\, dq_N}{(N-1)!\, h^{3N-3}}. \tag{2.29}$$

The function $f_1(p_1, q_1, t)$ has the normalization (2.26).

It is well known from thermodynamics that the entropy of an isolated system increases, or in the case of thermodynamic equilibrium, is constant. If $f_1(p_1, q_1, t)$ satisfies Boltzmann's kinetic equation [4], then the Boltzmann entropy increases, while in the case of statistical equilibrium it is constant. However, the Boltzmann definition (2.28) for the entropy gives the correct expression for the entropy as a thermodynamic function in the equilibrium state only for the ideal gas, and therefore S_{B} may not be identified with the entropy of a system and the Boltzmann definition of the entropy is inadequate in the general case.

The Gibbs definition (2.24) of the entropy is better than the Boltzmann definition, since in the equilibrium case it gives the correct expression for the entropy as a thermodynamic function (see §§3 and 5).

The Gibbs definition of the entropy raises no doubts for the equilibrium case. But if f depends on time, we have a different state of affairs. In this case, the Gibbs definition of the entropy has an intrinsic defect and needs to be made more precise (see Chapter IV).

It is easy to see that for an isolated system the Gibbs entropy does not depend on time and therefore cannot increase.

Indeed, let the distribution function be $f(p^0, q^0, 0)$ at $t = 0$ and $f(p, q, t)$ at time t, where (p, q) lies on the phase trajectory passing through (p^0, q^0) and determined by the equations of motion. According to Liouville's theorem (2.7), we have

$$f(p^0, q^0, 0) = f(p, q, t).$$

The Gibbs entropy at time t is

$$S = -\int f(p, q, t) \ln f(p, q, t) \frac{dp\, dq}{N!\, h^{3N}}$$
$$= -\int f(p^0, q^0, 0) \ln f(p^0, q^0, 0) \frac{dp^0\, dq^0}{N!\, h^{3N}},$$

since according to Liouville's theorem (2.2),

$$dp\, dq = dp^0\, dq^0.$$

Thus, the Gibbs entropy (2.24) for an isolated system does not change with time.

Gibbs attempted to prove that the entropy of an isolated system can increase in some sense. P. and T. Ehrenfest [5, 6] elucidated the sense in which this assertion is to be understood. For this, they suggested "coarsening" the Gibbs definition of entropy by introducing in place of the true distribution function $f(p, q, t)$, which may be called the fine-grained density of the distribution, a coarse-grained density

$$\tilde{f}(p, q, t) = \frac{1}{\omega} \int_{\omega} f(p, q, t)\, dp\, dq, \qquad (2.30)$$

where the integration is performed over fixed small cells ω of phase space. From a physical point of view, the coarse-grained averaging operation (2.30) corresponds to the fact that observed quantities are always averages over a certain region. Quantum mechanics establishes a lower bound for the cell ω, which cannot be smaller than h^{3N}.

The Gibbs entropy constructed from the coarsened density is, generally speaking, no longer constant in time and can increase; this is true however small the coarsening.

We shall compare the values of the Gibbs entropy calculated for the coarse-grained distribution function (2.30) at times t and t = 0; we shall assume that at the initial moment the true density coincides with the coarsened density:

$$f(p^0, q^0, 0) = \tilde{f}(p^0, q^0, 0). \tag{2.31}$$

We have

$$\begin{aligned} S_t - S_0 = & - \int \tilde{f}(p, q, t) \ln \tilde{f}(p, q, t)\, d\Gamma + \\ & + \int f(p^0, q^0, 0) \ln f(p^0, q^0, 0)\, d\Gamma_0 = \\ & = - \int \{f(p, q, t) \ln \tilde{f}(p, q, t) - f(p, q, t) \ln f(p, q, t)\}\, d\Gamma \end{aligned} \tag{2.31a}$$

(we have used Liouville's theorem and, as is always possible in the integrand, we have removed the tilde sign from the distribution function except when we take its logarithm).

For any two normalized distribution functions f and f' defined in the same phase space, the inequality

$$\int f \ln\left(\frac{f}{f'}\right) d\Gamma \geqslant 0. \tag{2.32}$$

holds. The equality sign is attained only when $f = f'$. Taking (2.32) into account, we obtain

$$S_t \geqslant S_0.$$

The inequality (2.32) is a consequence of the obvious inequality

$$\ln\left(\frac{f}{f'}\right) \geqslant 1 - \frac{f'}{f} \qquad (f' > 0,\ f > 0), \tag{2.32a}$$

where the equality sign holds only when $f = f'$. It is easy to see that (2.32a) is valid by noting that $\ln x - 1 + 1/x$ is a positive function equal to zero only when $x = 1$, and then putting $x = f/f'$.

We obtain the inequality (2.32) by multiplying (2.32a) by f and integrating over the whole phase space. In fact,

$$\int f \ln\left(\frac{f}{f'}\right) d\Gamma \geqslant \int f\left(1 - \frac{f'}{f}\right) d\Gamma = 0,$$

as we wished to prove.

We shall assume that $f(p^0, q^0, 0)$ does not correspond to an ensemble in statistical equilibrium; then at time t,

$$\bar{f}(p, q, t) \neq \tilde{\bar{f}}(p, q, t), \tag{2.33}$$

since, although $f(p, q, t)$ does not change along the phase trajectories, phase points from other cells will enter and leave the given cell ω, and these processes, generally speaking, do not cancel each other. A process of "intermixing" of space points over the phase cells occurs. Using (2.33), we obtain

$$S_t > S_0, \tag{2.34}$$

i.e., the entropy for the coarse-grained distribution function can now increase.

The introduction of coarse-grained averaging, however, still does not solve the question of the increase in entropy; we must convince ourselves that there really is intermixing. The finer the scale of the coarsening, the smaller the increase in entropy S_t becomes, and in the limit $\omega \to 0$ it tends to zero. But the increase in the physical entropy cannot depend on the scale of the coarsening.

To answer this objection, we note that in applying the coarse-grained averaging operation we are taking two limits: the usual statistical mechanical limit $V \to \infty$ ($N/V = \text{const}$), and making the cell ω go to zero. There are no reasons for assuming that the results will not depend on the order in which these limits are taken. The Ehrenfest coarsening of the distribution functions can be effective only if the limit $V \to \infty$ is taken first and then $\omega \to 0$, and if there is no uniform convergence in the limiting processes.

It is interesting to recall that Gibbs, in drawing an analogy between the tendency to equilibrium in a system and intermixing in an incompressible liquid, had already essentially introduced the coarsening procedure for the density of the distribution and had noted the absence of uniform convergence in the limiting processes (cf. [1], Chapter 12, pages 143-147).

Thus, the Ehrenfest coarse-grained averaging does not solve the problem of the entropy increase, although the idea of averaging distribution functions is of undoubted interest.

Smoothing of distribution functions is possible not only in phase space, but also in time. In fact, all observed quantities are average values over intervals of time T of the order of the time of observation. Therefore, we can introduce distribution functions averaged over the interval T

$$\overline{f(p, q, t)} = \frac{1}{T} \int_0^T f(p(t+t_1), q(t+t_1)) \, dt_1 \qquad (2.35)$$

or dynamic variables averaged in a corresponding manner.

This smoothing of distribution functions in time was widely applied by Kirkwood [7] and, apparently, is more effective than the coarse-grained smoothing of the Ehrenfests. It is analogous to the averaging of the equations of motion in nonlinear mechanics, which smooths out the rapid oscillations about the mean trajectory and helps us to determine the averaged motion [8]. In general, there is a deep relationship between the methods of nonlinear mechanics and statistical mechanics. A statistical system tends to a state of statistical equilibrium irrespective of the initial conditions, which are quickly "forgotten," just as a nonlinear system tends to its limiting cycle. The methods for eliminating the secular terms in the construction of kinetic equations are the same as those used in the solution of the equations of nonlinear mechanics [9].

§ 3. Gibbsian Statistical Ensembles

To construct statistical ensembles in the case of statistical equilibrium, we need to solve the problem of what integrals of motion the distribution function can depend on and what is its form for different external conditions macroscopically defining the ensemble. This problem was solved by Gibbs, although a rigorous justification of the distributions obtained is a complicated problem that is still not completely solved at the present time. It is not even clear to what extent this rigorous justification is possible.

For states differing from those of statistical equilibrium, the construction of statistical ensembles is a considerably more complicated problem, and progress in this direction has been contemplated only in recent years. Here the problem of actually constructing the ensembles is more pressing than their rigorous formal justification. We shall discuss these questions in Chapters III and IV.

According to Gibbs, the distribution function f in a state of statistical equilibrium depends only on single-valued additive integrals of motion.

The additivity of the integrals of motion implies that the integrals of motion of the complete system are additively composed of the integrals of motion of its subsystems.

Three such integrals of motion are known: the energy **H**, the total momentum **P** and the total angular momentum **M**. Consequently,

$$f = f(H, \boldsymbol{P}, \boldsymbol{M}). \tag{3.1}$$

The additivity property is fulfilled exactly for **P** and **M** (if we ignore interaction with the walls), and for H is fulfilled to within the surface energy at the interface of the subsystems, which arises from the interaction between particles in different subsystems.

If the total number of particles N in each system of the ensemble is not specified, we must regard N as being like a fourth integral of motion, since N does not change during the evolution of the systems. Thus, in this case,

$$f = f(H, N, \boldsymbol{P}, \boldsymbol{M}). \tag{3.1a}$$

The integral of motion **M** must be taken into account if the system rotates as a whole with constant angular velocity, and the integral of motion **P** must be taken into account for superfluid systems in quantum statistics.

The number of essential integrals of motion is reduced if we consider a system of particles in a motionless vessel. Then the total momentum **P** and angular momentum **M** are zero in a state of statistical equilibrium, and it is not necessary to take the additive integrals **P** and **M** into account. Consequently, for systems with a specified number of particles,

$$f = f(H) \tag{3.2}$$

or, if the number of particles is not specified,

$$f = f(H, N). \tag{3.2a}$$

In addition, the function f must depend on the parameters determining the ensemble macroscopically, e.g., the volume V and the total number of particles N, if the latter is specified; these parameters are assumed to be constant for all the copies of the system.

3.1. The Microcanonical Distribution

We shall consider a statistical ensemble of closed energetically isolated systems in a constant volume V, i.e., an ensemble of systems with constant particle number N, each system being situated in an adiabatic shell and having the same energy E to within $\Delta E \ll E$. Following Gibbs, we shall assume that the distribution function $f(p, q)$ for such an ensemble is constant in the layer ΔE and is equal to zero outside this layer:

$$f(p, q) = \begin{cases} [\Omega(E, N, V)]^{-1} & \text{for} \quad E \leqslant H(p, q) \leqslant E + \Delta E, \\ 0 & \text{outside this layer.} \end{cases} \tag{3.3}$$

Such a distribution is called microcanonical and the corresponding ensemble is a microcanonical ensemble. The macroscopic state of the systems in such an ensemble is determined by three extensive parameters E, V, and N. The constant $\Omega(E, N, V)$, which is called the statistical weight, is determined from the normalization condition (1.5a) and has the meaning of a dimensionless phase volume – the number of states in the layer ΔE:

$$\Omega(E, N, V) = \frac{1}{N!\, h^{3N}} \int\limits_{E \leqslant H(p, q) \leqslant E + \Delta E} dp\, dq. \tag{3.3a}$$

In the case of classical mechanics, we can go over to the limit $\Delta E \to 0$ and write f in the form

$$f = \Omega^{-1}(E, N, V)\, \delta(H(p, q) - E), \tag{3.4}$$

where

$$\begin{aligned} \Omega(E, N, V) &= \frac{1}{N!\, h^{3N}} \int \delta(H(p, q) - E)\, dp\, dq, \\ \Omega^{-1}(E, N, V) &= [\Omega(E, N, V)]^{-1}. \end{aligned} \tag{3.4a}$$

[In formulae (3.3a) and (3.4a), the number of states in the layer ΔE and the density of states at the constant-energy surface are given the same notation $\Omega(E, N, V)$.]

In quantum mechanics, the uncertainty relation between the time of observation t and the energy, $\Delta E \cdot t \sim \hbar$, prevents such a limiting process; making ΔE tend to zero would correspond to an infinite time of observation. Therefore, we shall use the representation (3.3) for f and only occasionally use (3.4) to simplify the calculations.

We shall calculate the entropy (2.24) for the microcanonical distribution:

$$S = \langle \eta \rangle = -\frac{1}{N!\,h^{3N}} \int f(p, q) \ln f(p, q)\, dp\, dq. \tag{3.5}$$

Putting the expression (3.3) into this, and taking (3.3a) into account, we obtain

$$S(E, N, V) = \ln \Omega (E, N, V). \tag{3.5a}$$

Thus, for a microcanonical ensemble, the entropy is equal to the logarithm of the statistical weight (3.3a). One can show that the entropy defined in this way does indeed possess the properties of the thermodynamic entropy. We shall return to this qeustion in §5.

The hypothesis that the microcanonical ensemble does in fact represent the macroscopic state of a closed energetically isolated system, i.e., that microcanonical averages coincide with the observed values of physical quantities, is one of the basic postulates of statistical mechanics.

The observed values of physical quantities A(p, q) are always average values over some observation-time interval τ. The problem of justifying the possibility of replacing time averages by averages over the microcanonical ensemble is called the e r g o d i c p r o b l e m. The problem consists in proving for closed energetically isolated systems that

$$\frac{1}{\tau} \int_0^{\tau} A(p(t), q(t))\, dt = \frac{1}{N!\,h^{3N}} \int f(p, q) A(p, q)\, dp\, dq, \tag{3.6}$$

where f is the microcanonical distribution. This problem is extremely complicated and, despite a number of important results that have been obtained, is still not resolved; we shall therefore not discuss it, but refer the reader to the literature [6, 11, 12].

The microcanonical distribution is sometimes useful for general investigations, since, of all the Gibbsian distributions, it is the one most directly connected with mechanics (all the parameters E, N, V have a mechanical meaning), but it is not convenient for practical application to concrete systems, since calculation of $\Omega(E, N, V)$ is very difficult.

Rather than consider energetically isolated systems, it is much more convenient to consider systems in thermal contact with the surroundings.

3.2. The Canonical Distribution

We now consider closed systems in thermal contact with a thermostat. By a thermostat, we shall mean a system with a large number of degrees of freedom, able to exchange energy with the given system and so large that its state is practically unchanged in this exchange.

A statistical ensemble of systems with a specified number of particles N and volume V in contact with a thermostat is called a *canonical ensemble*.

Such an ensemble is described by the *canonical distribution*†

$$f(p, q) = Q^{-1}(\theta, V, N) \exp\left(-\frac{H(p, q)}{\theta}\right), \tag{3.7}$$

where θ is the modulus of the canonical distribution, which, as will be shown below, plays the role of the temperature, and $Q(\theta, V, N)$ is the *statistical integral*, or *partition function*,‡ determined from the normalization condition (1.5a):

$$Q(\theta, V, N) = \int \exp\left(-\frac{H(p, q)}{\theta}\right) d\Gamma, \quad \text{where} \quad d\Gamma = \frac{dp\,dq}{N!\,h^{3N}}. \tag{3.8}$$

The partition function (3.8), constructed on the basis of the canonical distribution, is a function of the parameters θ, V, and N macroscopically defining the ensemble. Two of these, V and N,

---

†For a justification of the canonical distribution, see §3.3.

‡The term "partition function" is more usual in Western literature, and will be used hereafter. – Translation Editors.

are extensive parameters, i.e., are proportional to V when V/N = const, while the third, θ, is intensive, i.e., has a finite value as V → ∞, V/N = const. The partition function Q(θ, V, N) is a fundamental quantity, which determines the thermodynamic properties of the system.

The logarithm of the partition function (3.8) determines the free energy of the system:

$$F(\theta, V, N) = -\theta \ln Q(\theta, V, N). \tag{3.8a}$$

For real systems, when N is very large, we do not require the exact value of the function F(θ, V, N); it is sufficient to know the thermodynamic limit

$$\lim_{\substack{N\to\infty \\ V/N=v=\mathrm{const}}} \frac{F(\theta, V, N)}{N} = f(\theta, v),$$

i.e., the free energy per particle when the number of particles increases without limit at a given density. This function determines all the thermodynamic properties of the system.

A proof of the existence of the thermodynamic limit for the canonical ensemble was given by Van Hove [13] and by Bogolyubov and Khatset [34]. Van Hove proved the existence of the thermodynamic limit for the free energy under not very rigorous restrictions on the interaction potential between the particles. Bogolyubov and Khatset proved the existence of correlation functions in the limit, which determine the thermodynamic functions in the limit. Further developments in this direction were made by Yang and Lee [14], Ruelle [15], Dobrushin [16], and other authors [17a-17e]. These papers can be regarded as the beginning of a new field of mathematical physics, mathematic statistical physics. (See the bibliography in Ruelle's book [15].)

We shall obtain an expression for the average energy, by differentiating (3.8) with respect to θ and taking (2.19) into account:

$$\langle H\rangle = \theta^2 \frac{\partial}{\partial\theta} \ln Q(\theta, V, N) = -\theta^2 \frac{\partial}{\partial\theta}\left(\frac{F}{\theta}\right)_{V,N}. \tag{3.8b}$$

The entropy (2.24) of the Gibbs canonical ensemble is

$$S = \langle\eta\rangle = -\int f \ln f \, d\Gamma = \frac{\langle H\rangle - F}{\theta} = -\left(\frac{\partial F}{\partial\theta}\right)_{V,N} \tag{3.8c}$$

[here we have used the relations (3.8a) and (3.8b)].

The relations (3.8b) and (3.8c) have the form of thermodynamic equalities, indicating that F and S are indeed the free energy and entropy. We obtain a more complete thermodynamic analogy if we consider a canonical ensemble with slowly varying volume (cf. §5.4).

The canonical distribution (3.7) also enables us to calculate the fluctuations. Differentiating (3.8) twice with respect to $1/\theta$ and using (3.8b), we find an expression for the energy fluctuations in the Gibbsian ensemble:

$$\langle H^2\rangle - \langle H\rangle^2 = \langle (H - \langle H\rangle)^2\rangle = \frac{\partial^2 \ln Q}{\partial (1/\theta)^2} = \theta^2 \frac{\partial \langle H\rangle}{\partial \theta} = \theta^2 C_V, \qquad (3.8d)$$

where C_V is the heat capacity at constant volume.

The quantity $\langle H\rangle$ is proportional to the number of particles N, while θ does not depend on N; consequently, for large systems, the relative square fluctuations of the energy are proportional to 1/N, i.e., are very small. Therefore, the canonical and microcanonical ensembles differ little from each other. (For more details on the thermodynamic equivalence of the ensembles, see §13.)

We have assumed up to now that the system does not move as a whole and have considered a single integral of motion — the total energy H. In the case when, in addition to the energy, there exist other additive integrals of motion $\mathscr{P}_1, \ldots, \mathscr{P}_s$, the Gibbsian distribution has the form

$$f(p, q) = Q^{-1}(\theta, \mathscr{F}_1, \ldots, \mathscr{F}_s) \exp\left\{-\frac{H(p, q)}{\theta} - \sum_{1 \leqslant k \leqslant s} \mathscr{F}_k \mathscr{P}_k(p, q)\right\}, \qquad (3.9)$$

where $\mathscr{F}_k$ are the new thermodynamic parameters. We do not indicate the dependence of Q on V and N explicitly.

In the special case when the total momentum **P** and the total angular momentum **M** are taken into account, the distribution (3.9) has the form

$$f(p, q) = Q^{-1} \exp\left\{-\frac{1}{\theta}\left[H(p, q) - \boldsymbol{v} \cdot \boldsymbol{P}(p) - \boldsymbol{\omega} \cdot \boldsymbol{M}(p, q)\right]\right\}, \qquad (3.9a)$$

where **v** is the velocity of the system as a whole, and ω is the angular velocity of its rotation.

It is convenient to represent the expression (3.9) in a more symmetric form:

$$f(p, q) = \exp\left\{-\Phi(\mathcal{F}_0, \ldots, \mathcal{F}_s) - \sum_{0 \leqslant k \leqslant s} \mathcal{F}_k \mathcal{P}_k(p, q)\right\}, \tag{3.10}$$

where we have introduced the notation

$$\mathcal{P}_0(p, q) = H(p, q), \quad \mathcal{F}_0 = 1/\theta, \quad \Phi(\mathcal{F}_0, \ldots, \mathcal{F}_s) = \ln Q. \tag{3.10a}$$

The thermodynamic potential $\Phi(\mathcal{F}_0, \ldots, \mathcal{F}_s)$ is called the *Massieu – Planck thermodynamic function*.

For a distribution function of the form (3.10), the thermodynamic equalities and expressions for the fluctuations take a specially symmetric form:

$$\begin{gathered}
\langle \mathcal{P}_k \rangle = -\frac{\partial \Phi}{\partial \mathcal{F}_k}, \qquad \mathcal{F}_k = \frac{\partial S}{\partial \langle \mathcal{P}_k \rangle}, \\
S = \Phi + \sum_k \mathcal{F}_k \langle \mathcal{P}_k \rangle = \Phi - \sum_k \mathcal{F}_k \frac{\partial \Phi}{\partial \mathcal{F}_k}, \\
\langle \mathcal{P}_k^2 \rangle - \langle \mathcal{P}_k \rangle^2 = \frac{\partial^2 \Phi}{\partial \mathcal{F}_k^2},
\end{gathered} \tag{3.10b}$$

where S is the entropy.

3.3. Gibbs' Theorem on the Canonical Distribution

The postulates about the microcanonical distribution (3.3) and about the canonical distribution (3.7) are not independent. We shall now prove Gibbs' theorem on the canonical distribution, according to which a small part of a microcanonical ensemble of systems with many degrees of freedom is distributed canonically.

The combination of a given system with a thermostat that is assumed to be much greater than the given system (in its number of degrees of reedom) can always be regarded as one large closed isolated system. If we take the microcanonical distribution for this combined system, then it follows from Gibbs' theorem that the system in the thermostat is distributed canonically.

We shall now prove the Gibbs theorem.

Let a large system with Hamiltonian H consist of two subsystems (1) and (2) with Hamiltonians $H_1(p, q)$ and $H_2(p', q')$, where p, q and p', q' are the aggregates of the momenta and coordinates of the subsystems. We shall assume the interaction between the subsystems to be negligibly small; then

$$H = H_1(p, q) + H_2(p', q'). \tag{3.11}$$

We shall assume that the subsystem (1) is much smaller than subsystem (2), which we shall call the thermostat. We shall further assume that the combined system is distributed microcanonically. According to (3.3), its distribution function has the form

$$f(p, q;\ p', q') = \begin{cases} \Omega^{-1}(E) \text{ in the layer } E \leqslant H \leqslant E + \Delta E, \\ 0 \quad \text{outside this layer} \end{cases} \tag{3.12}$$

where the statistical weight $\Omega^{-1}(E)$ can also depend on the total particle number N and the volume V, although we shall omit this dependence for brevity.

In order to obtain the distribution function of subsystem (1), i.e., of the small subsystem, we must integrate the total distribution function over all the variables of the second subsystem (the thermostat), taking into account the normalization factors introduced in (1.5a), i.e.,

$$f_1(p, q) = \frac{1}{N_2!\, h^{3N_2}} \int f(p, q;\ p', q')\, dp'\, dq' =$$

$$= \Omega^{-1}(E) \frac{1}{N_2!\, h^{3N_2}} \int\limits_{E-H_1 \leqslant H_2 \leqslant E-H_1+\Delta E} dp'\, dq',$$

where the integration is performed over the variables p', q' lying in the layer

$$E - H_1(p, q) \leqslant H_2(p', q') \leqslant E - H_1(p, q) + \Delta E.$$

Taking into account the definition (3.3a) of the statistical weight, we note that the distribution function of the first subsystem is equal to the ratio of the statistical weight of the second subsystem with energy $E - H_1$ to the statistical weight of the whole system:

$$f_1(p, q) = \frac{\Omega_2(E - H_1(p, q))}{\Omega(E)}. \tag{3.13}$$

To calculate f_1, we must obtain the asymptotic limit of the ratio of the statistical weights of the thermostat and the whole system under the assumption that the thermostat is very large.

First, we shall give a very simple, but nonrigorous, derivation of the canonical distribution (3.7) from the equality (3.13).

We introduce the entropy $S_2(E)$ of the thermostat and the entropy $S(E)$ of the whole system using the relation (3.5a):

$$S_2(E) = \ln \Omega_2(E), \quad S(E) = \ln \Omega(E) \tag{3.14}$$

and rewrite (3.13) in the form

$$f_1(p, q) = \exp\{S_2(E - H_1(p, q)) - S(E)\}.$$

Because of the smallness of subsystem (1) compared with the thermostat ($H_1 \ll E$), we can expand the function $S_2(E - H_1)$ in a series in H_1, confining ourselves to two terms:

$$S_2(E - H_1(p, q)) \cong S_2(E) - \frac{\partial S_2}{\partial E} H_1(p, q). \tag{3.14a}$$

Using this expansion, we write f_1 in the form

$$f_1(p, q) = Q^{-1} \exp\left(-\frac{H_1(p, q)}{\theta}\right), \tag{3.15}$$

where Q is the normalizing factor (3.8), i.e., the partition function, and the quantity

$$\frac{1}{\theta} = \frac{\partial S_2(E)}{\partial E} = \frac{\partial \ln \Omega_2(E)}{\partial E} \tag{3.16}$$

plays the role of the inverse temperature. Thus, the system in the thermostat is distributed canonically, as we wished to prove.

To elucidate the nature of the asymptotic tendency of (3.13) toward the Gibbsian distribution as the dimensions of the thermostat increase, we shall give a beautiful and more rigorous proof of the Gibbs theorem, due to Krutkov [18].

We write expression (3.13) for f_1 in the form

$$f_1(p, q) = \frac{\omega_2(E - H_1(p, q))}{\omega(E)}, \tag{3.17}$$

where

$$\omega_2(E) = \frac{\Omega_2(E)}{\Delta E}, \quad \omega(E) = \frac{\Omega(E)}{\Delta E} \tag{3.18}$$

are respectively the densities of states of the thermostat and of the total system.

Since $H_1(p, q)$ and $H_2(p', q')$ depend only on variables referring to the different subsystems, the statistical weight $\Omega(E)$ of the total system can be represented in the form of successive integrations — first over the coordinates of the second subsystem at fixed $H_1(p, q) = E_1$, and then over the coordinates of the first subsystem with $0 \leq E_1 \leq E$, i.e.,

$$\Omega(E) = \frac{1}{N_1!\,h^{3N_1}} \int\limits_{0 \leqslant H_1(p,\,q) = E_1 \leqslant E} dp\,dq \times$$

$$\times \frac{1}{N_2!\,h^{3N_2}} \int\limits_{E-E_1 \leqslant H_2(p',\,q') \leqslant E-E_1+\Delta E} dp'\,dq'. \tag{3.18a}$$

We take the indistinguishability of the particles of the thermostat and of the system into account separately, since we are not permitting the possibility of exchange of particles between them.

It follows from (3.18a) that the density of states ω of the total system is related to the densities of states ω_1 and ω_2 of the first and second subsystems by an integral relation of the convolution type:

$$\omega(E) = \int_0^E \omega_1(E - E_2)\,\omega_2(E_2)\,dE_2. \tag{3.19}$$

The relation (3.19) can be regarded as an integral equation for $\omega_2(E)$. We shall solve it by means of a Laplace transformation. Multiplying (3.19) by $e^{-\lambda E}$, integrating over E from zero to infinity and going over to the variables $E_1 = E - E_2$ and E_2, we obtain

$$Q(\lambda) = Q_1(\lambda)\,Q_2(\lambda), \tag{3.20}$$

where

$$\begin{aligned} Q(\lambda) &= \int_0^\infty e^{-\lambda E}\,\omega(E)\,dE, \\ Q_k(\lambda) &= \int_0^\infty e^{-\lambda E}\,\omega_k(E)\,dE \qquad (k = 1,\ 2) \end{aligned} \tag{3.21}$$

are the Laplace transforms of the densities of states (3.18). We find the densities of states by inverting the Laplace transforms (3.21):

$$\omega_k(E) = \frac{1}{2\pi i} \int_{a-i\infty}^{a+i\infty} e^{\lambda E} Q_k(\lambda)\, d\lambda \qquad (k = 1, 2),$$
$$\omega(E) = \frac{1}{2\pi i} \int_{a-i\infty}^{a+i\infty} e^{\lambda E} Q(\lambda)\, d\lambda, \tag{3.22}$$

where $a > 0$ is a real positive constant.

We shall assume now that the second subsystem (the thermostat) consists of $n-1$ identical weakly interacting parts, each of which is the same as the first subsystem. Then, by virtue of (3.20),

$$Q(\lambda) = [Q_1(\lambda)]^n, \quad Q_2(\lambda) = [Q_1(\lambda)]^{n-1}, \tag{3.23}$$

and the solutions (3.22) take the form

$$\omega_2(E) = \frac{1}{2\pi i} \int_{a-i\infty}^{a+i\infty} e^{\lambda E} [Q_1(\lambda)]^{n-1}\, d\lambda,$$
$$\omega(E) = \frac{1}{2\pi i} \int_{a-i\infty}^{a+i\infty} e^{\lambda E} [Q_1(\lambda)]^n\, d\lambda. \tag{3.24}$$

We estimate the expressions (3.24) for the densities of states asymptotically as $n \to \infty$ using the method of steepest descents [19]. We write the second integral of (3.24) in the form

$$\omega(E) = \frac{1}{2\pi i} \int_{a-i\infty}^{a+i\infty} e^{n\chi(\lambda)}\, d\lambda, \tag{3.25}$$

where we have introduced the function

$$\chi(\lambda) = \lambda \frac{E}{n} + \ln Q_1(\lambda). \tag{3.25a}$$

The function $\chi(\lambda)$ is assumed to be analytic in the region $\operatorname{Re} \lambda > 0$ of complex values of λ. From the properties of analytic functions, the function $\chi(\lambda)$ can have neither a minimum nor a maximum, and its extremum corresponds to a saddle point. The saddle point λ_1

is determined from the conditions

$$\chi'(\lambda_1) = \frac{E}{n} + \frac{Q_1'(\lambda_1)}{Q_1(\lambda_1)} = 0, \tag{3.26}$$

$$\chi''(\lambda_1) = \frac{d^2}{d\lambda^2} \ln Q_1(\lambda)_{\lambda=\lambda_1} > 0, \tag{3.26a}$$

where λ_1 is the only real root of Eq. (3.26). We assume that the conditions for applying the method of steepest descents are fulfilled. Below, we shall see that the condition (3.26a) is fulfilled.

We introduce a new real variable ξ,

$$\lambda = \lambda_1 + i\xi,$$

and expand $\chi(\lambda)$ in a Taylor series about the saddle point up to terms quadratic in ξ:

$$\chi(\lambda_1 + i\xi) \cong \chi(\lambda_1) - \frac{\chi''(\lambda_1)}{2}\xi^2;$$

consequently,

$$\omega(E) = \frac{1}{2\pi}\int_{-\infty}^{\infty} \exp\left\{n\chi(\lambda_1) - \frac{n\chi''(\lambda_1)}{2}\xi^2\right\} d\xi = \frac{1}{2\pi} e^{n\chi(\lambda_1)} \sqrt{\frac{2\pi}{n\chi''(\lambda_1)}}$$

$$(\lambda_1 > 0),$$

and finally,

$$\omega(E) = e^{\lambda_1 E}[Q_1(\lambda_1)]^n \frac{1}{\sqrt{2\pi n\chi''(\lambda_1)}}. \tag{3.27}$$

For ω_2, we obtain completely analogously

$$\omega_2(E) = e^{\lambda_1 E}[Q_1(\lambda_1)]^{n-1} \frac{1}{\sqrt{2\pi(n-1)\chi''(\lambda_1)}}. \tag{3.27a}$$

Putting (3.27) and (3.27a) into (3.17), we find

$$f_1(p, q) = Q_1^{-1}(\lambda_1)\exp(-\lambda_1 H_1(p, q)), \tag{3.28}$$

which coincides with the canonical distribution (3.7) with modulus

$$\theta = \frac{1}{\lambda_1}. \tag{3.29}$$

The condition (3.26) at the saddle point coincides with the thermodynamic equality (3.8b), while the inequality (3.26a), as can easily be seen using (3.8d), means that the energy fluctuations are positive, and is consequently also fulfilled. Thus, the Gibbs theorem is proved.

In the derivation we have given of the canonical distribution, the system and thermostat are regarded as identical in nature. It is also possible not to make this assumption, but to assume that the thermostat and the system under consideration are described by different Hamiltonians, provided that the thermostat is sufficiently large and its interaction with the system is small. In this case, the Gibbs theorem remains valid, as follows from a simple argument [cf. (3.13)-(3.16)].

The Gibbs theorem can also be proved in other ways, e.g., by Khinchin's method [20, 21], which is based on applying the central limit theorem of probability theory. The construction of the distribution functions of statistical ensembles can also be approached in another way, based on information theory [22, 23], which we shall discuss later, in §4.

3.4. The Grand Canonical Distribution

Earlier we considered closed systems in contact with a thermostat. A more general case of contact of a system with its surroundings is also possible. We shall consider an open system in a thermostat, which can exchange not only energy with its surroundings, but also particles. For example, there may be permeable walls between the system and the reservoir of particles. Then the energy and number of particles in the system are not constant, but we assume the volume to be fixed. The statistical ensemble corresponding to an aggregate of such systems in thermal and material contact with their surroundings is called a *grand canonical ensemble*.

Such an ensemble is described by the *grand canonical distribution*†

$$f_N(p, q) = Q^{-1}(\theta, \mu, V) \exp\left\{-\frac{H(p, q) - \mu N}{\theta}\right\}, \tag{3.30}$$

†For a justification of the grand canonical distribution, see formulas (3.36)-(3.39) below, and §9.3.

where μ is the chemical potential, and $Q(\theta, \mu, V)$ is the partition function for the grand ensemble, determined from the normalization condition

$$\sum_{N \geqslant 0} \frac{1}{N!\,h^{3N}} \int f_N(p, q)\, d\boldsymbol{p}_1\, d\boldsymbol{q}_1 \ldots d\boldsymbol{p}_N\, d\boldsymbol{q}_N = 1, \tag{3.30a}$$

which is the natural generalization of the normalization (1.5a) to systems with a variable number of particles. Consequently,

$$Q(\theta, \mu, V) = \sum_{N \geqslant 0} \int \exp\left(-\frac{H(p, q) - \mu N}{\theta}\right) d\Gamma_N, \tag{3.31}$$

where

$$d\Gamma_N = \frac{d\boldsymbol{p}_1\, d\boldsymbol{q}_1 \ldots d\boldsymbol{p}_N\, d\boldsymbol{q}_N}{N!\,h^{3N}}.$$

The partition function for the grand canonical ensemble is a function of the parameters θ, μ, and V macroscopically determining the ensemble. One of these, V, is an extensive parameter, and two, θ and μ, are intensive. We have denoted the partition function for the grand ensemble by the same letter Q as the partition function of the canonical ensemble, but they should not be confused, since they are functions of different variables.†

Using (3.30), one can find the average value of any dynamic variable:

$$\langle A \rangle = \sum_{N \geqslant 0} \int A(p, q) f_N(p, q)\, d\Gamma_N. \tag{3.32}$$

The logarithm of the partition function (3.31) determines the thermodynamic potential $\Omega(\theta, \mu, V)$ for systems with a variable number of particles:

$$\Omega(\theta, \mu, V) = -\theta \ln Q(\theta, \mu, V). \tag{3.33}$$

†The partition function for the grand ensemble is sometimes denoted by $\Xi(\theta, \mu, V)$. We shall use the notation Q for partition functions for all the Gibbsian ensembles, indicating the type of ensemble by the arguments on which Q depends.

The existence of the thermodynamic limit

$$\lim_{\substack{V\to\infty \\ V/\langle N\rangle=\text{const}}} \frac{\Omega(\theta, \mu, V)}{V}$$

was proved by Yang and Lee [14], with certain limitations on the form of the interaction potential (cf. also [15-17d]).

We obtain expressions for the average energy and average particle number by differentiating (3.31) with respect to θ and μ:

$$\begin{aligned} \langle H\rangle - \mu\langle N\rangle &= \theta^2 \frac{\partial}{\partial\theta} \ln Q(\theta, \mu, V) = -\theta^2 \frac{\partial}{\partial\theta}\left(\frac{\Omega}{\theta}\right)_{\mu, V}, \\ \langle N\rangle &= \theta \frac{\partial}{\partial\mu} \ln Q(\theta, \mu, V) = -\left(\frac{\partial\Omega}{\partial\mu}\right)_{\theta, V}. \end{aligned} \tag{3.33a}$$

The average of minus the logarithm of the distribution function (3.30) is the entropy of the Gibbs grand canonical ensemble:

$$S = \langle\eta\rangle = -\sum_N \int f_N \ln f_N \, d\Gamma_N = \frac{\langle H\rangle - \Omega - \mu\langle N\rangle}{\theta} = -\left(\frac{\partial\Omega}{\partial\theta}\right)_{V, \mu} \tag{3.34}$$

(we have used the relations (3.33) and (3.33a)).

Using the grand canonical distribution, one can calculate the fluctuations in the energy and particle number in a grand ensemble. Differentiating (3.31) twice, with respect to μ and θ, we find

$$\begin{aligned} \langle N^2\rangle - \langle N\rangle^2 &= \theta \frac{\partial\langle N\rangle}{\partial\mu}, \\ \langle (H - \mu N)^2\rangle - \langle H - \mu N\rangle^2 &= \theta^2 \frac{\partial}{\partial\theta}(\langle H\rangle - \mu\langle N\rangle). \end{aligned} \tag{3.35}$$

The relative smallness of these fluctuations shows that the grand canonical ensemble differs little from the canonical and microcanonical ensembles, but it is considerably more convenient for calculations, since it is not necessary to take into account the supplementary conditions that the particle number and energy be constant.

For systems with a variable number of particles, a Gibbs theorem also holds, according to which a small part, with varying number of particles, of a microcanonical ensemble of systems with many degrees of freedom is distributed in accordance with the grand canonical ensemble.

A combination of a given system and a thermostat that is also a particle reservoir can be regarded as one large isolated and closed system. If the microcanonical distribution is used for the whole system, it follows from the Gibbs theorem that the open system in the thermostat is distributed in accordance with the grand canonical ensemble.

Let a large system with Hamiltonian H and particle number N consist of two subsystems (1) and (2) with Hamiltonians H_1 and H_2 and with particle numbers N_1 and N_2. If we neglect the interaction between the subsystems, then

$$H = H_1 + H_2, \quad N = N_1 + N_2.$$

We shall assume that subsystem (1) is much smaller than (2), which, as before, we shall call the thermostat.

We assume that the combined system is distributed microcanonically:

$$f_N = \begin{cases} \Omega^{-1}(E,\ N) \text{ in the layer } E \leqslant H \leqslant E + \Delta E, \\ 0 \quad \text{outside this layer} \end{cases} \tag{3.36}$$

We find the distribution function of subsystem (1) by integrating f over the variables of the second subsystem:

$$f_{N_1} = \int\limits_{E-H_1 \leqslant H_2 \leqslant E-H_1+\Delta E} f_{N_1+N_2}\, d\Gamma_{N_2}, \tag{3.37}$$

where

$$d\Gamma_{N_2} = \frac{dp_1'\, dq_1' \dots dp_{N_2}'\, dq_{N_2}'}{N_2!\, h^{3N_2}}.$$

Using (3.36), we find that the distribution function of the first subsystem is equal to the ratio of the statistical weight of the second subsystem with $E - H_1$ and $N - N_1$ to the statistical weight of the whole system:

$$f_{N_1}(p,\ q) = \frac{\Omega_2(E - H_1(p,\ q),\ N - N_1)}{\Omega(E,\ N)}. \tag{3.38}$$

Using the relation (3.5a), we introduce the entropy S_2 of the thermostat and the entropy S of the whole system

$$S_2(E,\ N) = \ln \Omega_2(E,\ N), \qquad S(E,\ N) = \ln \Omega(E,\ N)$$

and write (3.38) in the form

$$f_{N_1}(p,\ q) = \exp\{S_2(E - H_1(p,\ q),\ N - N_1) - S(E,\ N)\}.$$

Taking into account the smallness of the first subsystem in comparison with the thermostat ($H_1 \ll E$, $N_1 \ll N$), we expand the function $S_2(E - H_1, N - N_1)$ in a series in H_1 and N_1, confining ourselves to two terms:

$$S_2(E - H_1,\ N - N_1) \cong S_2(E,\ N) - \frac{\partial S_2}{\partial E} H_1 - \frac{\partial S_2}{\partial N} N_1.$$

Using this expansion, we write f_{N_1} in the form

$$f_N(p,\ q) = Q_1^{-1} \exp\left\{-\frac{H_1(p,\ q) - \mu N_1}{\theta}\right\}, \tag{3.39}$$

where

$$\frac{1}{\theta} = \frac{\partial S_2}{\partial E}, \qquad \frac{\mu}{\theta} = -\frac{\partial S_2}{\partial N}, \tag{3.39a}$$

i.e., θ is the temperature and μ the chemical potential.

One can also give a more rigorous proof of the Gibbs theorem for the grand canonical ensemble [24], analogous to the proof given in the previous section for the case of the canonical ensemble. In place of the integral relation (3.19), we shall have a similar relation, in which in addition to integrating over the energy, we perform a sum over the particle number N, a variable which takes only positive integral values. The problem thus reduces to the solution of an integral equation of the convolution type in continuous and discrete variables, and to the asymptotic estimation of the solution obtained. We shall not give this derivation here, but refer interested readers to the paper by Shubin [24]. Later, in §9 of Chapter II, we shall give a similar proof of the Gibbs theorem for the case of quantum statistics.

3.5. The Distribution for the Isobaric-Isothermal Ensemble

Until now, we have assumed the volume of the systems to be fixed; we shall now assume it to be variable, with the pressure and particle number fixed. This may be realized by means of a movable piston maintaining a constant pressure.

An ensemble of systems with constant particle number and given pressure in contact with a thermostat is called the Gibbs isobaric-isothermal ensemble. We shall discuss briefly the properties of this ensemble.

Together with its surroundings, the system can be regarded as one large closed system with constant energy and volume. Let E_1 and E_2 be the energies of the first and second subsystems, i.e., of the given system and the thermostat, and V_1 and V_2 their volumes, with

$$E = E_1 + E_2, \ V = V_1 + V_2 \qquad (E_1 \ll E_2, \ V_1 \ll V_2) \tag{3.40}$$

assumed to be constant. We can apply the microcanonical distribution to the combined system and then find the distribution function of the first subsystem by integrating over the coordinates of the particles of the second subsystem.

Repeating the arguments of §3.3, we obtain for the distribution function of the first subsystem

$$f_1(p, q) = \frac{\Omega_2(E - H_1, V - V_1)}{\Omega(E, V)} = \exp\{S_2(E - H_1, V - V_1) - S(E, V)\}. \tag{3.41}$$

Expanding S_2 in a series in E_1 and V_1, we obtain, since the first subsystem is small,

$$f_1(p, q) = Q^{-1}(\theta, p, N) \exp\left\{-\frac{H_1(p, q) + pV_1}{\theta}\right\}, \tag{3.42}$$

where

$$\frac{1}{\theta} = \frac{\partial S_2(E, V)}{\partial E}, \qquad \frac{p}{\theta} = \frac{\partial S_2(E, V)}{\partial V}. \tag{3.42a}$$

The parameter p, as we shall see below, plays the role of the pressure. (We are also using the symbol p for momenta, but this should lead to no confusion.)

Thus, we have shown that the distribution for an isobaric-isothermal ensemble has the form

$$f_V(p, q) = Q^{-1}(\theta, p, N) \exp\left\{-\frac{H(p, q) + pV}{\theta}\right\}, \tag{3.43}$$

where we have omitted the subscript 1.

Here $Q(\theta, p, N)$ is the partition function for the isobaric ensemble, determined from the normalization condition, which we can take, for example, in the form

$$\int f_V(p, q)\, d\Gamma\, dV = 1. \tag{3.43a}$$

Consequently,

$$Q(\theta, p, N) = \int \exp\left\{-\frac{H(p, q) + pV}{\theta}\right\} d\Gamma\, dV. \tag{3.43b}$$

With this normalization f_V has the dimensions of an inverse volume. A dimensionless normalization of f_V is also possible.

The partition function for the isobaric-isothermal ensemble is a function of the parameters θ, p, and N macroscopically determining the ensemble; two of these parameters, θ and p, are intensive, and one, N, is extensive. One must not confuse $Q(\theta, p, N)$ with $Q(\theta, V, N)$ and $Q(\theta, \mu, V)$ introduced earlier.

Using (3.43), we can calculate the average value of any dynamical variable $A(p, q)$:

$$\langle A\rangle = \int A(p, q) f_V(p, q)\, d\Gamma\, dV. \tag{3.44}$$

The logarithm of the partition function (3.43b) determines the thermodynamic potential $\Phi(\theta, p, N)$ for isobaric-isothermal systems, or, simply, the thermodynamic potential†

$$\Phi(\theta, p, N) = -\theta \ln Q(\theta, p, N). \tag{3.45}$$

Differentiating (3.43b) with respect to θ and p, we obtain expressions for the average energy and average volume:

$$\begin{aligned} \langle H\rangle + p\langle V\rangle &= \theta^2 \frac{\partial}{\partial\theta} \ln Q(\theta, p, N) = -\theta^2 \frac{\partial}{\partial\theta}\left(\frac{\Phi}{\theta}\right)_{p, N}, \\ \langle V\rangle &= -\theta \frac{\partial}{\partial p} \ln Q(\theta, p, N) = \left(\frac{\partial\Phi}{\partial p}\right)_{\theta, N}. \end{aligned} \tag{3.46}$$

The average of minus the logarithm of the distribution function (3.43) is the entropy of the isobaric-isothermal ensemble:

$$S = -\int f_V \ln f_V\, d\Gamma\, dV = \frac{\langle H\rangle + p\langle V\rangle - \Phi}{\theta} = -\left(\frac{\partial\Phi}{\partial\theta}\right)_{p, N}. \tag{3.47}$$

---

†$\Phi(\theta, p, N)$ should not be confused with the Massieu–Planck function (3.10a), since they depend on different variables. (Φ is often called the Gibbs free energy in Western literature. – Translation Editors.)

The relations (3.46) and (3.47) show that Φ plays the role of the thermodynamic potential, and the parameter p that of the pressure. The second derivative of Φ with respect to p determines the volume fluctuations:

$$\langle V^2\rangle - \langle V\rangle^2 = -\theta \frac{\partial^2\Phi}{\partial p^2} = -\theta \frac{\partial \langle V\rangle}{\partial p}. \tag{3.48}$$

Sometimes, one can also introduce a generalized Gibbsian ensemble for systems in a thermostat with variable volume and variable particle number [25, 26]. Then the ensemble is characterized by three intensive parameters, θ, μ, and p, i.e., by the temperature, the chemical potential and the pressure. This is inconvenient, since the parameters θ, μ, and p are not independent, but are connected by a relation $\mu = \mu(p, \theta)$. Therefore, we shall not use the generalized ensemble, which has no advantages over the other ensembles. To describe an ensemble of systems, it is always convenient to have at least one extensive parameter, as in the Gibbsian ensembles cited above.

The different Gibbsian ensembles are thermodynamically equivalent, i.e., thermodynamic functions calculated using them coincide for large systems in the limit $V \to \infty$, $N \to \infty$, $V/N =$ const. Therefore, whether one uses one ensemble or another is a question of practical convenience. As we have already said, the most convenient is the grand canonical ensemble, since it does not require us to take any supplementary conditions into account. For the calculation of fluctuations, the different Gibbsian ensembles are not equivalent, and lead, generally speaking, to different results. The reason for the thermodynamic equivalence of the statistical ensembles lies in the smallness of the fluctuations of energy, particle number and volume [(3.8d), (3.35), and (3.48)]. We can also give a more rigorous proof of the thermodynamic equivalence of the statistical ensembles by comparing the thermodynamic functions calculated for them [26, 27]. We shall return to this question in §13.

§4. The Connection between the Gibbsian Distributions and the Maximum of the Information Entropy

The concept of entropy in statistical mechanics is closely related to information theory. We shall examine this connection in the present section.

4.1. The Information Entropy

The information entropy, or, simply, the entropy, is a measure of the uncertainty in the information corresponding to a statistical distribution [22, 23, 28, 29].

Let p_k be a discrete distribution of probabilities of events. The quantity

$$H = -\sum_{k=1}^{n} p_k \ln p_k, \tag{4.1}$$

where

$$\sum_{k=1}^{n} p_k = 1, \tag{4.2}$$

is called the information entropy. It is also known as the Shannon entropy.

In fact, the quantity H is equal to zero if any of the p_k is equal to unity and the remaining p_k are zero, i.e., when the result of the experiment can be predicted with certainty and there is no indeterminacy in the information. The quantity H takes its largest value when all the p_k are equal, i.e., when $p_k = 1/n$. It is obvious that this limiting case possesses the greatest uncertainty. The entropy H is additive for a group of independent events with probabilities u_i and v_k, since if $p_{ik} = u_i v_k$, then

$$\begin{gathered} -\sum_{i,k} p_{ik} \ln p_{ik} = -\sum_{i} u_i \ln u_i - \sum_{k} v_k \ln v_k, \\ \sum_{k} u_k = \sum_{k} v_k = 1. \end{gathered} \tag{4.3}$$

It was proved by Shannon [28, 29] that the information entropy defined by (4.1) with the required continuity and additivity properties is unique within a constant factor.

For a probability distribution of a continuous variable x with density $f(x)$, the information entropy is

$$S_i = -\int f(x) \ln f(x)\, dx, \tag{4.4}$$

where

$$\int f(x)\, dx = 1.$$

The information entropy (4.4), like (4.1), is additive for independent events, i.e., if

$$f(x, y) = f_1(x) f_2(y),$$

then

$$-\int\int f(x, y) \ln f(x, y)\, dx\, dy = -\int f_1(x) \ln f_1(x)\, dx - \int f_2(y) \ln f_2(y)\, dy.$$

For the distribution $f(\mathrm{p}, \mathrm{q})$ in phase space, the Gibbs entropy (2.24) is also the information entropy, i.e.,

$$S_{\mathrm{i}} = \langle \eta \rangle = -\int f \ln f\, d\Gamma \tag{4.5}$$

with the normalization

$$\int f\, d\Gamma = 1 \tag{4.6}$$

for ensembles with a specified number of particles, or

$$S_{\mathrm{i}} = -\sum_{N \geqslant 0} \int f_N \ln f_N\, d\Gamma_N \tag{4.7}$$

with the normalization

$$\sum_{N \geqslant 0} \int f_N\, d\Gamma_N = 1 \tag{4.8}$$

for ensembles with a variable number of particles.

It is completely natural to treat the entropy in statistical mechanics as the information entropy, since statistical mechanics must not go beyond the limits of the restricted possibilities of measurements on macroscopic systems. If we use the language of information theory, we can say that the maximum quality of the information contained in the distribution function must be maintained. As we shall see below, the distribution functions for the Gibbsian ensembles satisfy this requirement.

The definition of entropy in the form (4.5) obviously only has meaning in the region of applicability of classical statistics; it can be regarded as the limiting case of the quantum-mechanical expression. Thus, in classical mechanics we introduce the concept of a probability density with invariant measure $d\Gamma$, since the concept of a probability and its measure are contained naturally in quantum mechanics.

We shall consider the extremal properties of the Gibbsian distributions, which Gibbs established long before the creation of information theory [1]. These are easily obtained from the auxiliary inequality (2.32)

$$\int f' \ln\left(\frac{f'}{f}\right) d\Gamma \geqslant 0, \tag{4.9}$$

where f and f' are any two normalized distributions, defined in the same phase space. The equality sign in (4.9) holds only when $f = f'$. In formula (4.9) and below, we drop the subscript N from $d\Gamma$.

4.2. Extremal Property of the Microcanonical Distribution

We shall now prove that of all the distributions with the same number of particles in the same energy layer, the microcanonical distribution (3.3) corresponds to the maximum value of the information entropy (4.5) [1].

Let f be the distribution function of the microcanonical ensemble, and f' an arbitrary distribution function defined in the same phase space and within the same energy layer, such that

$$\int f' \, d\Gamma = \int f \, d\Gamma = 1.$$

Putting f and f' into inequality (4.9), we obtain

$$-\int f' \ln f' \, d\Gamma \leqslant -\int f' \ln f \, d\Gamma = -\ln f \int f' \, d\Gamma = -\int f \ln f \, d\Gamma,$$

where we have used the constancy of f in the energy layer and the normalization condition for f' and f.

Thus, we have proved that of all the distributions with a given particle number in a given energy layer, the microcanonical distribution corresponds to the maximum information entropy. The other Gibbsian ensembles possess similar extremal properties, but under different conditions.

4.3. Extremal Property of the Canonical Distribution

If systems describable by a canonical ensemble are in contact with a thermostat, this means that they are characterized by a given average energy.

We shall show that the canonical distribution (3.7) corresponds to the maximum of the information entropy (4.5) for a given average energy

$$\langle H \rangle = \int Hf\, d\Gamma \tag{4.10}$$

and with conservation of the normalization

$$\int f\, d\Gamma = 1. \tag{4.11}$$

We shall find the extremum of the functional (4.5) under the supplementary conditions (4.10) and (4.11). Following the usual method, we seek the absolute extremum of the functional

$$-\int f \ln f\, d\Gamma - \beta \int fH\, d\Gamma - \lambda \int f\, d\Gamma,$$

where β and λ are Lagrange multipliers determined from the conditions (4.10) and (4.11). From the requirement that the first variation of this functional vanish, we find

$$f = Q^{-1}(\theta, V, N) \exp(-\beta H), \tag{4.12}$$

where

$$Q(\theta, V, N) = \int \exp(-\beta H)\, d\Gamma,$$

which coincides with the canonical distribution (3.7).

We have shown that (4.12) corresponds to the extremum of the functional (4.5). We now show that this extremum is a maximum.

Let f' be a normalized statistical distribution corresponding to the same average energy as the canonical distribution f

$$\int f'H\,d\Gamma = \int fH\,d\Gamma,$$

In other respects, f' is arbitrary. Putting (4.12) into the inequality (4.9), we obtain

$$-\int f'\ln f'\,d\Gamma \leqslant -\int f'\ln f\,d\Gamma = \ln Q + \beta\int f'H\,d\Gamma = \ln Q + \beta\int fH\,d\Gamma,$$

i.e.,

$$-\int f'\ln f'\,d\Gamma \leqslant -\int f\ln f\,d\Gamma.$$

Consequently, the canonical distribution corresponds to the maximum of the information entropy for a given average energy.

If the average values of any n quantities $\mathcal{P}_k$, which may also include the energy,

$$\langle\mathcal{P}_k\rangle = \int f\mathcal{P}_k\,d\Gamma \qquad (k = 0,\ 1,\ \ldots,\ n-1), \tag{4.13}$$

are specified, then from the condition for the extremum of the information entropy (4.5), using the same method as above, we immediately obtain the distribution

$$f = Q^{-1}\exp\left\{-\sum_{k=0}^{n-1}\mathcal{F}_k\mathcal{P}_k\right\}, \tag{4.14}$$

which coincides with (3.9), if the $\mathcal{P}_k$ are integrals of motion. If the $\mathcal{P}_k$ are not integrals of motion, then (4.14) does not satisfy Liouville's equation and cannot describe a statistical-equilibrium ensemble. We shall consider such distributions in Chapter IV.

4.4. Extremal Property of the Grand Canonical Distribution

If systems describable by the grand ensemble are in contact with a thermostat and a particle reservoir, this means that they are characterized by specifying the average energy and average number of particles. We shall show that the grand canonical distribution (3.30) corresponds to the maximum of the information entropy (4.7) for specified average energy

$$\langle H\rangle = \sum_N \int H f_N \, d\Gamma, \tag{4.15}$$

average number of particles,

$$\langle N\rangle = \sum_N \int N f_N \, d\Gamma \tag{4.16}$$

and with conservation of the normalization

$$\sum_N \int f_N \, d\Gamma = 1. \tag{4.17}$$

As above, we seek the absolute extremum of the functional

$$-\sum_N \int f_N \ln f_N \, d\Gamma - \beta \sum_N \int f_N H \, d\Gamma + \nu \sum_N \int f_N N \, d\Gamma - \lambda \sum_N \int f_N \, d\Gamma,$$

where β, ν, and λ are Lagrange multipliers. From the extremum condition, we find

$$f_N = Q^{-1} \exp\left\{-\frac{H - \mu N}{\theta}\right\}, \tag{4.18}$$

where

$$Q(\theta, \mu, V) = \sum_N \int \exp\left\{-\frac{H - \mu N}{\theta}\right\} d\Gamma,$$

which coincides with the grand canonical distribution (3.30).

It follows from the inequalities (4.9) that the extremum is a maximum. In fact, we have

$$\sum_N \int f'_N \ln \frac{f'_N}{f_N} \, d\Gamma \geqslant 0.$$

Putting (4.18) here in place of f_N, we obtain

$$\begin{aligned} -\sum_N \int f'_N \ln f'_N \, d\Gamma &\leqslant -\sum_N \int f'_N \ln f_N \, d\Gamma \\ &= -\sum_N \int f'_N \left(-\ln Q - \frac{H}{\theta} + \frac{\mu N}{\theta}\right) d\Gamma \\ &= -\sum_N \int f_N \left(-\ln Q - \frac{H - \mu N}{\theta}\right) d\Gamma, \end{aligned}$$

where we have used the conditions (4.15), (4.16), and (4.17). Consequently,

$$-\sum_N \int f'_N \ln f'_N \, d\Gamma \leqslant -\sum_N \int f_N \ln f_N \, d\Gamma,$$

i.e., the grand canonical distribution (4.18) does indeed correspond to the maximum of the information entropy for a specified average energy and average number of particles.

Finally, by the same method it is easy to show that the distribution (3.43) for an isobaric-isothermal ensemble corresponds to the maximum of the information entropy

$$S_u = -\int f_V \ln f_V \, d\Gamma \, dV \tag{4.19}$$

under the supplementary conditions that the average energy and average volume be constant:

$$\langle H \rangle = \int H f_V \, d\Gamma_V \, dV, \quad \langle V \rangle = \int V f_V \, d\Gamma \, dV. \tag{4.20}$$

The proof of this assertion is in no way different from the proofs given above.

The extremal properties that we have considered in this section for the Gibbsian ensembles enable us to introduce these ensembles in a somewhat different way. Information theory has borrowed many of its ideas from statistical mechanics. Now, when information theory is a well developed theory, one can, following Jaynes [22, 23], regard its concepts as primary concepts and use them in statistical mechanics. We can postulate the existence of the invariant probability measure (1.5b), and then, regarding statistical mechanics as information theory, obtain all the Gibbsian distributions from the condition that the information entropy be a maximum [22, 23]; all the calculations that we have given remain valid.

This method of deriving the statistical distributions must not, however, be regarded as a rigorous foundation for statistical mechanics; fundamental questions are simply not considered. But, in any case, the use of the extremal properties of the information entropy is a very convenient heuristic method for finding the different distribution functions. This method is suitable in both classical and quantum statistical mechanics. It is especially convenient for the nonequilibrium case, and we shall use it frequently (cf. Chapter IV, §§20 and 27).

§ 5. Thermodynamic Equalities

5.1. Quasi-Static Processes

Up to now, we have obtained thermodynamic relations by simply differentiating the partition functions for the various ensembles with respect to the variables on which they depend. To construct a complete system of thermodynamic equalities, we must consider the process of infinitely slow variation of the external parameters determining the given ensemble, i.e., a quasi-static process, since equilibrium thermodynamics studies precisely such processes. Meanwhile, without any theoretical justification, we shall take it that quasi-static processes exist. In Chapter IV, we shall consider the influence on statistical ensembles of varying the external parameters, and shall make the concept of a quasi-static process more precise.

Let the external parameters $a_1, a_2, \ldots, a_s$ characterize macroscopically a statistical-equilibrium state of the dynamical systems under consideration. Such parameters may be the volume of the vessel, the intensity of the external electric or magnetic field, etc.

We assume that the ensemble is in a state of statistical equilibrium. If the external parameters change, then, generally speaking, the distribution function of the ensemble will also change. We imagine that the external parameters $a_1, a_2, \ldots, a_s$ change so slowly that they can be assumed to be practically constant over a period of the order of the time it takes for the system to relax to the equilibrium distribution. It can then be assumed that at each moment of time, the system is in a state of statistical equilibrium. We shall call such a process of variation of the external parameters quasi-static.

If the parameters $a_1, \ldots, a_s$ are regarded as generalized coordinates, the corresponding generalized forces are

$$A_i = -\frac{\partial H}{\partial a_i}. \tag{5.1}$$

For a quasi-static process, the observed value of the generalized forces is equal to the average value over the equilibrium statistical ensemble:

$$\langle A_i \rangle = -\left\langle \frac{\partial H}{\partial a_i} \right\rangle = -\int f(p, q) \frac{\partial H(p, q)}{\partial a_i} d\Gamma. \tag{5.2}$$

If, for example, the volume V of the system is chosen as one of the external parameters, then the corresponding generalized force is the pressure,

$$p = -\left\langle \frac{\partial H(p, q)}{\partial V} \right\rangle. \tag{5.3}$$

Below, in §5.3, we shall give the explicit form of the dynamic variable $\partial H(p, q)/\partial V$.

5.2. Thermodynamic Equalities for the Microcanonical Ensemble

In systems describable by the microcanonical distribution (3.3), the pressure can be calculated by differentiating the statistical weight (3.3a) with respect to the volume, i.e., by differentiating the corresponding phase integral with respect to variable limits of integration. For this, it is more convenient to use the δ-function form (3.4a) of the statistical weight:

$$\Omega(E, N, V) = \int \delta(H(p, q) - E) \frac{dp\, dq}{N!\, h^{3N}}. \tag{5.4}$$

Here, we assume that H(p, q) depends on V through the potential $U_V(q)$, which represents the influence of the walls and increases sharply toward the boundaries of the volume V, and that the limits of integration are infinite. Differentiating (5.4) with respect to V, we obtain

$$\frac{\partial \Omega}{\partial V} = \int \frac{\partial}{\partial V} \delta(H(p, q) - E)\, d\Gamma = -\int \frac{\partial}{\partial E} \delta(H(p, q) - E) \frac{\partial H(p, q)}{\partial V} d\Gamma$$

or, since $\partial H(p, q)/\partial V$ does not depend on E,

$$\frac{\partial \Omega}{\partial V} = -\frac{\partial}{\partial E} \int \delta(H(p, q) - E) \frac{\partial H(p, q)}{\partial V} d\Gamma.$$

Using (3.4), we can rewrite this equality in the form

$$\frac{\partial \Omega (E, N, V)}{\partial V} = -\frac{\partial}{\partial E}\left(\Omega (E, N, V)\left\langle\frac{\partial H (p, q)}{\partial V}\right\rangle\right), \tag{5.5}$$

whence it follows that

$$\frac{\partial}{\partial V}\ln \Omega (E, N, V) = -\left\langle\frac{\partial H (p, q)}{\partial V}\right\rangle\frac{\partial}{\partial E}\ln \Omega (E, N, V) - \frac{\partial}{\partial E}\left\langle\frac{\partial H (p, q)}{\partial V}\right\rangle. \tag{5.5a}$$

The first term in the right-hand side of (5.5a) is finite as $N \to \infty$ ($V/N = \text{const}$), since the entropy $S = \ln \Omega$ is proportional to the volume, while the second term falls off like $1/N$ and can therefore be omitted. Consequently,

$$\frac{\partial}{\partial V}\ln \Omega (E, N, V) = -\left\langle\frac{\partial H (p\ q)}{\partial V}\right\rangle\frac{\partial}{\partial E}\ln \Omega (E, N, V), \tag{5.5b}$$

which, using (5.3), can be written in the form

$$\frac{p}{\theta} = \frac{\partial}{\partial V}\ln \Omega (E, N, V) = \frac{\partial S (E, N, V)}{\partial V}, \tag{5.6}$$

where the quantity

$$\frac{1}{\theta} = \frac{\partial}{\partial E}\ln \Omega (E, N, V) = \frac{\partial S (E, N, V)}{\partial E} \tag{5.6a}$$

plays the role of the inverse temperature.

From the relations (5.6) and (5.6a), we can obtain a complete system of thermodynamic equalities for the microcanonical ensemble.

The differential of the entropy

$$S (E, N, V) = \ln \Omega (E, N, V)$$

is equal to

$$dS = \frac{\partial S}{\partial E} dE + \frac{\partial S}{\partial V} dV + \frac{\partial S}{\partial N} dN \tag{5.7}$$

or, taking (5.6) and (5.6a) into account,

$$\theta\, dS = dE + p\, dV - \mu\, dN \tag{5.8}$$

where

$$-\frac{\mu}{\theta}=\frac{\partial S(E, N, V)}{\partial N}, \tag{5.8a}$$

and μ is the chemical potential.

Consequently, $1/\theta$ is the integrating factor for the left-hand side of Eq. (5.8); therefore, in agreement with macroscopic thermodynamics, the quantity $\theta = kT$, where k is Boltzmann's constant, can be identified with the temperature on the absolute scale, and S with the entropy. Equation (5.8) has the form of the usual thermodynamic identity, expressing the first and second laws of thermodynamics. Thus, all the thermodynamic relations can be derived from the microcanonical distribution.

5.3. The Virial Theorem

We have defined the pressure as the average value of the generalized force $\partial H(p, q)/\partial V$, which is a dynamic variable, i.e., a function of the momenta and coordinates of all the particles. We shall find a more explicit form of this dynamic variable. As before, we shall start from the microcanonical ensemble, although all the derivations can also be carried through analogously for any of the Gibbsian ensembles.

We write the statistical weight (5.4) in the form

$$\Omega(E, N, V)=\int_{\{\ldots V \ldots\}} \delta(H(p, q)-E) \frac{dp\, dq}{N!\, h^{3N}}, \tag{5.9}$$

where we assume that H(p, q) does not depend explicitly on V, while the term $U_V(q)$ describing the influence of the walls is taken into account by restricting the range of integration in such a way that each of the q_i lies in the volume V (we have indicated this by the symbol $\{\ldots V \ldots\}$).

It is convenient to describe the variability of the volume by introducing the parameter λ^3 multiplying V:

$$\Omega(E, N, \lambda^3 V)=\int_{\{\ldots \lambda^3 V \ldots\}} \delta(H(p, q)-E) \frac{dp\, dq}{N!\, h^{3N}}. \tag{5.9a}$$

We change the integration variables, scaling up by a factor of λ:

$$q=\lambda q', \qquad p=\lambda^{-1} p'. \tag{5.9b}$$

This is a canonical transformation that leaves the phase volume dpdq unchanged and makes the limits of integration independent of λ:

$$\Omega(E, N, \lambda^3 V) = \int\limits_{\{\ldots V \ldots\}} \delta\left(H\left(\frac{p}{\lambda}, \lambda q\right) - E\right) \frac{dp\, dq}{N!\, h^{3N}}.$$

Differentiating this expression with respect to λ, we obtain

$$\frac{\partial \Omega}{\partial \lambda} = -\int\limits_{\{\ldots V \ldots\}} \frac{\partial}{\partial E} \delta\left(H\left(\frac{p}{\lambda}, \lambda q\right) - E\right) \frac{\partial H\left(\frac{p}{\lambda}, \lambda q\right)}{\partial \lambda} d\Gamma,$$

or, putting $\lambda = 1$,

$$\left(\frac{\partial \Omega}{\partial \lambda}\right)_{\lambda=1} = -\frac{\partial}{\partial E} \Omega \left\langle \frac{\partial}{\partial \lambda} H\left(\frac{p}{\lambda}, \lambda q\right)\right\rangle_{\lambda=1}. \tag{5.10}$$

On the other hand,

$$\left(\frac{\partial \Omega}{\partial \lambda}\right)_{\lambda=1} = \frac{\partial \Omega}{\partial V} 3V; \tag{5.10a}$$

consequently,

$$\left(\frac{\partial \Omega}{\partial V}\right)_{\lambda=1} = -\frac{\partial}{\partial E} \Omega \frac{1}{3V} \left\langle \frac{\partial}{\partial \lambda} H\left(\frac{p}{\lambda}, \lambda q\right)\right\rangle_{\lambda=1}. \tag{5.11}$$

Comparing (5.11) with (5.5), we find an explicit form for the dynamic variable $-\partial H(p, q)/\partial V$ corresponding to the pressure:

$$-\frac{\partial H(p, q)}{\partial V} = -\frac{1}{3V} \frac{\partial}{\partial \lambda} H\left(\frac{p}{\lambda}, \lambda q\right)_{\lambda=1}. \tag{5.12}$$

For example, if H has the form of the Hamiltonian (1.2) of a system of particles with pair interaction (not necessarily radially symmetric), we obtain directly from (5.12)

$$-\frac{\partial H(p, q)}{\partial V} = \frac{2}{3V} \sum_i \frac{p_i^2}{2m} + \frac{1}{6V} \sum_{i \neq j} (\boldsymbol{q}_i - \boldsymbol{q}_j) \cdot \boldsymbol{F}_{ij}, \tag{5.13}$$

where

$$\boldsymbol{F}_{ij} = -\frac{\partial \phi(\boldsymbol{q}_i - \boldsymbol{q}_j)}{\partial \boldsymbol{q}_i}$$

is the force of the pair interaction between the i-th and j-th particles.

The formula (5.13) gives the required explicit representation of the dynamical variable describing the pressure.

The average value of the dynamical variable (5.13) leads to an expression for the pressure

$$p = \frac{2}{3V} \sum_i \frac{\langle p_i^2 \rangle}{2m} + \frac{1}{6V} \sum_{i \neq j} \langle (\boldsymbol{q}_i - \boldsymbol{q}_j) \cdot \boldsymbol{F}_{ij} \rangle, \tag{5.14}$$

which is called the virial theorem. The quantity

$$\frac{1}{2} \sum_{i \neq j} \langle (\boldsymbol{q}_i - \boldsymbol{q}_j) \cdot \boldsymbol{F}_{ij} \rangle$$

is called the virial.

Consequently, according to the virial theorem, the pressure is equal to two-thirds of the kinetic energy density plus one third of the virial density. This theorem remains valid if by $\langle \dots \rangle$ we mean averaging over any of the Gibbsian ensembles, and not only over the microcanonical ensemble. It is also valid for the case of quantum statistics, if by $\langle \dots \rangle$ we mean averaging over a quantum ensemble (cf. §11.3 of Chapter II). For a classical canonical ensemble, it is easy to calculate the average kinetic energy:

$$\sum_i \frac{\langle p_i^2 \rangle}{2m} = \frac{3N}{2} \theta, \tag{5.14a}$$

and the virial theorem gives

$$p = \frac{N\theta}{V} + \frac{1}{6V} \sum_{i \neq j} \langle (\boldsymbol{q}_i - \boldsymbol{q}_j) \cdot \boldsymbol{F}_{ij} \rangle. \tag{5.14b}$$

This form of the virial theorem is valid only in classical statistics.

The relation (5.14a) can be rewritten in the form

$$\frac{\langle p_i^2 \rangle}{2m} = \frac{3}{2} \theta = \frac{3}{2} kT, \tag{5.14c}$$

i.e., in classical statistics the average kinetic energy associated with one degree of freedom is the same for all degrees of freedom and is equal to

$$\frac{\theta}{2} = \frac{kT}{2}.$$

For the harmonic oscillator in classical statistics, the potential energy is also uniformly distributed over the degrees of freedom, and we also have $\theta/2$ for each degree of freedom.

5.4. Thermodynamic Equalities for the Canonical Ensemble

In the case when the ensemble is described by the canonical distribution (3.7), the average value of the derivative of the Hamiltonian with respect to the parameter a_i is equal to

$$\left\langle \frac{\partial H}{\partial a_i} \right\rangle = e^{F/\theta} \int e^{-H/\theta} \frac{\partial H}{\partial a_i} \, d\Gamma = -\theta e^{F/\theta} \frac{\partial}{\partial a_i} \int e^{-H/\theta} \, d\Gamma =$$

$$= -\theta e^{F/\theta} \frac{\partial}{\partial a_i} e^{-F/\theta} = \left(\frac{\partial F}{\partial a_i} \right)_{\theta, N}.$$

Consequently, the observed value of the average generalized force corresponding to change of the parameter a_i in a quasi-static process at constant θ and N is equal to

$$\langle A_i \rangle = -\left(\frac{\partial F}{\partial a_i} \right)_{\theta, N}, \tag{5.15}$$

or, in the particular case when $a_i = V$,

$$p = -\left(\frac{\partial F}{\partial V} \right)_{\theta, N}. \tag{5.16}$$

It is now easy to obtain all the thermodynamic relations for the case of a canonical ensemble. The free energy F of a system is a function of θ, a_1, ..., a_s and N. Consequently,

$$dF = \left(\frac{\partial F}{\partial \theta} \right)_{a_i, N} d\theta + \sum_{i=1}^{s} \left(\frac{\partial F}{\partial a_i} \right)_{\theta, N} da_i + \left(\frac{\partial F}{\partial N} \right)_{a_i, \theta} dN, \tag{5.17}$$

or

$$dF = -S \, d\theta - \sum_{i=1}^{s} \langle A_i \rangle \, da_i + \mu \, dN; \tag{5.17a}$$

here we have used the relations (3.8c) and (5.15) and have introduced the **chemical potential**

$$\mu = \left(\frac{\partial F}{\partial N}\right)_{a_i,\,\theta}. \tag{5.18}$$

In §13, it will be proved that the μ from (5.18) and from (5.8a) coincide in the thermodynamic limit.

Using (3.8c), we can write the relation (5.17) in the form

$$d\langle H\rangle = d(F+\theta S) = \theta\, dS - \sum_{i=1}^{s} \langle A_i\rangle\, da_i + \mu\, dN. \tag{5.19}$$

Consequently, $1/\theta$ is the integrating factor for

$$d\langle H\rangle + \sum_i \langle A_i\rangle\, da_i - \mu\, dN;$$

therefore $\theta = kT$ can be identified with the temperature of the thermostat on the absolute scale and S with the entropy.

Equation (5.17a) contains the complete system of thermodynamic relations, which can be expressed not only in terms of the free energy F, but also in terms of other thermodynamic potentials.

The thermodynamic equality (5.17a) can be rewritten in the form

$$d\left(F + \sum_i \langle A_i\rangle\, a_i\right) = d\Phi = -S\, d\theta + \sum_i a_i\, d\langle A_i\rangle + \mu\, dN, \tag{5.20}$$

where

$$\Phi = F + \sum_i \langle A_i\rangle\, a_i \tag{5.21}$$

is the thermodynamic potential for isobaric-isothermal systems (a function of the variables θ, A_i and N), which we introduced earlier for the particular case $a_i = V$, $\langle A_1\rangle = p$. In this case,

$$\Phi = F + pV. \tag{5.22}$$

The thermodynamic potential Φ is often called, simply, the thermodynamic potential (in a restricted meaning of the words) or the

Gibbs potential, and denoted by $G(\theta, p, N)$. The passage from F to Φ is a Legendre transformation for thermodynamic functions.

It follows from (5.20) that $\mu = \partial\Phi/\partial N$. On the other hand, the function Φ depends on only one extensive variable N, and since the thermodynamic limit

$$\lim_{\substack{N\to\infty \\ \langle V\rangle/N=\mathrm{const}}} \frac{\Phi}{N}$$

must be finite, Φ must be proportional to N and have the form $\Phi = Nf(\theta, p)$. Consequently,

$$\mu = \frac{\Phi}{N}. \tag{5.23}$$

The thermodynamic equality (5.17a) can also be written in the form

$$d(F - \mu N) = d\Omega = -S\,d\theta - \sum_i \langle A_i\rangle\,da_i - N\,d\mu, \tag{5.24}$$

where

$$\Omega = F - \mu N \tag{5.25}$$

is the thermodynamic potential in the variables θ, a_i, and μ.† The passage from F to Ω is also a Legendre transformation.

Taking (5.21) and (5.23) into account, we obtain for Ω the expression

$$\Omega = F - \Phi = -\sum_i \langle A_i\rangle a_i. \tag{5.26}$$

In the particular case when there is only one external parameter V,

$$\Omega = -pV. \tag{5.27}$$

Consequently, after introducing quasi-static processes, we can obtain all the thermodynamic functions from one canonical distribution.

†Do not confuse (5.25) with the statistical weight (3.39a), which is denoted by the same symbol.

5.5. Thermodynamic Equalities for the Grand Canonical Ensemble

A similar derivation of the thermodynamic equalities can also be performed for other ensembles, e.g., for the grand canonical ensemble. In this case,

$$\left\langle \frac{\partial H}{\partial a_i} \right\rangle = e^{\Omega/\theta} \sum_N \int e^{-(H-\mu N)/\theta} \frac{\partial H}{\partial a_i} \, d\Gamma = \left(\frac{\partial \Omega}{\partial a_i} \right)_{\theta, \mu}.$$

Consequently, the average generalized force is equal to

$$\langle A_i \rangle = -\left(\frac{\partial \Omega}{\partial a_i} \right)_{\theta, \mu}, \tag{5.28}$$

or, in a particular case,

$$p = -\left(\frac{\partial \Omega}{\partial V} \right)_{\theta, \mu}. \tag{5.29}$$

The thermodynamic potential Ω is a function of θ, μ, $a_1, \ldots, a_s$. Therefore,

$$d\Omega = \left(\frac{\partial \Omega}{\partial \theta} \right)_{a_i, \mu} d\theta + \sum_i \left(\frac{\partial \Omega}{\partial a_i} \right)_{\theta, \mu} da_i + \left(\frac{\partial \Omega}{\partial \mu} \right)_{a_i, \theta} d\mu, \tag{5.30}$$

or, if we use the relations (3.33a), (3.34), and (5.28)

$$d\Omega = -S\, d\theta - \sum_i \langle A_i \rangle \, da_i - \langle N \rangle \, d\mu. \tag{5.31}$$

Thus, we have obtained a thermodynamic relation that coincides with (5.24) if we put $\langle N \rangle = N$.

§ 6. Fluctuations

6.1. Quasi-Thermodynamic Theory of Fluctuations

The Gibbsian statistical ensembles enable us to calculate the fluctuations of any dynamical variable in a state of statistical equilibrium. We have already considered the fluctuations of certain quantities, e.g., the energy fluctuations (3.8d) in a canonical ensemble, the fluctuations of particle number and energy (3.35) in a grand canonical ensemble, and the volume fluctuations (3.48) in an isobaric-isothermal ensemble. In all these cases, we were interested in fluctuations of quantities on which the distribution function of the ensemble depended explicitly. In a microcanonical ensemble, the energy and particle number are specified and, consequently, do not fluctuate, although pressure fluctuations occur in this ensemble.

The calculation of fluctuations for an arbitrary dynamical quantity is a problem that is no less complicated than the calculation of its average value; we must therefore set limits on the problem. One might be interested in the probability distribution of the fluctuations of different quantities, with the thermodynamic functions of the system assumed known. This is the problem posed in the quasi-thermodynamic theory of fluctuations; we shall now give an account of this theory, following the work of Green and Callen [30]. A simplifying feature in the theory of fluctuations is their relative smallness.

Let $\xi_1, \xi_2, \ldots, \xi_s$ be physical variables characterizing the system; they are not necessarily integrals of motion.† We assume, however, that the average values $\langle \xi_k \rangle$ can characterize some state of incomplete statistical equilibrium. We must determine the thermodynamic functions of this state.

Following Leontovich [2], we define the free energy of a nonequilibrium state characterized by specifying the average values $\langle \xi_k \rangle$ as the free energy of the equilibrium state in the auxiliary fields that constrain the system to be in equilibrium at the given values of $\langle \xi_k \rangle$. We shall use this device repeatedly below.

We shall consider the most complete possible equilibrium at given values of $\langle \xi_k \rangle$. This means that we must use a statistical ensemble in which the average values of the quantities ξ_k

$$\langle \xi_k \rangle = \int \xi_k f(p, q) \, d\Gamma \tag{6.1}$$

are specified, and in which the information entropy (4.5) is a maxi-

†In [30], the quantities ξ_k are assumed to be integrals of motion.

mum. For ensembles with unspecified N, (6.1) implies summation over N as well as integration.

In our case, the conditions (6.1) must be added to the usual conditions of constancy of the average energy (4.15), constancy of the average particle number (4.16), and conservation of the normalization.

Repeating the arguments of §4, we obtain

$$f = Q^{-1}\exp\left\{-\beta\left(H - \mu N - \sum_k a_k \xi_k\right)\right\}, \tag{6.2}$$

where the a_k are parameters thermodynamically conjugate to $\langle \xi_k \rangle$ and determined from the equations (6.1).

The partition function Q, which can be determined from the normalization condition for the distribution function (6.2), is a function of θ, μ, and a_k or of θ, μ, and $\langle \xi_k \rangle$ and determines the thermodynamic functions in a state of incomplete statistical equilibrium with specified $\langle \xi_k \rangle$ as a function of θ, μ, and $\langle \xi_k \rangle$. Below we shall not indicate the possible dependence of Q on the volume V.

Generally speaking the distribution function (6.2) does not satisfy Liouville's equation if the $\langle \xi_k \rangle$ are not integrals of motion, but it turns out to be suitable for the calculation of fluctuations.

The physical meaning of the distribution function (6.2) of systems in incomplete statistical equilibrium is that we are regarding such a state as a state of statistical equilibrium, although in some auxiliary fields a_k that constrain it to be an equilibrium state [2].

It is convenient to rewrite the distribution function (6.2) in a more symmetric form, similar to (4.14):

$$f = \exp\left\{-\Phi(\mathcal{F}_0, \ldots, \mathcal{F}_n) - \sum_{k=0}^{n} \mathcal{F}_k \mathcal{P}_k\right\}, \tag{6.3}$$

where

$$\begin{aligned} &\mathcal{F}_0 = \beta, && \mathcal{P}_0 = H, \\ &\mathcal{F}_1 = -\beta\mu, && \mathcal{P}_1 = N, \\ &\mathcal{F}_k = -\beta a_{k-1}, && \mathcal{P}_k = \xi_{k-1} \qquad (k = 2, 3, \ldots, n). \end{aligned} \tag{6.3a}$$

The function

$$\Phi(\mathcal{F}_0, \ldots, \mathcal{F}_n) = \ln Q(\mathcal{F}_0, \ldots, \mathcal{F}_n). \tag{6.4}$$

determined from the normalization condition for (6.3)

$$e^{\Phi} = \sum_N \int \exp\left\{-\sum_k \mathcal{F}_k \mathcal{P}_k\right\} d\Gamma, \tag{6.5}$$

is called the Massieu–Planck thermodynamic function. It is connected with the entropy (4.7) by the relation

$$S = \Phi + \sum_{k=0}^{n} \mathcal{F}_k \langle \mathcal{P}_k \rangle \tag{6.6}$$

and makes it possible to represent all the thermodynamic relations in a specially symmetric form.

For example, the average values and fluctuations of the quantities $\mathcal{P}_k$ are

$$\begin{gathered}\langle \mathcal{P}_k \rangle = -\frac{\partial \Phi}{\partial \mathcal{F}_k}, \\ \langle \mathcal{P}_i \mathcal{P}_k \rangle - \langle \mathcal{P}_i \rangle \langle \mathcal{P}_k \rangle = \frac{\partial^2 \Phi}{\partial \mathcal{F}_i \, \partial \mathcal{F}_k} = -\frac{\partial \langle \mathcal{P}_i \rangle}{\partial \mathcal{F}_k} = -\frac{\partial \langle \mathcal{P}_k \rangle}{\partial \mathcal{F}_i}.\end{gathered} \tag{6.7}$$

It is also easy to calculate the higher correlations in an analogous way [30].

Using the entropy (6.6), we can write the distribution function (6.3) in the form

$$f = \exp\left\{-S - \sum_{k=0}^{n} \mathcal{F}_k (\mathcal{P}_k - \langle \mathcal{P}_k \rangle)\right\}, \tag{6.8}$$

where

$$e^S = \sum_N \int \exp\left\{-\sum_k \mathcal{F}_k (\mathcal{P}_k - \langle \mathcal{P}_k \rangle)\right\} d\Gamma. \tag{6.9}$$

Differentiating (6.9) with respect to the $\langle \mathcal{P}_k \rangle$, we obtain another form of the thermodynamic equalities

$$\frac{\partial S}{\partial \langle \mathcal{P}_k \rangle} = \mathcal{F}_k, \tag{6.10}$$

since the terms arising from the differentiation of $\mathscr{F}_i$ with respect to the $\langle \mathscr{P}_k \rangle$, vanish. The formulas (6.7)-(6.10) serve as a starting point for the calculation of the probability distribution of fluctuations of the quantities $\mathscr{P}_k$.

6.2. The Gaussian Distribution for the Probability of Fluctuations

We turn now to the calculation of the probability of a fluctuation of the quantity $\mathscr{P}_k$. In the formulas (6.3) and (6.8), the quantity $\mathscr{P}_k$ is a dynamic variable, i.e., a function of the momenta and coordinates p, q of all the particles.

Following Einstein [31], from (6.3) we construct a macroscopic distribution function, assuming that the $\mathscr{P}_k$ have certain fixed values.

We introduce the macroscopic distribution function W:

$$W\, d\mathscr{P}_0 \ldots d\mathscr{P}_n = \Omega\, d\mathscr{P}_0 \ldots d\mathscr{P}_n \exp\left\{-\Phi(\mathscr{F}_0, \ldots, \mathscr{F}_n) - \sum_k \mathscr{F}_k \mathscr{P}_k\right\}, \tag{6.11}$$

which gives the probability that the parameters $\mathscr{P}_0, \ldots, \mathscr{P}_n$ lie in the regions $d\mathscr{P}_0, \ldots, d\mathscr{P}_n$ close to the values $\mathscr{P}_0, \ldots, \mathscr{P}_n$. We now treat the $\mathscr{P}_k$ not as dynamic variables, but as ordinary quantities, while keeping the previous notation. The quantity Ω has the meaning of the number of states in the regions $d\mathscr{P}_0, \ldots, d\mathscr{P}_n$ close to $\mathscr{P}_0, \ldots, \mathscr{P}_n$. We must now normalize the function W not in the phase space, but in the space of the values $\mathscr{P}_0, \ldots, \mathscr{P}_n$:

$$\int W(\mathscr{P}_0, \ldots, \mathscr{P}_n)\, d\mathscr{P}_0 \ldots d\mathscr{P}_n = 1. \tag{6.12}$$

The quantity Ω can be estimated with the help of the entropy S of the microcanonical ensemble in which the parameters $\mathscr{P}_0, \ldots, \mathscr{P}_n$ are specified to be in the regions $d\mathscr{P}_0, \ldots, d\mathscr{P}_n$ close to the values $\mathscr{P}_0, \ldots, \mathscr{P}_n$:

$$S = \ln \frac{\Omega}{\Omega_0}, \tag{6.13}$$

where Ω_0 is a normalization constant, unimportant for the present, which we shall determine later.

Taking (6.6) and (6.13) into account, we write the macroscopic distribution function (6.11) in the form

$$W = \Omega_0 \exp\left\{s - S - \sum_k \mathcal{F}_k(\mathcal{P}_k - \langle \mathcal{P}_k \rangle)\right\}, \tag{6.14}$$

where S is the entropy in the quasi-equilibrium grand canonical ensemble (6.3).

We shall make a comment that is important for what follows. Because of the thermodynamic equivalence of the statistical ensembles, the entropy in the ensemble (6.3) is the same function of $\langle \mathcal{P}_0 \rangle, \ldots, \langle \mathcal{P}_n \rangle$, as the entropy in the corresponding microcanonical ensemble is of $\mathcal{P}_0, \ldots, \mathcal{P}_n$, i.e., S and s are the same functions, but of different variables. Therefore, s − S can be expanded in a series in $\mathcal{P}_k - \langle \mathcal{P}_k \rangle$ and, because of the smallness of the fluctuations, we can confine ourselves to the second-order terms. Using (6.10), we obtain

$$\begin{gathered} s - S = \sum_k \mathcal{F}_k \Delta\mathcal{P}_k + \frac{1}{2} \sum_{i,k} \frac{\partial^2 S}{\partial \langle \mathcal{P}_i \rangle \partial \langle \mathcal{P}_k \rangle} \Delta\mathcal{P}_i \Delta\mathcal{P}_k, \\ \Delta\mathcal{P}_k = \mathcal{P}_k - \langle \mathcal{P}_k \rangle. \end{gathered} \tag{6.15}$$

Putting (6.15) into (6.14) and noting that the linear terms cancel, we obtain the distribution function for the fluctuations:

$$W = A \exp\left\{\frac{1}{2} \sum_{i,k} \frac{\partial^2 S}{\partial \langle \mathcal{P}_i \rangle \partial \langle \mathcal{P}_k \rangle} \Delta\mathcal{P}_i \Delta\mathcal{P}_k\right\} \tag{6.16}$$

[the constant A is determined from the normalization condition (6.12)]. Consequently, the probability of fluctuations of the quantities $\mathcal{P}_k$ is determined by the **Gaussian distribution** (6.16).

We write the Gaussian distribution (6.16) in the form

$$W = A \exp\left\{-\frac{1}{2} \sum_{i,k} \lambda_{ik} x_i x_k\right\}, \tag{6.17}$$

where

$$\lambda_{ik} = \lambda_{ki} = -\frac{\partial^2 S}{\partial \langle \mathcal{P}_i \rangle \partial \langle \mathcal{P}_k \rangle}, \qquad x_i = \Delta\mathcal{P}_i = \mathcal{P}_i - \langle \mathcal{P}_i \rangle. \tag{6.17a}$$

or, after calculation of the normalization constant A,

$$W = \frac{\sqrt{\lambda}}{(2\pi)^{(n+1)/2}} \exp\left\{-\frac{1}{2}\sum_{i,k} \lambda_{ik} x_i x_k\right\}, \tag{6.18}$$

where λ is a determinant with elements λ_{ik}, which we assume to be positive and nonzero, and n + 1 is the number of the variables x_i.

Using the Gaussian distribution function (6.18), we can calculate all fluctuations. We write (6.18) in the form

$$W = Ae^{-K(x_0, \ldots, x_n)}, \tag{6.18a}$$

where

$$K(x_0, \ldots, x_n) = \frac{1}{2}\sum_{i,k} \lambda_{ik} x_i x_k = \frac{1}{2}\boldsymbol{\lambda} : \boldsymbol{xx} \tag{6.18b}$$

is a quadratic form; $\boldsymbol{\lambda}$ is the tensor λ_{ik}, $\mathbf{x}$ is a vector with components x_i and the symbol : denotes the complete contraction of two tensors.

We shall calculate the average value over the Gaussian distribution of the product of x_i with X_k,

$$X_k = \frac{\partial K}{\partial x_k}. \tag{6.19}$$

Integrating by parts, we obtain

$$A \int e^{-K} x_i \frac{\partial K}{\partial x_k} dx_0 \ldots dx_n = -A \int x_i \frac{\partial}{\partial x_k} e^{-K} dx_0 \ldots dx_n = \delta_{ik}.$$

Consequently,

$$\left\langle x_i \frac{\partial K}{\partial x_k}\right\rangle = \langle x_i X_k \rangle = \delta_{ik}, \tag{6.20}$$

or

$$\sum_m \langle x_i x_m \rangle \lambda_{mk} = \delta_{ik}, \tag{6.21}$$

i.e., the product of the matrix $\langle x_i x_k \rangle$ with λ_{mk} is equal to the unit matrix.

Thus, the mean square fluctuations are determined by the matrix that is the reciprocal of λ_{mk}:

$$\langle x_i x_m \rangle = (\boldsymbol{\lambda}^{-1})_{im}. \tag{6.22}$$

We shall show that $\boldsymbol{\lambda}$ is a positive-definite matrix. We reduce (6.21) to diagonal form by going over from x to x' by means of a canonical transformation; then

$$\langle (x_i')^2 \rangle = \frac{1}{\lambda_{ii}}, \tag{6.22a}$$

and, consequently, $\lambda_{ii} > 0$, as we assumed earlier.

The formula (6.22) gives the squared fluctuations calculated by means of the Gaussian distribution. On the other hand, the formula (6.7) gives the exact value of the fluctuations calculated by means of the Gibbsian distribution (6.3):

$$\langle x_i x_k \rangle = \frac{\partial^2 \Phi}{\partial \mathcal{F}_i \, \partial \mathcal{F}_k}. \tag{6.23}$$

It is easily seen that these values coincide. Indeed, the matrices

$$\frac{\partial^2 \Phi}{\partial \mathcal{F}_i \, \partial \mathcal{F}_k} \quad \text{and} \quad -\frac{\partial^2 S}{\partial \langle \mathcal{P}_i \rangle \, \partial \langle \mathcal{P}_k \rangle}$$

are mutually reciprocal. Using (6.7) and (6.10), we obtain

$$-\sum_k \frac{\partial^2 \Phi}{\partial \mathcal{F}_i \, \partial \mathcal{F}_k} \frac{\partial^2 S}{\partial \langle \mathcal{P}_m \rangle \, \partial \langle \mathcal{P}_k \rangle} = \sum_k \frac{\partial \langle \mathcal{P}_i \rangle}{\partial \mathcal{F}_k} \frac{\partial \mathcal{F}_k}{\partial \langle \mathcal{P}_m \rangle} = \frac{\partial \langle \mathcal{P}_i \rangle}{\partial \langle \mathcal{P}_m \rangle} = \delta_{im}. \tag{6.24}$$

This means that the Gaussian distribution (6.18) gives the exact value of the square fluctuations of the quantities x_i. For fluctuations of higher order, this is no longer true, and higher terms of the expansion (6.15) of the entropy must be taken into account.

We shall go no deeper into the theory of fluctuations in this chapter, but refer the interested reader to the literature [2, 26, 32, 33]. We shall repeatedly return to questions concerning fluctuations, since they are intimately connected with irreversible processes.

This section concludes our brief survey of the basic concepts of classical statistical mechanics.

Classical statistical mechanics is applicable at sufficiently high temperatures, when quantum effects can be neglected. In other circumstances, it can lead to erroneous conclusions. For example, the equipartition of energy between the degrees of freedom, which follows from classical statistics, is not valid at low temperatures. And even in the region where it is applicable classical statistical mechanics, as we have seen in §1, borrows some of its postulates from quantum statistics. For example, the assumption of the existence of a minimum cell h^{3N} in phase space, and the factor $1/N!$ that takes into account that states differing only by a permutation of the particles are identical, are introduced into classical statistics from outside. These effects are taken into account completely naturally in quantum statistical mechanics, the basic principles of which we describe in Chapter II.

Chapter II

Equilibrium Statistical Thermodynamics of Quantum Systems

We shall give a brief survey of the basic principles of the statistical mechanics of quantum systems for the equilibrium case, to the extent that this will be required for the following development (cf. [1-4]).

§ 7. The Statistical Operator

7.1. The Pure Ensemble

Until now, we have considered classical statistical mechanics, in which the state of systems are described by a point (p, q) in the 6N-dimensional phase space, and the evolution of the state in time by Hamilton's equations (1.1). The dynamic variables, e.g., the energy (1.2) and the total momentum, were functions of the coordinates and momenta q, p, i.e., functions of the state of the dynamic system.

Quantum statistical mechanics originates from the basic concepts of quantum mechanics, in which the situation is completely different. In quantum mechanics, the state of a dynamic system is described by a wavefunction $\Psi(x_1, \ldots, x_N, t)$, or, more briefly, $\Psi(x, t)$, which depends on the time and on the coordinates $x_1, \ldots, x_N$ of the particles, or on another system of simultaneously measurable quantities.

The evolution of the state in time is determined by the Schrödinger equation

$$i\hbar \frac{\partial \Psi}{\partial t} = H\Psi, \tag{7.1}$$

where H is a self-adjoint operator acting on the wave function Ψ, and $\hbar$ is Planck's constant.

For example, for a system of N identical particles of mass m, with no internal degrees of freedom and interacting through a potential $\phi(|\mathbf{x}|)$, the Schrödinger equation has the form

$$i\hbar \frac{\partial \Psi}{\partial t} = \left\{ -\frac{\hbar^2}{2m} \sum_{1 \leqslant j \leqslant N} \nabla_j^2 + \frac{1}{2} \sum_{j \neq k} \phi(|x_j - x_k|) \right\} \Psi, \tag{7.2}$$

where

$$\nabla_j^2 = \sum_{1 \leqslant \alpha \leqslant 3} \frac{\partial^2}{(\partial x_j^{\alpha})^2}$$

is the Laplacian.

The Schrödinger equation completely determines Ψ at any time t, if Ψ is known at the initial time t = 0. For example, for an isolated system, when H does not depend explicitly on time,

$$\Psi(t) = e^{\frac{1}{i\hbar} Ht} \Psi(0) \tag{7.3}$$

is the formal solution of the Schrödinger equation.

In quantum mechanics, the dynamic variables are not functions of the state of the dynamic system, but are linear self-adjoint operators, acting in the space of the wave functions. Their spectra determine the possible observable values of the physical quantities. Therefore, specifying the state of the system, i.e., Ψ, does not imply exact knowledge of the dynamic variables. The wave function Ψ enables us to calculate only the mean value in the state Ψ of any dynamic variable represented by the operator A:

$$\bar{A} = (\Psi^*, A\Psi), \tag{7.4}$$

where the wave functions are normalized to unity

$$(\Psi^*, \Psi) = 1, \tag{7.5}$$

and the brackets denote a scalar product of the functions in Hilbert space, i.e.,

$$(\Psi^*, \Phi) = \int \Psi^*(x)\,\Phi(x)\,dx; \tag{7.6}$$

x denotes the set of coordinates $\mathbf{x}_1, \mathbf{x}_2, \ldots, \mathbf{x}_N$.

The function $\Psi(x)$, generally speaking, depends also on the time t, i.e., we should write $\Psi(x, t)$, but we omit the argument t. If the state is also characterized by spin variables $\sigma_1, \ldots, \sigma_N$, then in (7.6) in addition to the integration we must also sum over the spin variables.

The formula (7.4) gives only probabilistic predictions about the observable values of any physical quantities. Only in the special case when Ψ is an eigenfunction of the operator A does formula (7.4) give the exact value of the quantity A in the state Ψ.

A state which can be described by a wave function is called a *pure state*. The corresponding statistical ensemble, i.e., a large number of noninteracting "copies" of the given system, each in the given quantum state under the condition that averages are calculated from formula (7.4), is called a *pure ensemble*. A pure state is usually simply called a quantum-mechanical state. It corresponds to complete, maximum possible information on the quantum-mechanical system. The whole of quantum mechanics, with the exception of certain problems in the theory of measurement [2, 5-9], is based on the application of pure ensembles.

The expressions for the mean values of dynamic quantities in a pure ensemble are conveniently represented by means of a projection operator.

We write the linear operator A in the matrix x representation, defining it by means of matrix elements,

$$A\Psi(x) = \int A(x, x')\,\Psi(x')\,dx'. \tag{7.7}$$

Putting (7.7) into (7.4) we obtain

$$\bar{A} = \int A(x, x')\,\mathscr{P}(x', x)\,dx\,dx' = \mathrm{Tr}\,(A\mathscr{P}), \tag{7.8}$$

where

$$\mathscr{P}(x, x') = \Psi(x)\,\Psi^*(x') \tag{7.8a}$$

is called a **projection operator** which thus represents the pure ensemble.

This name is connected with the fact that the effect of the operator $\mathcal{P}$ on any function φ projects it on to the direction of Ψ in Hilbert space. In fact,

$$\mathcal{P}\varphi = \int \mathcal{P}(x, x')\varphi(x')\,dx' = (\Psi^*, \varphi)\,\Psi(x). \tag{7.9}$$

The functions Ψ are assumed to be normalized.

The projection operator is Hermitian, as follows from (7.8a):

$$\mathcal{P}^*(x, x') = \mathcal{P}(x', x).$$

In addition, it has the property

$$\mathcal{P}^2 = \mathcal{P}, \tag{7.10}$$

which follows from (7.9). This property is obvious, since after one projection operation, all the subsequent projections on to the same direction give the same result.

Also,

$$\mathrm{Tr}\,\mathcal{P} = 1, \tag{7.11}$$

always, which follows from (7.8) if we replace A by the unit operator, or from (7.8a) using the normalization (7.5).

We shall show that all the eigenvalues of the projection operator are zero except one, which is equal to unity.

The Hermitian operator $\mathcal{P}$ can always be brought to diagonal form. Then its eigenvalues will also satisfy Eq. (7.10), and, consequently, will be zero or unity. But by virtue of the normalization condition (7.11), the projection operator can then have only one eigenvalue, equal to unity. Therefore, all the eigenvalues of the projection operator are equal to zero, except one which is equal to unity.

The condition (7.10) together with the Hermiticity condition can be regarded as a definition of the projection operator, and consequently also of a pure state.

Knowledge of the wave function $\Psi_1(x, t)$ enables us to calculate the probability of a transition from a state $\Psi_1(x, t)$ to any state

$\Psi_2(x, t)$ in time t:

$$W_{12}(t) = |(\Psi_2^*(t), \Psi_1(t))|^2,$$

which can be written using projection operators:

$$W_{12}(t) = \int \mathcal{P}_1(x, x', t)\mathcal{P}_2(x', x, t)\,dx\,dx' = \mathrm{Tr}\,(\mathcal{P}_1(t)\mathcal{P}_2(t)),$$

where

$$\mathcal{P}_\alpha(x, x', t) = \Psi_\alpha(x, t)\Psi_\alpha^*(x', t) \qquad (\alpha = 1, 2)$$

are the projection operators corresponding to the states Ψ_1 and Ψ_2.

For a pure ensemble, minus the mean of the logarithm of $\mathcal{P}$, which vanishes:

$$-\langle \ln \mathcal{P} \rangle = -\mathrm{Tr}\,(\mathcal{P} \ln \mathcal{P}) = 0.$$

corresponds to the information entropy (4.1). In fact, if we diagonalize $\mathcal{P}$, the product $\mathcal{P}_{nn} \ln \mathcal{P}_{nn}$ is zero, since $\mathcal{P}_{nn}$ is either zero or unity, and $x \ln x$ is taken to be zero at $x = 0$. Thus, for a pure ensemble, the degree of indeterminacy in the information is zero, i.e., it corresponds to complete, maximum possible information about quantum-mechanical systems.

Quantum statistical mechanics is in some sense simpler than classical, since it already contains the concept of probability, but the quantum-mechanical pure ensemble is found to be inadequate in quantum statistics, since, as a rule, we do not have at our disposal complete information about the systems under study, which consist of a large number of particles.

7.2. The Mixed Ensemble and the Statistical Operator

Quantum statistical mechanics used a statistical ensemble of a more general type than the "pure" ensemble considered above, namely a *mixed ensemble* (or "mixture"), which is based on an incomplete set of data on the system (cf. [1-10]).

We shall consider a large number of identical noninteracting copies of the given system; these copies can be in different quantum states.

In a mixed ensemble, only the probabilities $w_1, w_2, \ldots$ of finding the system in the different quantum states $\Psi_1, \Psi_2, \ldots$ are defined. The mean value of any physical quantity represented by the operator A is defined in a mixed state by the expression

$$\langle A \rangle = \sum_k w_k (\Psi_k^*, A\Psi_k), \tag{7.12}$$

with

$$\sum_k w_k = 1, \quad w_k \geqslant 0. \tag{7.12a}$$

Here, $(\Psi_k^*, A\Psi_k)$ is the quantum-mechanical mean value of the operator A in the state Ψ_k. The supplementary conditions (7.12a) mean that the total probability for all quantum states is unity, and that a probability cannot be a negative quantity.

The pure ensemble is a particular case of the mixed ensemble, in which all the probabilities w_k are zero except one which is unity. Then (7.12) goes over to (7.4).

In a mixed ensemble, in contrast to the pure ensemble, the different quantum states do not interfere, since in the definition of the averages (7.12) for the mixture, the mean values and not the wave functions are combined. If the system were described by a wave function in the form of a superposition of states Ψ_k, interference cross-terms connecting the different quantum states, of which there are none in (7.12), would be present in the expression for the averages (7.4).

To study mixed ensembles, it is convenient to introduce the statistical operator proposed by Neumann [2, 3] and also, for a special case, by Landau [11]. We write the linear operator A in the matrix x-representation (7.7). Putting (7.7) into (7.12), we obtain

$$\langle A \rangle = \int A(x, x') \rho(x', x) \, dx \, dx', \tag{7.13}$$

or

$$\langle A \rangle = \mathrm{Tr}\,(A\rho), \tag{7.14}$$

where

$$\rho(x, x') = \sum_k w_k \Psi_k(x) \Psi_k^*(x') \tag{7.15}$$

is the statistical operator in the matrix x-representation, or **density matrix**. The density matrix (7.15) depends on 2N variables $x_1, \ldots, x_N; x'_1, \ldots, x'_N$, i.e., on the same number of variables as the distribution function in classical statistical mechanics, which depends on the 2N coordinates and momenta $q_1, \ldots, q_N; p_1, \ldots, p_N$.

The statistical operator ρ obeys the normalization condition

$$\operatorname{Tr} \rho = 1, \tag{7.16}$$

since

$$\operatorname{Tr} \rho = \int \rho(x, x)\, dx = \sum_k w_k (\Psi_k^*, \Psi_k),$$

and from the normalization conditions for the wave functions and the probabilities w_k, it follows that

$$(\Psi_k^*, \Psi_k) = 1, \qquad \sum_k w_k = 1.$$

The normalization condition (7.16) also follows from (7.14), if we replace A by the unit operator. This normalization condition is the quantum analog of the normalization (1.5a) of the distribution function.

Formula (7.14) is convenient in that the trace of a matrix is invariant under unitary transformations of the operators. Therefore, formula (7.14) does not depend on the representation of the operators A and ρ; it is valid for any representation, and not only the matrix x-represention, of the operators. In practical problems, other more convenient representations are usually used for the operators.

For example, in a discrete matrix n-representation,

$$\langle A \rangle = \sum_{m, n} A_{mn} \rho_{nm},$$

where A_{mn} are the matrix elements of the operator in the n-representation, and ρ_{nm} is the density matrix in the n-representation.

The density matrix (7.15) is Hermitian:

$$\rho^*(x, x') = \rho(x', x), \tag{7.17}$$

which follows directly from its definition (7.15).

Using the projection operator (7.8a), we can write the density matrix (7.15) in the form

$$\rho = \sum_k w_k \mathscr{P}_{\Psi_k}, \quad \sum_k w_k = 1, \quad w_k \geqslant 0, \tag{7.18}$$

where $\mathscr{P}_{\Psi_k}$ is the projection operator onto the state Ψ_k.

We shall show that the density matrix is positive-definite, i.e., has no negative eigenvalues. This property follows from (7.18), since a sum of positive-definite matrices is also positive-definite, and the projection operator, as we saw earlier, is positive-definite. Incidentally, the fact that the eigenvalues of ρ are positive can easily be proved directly.

Since ρ is Hermitian, we can write the condition that its eigenvalues be positive-definite in the form

$$\langle A^2 \rangle = \mathrm{Tr}\,(\rho A^2) \geqslant 0, \tag{7.19}$$

where A^2 is an arbitrary Hermitian operator. In fact, by diagonalizing ρ, which is possible since it is Hermitian, we can write (7.19) in the form

$$\sum_{n,\,k} \rho_{nn} A_{nk} A_{kn} = \sum_{n,\,k} \rho_{nn} \mid A_{nk} \mid^2 \geqslant 0,$$

whence it follows that $\rho_{nn} \geq 0$. The property (7.19) is fulfilled for the density matrix (7.15), since

$$\langle A^2 \rangle = \sum_k w_k \, (A^2)_{kk} = \sum_{k,\,m} w_k A_{km} A_{mk} = \sum_{k,\,m} w_k \mid A_{km} \mid^2 \geqslant 0, \tag{7.19a}$$

and, consequently, the density matrix is positive-definite.

It is not difficult to see that any positive-definite Hermitian operator satisfying the normalization condition (7.16) can be represented in the form (7.18). For this, we must bring it to diagonal form, and then represent it as a sum of matrices of which all the diagonal elements are zero except one. The positive eigenvalues of the operator, whose sum is equal to unity by virtue of the normalization, will play the role of the w_k, while the remaining matrices will be projection matrices.

We shall show that all matrix elements of the statistical operator are bounded. The trace of the square of the statistical operator is

$$\operatorname{Tr}\rho^2 = \sum_{m,\,n} |\rho_{mn}|^2.$$

We note that in a diagonal representation this quantity is less than unity, since in this case, because the eigenvalues ρ_{nn} of the statistical operator are positive,

$$\sum_n \rho_{nn}^2 \leqslant \Big(\sum_n \rho_{nn}\Big)^2 = 1.$$

Taking into account that the trace is independent of the representation, we obtain

$$\operatorname{Tr}\rho^2 = \sum_{m,\,n} |\rho_{mn}|^2 \leqslant 1.$$

This inequality means that all the matrix elements of the statistical operator are bounded.

§ 8. The Quantum Liouville Equation

8.1. The Liouville Equation in the Quantum Case

We shall consider the time evolution of the statistical operator for an ensemble of systems with Hamiltonian H, which can depend on the time. The density matrix at time t has the form (7.15), but now the Ψ_k depend on time:

$$\rho(x, x', t) = \sum_k w_k \Psi_k(x, t)\, \Psi_k^*(x', t), \tag{8.1}$$

where the w_k do not depend on t, since they correspond to the distribution of probabilities at $t = 0$. The function $\Psi_k(x, t)$ are solutions of the Schrödinger equation satisfying the initial condition

$$\Psi_k(x, t)\,|_{t=0} = \Psi_k(x),$$

where the $\Psi_k(x)$ are some set of wave functions defining the statistical operator at $t = 0$:

$$\rho(x, x') = \sum_k w_k \Psi_k(x)\, \Psi_k^*(x').$$

Because of this initial condition, the nonstationary solutions of the Schrödinger equation depend on the quantum number k.

If a fraction w_k of the dynamic systems were in the state $\Psi_k(x, 0)$ at the initial moment of time, then at time t there will be the same fraction of systems in the state $\Psi_k(x, t)$.

The change of the state $\Psi_k(x, t)$ in time is determined by the Schrödinger equation (7.1)

$$i\hbar \frac{\partial \Psi_k(x, t)}{\partial t} = H\Psi_k(x, t), \tag{8.2}$$

which, using (7.7), we can write in the matrix form

$$i\hbar \frac{\partial \Psi_k(x, t)}{\partial t} = \int H(x, x') \Psi_k(x', t)\, dx'. \tag{8.3}$$

Consequently, the density matrix (8.1) satisfies the equation

$$\begin{aligned} i\hbar \frac{\partial \rho(x, x', t)}{\partial t} &= \int \sum_k \big(H(x, x'') w_k \Psi_k(x'', t) \Psi_k^*(x', t) - \\ &\quad - w_k \Psi_k(x, t) \Psi_k^*(x'', t) H(x'', x')\big)\, dx'' = \\ &= \int \big(H(x, x'') \rho(x'', x', t) - \rho(x, x'', t) H(x'', x')\big)\, dx'', \end{aligned} \tag{8.4}$$

where we have used the Hermiticity property of the Hamiltonian

$$H^*(x, x') = H(x', x). \tag{8.5}$$

Thus, we have obtained the equation of motion of the density matrix – a quantum Liouville equation – in the matrix form (8.4). This can be written conveniently in the operator form

$$i\hbar \frac{\partial \rho}{\partial t} = [H, \rho], \tag{8.6}$$

where

$$\frac{1}{i\hbar}[H, \rho] = \frac{1}{i\hbar}(H\rho - \rho H) \tag{8.7}$$

is the quantum Poisson bracket.

The quantum Liouville equation (8.6) is analogous to the classical Liouville equation (2.11) for the distribution function $f(p, q, t)$.

In place of the classical Poisson bracket (2.10), it contains the quantum Poisson bracket (8.7). There is, however, an essential difference. The density matrix $\rho(x, x', t)$ is a complex function of the set of coordinates $x_1, \ldots, x_N$ and $x'_1, \ldots, x'_N$ of the particles, while $f(p, q, t)$ is a real function of the set of coordinates and momenta. As shown by Wigner [12], there exists a closer analogy between the density matrix $\rho(x, p, t)$ in the mixed coordinate-momentum representation and the classical distribution function (cf. §14).

In the case of statistical equilibrium, ρ and H do not depend explicitly on time, and the quantum Liouville equation has the form

$$[H, \rho] = 0, \tag{8.8}$$

i.e., in this case, the statistical operator ρ commutes with the Hamiltonian and, consequently, is an integral of motion. In classical statistical mechanics, the equilibrium distribution function, as we saw in §2, is also an integral of motion, as can be seen from (2.13).

The commutativity of the operators ρ and H and their Hermiticity show that they have a common set of eigenfunctions. Therefore, in the case of statistical equilibrium, the density matrix can be represented in the form

$$\rho(x, x') = \sum_k w_k \Psi_k(x) \Psi_k^*(x'), \tag{8.9}$$

where the $\Psi_k(x)$ are the eigenfunctions of the Hamiltonian

$$H\Psi_k = E_k \Psi_k. \tag{8.10}$$

In quantum mechanics, not all the eigenfunctions are permissible wave functions of the system, but only those which satisfy the necessary symmetry properties.

For a system of particles with zero or integral spin (in multiples of $\hbar$), only the wave functions that are symmetric with respect to simultaneous interchange of the coordinates and spins of the particles are permissible. In this case, we say that the particles obey Bose statistics.

For a system of particles with half-integral spin (in units $\hbar$), only wave functions that are antisymmetric with respect to interchange of the coordinates and spins are permissible. In this case, we say that the particles obey Fermi statistics.

In the expression (8.9) for the density matrix, the summation is assumed to be only over the permissible, and not over all, quantum states of the system.

The Liouville equation (8.6) enables us to find the statistical operator for any moment of time, if it is known at the initial moment.

Let the statistical operator $\rho(0)$ at $t = 0$ be given. Then at time t, the statistical operator has the form

$$\rho(t) = e^{-iHt/\hbar}\rho(0)\,e^{iHt/\hbar}, \tag{8.11}$$

if the Hamiltonian H does not depend on t. In fact, differentiating (8.11) with respect to time, we see that $\rho(t)$ satisfies the Liouville equation (8.6). In addition, $\rho(t)$ satisfies the initial condition

$$\rho(t)\,|_{t=0} = \rho(0). \tag{8.11a}$$

The expression (8.11) is a formal solution of the Liouville equation (8.6). It is analogous to the expression (2.17) in classical statistical mechanics. We shall use this formal device for integrating the Liouville equation frequently later.

If the Hamiltonian H_t depends explicitly on time, the Liouville equation can be formally integrated with the help of the *evolution operator* U(t, 0), a unitary operator satisfying the equation

$$i\hbar\,\frac{\partial U(t, 0)}{\partial t} = H_t U(t, 0), \text{ where } U^{+}(t_1, t_2) = U^{-1}(t_1, t_2), \tag{8.12}$$

and the initial condition

$$U(0, 0) = 1. \tag{8.12a}$$

The statistical operator at time t has the form

$$\rho(t) = U(t, 0)\,\rho(0)\,U^{-1}(t, 0). \tag{8.13}$$

In fact, $\rho(t)$ satisfies the Liouville equation

$$\frac{\partial \rho(t)}{\partial t} = \frac{1}{i\hbar}\,[H_t, \rho(t)] \tag{8.14}$$

and the initial condition (8.11a).

8.2. The Schrödinger and Heisenberg Pictures for Statistical Operators

Until now, we have used a representation in which the statistical operator ρ depends on time, while the dynamic variables do not depend on time (through the coordinates and momenta); they can depend on time only through external fields. This corresponds to the Schrödinger picture in quantum mechanics.

It is sometimes more convenient to use the Heisenberg picture, in which ρ does not depend on time, but the dynamic variables depend on time through the coordinates and momenta, in addition to the possible dependence on time as a parameter through external fields.

The mean value of any dynamic variable is equal to

$$\langle A\rangle = \mathrm{Tr}(\rho(t)\, A). \tag{8.15}$$

Substituting ρ (t) into this from (8.11) [or (8.13)] and using the cyclic invariance of the trace, we obtain

$$\langle A\rangle = \mathrm{Tr}(\rho(0)\, A(t)), \tag{8.16}$$

where

$$A(t) = e^{iHt/\hbar} A\, e^{-iHt/\hbar} \tag{8.17}$$

or

$$A(t) = U^{-1}(t, 0)\, AU(t, 0) \tag{8.17a}$$

is the operator A in the Heisenberg picture; U(t, 0) is the evolution operator (8.12). The formula (8.15) corresponds to the Schrödinger picture, and (8.17) and (8.17a) to the Heisenberg picture for the operator A.

We shall obtain expressions for the time derivative of a dynamic variable in the Heisenberg picture in the general case.

Differentiating the identity (8.15) with respect to time, we find

$$\frac{d}{dt}\langle A\rangle = \mathrm{Tr}\left(\frac{\partial\rho}{\partial t} A + \rho\frac{\partial A}{\partial t}\right).$$

Substituting $\partial\rho/\partial t$ from Liouville's equation (8.6) into this, we obtain

$$\frac{d}{dt}\langle A\rangle = \mathrm{Tr}\left\{\left(\frac{\partial A}{\partial t} + \frac{1}{i\hbar}[A, H]\right)\rho\right\}$$

or

$$\frac{d}{dt}\langle A\rangle = \mathrm{Tr}\left(\frac{dA}{dt}\rho\right) = \left\langle\frac{dA}{dt}\right\rangle, \tag{8.18}$$

where

$$\frac{dA}{dt} = \frac{\partial A}{\partial t} + \frac{1}{i\hbar}[A, H] \tag{8.19}$$

is the time derivative of the dynamic variable A. This same relation can be obtained by differentiating (8.17) and (8.17a) with respect to time. The formulas (8.18) and (8.19) are analogous to the formulas (2.19a) and (2.18) of classical statistical mechanics.

If the dynamic variable A does not depend explicitly on time, its derivative is equal to

$$\frac{dA}{dt} = \frac{1}{i\hbar}[A, H]. \tag{8.20}$$

Below, we shall make wide use of the equations of motion for dynamic variables.

8.3. The Entropy Operator

In quantum statistical mechanics we can introduce an entropy operator

$$\eta = -\ln\rho, \tag{8.21}$$

analogous to minus the logarithm of the distribution function in classical statistical mechanics, (2.22).

As we saw earlier, the statistical operator ρ is Hermitian and positive-definite. Consequently, its logarithm is Hermitian, and the entropy operator η is positive-definite. In fact, if w_1, w_2, ... are the eigenvalues of the operator ρ, with $0 \le w_k \le 1$, then $-\ln w_1$, $-\ln w_2$, ... are the eigenvalues of η, since the eigenvalues of a function of an operator are equal to the same function of the eigenvalues.

It follows from the inequality $w_k \le 1$ that $-\ln w_k \ge 0$, i.e., that the eigenvalues of η are positive, but not necessarily bounded, although $\sum w_k \ln w_k$ is always bounded.

The entropy operator η has the property of additivity, i.e., if the operator ρ is a direct product† of operators ρ_1 and ρ_2

$$\rho = \rho_1 \otimes \rho_2, \tag{8.22}$$

†The direct product $A \otimes B$ of a matrix $A \equiv [a_{ik}]$ and $B \equiv [b_{i'k'}]$ is the matrix $A \otimes B \equiv [c_{jl}]$ where $c_{jl} = a_{ik}b_{i'k'}$ ($j \equiv ii'$ and $l \equiv kk'$).

which denotes the direct product of the corresponding matrices, then

$$\eta = \eta_1 + \eta_2, \tag{8.23}$$

where $\eta = -\ln \rho$, $\eta_1 = -\ln \rho_1$, and $\eta_2 = -\ln \rho_2$.

The entropy operator η, like ρ, satisfies Liouville's equation

$$i\hbar \frac{d\eta}{dt} = [H, \eta], \tag{8.24}$$

as can easily be seen directly. For example, if ρ satisfies Eq. (8.6), then ρ^2 satisfies the same equation

$$i\hbar \frac{d}{dt} \rho\rho = [H, \rho\rho], \tag{8.25}$$

since the Poisson bracket has the property

$$[H, \rho\rho] = [H, \rho]\rho + \rho[H, \rho].$$

Equation (8.24) sometimes turns out to be very convenient, since H and η are Hermitian additive operators.

8.4. Entropy

Minus the mean of the logarithm of the statistical operator, i.e., the mean value of the entropy operator, is called the Gibbs entropy,

$$S = \langle \eta \rangle = -\langle \ln \rho \rangle = -\operatorname{Tr}(\rho \ln \rho). \tag{8.26}$$

This definition corresponds to the Gibbs definition (2.24) of the entropy in classical statistical mechanics, and is its quantum generalization.

From the properties of the statistical operator considered in §7.2, it follows that the entropy (8.26) is a positive-definite quantity. In fact, in a diagonal representation, it has the form

$$S = -\sum_n \rho_{nn} \ln \rho_{nn} \geqslant 0, \tag{8.27}$$

since, according to (7.19), the eigenvalues of the statistical operator cannot be negative, $\rho_{nn} \geq 0$.

Only in the particular case when the statistical operator describes a pure state do we have S = 0.

The entropy (8.26) possesses the property of additivity. If ρ describes statistically independent ensembles, and is the direct product (8.22) of ρ_1 and ρ_2, then

$$S = S_1 + S_2, \tag{8.28}$$

where

$$S = -\langle \ln \rho \rangle, \quad S_1 = -\langle \ln \rho_1 \rangle, \quad S_2 = -\langle \ln \rho_2 \rangle.$$

The entropy defined by means of formula (8.26) for an isolated system does not depend on time.

In fact, the statistical operator at time t is connected with its value at t = 0 by the unitary transformation (8.13):

$$\rho(t) = U(t, 0)\,\rho(0)\,U^{-1}(t, 0). \tag{8.29}$$

For example, $U(t, 0) = e^{-iHt/\hbar}$, if the Hamiltonian does not depend on time.

Then we have

$$\begin{aligned} S(t) &= -\operatorname{Tr}\left\{U(t, 0)\,\rho(0)\,U^{-1}(t, 0)\ln\left(U(t, 0)\,\rho(0)\,U^{-1}(t, 0)\right)\right\} \\ &= -\operatorname{Tr}\left\{U(t, 0)\,\rho(0)\,U^{-1}(t, 0)\,U(t, 0)\ln(\rho(0))\,U^{-1}(t, 0)\right\}, \end{aligned}$$

since

$$\ln\left(U(t, 0)\,\rho(0)\,U^{-1}(t, 0)\right) = U(t, 0)\ln(\rho(0))\,U^{-1}(t, 0),$$

which is true in general for any function of the operator and can be proved by expanding in a Taylor series. Taking into account that

$$U(t, 0)\,U^{-1}(t, 0) = 1$$

and that operators in the trace can be cyclically permuted, we obtain

$$S(t) = -\operatorname{Tr}\{\rho(0)\ln\rho(0)\} = S(0). \tag{8.30}$$

On the other hand, it is well known from thermodynamics that the entropy of an isolated system can increase. Therefore, for nonequilibrium processes it is sometimes suggested [14] that the entropy to be associated with the thermodynamic entropy is not (8.26), but the entropy calculated with the help of the "coarsened" statistical operator $\tilde{\rho}$, averaged over a small region $\Delta\Gamma$ of quantum-mechanical states, the coarse-grained statistical operator

$$\tilde{\rho} = \frac{1}{\Delta\Gamma}\operatorname{Tr}_{\Delta\Gamma}\rho. \tag{8.31}$$

The coarse-graining operation (8.31) for the statistical operator is analogous to the coarse-graining operation (2.30) for the distribution function in classical statistical mechanics. For the coarse-grained statistical operator (8.31), the entropy

$$S_t = -\operatorname{Tr}(\tilde{\rho}(t)\ln\tilde{\rho}(t)) \tag{8.32}$$

can now increase.

Let the state at t = 0 be described by the coarse-grained statistical operator

$$\rho(0) = \tilde{\rho}(0).$$

The corresponding entropy is

$$S_0 = -\operatorname{Tr}(\tilde{\rho}(0)\ln\tilde{\rho}(0)). \tag{8.33}$$

At time t, the entropy calculated using the coarsened statistical operator is given by (8.32), so that

$$\begin{aligned} S_t - S_0 &= -\operatorname{Tr}(\tilde{\rho}(t)\ln\tilde{\rho}(t)) + \operatorname{Tr}(\tilde{\rho}(0)\ln\tilde{\rho}(0)) \\ &= -\operatorname{Tr}(\rho(t)\ln\tilde{\rho}(t)) + \operatorname{Tr}(\rho(t)\ln\rho(t)) \end{aligned} \tag{8.34}$$

[cf. (2.31a)], since, according to Liouville's theorem,

$$\operatorname{Tr}(\rho(t)\ln\rho(t)) = \operatorname{Tr}(\rho(0)\ln\rho(0)).$$

For any two statistical operators, the inequality

$$\operatorname{Tr}(\rho\ln\rho) \geqslant \operatorname{Tr}(\rho\ln\rho_1), \tag{8.35}$$

holds, the equality sign being attained only when $\rho = \rho_1$. The inequality (8.35) follows from the obvious inequality

$$\ln x \geqslant 1 - \frac{1}{x}, \qquad x > 0, \tag{8.36}$$

where the equality sign holds only when x = 1.

Putting $x = \rho\rho_1^{-1}$ into (8.36) (ρ and ρ_1 are positive-definite operators) and averaging the inequality over ρ, we obtain

$$\operatorname{Tr}\{\rho\ln(\rho\rho_1^{-1})\} \geqslant \operatorname{Tr}\{\rho(1 - \rho_1\rho^{-1})\} = 0, \tag{8.37}$$

since both operators are normalized and the operators in the trace can be permuted cyclically. The inequality (8.37) coincides with (8.35), as we wished to prove.

If we put $\rho = \rho(t)$ and $\rho_1 = \tilde{\rho}(t)$, it follows from (8.34) and (8.35) that

$$S_t \geqslant S_0.$$

We assume that $\rho(t)$ does not describe a state of statistical equilibrium; then, generally speaking,

$$\tilde{\rho}(t) \neq \rho(t) \tag{8.38}$$

and

$$S_t > S_0, \tag{8.39}$$

i.e., the entropy S_t can increase.

The above comments on the coarsening of the statistical operator do not solve the problem of defining the entropy of a nonequilibrium state. The concept of the entropy of a nonequilibrium state will be examined in Chapter IV.

§ 9. Gibbsian Statistical Ensembles in the Quantum Case

The basic ideas, expounded in §3, of the theory of Gibbsian statistical ensembles can be carried over directly to quantum statistical mechanics [1-3].

In a state of statistical equilibrium, the statistical operator can depend only on the additive integrals of motion of the quantum Liouville equation (8.6). Three such integrals of motion are known: the total energy, represented by the Hamiltonian operator H (independent of time), the total momentum **P**, and the total angular momentum **M**. All these quantities are dynamic variables in the quantum-mechanical sense, i.e., Hermitian operators acting in the wave function space.

Consequently, in accordance with the basic idea of the Gibbsian ensembles, ρ is a function H, **P**, and **M**:

$$\rho = \rho(H, \boldsymbol{P}, \boldsymbol{M}). \tag{9.1}$$

If the number of particles N in the ensemble is not specified, it must be treated as the fourth integral of motion:

$$[N, H] = 0,$$

where N is an operator taking the integral positive values 0, 1, 2, ... Then

$$\rho = \rho(H, N, \boldsymbol{P}, \boldsymbol{M}). \tag{9.2}$$

If we consider systems in a motionless vessel, then $\mathbf{P} = \mathbf{M} = 0$, and it is not necessary to take account of these integrals of motion. Consequently, for systems with a specified number of particles,

$$\rho = \rho(H), \tag{9.3}$$

and for systems with an unspecified number of particles,

$$\rho = \rho(H, N). \tag{9.4}$$

In addition, the statistical operator can depend parametrically on the quantities which are specified for the systems in the ensemble, e.g., on the volume V and the particle number N, in the case (9.3), or in the case (9.4), on V.

9.1. The Microcanonical Distribution

In quantum statistical mechanics, the microcanonical distribution can be introduced exactly as in classical statistical mechanics (cf. 3.1). For this, we shall consider an ensemble of closed, energetically isolated systems with constant volume V and total particle number N, each having the same energy E to within $\Delta E \ll E$. We assume that for such systems all the quantum-mechanical states in the layer E, E + ΔE are equally probable. Such a distribution, when

$$w(E_k) = \begin{cases} \Omega^{-1}(E, N, V) & \text{for} \quad E \leqslant E_k \leqslant E + \Delta E, \\ 0 & \text{outside this layer,} \end{cases} \tag{9.5}$$

is called a microcanonical distribution, and the corresponding ensemble is called the microcanonical ensemble of quantum statistics.

The microcanonical distribution (9.5) is the quantum generalization of the distribution (3.3) of classical statistical mechanics. The difference lies in the fact that the statistical weight Ω(E, N, V) is no longer simply equal to the phase volume (3.3a), but is the number of quantum-mechanical states in the layer E, E + ΔE for a system with particle number N and volume V. This follows from the fact that the probability $w(E_k)$ must be normalized to unity:

$$\sum_k w(E_k) = 1. \tag{9.6}$$

As in Chapter I, we assume the quantity ΔE to be a small, but finite, quantity, since in quantum mechanics fixing the energy exactly would require an infinite observation time, in accordance with the energy–time uncertainty relation. ΔE can be chosen, for example, to be the average magnitude of the energy fluctuations of the system.

A theoretical treatment of an ensemble of completely isolated systems, an idealized limiting case, can be given. This model is convenient in that all the discussion given in §8.1 and at the beginning of this section on the properties of isolated systems is exactly applicable to it. For completely isolated systems, $\Omega(E, N, V)$ is equal to the multiplicity of the level of energy E in a system with particle number N and volume V. If N is large, the number $\Omega(E, N, V)$ is very great.

The statistical operator (8.9) corresponds to the microcanonical distribution (9.5). In matrix form, it is

$$\rho(x, x') = \Omega^{-1}(E, N, V) \sum_{1 \leqslant k \leqslant \Omega} \Psi_k(x)\, \Psi_k^*(x'), \tag{9.7}$$

where x is the set of coordinates (and spins) of the N particles, and $\Psi_1, \ldots, \Psi_\Omega$ are the eigenfunctions of the Hamiltonian operator H, corresponding to the energy E. The statistical operator (9.7) can be written in operator form:

$$\rho = \Omega^{-1}(E, N, V)\, \Delta(H - E), \tag{9.7a}$$

where H is the Hamiltonian of the system, and $\Delta(x)$ is a function differing from zero only in the thin energy layer $0 \leq x \leq \Delta E$, where it is equal to unity, and is zero outside this layer.

From the quantum Liouville theorem (8.6), it follows that the microcanonical distribution is stationary. It must, however, be emphasized that the assumption of equal probability of quantum states with the same energy for a closed isolated system, though by no means self-evident, is the simplest assumption. The problem of justifying this hypothesis is known as the quantum-mechanical ergodic problem. We shall not discuss these questions here, but refer the reader to the literature [13, 14].

An extremal property of the microcanonical distribution can serve as an argument in its favor. Of all the distributions in the same energy layer, the microcanonical distribution is the one corresponding to the maximum entropy (cf. §10.1). The extremal property of the microcanonical distribution in classical statistical mechanics has already been discussed in §4.

We shall calculate the entropy for the microcanonical distribution. In a diagonal representation,

$$S = \langle \eta \rangle = - \operatorname{Tr} (\rho \ln \rho) = - \sum_k w_k \ln w_k, \qquad (9.8)$$

or, since all the w_k in the layer E, $E + \Delta E$ are the same and equal to $\Omega^{-1}(E, N, V)$, we obtain

$$S = \ln \Omega (E, N, V), \qquad (9.9)$$

i.e., the entropy for the microcanonical ensemble is equal to the logarithm of the statistical weight. Formula (9.9) corresponds to the Planck definition of entropy, which is valid, generally speaking, for the equilibrium state and the microcanonical ensemble.

The microcanonical distribution is inconvenient for practical application, since to calculate the statistical weight it is necessary to investigate the distribution of eigenvalues of the Hamiltonian H, and this is a very complicated problem. Rather than consider energetically isolated systems, it is more convenient to consider systems in thermal contact with the surroundings.

9.2. The Canonical Distribution

We shall consider quantum-mechanical systems with constant particle number and constant volume in contact with a thermostat. The thermostat is assumed to be so large that its state is practically unchanged by exchange of energy with the systems of the ensemble. A statistical ensemble of quantum-mechanical systems with specified particle number N and constant volume V in contact with a thermostat is called a *canonical ensemble* in quantum statistics. Such an ensemble is described by the *canonical distribution*

$$w(E_k) = Q^{-1}(\theta, V, N) \exp(- E_k/\theta), \qquad (9.10)$$

where θ is the modulus of the canonical distribution, playing the role of the temperature, and $Q(\theta, V, N)$ is the statistical sum (partition function), determined from the normalization condition (7.12a):

$$Q(\theta, V, N) = \sum_k \exp(-E_k/\theta). \tag{9.11}$$

In the partition function (9.11), the summation is performed over all the quantum-mechanical states permitted by symmetry; states belonging to a degenerate level are assumed to be different. For the justification of the canonical distribution in the quantum case, see §9.3.

The logarithm of the partition function (9.11) determines the free energy

$$F(\theta, V, N) = -\theta \ln Q(\theta, V, N) \tag{9.12}$$

as a function of the parameters θ, V, and N.

The canonical distribution (9.10) is much more convenient than the microcanonical one, since the sums (9.11) over the eigenvalues can sometimes be calculated without knowledge of the eigenvalues themselves. In calculating the partition function, we need only take into account the supplementary condition that the particle number be constant, and not, as in the calculation of the statistical weight in the microcanonical distribution, the condition that the particle number and energy be constant; it is therefore much simpler to work with the canonical, rather than the microcanonical distribution.

The density matrix

$$\rho(x, x') = Q^{-1}(\theta, V, N) \sum_k e^{-E_k/\theta}\, \Psi_k(x)\, \Psi_k^*(x'), \tag{9.13}$$

where x is the set of coordinates (and, possibly, spins) $x_1, \ldots, x_N$ of the particles and the $\Psi_k(x)$ are the eigenfunctions of the Hamiltonian H, corresponds to the canonical distribution (9.10).

We shall introduce the operator $\exp(-H/\theta)$, stipulating that it acts not in the whole wave function space, but only in the space of the wave functions permitted by symmetry. Then (9.13) and (9.11) can be rewritten in a more compact operator form:

$$\rho = Q^{-1}(\theta, V, N)\, e^{-H/\theta} = e^{(F-H)/\theta}, \tag{9.14}$$

$$Q(\theta, V, N) = \mathrm{Tr}\, e^{-H/\theta} = \sum_k \int \Psi_k^*(x)\, e^{-H/\theta}\, \Psi_k(x)\, dx. \tag{9.15}$$

The expression (9.15) for the partition function is very convenient since, because of the invariance of the trace with respect to the representation of the matrices, it does not depend on the choice of the functions $\Psi_k(x)$, which may even be noneigenfunctions of H.

Until now, we have assumed that the system is not moving as a whole and has a single additive integral, the energy H. In the case when there exist integrals of motion $\mathcal{P}_1, \ldots, \mathcal{P}_s$, in addition to the total energy H, the statistical operator has the form

$$\rho = Q^{-1}(\theta, \mathcal{F}_1, \ldots, \mathcal{F}_s) \exp\left\{-\frac{H}{\theta} - \sum_{1\leqslant k\leqslant s} \mathcal{F}_k \mathcal{P}_k\right\}, \tag{9.16}$$

where $\mathcal{F}_1, \ldots, \mathcal{F}_s$ are new thermodynamic parameters, determined from the conditions

$$\langle \mathcal{P}_k \rangle = \mathrm{Tr}\,(\rho \mathcal{P}_k). \tag{9.16a}$$

9.3. Gibb's Theorem on the Canonical Distribution

In quantum, as in classical statistical mechanics, the postulates on the microcanonical distribution (9.5) and on the canonical distribution (9.10) are not independent. Here also, there is a Gibbs theorem on the canonical distribution, according to which a small part of a microcanonical ensemble of quantum systems is distributed canonically. The proof of this theorem is very similar to the corresponding proof of the Gibbs theorem in the classical case, given in §3.3.

We shall treat the combination of the given system and the thermostat as a single energetically isolated closed system with Hamiltonian

$$H = H_1 + H_2, \tag{9.17}$$

where H_1 is the Hamiltonian of the given system, and H_2 is the Hamiltonian of the thermostat, which is assumed to be considerably larger than the given system, i.e., to have a much larger number of degrees of freedom. We assume the interaction between the system and the thermostat to be very small, but nonzero, since it must maintain a constant energy for the combined system. In fact, the thermal contact with the thermostat is effected through the walls of the vessel and is therefore a small surface effect.

The wave function of the Hamiltonian (9.17) of the combined system can be decomposed into a product of the wave functions of the thermostat (system 2) and of the system 1 under consideration:

$$\Psi_{ik}(x, y) = \Psi_k(x)\Psi_i(y), \tag{9.18}$$

where $\Psi_k(x)$ are the eigenfunctions of H_1, and $\Psi_i(y)$ are the eigenfunctions of H_2; x and y are respectively the sets of coordinates of the system under consideration and of the thermostat.

The energy levels of the combined system are equal to the sum of the levels of the systems 1 and 2:

$$E_{ik} = E_i + E_k, \tag{9.18a}$$

where E_k are the energy levels of system 1, and E_i are the energy levels of the thermostat.

In accordance with (7.15), the density matrix of the combined system has the form

$$\rho(xy, x'y') = \sum_{i,k} w_{ik}\Psi_{ik}(x, y)\Psi^*_{ik}(x', y'), \tag{9.19}$$

where w_{ik} is given by the expression (9.5).

We shall obtain the density matrix of system 1 by calculating the trace of the total statistical operator over the coordinates of the thermostat:

$$\rho(x, x') = \mathrm{Tr}_{(2)}\rho(xy, x'y') = \sum_{i,k} w_{ik}\int \Psi_{ik}(x, y)\Psi^*_{ik}(x', y)\,dy,$$

whence, using (9.18) and assuming the eigenfunctions to be normalized, we obtain

$$\rho(x, x') = \sum_k w_k\Psi_k(x)\Psi^*_k(x'), \tag{9.20}$$

where

$$w_k = \sum_i w_{ik}. \tag{9.20a}$$

Consequently, in order to find the probability distribution of the states in system 1, we must sum the probability distribution in the combined system over all the states of the thermostat:

$$w(E_k) = \sum_{\substack{i \\ (E_i+E_k=E)}} w(E_i + E_k) = \frac{1}{\Omega(E)} \sum_{\substack{i \\ (E_i=E-E_k)}} 1. \tag{9.21}$$

[We omit the arguments N and V of $\Omega(E)$ for brevity.] We can write this expression for the probability distribution of states in the given system in the form

$$w(E_k) = \frac{\Omega_2(E - E_k)}{\Omega(E)}, \tag{9.21a}$$

where $\Omega_2(E - E_k)$ is the number of quantum states of the thermostat corresponding to the level $E - E_k$, and $\Omega(E)$ is the number of states of the combined system corresponding to the level E. Formula (9.21a) is the quantum analog of formula (3.13).

To calculate $w(E_k)$, we must obtain an asymptotic estimate for the ratio of the statistical weights of the thermostat and of the combined system by assuming that the thermostat is very large, as was done in the case of classical statistics for the ratio (3.13).

First, we shall give a simple, though not rigorous, derivation of the canonical distribution.

Introducing the entropy $S_2(E)$ of the thermostat and the entropy $S(E)$ of the combined system using the relation (9.9), we write (9.21a) in the form

$$w(E_k) = \exp\{S_2(E - E_k) - S(E)\}. \tag{9.22}$$

Taking into account that the system 1 is small compared with the thermostat, i.e., $E_k \ll E$, we expand $S_2(E - E_k)$ in a series in E_k and confine ourselves to two terms:

$$S_2(E - E_k) \cong S_2(E) - \frac{\partial S_2}{\partial E} E_k.$$

Taking this expansion into account, we rewrite (9.22) in the form

$$w(E_k) = Q^{-1} \exp\left(-\frac{E_k}{\theta}\right), \tag{9.23}$$

where Q is the partition function (9.11), and

$$\frac{1}{\theta} = \frac{\partial S_2(E)}{\partial E} = \frac{\partial \ln \Omega_2(E)}{\partial E} \tag{9.23a}$$

is the inverse temperature. Thus, a small part of a microcanonical ensemble is distributed canonically.

We shall now give a more rigorous proof of the Gibbs theorem, analogous to the proof by Krutkov given in §3.3.

We calculate the number of eigenfunctions of the whole system with energy E. Each eigenfunction of system 1 with energy E_1 can be combined with any of the eigenfunctions of system 2 (the thermostat) with energy $E - E_1$. $\Omega_1(E_1)$ eigenfunctions of system 1 correspond to the level of energy E_1. $\Omega_2(E - E_1)$ eigenfunctions of system 2 correspond to the level of energy $E - E_1$. Consequently, the total number $\Omega(E)$ of eigenfunctions of the system, corresponding to energy E, is

$$\Omega(E) = \sum_{E_1 \leqslant E} \Omega_1(E_1)\,\Omega_2(E - E_1). \tag{9.24}$$

The relation (9.24) can be regarded as an equation for Ω_2, if Ω and Ω_1 are assumed known. This equation is the quantum analog of the integral equation (3.19).

Multiplying (9.24) by $\exp(-\lambda E)$ and summing over all E, we obtain

$$\sum_{0 \leqslant E < \infty} e^{-\lambda E}\,\Omega(E) = \sum_{0 \leqslant E < \infty} e^{-\lambda E} \sum_{0 \leqslant E_2 \leqslant E} \Omega_2(E_2)\,\Omega_1(E - E_2).$$

Changing the order of summation in the right-hand side of the equation, we have

$$\sum_{0 \leqslant E < \infty} e^{-\lambda E}\,\Omega(E) = \sum_{0 \leqslant E_2 < \infty}\ \sum_{E_2 \leqslant E < \infty} e^{-\lambda E}\,\Omega_2(E_2)\,\Omega_1(E - E_2).$$

Making a change of variables $E_1 = E - E_2$, we obtain

$$Q(\lambda) = Q_1(\lambda)\,Q_2(\lambda), \tag{9.25}$$

where

$$Q(\lambda) = \sum_{0 \leqslant E < \infty} e^{-\lambda E}\,\Omega(E), \qquad Q_\alpha(\lambda) = \sum_{0 \leqslant E < \infty} e^{-\lambda E}\,\Omega_\alpha(E) \tag{9.26}$$

$$(\alpha = 1,\ 2).$$

To estimate the statistical weight, it is necessary to invert Eqs. (9.26).

We shall show that for a discrete spectrum, the number of eigenvalues with energy in the interval from 0 to E is equal to

$$\Gamma(E) = \frac{1}{2\pi i} \int_{a - i\infty}^{a + i\infty} e^{\lambda E}\,Q(\lambda)\,\frac{d\lambda}{\lambda}, \tag{9.27}$$

where a is a positive constant, and

$$Q(\lambda)=\sum_{0\leqslant E<\infty} e^{-\lambda E}\Omega(E)=\sum_k e^{-\lambda E_k}. \tag{9.28}$$

Formula (9.27) expresses a theorem on the inversion of the partition function.

Putting (9.28) into (9.27), we obtain

$$\Gamma(E)=\sum_k \mathcal{F}(E-E_k), \tag{9.29}$$

where $\mathcal{F}(x)$ is the discontinuous function:

$$\mathcal{F}(x)=\frac{1}{2\pi i}\int_{a-i\infty}^{a+i\infty}\frac{e^{\lambda x}}{\lambda}d\lambda=\begin{cases}0 & \text{for} \quad x<0,\\ {}^1/_2 & \text{for} \quad x=0,\\ 1 & \text{for} \quad x>0.\end{cases} \tag{9.29a}$$

Thus, Γ(E) gives with very good accuracy the total number of eigenvalues less than E, since for the zero of energy we can always choose the lowest eigenvalue. In applying (9.29) to calculate the number of eigenvalues, we can certainly neglect the error introduced by the value of $\mathcal{F}(x)$ at x = 0, since the number of eigenvalues is very great.

The number of eigenvalues in the energy interval (E, E + ΔE) is, clearly,

$$\Omega(E)=\Gamma(E+\Delta E)-\Gamma(E), \tag{9.30}$$

and, consequently, using the expression (9.27) for Γ(E), we obtain

$$\Omega(E)=\frac{1}{2\pi i}\int_{a-i\infty}^{a+i\infty}\{e^{\lambda(E+\Delta E)}-e^{\lambda E}\}Q(\lambda)\frac{d\lambda}{\lambda}. \tag{9.31}$$

Formula (9.31) gives the required inversion of the sum (9.28).

Using (9.31), we obtain the inversions of the sums (9.26):

$$\begin{aligned}\Omega(E)&=\frac{1}{2\pi i}\int_{a-i\infty}^{a+i\infty}e^{\lambda E}\frac{e^{\lambda\Delta E}-1}{\lambda}Q(\lambda)\,d\lambda,\\ \Omega_2(E)&=\frac{1}{2\pi i}\int_{a-i\infty}^{a+i\infty}e^{\lambda E}\frac{e^{\lambda\Delta E}-1}{\lambda}\,\frac{Q(\lambda)}{Q_1(\lambda)}\,d\lambda.\end{aligned} \tag{9.32}$$

In accordance with the basic idea of the Gibbsian ensembles, we now assume that system 2 (the thermostat) consists of $n-1$ identical weakly interacting parts, each of which is identical with system 1. The combined system consists of n such systems, n being assumed to be very large; in the limit, $n \to \infty$. We recall that system 1 consists, in turn, of a large number of particles.

It follows from (9.25) that $Q(\lambda)$ can be factorized into a product of n equal factors $Q_1(\lambda)$:

$$Q(\lambda) = [Q_1(\lambda)]^n, \tag{9.33}$$

and analogously for the thermostat:

$$Q_2(\lambda) = [Q_1(\lambda)]^{n-1}. \tag{9.33a}$$

Therefore, the expressions (9.32) for the statistical weight take the form

$$\begin{aligned} \Omega(E) &= \frac{1}{2\pi i} \int_{a-i\infty}^{a+i\infty} e^{\lambda E} \frac{e^{\lambda \Delta E} - 1}{\lambda} [Q_1(\lambda)]^n \, d\lambda, \\ \Omega_2(E) &= \frac{1}{2\pi i} \int_{a-i\infty}^{a+i\infty} e^{\lambda E} \frac{e^{\lambda \Delta E} - 1}{\lambda} [Q_1(\lambda)]^{n-1} \, d\lambda. \end{aligned} \tag{9.34}$$

We shall estimate these expressions asymptotically as $n \to \infty$ by the method of steepest descents. It is convenient to write the first integral in (9.34) in the form

$$\Omega(E) = \frac{1}{2\pi i} \int_{a-i\infty}^{a+i\infty} e^{n\chi(\lambda)} \frac{e^{\lambda \Delta E} - 1}{\lambda} d\lambda, \tag{9.35}$$

where

$$\chi(\lambda) = \lambda \frac{E}{n} + \ln Q_1(\lambda). \tag{9.35a}$$

Repeating the same arguments as were used in the classical case [cf. (3.25)], we find asymptotic estimates for the statistical weights:

$$\begin{aligned} \Omega(E) &= \frac{e^{\lambda_1 \Delta E} - 1}{\lambda_1 \sqrt{2\pi n \chi''(\lambda_1)}} e^{\lambda_1 E} [Q_1(\lambda_1)]^n, \\ \Omega_2(E) &= \frac{e^{\lambda_1 \Delta E} - 1}{\lambda_1 \sqrt{2\pi (n-1) \chi''(\lambda_1)}} e^{\lambda_1 E} [Q_1(\lambda_1)]^{n-1}, \end{aligned} \tag{9.36}$$

where the parameter λ_1 is determined from the condition (3.26) for the existence of a saddle point.

Substituting the resulting estimates (9.36) into (9.21a), we see that the expression for $w(E_k)$ does not depend on the quantity ΔE and has the form

$$w(E_k) = Q^{-1}(\theta, V, N)\, e^{-E_k/\theta}, \tag{9.37}$$

where

$$\theta = \frac{1}{\lambda_1} \tag{9.37a}$$

is the temperature, and

$$Q(\theta, V, N) = \sum_E e^{-E/\theta}\, \Omega_1(E) = \sum_k e^{-E_k/\theta} \tag{9.37b}$$

is the partition function.

Thus, if the combined system is distributed microcanonically then a small subsystem of it is distributed in accordance with the canonical distribution.

9.4. The Grand Canonical Distribution

We shall consider quantum-mechanical systems with constant volume, in contact with a thermostat which also serves as a reservoir of particles. The thermostat is assumed to be so large that its state is practically unchanged by exchange of energy and particles with the systems of the ensemble. A statistical ensemble of quantum-mechanical systems with given volume V in contact with a thermostat and a particle reservoir is called the *grand canonical ensemble* in quantum statistics. Such an ensemble is described by the *grand canonical distribution*

$$w_N(E_k) = Q^{-1}(\theta, \mu, V) \exp\left(-\frac{E_k - \mu N}{\theta}\right), \tag{9.38}$$

where θ is the absolute temperature, and $Q(\theta, \mu, V)$ is the partition function for the grand ensemble, determined from the normalization condition

$$\sum_{k, N} w_N(E_k) = 1 \tag{9.38a}$$

and equal to

$$Q(\theta, \mu, V) = \sum_{N, k} \exp\left(-\frac{E_k - \mu N}{\theta}\right). \tag{9.39}$$

Here, we assume throughout that E_k depends on N, i.e., $E_k = E_{k,N}$, although we do not indicate this explicitly. For the justification of the grand canonical distribution see §9.5.

In the partition function (9.39), the summation is performed over all the accessible quantum-mechanical states and over all positive integer values of $N \geq 0$. We have denoted the partition functions of the grand ensemble (9.39) and of the canonical ensemble (9.11) by the same letter Q, but we indicate the difference between them by the variables on which they depend.†

The logarithm of the partition function (9.39) determines the thermodynamic potential $\Omega(\theta, \mu, V)$:

$$\Omega(\theta, \mu, V) = -\theta \ln Q(\theta, \mu, V). \tag{9.40}$$

Corresponding to the grand canonical distribution (9.38), we have the density matrix

$$\rho(x, x') = \sum_{N, k} \exp\left(\frac{\Omega - E_k + \mu N}{\theta}\right) \Psi_k(x) \Psi_k^*(x'), \tag{9.41}$$

where x is the aggregate of the coordinates and spins of the particles, and $\Psi_k(x)$ are eigenfunctions of the Hamiltonian H and of the operator N, i.e., $\Psi_k(x) = \Psi_{k,N}(x)$. Since the operator H commutes with the operator for the total particle number N, the functions $\Psi_k(x)$ can be simultaneously eigenfunctions of the operator N also.

We introduce the operator $e^{-(H-\mu N)/\theta}$, acting in the space of the permissible wave functions of the system; here we are regarding N as an operator, although we are retaining the previous notation for it. Then the formulas (9.41) and (9.39) can be written in a more compact operator form:

$$\rho = e^{(\Omega - H + \mu N)/\theta}, \tag{9.42}$$

$$e^{-\Omega/\theta} = \operatorname{Tr} e^{-(H-\mu N)/\theta} = \sum_{k, N} \int \Psi_k^*(x) e^{-(H-\mu N)/\theta} \Psi_k(x)\, dx, \tag{9.42 a}$$

†The partition function of the grand ensemble is sometimes denoted by $\Xi(\theta, \mu, V)$.

where $\Psi_k(x)$ is an arbitrary complete set of the functions permitted by the symmetry of antisymmetry requirements, but not necessarily satisfying the Schrödinger equation. Formula (9.42a) is convenient because of the invariance of the trace with respect to the representation of the operators.

Until now, we have considered systems consisting of only one sort of particle. It is easy to generalize the grand canonical distribution to systems consisting of several sorts of particles. We may imagine that the system is in thermal and material contact with s large reservoirs of particles and energy, with semipermeable partitions allowing only one sort of molecule to pass. The statistical operator of such an ensemble will have the form

$$\rho = \exp\left\{\frac{\Omega + \sum\limits_{1 \leqslant \alpha \leqslant s} \mu_\alpha N_\alpha - H}{\theta}\right\}, \tag{9.43}$$

where μ_α is the chemical potential for particles of type α.

9.5. Gibbs' Theorem on the Grand Canonical Distribution

For an ensemble of quantum-mechanical systems with variable particle number, a Gibbs theorem analogous to the corresponding theorem in classical statistics holds: a small part of a microcanonical ensemble of quantum-mechanical systems with many degrees of freedom, if the number of particles in it is not constant, is distributed in accordance with the grand canonical ensemble (9.38).

We shall give a proof of this theorem.

Let a system with energy E and particle number N consist of two weakly interacting subsystems with energies E_1, E_2 and particle numbers N_1, N_2 respectively; then

$$E = E_1 + E_2, \qquad N = N_1 + N_2. \tag{9.44}$$

We assume that the second subsystem (the thermostat and particle reservoir) is much larger than the first,

$$E_1 \ll E_2, \qquad N_1 \ll N_2.$$

Since the combined system is isolated and closed, we can apply the microcanonical distribution (9.5) to it. Repeating the arguments used in deriving the canonical distribution in §3.3, we find the distribution of probabilities $w_{N_1}(E_1)$ in the first, small subsystem by summing the microcanonical distribution of the combined system over all the states of the second subsystem. Thus, we obtain, in complete analogy with formula (9.21a),

$$w_{N_1}(E_1) = \frac{\Omega_2(E - E_1, N - N_1)}{\Omega(E, N)}, \tag{9.45}$$

where Ω_2 is the statistical weight of the second subsystem and Ω is the statistical weight of the whole system. An elementary proof of the Gibbs theorem follows immediately from this. Expressing the statistical weights in (9.45) in terms of the entropies of the second subsystem and of the whole system by the relation (9.9) and expanding the exponent in $E_1 \ll E$, $N_1 \ll N$, we immediately obtain the grand canonical distribution (9.38). A similar derivation was performed for the classical case in §3.4.

We shall give below a more rigorous proof of the Gibbs theorem, based on inversion of the partition functions.

The statistical weight Ω of the combined system is connected with the statistical weights Ω_1 and Ω_2 of the subsystems by the relation

$$\Omega(E, N) = \sum_{\substack{0 \leqslant N_1 \leqslant N \\ 0 \leqslant E_1 \leqslant E}} \Omega_1(E_1, N_1)\,\Omega_2(E - E_1, N - N_1), \tag{9.46}$$

which is analogous to the relation (9.24), but allows for the possibility of different distributions of the particles between the subsystems 1 and 2.

The relation (9.46) can be regarded as an equation for Ω_2, if Ω and Ω_1 are assumed known. It has the form of a finite-difference equation with respect to the variables E and N. In the case of classical statistics, it goes over into the integral equation in the variable E and difference equation in the variable N investigated by Shubin [15].

We shall solve Eq. (9.46) for Ω_2. For this, we multiply both sides of the equation by $e^{-\lambda E + \nu N}$ and sum it over all values of E and N from 0 to ∞:

$$\sum_{\substack{0 \leqslant N < \infty \\ 0 \leqslant E < \infty}} e^{-\lambda E + \nu N}\,\Omega(E, N) =$$

$$= \sum_{\substack{0 \leqslant N < \infty \\ 0 \leqslant E < \infty}} \sum_{\substack{0 \leqslant N_1 \leqslant N \\ 0 \leqslant E_1 \leqslant E}} e^{-\lambda E + \nu N}\,\Omega_1(E_1, N_1)\,\Omega_2(E - E_1, N - N_1).$$

Changing the order of summation in the right-hand side of the equation and making a change of variables $E - E_1 = E_2$, $N - N_1 = N_2$, we transform this equation to the form

$$Q(\lambda, \nu) = Q_1(\lambda, \nu)\, Q_2(\lambda, \nu), \tag{9.47}$$

where

$$\begin{aligned} Q(\lambda, \nu) &= \sum_{\substack{0 \leqslant N < \infty \\ 0 \leqslant E < \infty}} e^{-\lambda E + \nu N}\, \Omega(E, N), \\ Q_\alpha(\lambda, \nu) &= \sum_{\substack{0 \leqslant N < \infty \\ 0 \leqslant E < \infty}} e^{-\lambda E + \nu N}\, \Omega_\alpha(E, N) \qquad (\alpha = 1, 2). \end{aligned} \tag{9.48}$$

We note that if ν is purely imaginary, the right-hand sides of (9.48) go over into Fourier series in the variable ν.

If the relations (9.48) are inverted for the statistical weights, in the same way as was done for the relations (9.26), we obtain

$$\begin{aligned} \Omega_2(E, N) &= \frac{1}{(2\pi i)^2} \int\limits_{c-2\pi i}^{c+2\pi i} d\nu \int\limits_{a-i\infty}^{a+i\infty} \frac{Q(\lambda, \nu)}{Q_1(\lambda, \nu)}\, e^{\lambda E - \nu N} \left(e^{\lambda\, \Delta E - \nu\, \Delta N} - 1\right) \frac{d\lambda}{\lambda}, \\ \Omega(E, N) &= \frac{1}{(2\pi i)^2} \int\limits_{c-2\pi i}^{c+2\pi i} d\nu \int\limits_{a-i\infty}^{a+i\infty} Q(\lambda, \nu)\, e^{\lambda E - \nu N} \left(e^{\lambda\, \Delta E - \nu\, \Delta N} - 1\right) \frac{d\lambda}{\lambda}, \end{aligned} \tag{9.49}$$

where $a > 0$, $c > 0$. Thus, (9.49) gives the solution of Eq. (9.46).

Let the second subsystem, i.e., the reservoir of energy and particles, consist of $n - 1$ subsystems, identical with the first subsystem. Then the combined system consists of n such subsystems. On the basis of (9.47), we shall have

$$Q_2(\lambda, \nu) = [Q_1(\lambda, \nu)]^{n-1}, \qquad Q(\lambda, \nu) = [Q_1(\lambda, \nu)]^n. \tag{9.50}$$

Using these relations, we write (9.49) in the form

$$\begin{aligned} \Omega_2(E, N) &= \frac{1}{(2\pi i)^2} \int\limits_{c-2\pi i}^{c+2\pi i} d\nu \int\limits_{a-i\infty}^{a+i\infty} [Q_1(\lambda, \nu)]^{n-1}\, e^{\lambda E - \nu N} \left(e^{\lambda\, \Delta E - \nu\, \Delta N} - 1\right) \frac{d\lambda}{\lambda}, \\ \Omega(E, N) &= \frac{1}{(2\pi i)^2} \int\limits_{c-2\pi i}^{c+2\pi i} d\nu \int\limits_{a-i\infty}^{a+i\infty} [Q_1(\lambda, \nu)]^{n}\, e^{\lambda E - \nu N} \left(e^{\lambda\, \Delta E - \nu\, \Delta N} - 1\right) \frac{d\lambda}{\lambda}. \end{aligned} \tag{9.51}$$

Putting (9.51) into (9.45), we obtain

$$w_{N_1}(E_1) = \frac{\int\limits_{c-2\pi i}^{c+2\pi i} d\nu \int\limits_{a-i\infty}^{a+i\infty} [Q_1(\lambda, \nu)]^{n-1} e^{\lambda(E-E_1)-\nu(N-N_1)} (e^{\lambda\Delta E-\nu\Delta N}-1)\frac{d\lambda}{\lambda}}{\int\limits_{c-2\pi i}^{c+2\pi i} d\nu \int\limits_{a-i\infty}^{a+i\infty} [Q_1(\lambda, \nu)]^{n} e^{\lambda E-\nu N} (e^{\lambda\Delta E-\nu\Delta N}-1)\frac{d\lambda}{\lambda}}. \tag{9.52}$$

The integrals in formula (9.52) can be evaluated by the method of steepest descents, by using the fact, as we did in §9.3, that n is large.

The saddle point for the integrand in the denominator of (9.52) corresponds to the minimum of the function $\chi(\lambda, \nu)$ [cf. (9.35) and (9.35a)] at real values of the variables λ and ν:

$$\chi(\lambda, \nu) = \lambda\frac{E}{n} - \nu\frac{N}{n} + \ln Q_1(\lambda, \nu).$$

Consequently, on the basis of properties of functions of a complex variable, the function $\chi(\lambda, \nu)$ has a sharp maximum in the direction parallel to the imaginary axis, since n is large (E/n = const, N/n = const).

We find the minimum of $\chi(\lambda, \nu)$ from the equations

$$\begin{aligned} \frac{E}{n} + \frac{\partial}{\partial\lambda}\ln Q_1(\lambda, \nu) &= 0, \\ -\frac{N}{n} + \frac{\partial}{\partial\nu}\ln Q_1(\lambda, \nu) &= 0. \end{aligned} \tag{9.53}$$

Let the roots of equations (9.53) be λ_1 and ν_1. We take the integration path in (9.52) through these points. We note that an asymptotic expression for $w_{N_1}(E_1)$ as $n \to \infty$ can be written down immediately, if we take the slowly varying function

$$Q_1^{-1}(\lambda, \nu)\, e^{-\lambda E_1 + \nu N_1}$$

out of the integral in the numerator at the saddle point $\lambda = \lambda_1$, $\nu = \nu_1$. Then the integral that remains cancels with the integral in the denominator, and so we obtain

$$w_{N_1}(E_1) = Q_1^{-1}(\lambda_1, \nu_1)\, e^{-\lambda_1 E_1 + \nu_1 N_1}, \tag{9.54}$$

which coincides with (9.38), if we put

$$\frac{1}{\theta} = \lambda_1, \quad \frac{\mu}{\theta} = \nu_1. \tag{9.55}$$

Thus, the probabilities of states in the small subsystem are distributed in accordance with the grand canonical ensemble.

9.6. The Distribution for the Isobaric-Isothermal Ensemble

We shall consider quantum-mechanical systems with constant particle number but variable volume in contact with a thermostat. A statistical ensemble of quantum-mechanical systems with given particle number N and pressure p in contact with a thermostat is called an *isobaric-isothermal ensemble*.

Let a system with energy E and volume V consists of two weakly interacting parts, with energies E_1, E_2 and volumes V_1, V_2:

$$E = E_1 + E_2, \quad V = V_1 + V_2, \tag{9.56}$$

the first subsystem being much smaller than the second (the thermostat),

$$E_1 \ll E_2, \quad V_1 \ll V_2.$$

We assume that the combined system is distributed microcanonically. Then, repeating the arguments of the preceding subsection, we find the distribution of probabilities in the first subsystem:

$$w_{V_1}(E_1) = \frac{\Omega_2(E - E_1, V - V_1)}{\Omega(E, V)} = \exp\{S_2(E - E_1, V - V_1) - S(E, V)\}, \tag{9.57}$$

where $\Omega_2(E, V)$ and $\Omega(E, V)$ are the numbers of quantum-mechanical states with E and V for the thermostat and the whole system respectively, and S_2 and S are the entropies of the thermostat and of the whole system.

Taking into account the smallness of the first subsystem, we expand the entropy in (9.57) in powers of E_1 and V_1. Confining ourselves to the linear terms, we obtain

$$w_V(E_k) = Q^{-1}(\theta, p, N) \exp\left\{-\frac{E_k + pV}{\theta}\right\} = \exp\left\{\frac{\Phi - pV - E_k}{\theta}\right\}, \tag{9.58}$$

where

$$\frac{1}{\theta}=\frac{\partial S_2(E, V)}{\partial E}, \qquad \frac{p}{\theta}=\frac{\partial S_2(E, V)}{\partial V}, \tag{9.58 a}$$

θ is the temperature, p is the pressure and $\Phi(\theta, p, N)$ is the Gibbs thermodynamic potential, in complete analogy with the corresponding formulas (3.42)-(3.43) of classical statistical mechanics.

To the distribution (9.58) corresponds the statistical operator

$$\rho=\exp\left\{\frac{\Phi-pV-H}{\theta}\right\}. \tag{9.59}$$

We have considered four types of Gibbsian statistical ensembles: microcanonical, canonical, grand canonical, and isobaric-isothermal ensembles. Sometimes a generalized Gibbsian ensemble is introduced, in which the energy, particle number and volume are variable [16, 17]. But, as we have already noted in Chapter I, such an ensemble is inconvenient, since it becomes necessary to introduce the intensive variables θ, μ, and p, which are not independent. In constructing statistical ensembles, it is convenient to retain at least one extensive thermodynamic variable.

§ 10. The Connection between the Gibbsian Distributions and the Maximum of the Information Entropy (Quantum Case)

In §4, we considered the connection between the Gibbsian distributions of classical statistical mechanics and the maximum of the information entropy. Completely analogous relations also hold in quantum statistical mechanics.

The information entropy (4.1) is defined for a discrete distribution of probabilities, and quantum statistics studies distributions over discrete quantum states; therefore, the analogy between the information entropy (4.1) and the entropy (8.26) in quantum statistical mechanics is even closer than in classical statistical mechanics. In quantum statistics, we do not have the difficulty with the choice of an invariant probability measure which arises for continuous distributions.

We have already introduced the entropy for quantum ensembles in §8.4:

$$S_i = -\langle \ln \rho \rangle = -\operatorname{Tr}(\rho \ln \rho), \tag{10.1}$$

or, if the statistical operator is represented in a diagonal form,

$$S_i = -\sum_k w_k \ln w_k. \tag{10.1a}$$

We have denoted the entropy not by S, but by S_i, in order to emphasize that we are regarding the information entropy as a functional of an arbitrary statistical operator ρ.

We shall examine the extremal properties of the Gibbsian quantum statistical ensembles. The extremal properties of all the Gibbsian ensembles can be obtained from the inequality (8.35):

$$\operatorname{Tr}(\rho' \ln \rho') \geqslant \operatorname{Tr}(\rho' \ln \rho), \tag{10.2}$$

where ρ and ρ' are arbitrary statistical operators. We have already used this inequality in §8.4.

10.1. Extremal Property of the Microcanonical Distribution

We shall prove that, of all the distributions with the same number of particles in the same energy layer, the microcanonical distribution (9.7a) corresponds to the maximum value of the information entropy (10.1).

Let ρ be the statistical operator (9.7a) of the microcanonical distribution, and ρ' be an arbitrary statistical operator acting in the same space and differing from zero in the same energy layer as ρ. It follows from the normalization condition for statistical operators that

$$\operatorname{Tr} \rho = \operatorname{Tr} \rho' = 1.$$

Putting ρ and ρ' into the inequality (10.2), we obtain

$$-\operatorname{Tr}(\rho' \ln \rho') \leqslant -\operatorname{Tr}(\rho' \ln \rho) = \operatorname{Tr} \rho' \ln \Omega = \ln \Omega(E, N, V),$$

i.e., taking (9.8) and (9.9) into account,

$$-\operatorname{Tr}(\rho' \ln \rho') \leqslant -\operatorname{Tr}(\rho \ln \rho). \tag{10.3}$$

Thus, it is proved that, of all the distributions in a given layer, the microcanonical distribution (9.7a) corresponds to the maximum of the information entropy.

10.2. Extremal Property of the Canonical Distribution

We shall prove that the canonical distribution corresponds to the maximum of the information entropy (10.1) for a given average energy

$$\langle H\rangle = \mathrm{Tr}\,(\rho H) \tag{10.4}$$

when the normalization is conserved

$$\mathrm{Tr}\,\rho = 1. \tag{10.5}$$

We shall seek the extremum of the functional (10.1) under the auxiliary conditions (10.4) and (10.5). For this, we must find the absolute extremum of the functional

$$-\mathrm{Tr}\,(\rho\ln\rho) - \beta\,\mathrm{Tr}\,(\rho H) - \lambda\,\mathrm{Tr}\,\rho,$$

where β and λ are Lagrange multipliers determined from the conditions (10.4) and (10.5). From the condition that the first variation of this functional vanish, we find

$$\rho = Q^{-1}\exp(-\beta H), \tag{10.6}$$

where

$$Q(\theta, V, N) = \mathrm{Tr}\,\exp(-\beta H), \qquad \beta = \frac{1}{\theta}, \tag{10.6a}$$

which coincides with the canonical distribution (9.14). Thus, (10.6) corresponds to an extremum of (10.1).

We now prove that (10.6) corresponds to the maximum of (10.1).

Let ρ' be a normalized statistical operator corresponding to the same average energy as (10.6),

$$\mathrm{Tr}\,(\rho' H) = \mathrm{Tr}\,(\rho H),$$

but arbitrary in other respects. Putting (10.6) into the inequality (10.2), we obtain

$$-\mathrm{Tr}\,(\rho'\ln\rho') \leqslant -\mathrm{Tr}\,(\rho'\ln\rho) = \mathrm{Tr}\,\rho'\ln Q + \beta\,\mathrm{Tr}\,(\rho' H) = \ln Q + \beta\,\mathrm{Tr}\,(\rho H),$$

i.e.,

$$-\operatorname{Tr}(\rho' \ln \rho') \leqslant -\operatorname{Tr}(\rho \ln \rho),$$

where ρ is the canonical distribution (10.6).

Consequently, of all distributions with the same average energy, the canonical distribution corresponds to the maximum information entropy.

In the case when the average values of any n quantities are given

$$\langle \mathscr{P}_k \rangle = \operatorname{Tr}(\rho \mathscr{P}_k) \qquad (k = 0, 1, 2, \ldots, n-1), \tag{10.7}$$

we obtain from the extremum condition for the information entropy (10.1)

$$\rho = \exp\left\{ -\Phi(\mathscr{F}_0, \ldots, \mathscr{F}_{n-1}) - \sum_{k=0}^{n-1} \mathscr{F}_k \mathscr{P}_k \right\}, \tag{10.8}$$

which corresponds to its maximum.

10.3. Extremal Property of the Grand Canonical Distribution

We shall prove that the grand canonical distribution (9.42) corresponds to the maximum of the information entropy (10.1) for a given average energy

$$\langle H \rangle = \operatorname{Tr}(\rho H) \tag{10.9}$$

and given average particle number

$$\langle N \rangle = \operatorname{Tr}(\rho N) \tag{10.10}$$

for conservation of the normalization

$$\operatorname{Tr} \rho = 1. \tag{10.11}$$

We seek the absolute extremum of the functional

$$-\operatorname{Tr}(\rho \ln \rho) - \beta \operatorname{Tr}(\rho H) + \nu \operatorname{Tr}(\rho N) - \lambda \operatorname{Tr} \rho,$$

where β, ν, and λ are Lagrange multipliers. From the extremum of the functional, we find

$$\rho = \exp\left\{ \frac{\Omega - H + \mu N}{\theta} \right\}, \tag{10.12}$$

where

$$e^{-\Omega/\theta} = \mathrm{Tr}\, \exp\left\{-\frac{H-\mu N}{\theta}\right\}, \quad \beta = \frac{1}{\theta}, \quad \nu = \frac{\mu}{\theta}, \tag{10.12a}$$

which coincides with the statistical operator (9.42) of the grand canonical distribution.

It follows from inequality (10.2) that the extremum corresponds to a maximum:

$$-\mathrm{Tr}\,(\rho' \ln \rho') \leqslant -\mathrm{Tr}\,(\rho' \ln \rho) = -\mathrm{Tr}\left\{\rho' \left(\frac{\Omega}{\theta} - \frac{H}{\theta} + \frac{\mu N}{\theta}\right)\right\} = -\mathrm{Tr}\,(\rho \ln \rho), \tag{10.13}$$

where we have used the conditions (10.9)-(10.11) for ρ and ρ', i.e., that

$$\mathrm{Tr}\,(\rho' H) = \mathrm{Tr}\,(\rho H), \qquad \mathrm{Tr}\,(\rho' N) = \mathrm{Tr}\,(\rho N).$$

Thus, the statistical operator (10.12) corresponds to the maximum information entropy for a given average energy and average particle number.

By an analogous method, it is easy to convince oneself that the statistical operator (9.59) for the isobaric-isothermal ensemble corresponds to the maximum of the information entropy

$$S_i = -\int \mathrm{Tr}\,(\rho \ln \rho)\, dV \tag{10.14}$$

under the auxiliary conditions that the average energy and average volume be constant:

$$\langle H \rangle = \int \mathrm{Tr}\,(\rho H)\, dV, \qquad \langle V \rangle = \int \mathrm{Tr}\,(\rho V)\, dV. \tag{10.15}$$

The extremal properties of the Gibbsian quantum ensembles were noticed a very long time ago [2, 3, 9]. Indeed, in generalizing the Gibbsian ensembles to the case of quantum statistics, von Neumann started from the extremal properties of the entropy [2, 3].

It would be possible to make the extremal properties of the Gibbsian statistical ensembles, considered above, the basis of their definition, as was done by Jaynes [18, 19]. Below, we shall use the extremal properties of the entropy frequently to construct ensembles in nonequilibrium statistical thermodynamics (cf. Chapter IV).

§ 11. Thermodynamic Equalities

11.1. Quasi-Static Processes

To obtain thermodynamic equalities in quantum statistical mechanics, it is necessary, as in classical statistical mechanics, to consider a quasi-static process with infinitely slow variation of the external parameters defining the ensemble. We assume that, in a quasi-static process, the external parameters $a_1, a_2, \ldots, a_s$ vary so slowly that the ensemble of quantum-mechanical systems can be assumed to be in statistical equilibrium at each moment of time. To the parameters $a_1, \ldots, a_s$ correspond the generalized forces

$$A_i = -\frac{\partial H}{\partial a_i}. \tag{11.1}$$

Their observed values in a quasi-static process are equal to the average values calculated by means of the equilibrium statistical operator:

$$\langle A_i \rangle = \mathrm{Tr}\,(\rho A_i) = -\left\langle \frac{\partial H}{\partial a_i} \right\rangle. \tag{11.2}$$

In the particular case when we take the volume V of the system as the generalized parameter, the generalized force is the pressure:

$$p = -\left\langle \frac{\partial H}{\partial V} \right\rangle. \tag{11.3}$$

The quantity $\partial H/\partial V$ is a dynamic variable, the form of which we shall make more precise below.

11.2. Thermodynamic Equalities for the Microcanonical Ensemble

For the microcanonical ensemble, according to (9.9), the entropy is equal to the logarithm of the statistical weight:

$$S(E, N, V) = \ln \Omega(E, N, V). \tag{11.4}$$

The total increment in entropy on change of the energy, particle number and volume, is equal to

$$dS = \frac{\partial \ln \Omega}{\partial E}\, dE + \frac{\partial \ln \Omega}{\partial V}\, dV + \frac{\partial \ln \Omega}{\partial N}\, dN, \tag{11.5}$$

which can be written in the form of the usual thermodynamic identity

$$dS = \frac{1}{\theta} dE + \frac{p}{\theta} dV - \frac{\mu}{\theta} dN, \tag{11.6}$$

where

$$\frac{1}{\theta} = \frac{\partial \ln \Omega}{\partial E}, \quad \frac{p}{\theta} = \frac{\partial \ln \Omega}{\partial V}, \quad \frac{\mu}{\theta} = - \frac{\partial \ln \Omega}{\partial N}, \tag{11.7}$$

θ is the temperature, p is the pressure, and μ is the chemical potential.

It is easily verified that the pressure defined by formula (11.7) is indeed the same as the average value of the generalized force $-\partial H/\partial V$. In fact, the statistical weight of the microcanonical distribution (9.7a) can be written in the form

$$\Omega(E, N, V) = \mathrm{Tr}\,(\Delta(H - E)). \tag{11.8}$$

For calculational convenience, it can be assumed that Δ is a continuous function approximating the step function. We shall calculate the partial derivative of Ω with respect to V, from the identity (11.8):

$$\frac{\partial \Omega}{\partial V} = \mathrm{Tr}\left(\frac{\partial}{\partial V} \Delta(H - E)\right) = -\,\mathrm{Tr}\left\{\frac{\partial}{\partial E} \Delta(H - E) \frac{\partial H}{\partial V}\right\} =$$

$$= - \frac{\partial}{\partial E} \mathrm{Tr}\left\{\Delta(H - E) \frac{\partial H}{\partial V}\right\},$$

or

$$\frac{\partial \Omega}{\partial V} = - \frac{\partial}{\partial E} \Omega \left\langle \frac{\partial H}{\partial V} \right\rangle \cong - \frac{\partial \Omega}{\partial E} \left\langle \frac{\partial H}{\partial V} \right\rangle \tag{11.9}$$

(we have discarded small terms of the order of the fluctuations, which is permissible in the thermodynamic limit). Consequently,

$$\frac{\partial \ln \Omega(E, N, V)}{\partial V} = - \frac{\partial \ln \Omega(E, N, V)}{\partial E} \left\langle \frac{\partial H}{\partial V} \right\rangle, \tag{11.9a}$$

and p, defined by Eq. (11.7), coincides with the average value of the generalized force $-\partial H/\partial V$, as we wished to prove.

The physical meaning of θ as the temperature is obvious from the fact that this quantity coincides with the integrating factor for $dE + pdV - \mu dN$.

11.3. The Virial Theorem for Quantum Systems

We shall consider the virial theorem for the case of quantum statistics, analogous to the classical virial theorem considered in §5.3.

Earlier, we defined the pressure as the average value of the generalized force operator $-\partial H(p, x)/\partial V$, which is a function of the momentum and coordinate operators p and x. We shall determine the explicit form of this operator.

We shall start from the statistical distribution (9.7a). We write the statistical weight (11.8) in the form

$$\Omega(E, N, V) = \mathrm{Tr}_V \{\Delta(H(p, x) - E)\}, \tag{11.10}$$

where we assume that H(p, x) does not depend on the volume V, but the volume dependence appears only through the dimensions of the basic region $V = L^3$ of normalization of the wave functions over which the trace in (11.10) is calculated; we indicate this by the subscript V in the trace. We shall describe the variability of the volume by introducing the parameter λ^3 multiplying V:

$$\Omega(E, N, \lambda^3 V) = \mathrm{Tr}_{\lambda^3 V} \{\Delta(H(p, x) - E)\}. \tag{11.10a}$$

We make the change of variables

$$x = \lambda x', \quad p = \lambda^{-1} p', \tag{11.10b}$$

i.e., we perform a canonical transformation which does not change the phase of the wave functions

$$(xp) = (x'p')$$

but makes the basic region of normalization independent of λ:

$$\Omega(E, N, \lambda^3 V) = \mathrm{Tr}_V \left\{\Delta\left(H\left(\frac{p}{\lambda}, \lambda x\right) - E\right)\right\}. \tag{11.11}$$

Differentiating Ω with respect to λ, we obtain

$$\frac{\partial \Omega}{\partial \lambda} = -\mathrm{Tr}_V \left\{ \frac{\partial}{\partial E} \Delta\left(H\left(\frac{p}{\lambda}, \lambda x\right) - E\right) \frac{\partial H\left(\frac{p}{\lambda}, \lambda x\right)}{\partial \lambda} \right\}$$

or, putting $\lambda = 1$,

$$\left(\frac{\partial \Omega}{\partial \lambda}\right)_{\lambda=1} = -\frac{\partial}{\partial E}\left(\Omega \left\langle \frac{\partial}{\partial \lambda} H\left(\frac{p}{\lambda}, \lambda x\right)\right\rangle\right)_{\lambda=1}. \tag{11.12}$$

Taking (5.10a) into account, we obtain

$$\frac{\partial \Omega}{\partial V} = -\frac{\partial}{\partial E}\Omega \frac{1}{3V}\left\langle \frac{\partial}{\partial \lambda} H\left(\frac{p}{\lambda}, \lambda x\right)\right\rangle_{\lambda=1}. \tag{11.12a}$$

Comparing (11.12a) with (11.9), we find an explicit form for the operator $-\partial H(p, x)/\partial V$ corresponding to the pressure:

$$-\frac{\partial H(p, x)}{\partial V} = -\frac{1}{3V}\frac{\partial}{\partial \lambda} H\left(\frac{p}{\lambda}, \lambda x\right)_{\lambda=1}. \tag{11.13}$$

The formula obtained differs from (5.12) only in the fact that p and x are now noncommuting operators. In other respects, it has the same form.

In the particular case of a Hamiltonian of particles with pair interaction, we obtain

$$-\frac{\partial H(p, x)}{\partial V} = -\frac{2}{3V}\sum_i \frac{\hbar^2\nabla_i^2}{2m} - \frac{1}{6V}\sum_{i\neq j}(\boldsymbol{x}_i - \boldsymbol{x}_j)\cdot\frac{\partial \phi(\boldsymbol{x}_i - \boldsymbol{x}_j)}{\partial \boldsymbol{x}_i}. \tag{11.14}$$

This formula gives the required representation for the pressure operator. The average value of the operator (11.14) gives an expression for the pressure:

$$p = -\frac{2}{3V}\sum_i \frac{1}{2m}\langle \hbar^2\nabla_i^2\rangle + \frac{1}{6V}\sum_{i\neq j}\langle(\boldsymbol{x}_i - \boldsymbol{x}_j)\cdot \boldsymbol{F}_{ij}\rangle,$$
$$\boldsymbol{F}_{ij} = -\frac{\partial \phi(\boldsymbol{x}_i - \boldsymbol{x}_j)}{\partial \boldsymbol{x}_i}, \tag{11.15}$$

i.e., a generalization of the virial theorem for the case of quantum statistics. Thus, in quantum statistics, as in classical statistics, the pressure is equal to two-thirds of the average kinetic energy density plus one-third of the virial.

The difference from the classical case consists in the fact that the operators for the kinetic energy and for the virial occurring in (11.15) do not commute. Therefore, formula (5.14c), which expresses the law of equipartition of energy among the degrees of freedom, no longer holds in the quantum case.

11.4. Thermodynamic Equalities for the Canonical Ensemble

The entropy for the canonical ensemble (9.14) is equal to

$$S = -\operatorname{Tr}(\rho \ln \rho) = \frac{\langle H\rangle - F}{\theta}. \tag{11.16}$$

Differentiating the identity (9.15)

$$e^{-F/\theta} = \mathrm{Tr}\ e^{-H/\theta}$$

with respect to θ, we find an expression for the average energy:

$$\langle H \rangle = -\theta^2 \frac{\partial}{\partial \theta} \left(\frac{F}{\theta} \right)_{a_i,\, N}. \tag{11.17}$$

Putting (11.17) into (11.16), we obtain another expression for the entropy:

$$S = -\left(\frac{\partial F}{\partial \theta} \right)_{a_i,\, N}. \tag{11.18}$$

The average value of the generalized force (11.1) over the canonical distribution is equal to

$$\langle A_i \rangle = -\left\langle \frac{\partial H}{\partial a_i} \right\rangle = -e^{F/\theta}\ \mathrm{Tr} \left(e^{-H/\theta} \frac{\partial H}{\partial a_i} \right), \tag{11.19}$$

or

$$\langle A_i \rangle = \theta\, e^{F/\theta} \frac{\partial}{\partial a_i}\ \mathrm{Tr}\, e^{-H/\theta} = -\theta\, e^{F/\theta} \frac{\partial}{\partial a_i} e^{-F/\theta}.$$

Consequently, the observed value of the average generalized force corresponding to change of the parameter a_i is equal to

$$\langle A_i \rangle = -\left(\frac{\partial F}{\partial a_i} \right)_{\theta,\, N} \tag{11.20}$$

or, in the particular case when $a_i = V$,

$$p = -\left(\frac{\partial F}{\partial V} \right)_{\theta,\, N}. \tag{11.20a}$$

We write the system of thermodynamic equalities obtained in the form of one relation, by calculating the increment in the free energy F on variation of the parameters θ, a_1, ..., a_s and N:

$$dF = \left(\frac{\partial F}{\partial \theta} \right)_{a_i,\, N} d\theta + \sum_{i=1}^{s} \left(\frac{\partial F}{\partial a_i} \right)_{\theta,\, N} da_i + \left(\frac{\partial F}{\partial N} \right)_{\theta,\, a_i} dN, \tag{11.21}$$

or, taking (11.18) and (11.20) into account,

$$dF = -S\, d\theta - \sum_{i=1}^{s} \langle A_i \rangle\, da_i + \mu\, dN, \tag{11.22}$$

where

$$\mu = \left(\frac{\partial F}{\partial N}\right)_{\theta,\, a_i}. \tag{11.23}$$

Equation (11.122) contains a complete system of thermodynamic relations, which can be expressed not only in terms of F, but also in terms of other thermodynamic functions, as in §5.4.

11.5. Thermodynamic Equalities for the Grand Canonical Ensemble

The thermodynamic equalities for the grand canonical ensemble are obtained in exactly the same way as for the canonical ensemble.

The entropy for the grand canonical ensemble (9.42) is equal to

$$S = -\operatorname{Tr}(\rho \ln \rho) = \frac{\langle H\rangle - \Omega - \mu \langle N\rangle}{\theta}. \tag{11.24}$$

Differentiating the identity (9.42a) with respect to θ and μ, we obtain expressions for the average energy and average particle number:

$$\langle H\rangle - \mu\langle N\rangle = -\theta^2 \frac{\partial}{\partial\theta}\left(\frac{\Omega}{\theta}\right)_{\mu,\, a_i}, \qquad \langle N\rangle = -\left(\frac{\partial\Omega}{\partial\mu}\right)_{\theta,\, a_i}. \tag{11.25}$$

Putting (11.25) into (11.24), we write the entropy in the form

$$S = -\left(\frac{\partial\Omega}{\partial\theta}\right)_{a_i,\, \mu}. \tag{11.26}$$

The generalized for (11.1) averaged over the grand canonical ensemble is equal to

$$\langle A_i\rangle = -\left\langle\frac{\partial H}{\partial a_i}\right\rangle = -e^{\Omega/\theta}\operatorname{Tr}\left(e^{-(H-\mu N)/\theta}\frac{\partial H}{\partial a_i}\right) =$$

$$= \theta\, e^{\Omega/\theta}\left(\frac{\partial}{\partial a_i}\operatorname{Tr} e^{-(H-\mu N)/\theta}\right)_{\theta,\, \mu} = \theta\, e^{\Omega/\theta}\left(\frac{\partial}{\partial a_i} e^{-\Omega/\theta}\right)_{\theta,\, \mu}$$

Consequently,

$$\langle A_i\rangle = -\left(\frac{\partial\Omega}{\partial a_i}\right)_{\theta,\, \mu}. \tag{11.27}$$

In the particular case when $a_i = V$,

$$p = -\left(\frac{\partial\Omega}{\partial V}\right)_{\theta,\ \mu}. \tag{11.27a}$$

To obtain the complete system of thermodynamic equalities, we calculate the increment in the thermodynamic potential $\Omega(\theta, \mu, a_1, \ldots, a_s)$ on change of the parameters θ, μ, a_1, ..., and a_s:

$$d\Omega = \left(\frac{\partial\Omega}{\partial\theta}\right)_{\mu,\ a_i} d\theta + \sum_{i=1}^{s}\left(\frac{\partial\Omega}{\partial a_i}\right)_{\theta,\mu} da_i + \left(\frac{\partial\Omega}{\partial\mu}\right)_{\theta,\mu} d\mu, \tag{11.28}$$

or, taking (11.26), (11.27), and (11.25) into account,

$$d\Omega = -S\,d\theta - \sum_{i=1}^{s}\langle A_i\rangle\,da_i - \langle N\rangle\,d\mu. \tag{11.29}$$

Equation (11.29) contains a complete system of thermodynamic relations.

11.6. Nernst's Theorem

In the preceding subsections of this section, we have obtained thermodynamic equalities expressing the first and second laws of thermodynamics, by starting from different Gibbsian ensembles. We now discuss **Nernst's theorem**, or the **third law of thermodynamics**.

Nernst's theorem establishes the behavior of the thermodynamic functions as the temperature tends to zero, and is related to the quantum properties of systems at low temperatures. Nernst established experimentally that, as the temperature tends to zero, for all substances, the difference in their entropies $S(\theta, a_i)$ (which is the only measurable quantity) tends to zero, along with its derivatives with respect to the external parameters, i.e.,

$$S(0, a_i) = S(0, a_i'), \quad \left(\frac{\partial S}{\partial a_i}\right)_{\theta=0} = 0 \tag{11.30}$$

for all values of the parameters a_i, a_i'. For example, if the external parameter is the volume, then

$$S(0, V_1) = S(0, V_2), \quad \left(\frac{\partial S}{\partial V}\right)_{\theta=0} = 0. \tag{11.30a}$$

Since the limiting value of the entropy does not depend on the parameters defining the system, it is convenient, following Planck, to put it equal to zero and obtain an absolute scale for the entropy of any substance:

$$S(0, a_i) = 0. \tag{11.30b}$$

These features of the behavior of the entropy at low temperatures are known as Nernst's theorem [20-22].

Nernst's theorem is not applicable to substances which are not in a state of statistical equilibrium, e.g., amorphous materials or disordered alloys, which can exist at very low temperatures as "frozen" metastable states with very long relaxation times. As a result of the incorrect application of Nernst's theorem to such substances, doubts used to be expressed about its validity.

Unlike the case of the first and second laws of thermodynamics, which follow directly from the Gibbsian distributions, there is no general statistical proof of Nernst's theorem, although for all known, physically reasonable models, it can be shown, using quantum statistical mechanics, that Nernst's theorem is fulfilled.

We shall examine the limit to which the Gibbsian distribution

$$w_k = e^{(F-E_k)/\theta} \tag{11.31}$$

tends as the temperature tends to zero.

It is convenient to express the free energy in (11.31) in terms of the entropy, using the relation (11.16),

$$w_k = \exp\left\{-S + \frac{\langle H\rangle - E_k}{\theta}\right\},$$

or

$$w_k = \exp\left\{-S + \frac{\langle H\rangle - E_0}{\theta} + \frac{E_0 - E_k}{\theta}\right\}, \tag{11.31a}$$

where E_0 is the energy of the ground level, with $E_k > E_0$ for $k \neq 0$ since the excited levels lie above the ground level. As $\theta \to 0$, the average energy $\langle H\rangle$ tends to E_0. Calculating the limit of the expression (11.31a) as $\theta \to 0$ using l'Hôpital's rule, we obtain

$$\lim_{\theta\to 0} w_k = w_k(0) = \exp(-S(0) + C_V(0))\,\delta_{E_k - E_0}, \tag{11.32}$$

where

$$\delta_{E_k - E_0} = \begin{cases} 1 & \text{for} \quad E_k = E_0, \\ 0 & \text{for} \quad E_k \neq E_0, \end{cases}$$

and $C_V(0) = \left(\frac{\partial \langle H \rangle}{\partial \theta}\right)_{\theta=0}$ is the heat capacity at constant volume and $\theta = 0$. From the normalization condition for the probability (11.32), it follows that

$$w_k(0) = \frac{1}{\Omega_0} \delta_{E_k - E_0}, \tag{11.33}$$

where Ω_0 is the multiplicity of the ground level E_0. But from the expression

$$S = \frac{\langle H \rangle - F}{\theta}$$

it follows from l'Hôpital's rule that

$$S(0) = C_V(0) + S(0),$$

i.e., that

$$C_V(0) = 0.$$

Thus, the limiting value of the entropy as $\theta \to 0$ is

$$S(0) = \ln \Omega_0. \tag{11.34}$$

It follows from (11.33) that as $\theta \to 0$, the canonical ensemble becomes a microcanonical ensemble with entropy (11.34).

For all known systems (crystal lattices, quantum gases, etc.), the ground level is nondegenerate, i.e.,

$$\Omega_0 = 1,$$

and, consequently, for these systems the entropy tends to zero as $\theta \to 0$. Even when $\Omega_0 \gg 1$, but the thermodynamic limit

$$\lim_{N \to \infty} \frac{\ln \Omega_0}{N} = 0,$$

we can assume that

$$S(0) = 0. \tag{11.35}$$

Occasionally in textbooks, Nernst's theorem is incorrectly associated entirely with absence of degeneracy of the ground level. In fact, the essence of Nernst's theorem does not lie in this, but in the features of the energy spectrum for small excitations.

If we associate Nernst's theorem only with absence of degeneracy of the ground level, then the features in the behavior of the thermodynamic functions which follow from Nernst's theorem would begin to appear only at very low temperatures θ_1 of the order of the energy difference between the first excited level and the ground level,

$$\theta_1 = E_1 - E_0,$$

and since the spectra of macroscopic bodies are practically continuous, these are very low, unobservable temperatures. For example, for an ideal gas of atoms with mass m in a volume $V = L^3$,

$$E_1 - E_0 = \frac{\hbar^2}{2m} k_{\min}^2 = \frac{h^2}{2mV^{2/3}},$$

where $k_{\min} = 2\pi/L$ is the minimum value of the wave vector. For a crystal lattice,

$$E_1 - E_0 = \hbar s k_{\min} = \frac{hs}{V^{1/3}},$$

where s is the speed of sound.

In fact, the behavior of the entropy required by Nernst's theorem begins to appear at much higher temperatures. For example, for ideal quantum gases, Nernst's theorem follows from degeneracy effects. For an ideal Bose gas, the behavior of the entropy corresponding to Nernst's theorem begins to appear at temperatures of the order of the degeneracy temperature θ_0

$$\theta_0 \approx \frac{\hbar^2}{m}\left(\frac{N}{V}\right)^{2/3}, \tag{11.36}$$

and, for an ideal Fermi gas, at temperatures lower than the temperature corresponding to the Fermi energy; this latter temperature is given, in order of magnitude, by the same expression (11.36) but for electrons in a metal can be very large, because of their small mass.

The degeneracy temperature of ideal gases is much higher than the θ_1 defined by the position of the first level. For crystal lattices, Nernst's theorem begins to be manifested at temperatures of the order of the Debye temperature θ_D, determined by the energy of the elementary excitations with maximum wave vector k_D:

$$\theta_D = \hbar s k_D = \hbar s \left(\frac{6\pi^2 N}{V} \right)^{1/3}.$$

The fact that the degeneracy temperature and the Debye temperature are proportional to Planck's constant $\hbar$ shows that Nernst's theorem is connected with the quantum properties of the system. In order to prove Nernst's theorem for the general case, it would be necessary to investigate the distribution of eigenvalues E_k close to the ground level, i.e., to investigate the function $\Omega(E, N, V)$ close to $E = E_0$. Up to the present time, it has been possible to do this only for certain models. For all the models of physical interest investigated, the distribution of eigenvalues close to the ground level is such that Nernst's theorem is fulfilled. It can be stated that in all cases in which the lower part of the spectrum of the system can be represented in the form of an ideal gas of quasi-particles (of the Fermi or Bose type), Nernst's theorem is found to be fulfilled.

§ 12. Fluctuations in Quantum Systems

We shall consider fluctuations for the Gibbsian quantum statistical ensembles. It is especially simple to calculate the fluctuations of quantities of which the statistical operator describing the ensemble depends, such as, for example, the energy fluctuations in a canonical ensemble.

12.1. Fluctuations in the Canonical Ensemble

The average value of the energy for the canonical ensemble is

$$\langle H \rangle = \mathrm{Tr}\, (e^{(F-H)/\theta} H). \tag{12.1}$$

Differentiating this identity with respect to θ at constant V and N and taking (11.17) into account, we obtain an expression for the energy fluctuations in the canonical ensemble

$$\langle H^2 \rangle - \langle H \rangle^2 = \theta^2 \frac{\partial \langle H \rangle}{\partial \theta}, \tag{12.2}$$

which has the same form as in the classical case (3.8d), except that the averaging is performed not by means of the classical distribution function, but by means of the statistical operator.

It follows from (12.2) that the energy fluctuations in the canonical ensemble are relatively small, since the average energy is proportional to the particle number N, while θ does not depend on N.

12.2. Fluctuations in the Grand Canonical Ensemble

In an analogous way, we calculate the fluctuations of energy and particle number in the grand canonical ensemble. Differentiating the expressions

$$\begin{aligned} \langle H-\mu N\rangle &= \mathrm{Tr}\,(e^{(\Omega-H+\mu N)/\theta}(H-\mu N)), \\ \langle N\rangle &= \mathrm{Tr}\,(e^{(\Omega-H+\mu N)/\theta}\,N) \end{aligned} \tag{12.3}$$

with respect to θ and μ with the remaining parameters held constant, using (11.25) we obtain expressions for the fluctuations of energy and particle number in the grand canonical ensemble:

$$\begin{aligned} \langle (H-\mu N)^2\rangle - \langle H-\mu N\rangle^2 &= \theta^2 \frac{\partial}{\partial\theta}(\langle H\rangle - \mu\langle N\rangle), \\ \langle N^2\rangle - \langle N\rangle^2 &= \theta\,\frac{\partial\langle N\rangle}{\partial\mu}, \end{aligned} \tag{12.4}$$

i.e., the same expressions as in the classical case (3.35).

It follows from (12.4) that the fluctuations of energy and particle number in the grand canonical ensemble are relatively small.

12.3. Fluctuations in a Generalized Ensemble

We shall consider fluctuations for the distribution described by the statistical operator (10.8)

$$\rho = \exp\left\{-\Phi(\mathcal{F}_0, \ldots, \mathcal{F}_n) - \sum_{k=0}^{n} \mathcal{F}_k \mathcal{P}_k\right\}, \tag{12.5}$$

in which the average values

$$\langle \mathcal{P}_k\rangle = \mathrm{Tr}\,(\rho\mathcal{P}_k) \qquad (k=0,\ 1,\ \ldots,\ n). \tag{12.6}$$

are given. In (12.5), $\Phi(\mathcal{F}_0, \ldots, \mathcal{F}_n)$ is the Massieu–Planck function, determined from the condition that the trace be normalized to unity:

$$e^{\Phi} = \operatorname{Tr} e^{-\sum \mathcal{F}_k \mathcal{P}_k}. \tag{12.7}$$

Differentiating this identity with respect to $\mathcal{F}_k$, we obtain the average value of $\mathcal{P}_k$:

$$\langle \mathcal{P}_k \rangle = \operatorname{Tr} \left(e^{-\Phi - \sum \mathcal{F}_i \mathcal{P}_i} \mathcal{P}_k \right) = -\frac{\partial \Phi}{\partial \mathcal{F}_k}. \tag{12.8}$$

In calculating the fluctuations, it is necessary to distinguish two cases: when all the $\mathcal{P}_k$ are integrals of motion, and when not all the $\mathcal{P}_k$ are integrals of motion.

We now consider the first case. Since it is assumed that all the $\mathcal{P}_k$ commute, the exponential of the sum of operators can be differentiated as an ordinary function, even outside the trace. Differentiating the identity (12.8) with respect to $\mathcal{F}_i$, we obtain an expression for the fluctuations of the quantities

$$\langle \mathcal{P}_k \mathcal{P}_i \rangle - \langle \mathcal{P}_k \rangle \langle \mathcal{P}_i \rangle = -\frac{\partial \langle \mathcal{P}_k \rangle}{\partial \mathcal{F}_i} = -\frac{\partial \langle \mathcal{P}_i \rangle}{\partial \mathcal{F}_k} = \frac{\partial^2 \Phi}{\partial \mathcal{F}_i \partial \mathcal{F}_k}, \tag{12.9}$$

analogous to the classical expression (6.7).

In the second case, when not all the $\mathcal{P}_k$ are integrals of motion and the $\mathcal{P}_k$ may not commute, we must exercise care in differentiating the exponential containing the sum of noncommuting operators.

We shall derive a formula for the differentiation of the exponential $e^{A(\alpha)}$ with respect to the parameter α. For this, we must expand the exponential e^{A+B}, where $B = \delta A$ and does not commute with A, in a series in B. It is convenient to introduce an auxiliary operator $\mathcal{K}(\tau)$

$$e^{(A+B)\tau} = \mathcal{K}(\tau) e^{A\tau}, \tag{12.10}$$

satisfying the condition

$$\mathcal{K}(0) = 1. \tag{12.10 a}$$

The expression (12.10) can be differentiated simply with respect to τ, since A + B does not depend on τ, and the increment $(A + B)\delta\tau$

commutes with A + B. The relation (12.10) is equivalent to a differential equation for $\mathcal{K}$

$$\frac{\partial \mathcal{K}}{\partial \tau} = \mathcal{K}\, e^{A\tau} B\, e^{-A\tau} \tag{12.11}$$

with the initial condition (12.10a), since on differentiation of (12.10) with respect to τ, the terms $\mathcal{K} e^{A\tau} A$ in the left- and right-hand sides cancel. The differential equation (12.11) and initial condition (12.10a) are equivalent to an operator integral equation for $\mathcal{K}$:

$$\mathcal{K}(\tau) = 1 + \int_0^\tau \mathcal{K}(\tau_1)\, e^{A\tau_1} B e^{-A\tau_1}\, d\tau_1. \tag{12.12}$$

Iteration of Eq. (12.12) gives an expansion of $\mathcal{K}$ in powers of B. Confining ourselves to the first approximation in B, we obtain

$$\mathcal{K}(\tau) \cong 1 + \int_0^\tau e^{A\tau_1} B e^{-A\tau_1}\, d\tau_1. \tag{12.12a}$$

Putting $\tau = 1$ in (12.12a) and substituting this expression in (12.10), we obtain

$$e^{A+B} \cong \left(1 + \int_0^1 e^{A\tau} B e^{-A\tau}\, d\tau\right) e^A,$$

i.e.,

$$\delta e^A = \int_0^1 e^{A\tau}\, \delta A\, e^{-A\tau} e^A\, d\tau, \tag{12.13}$$

or

$$\frac{d}{d\alpha} e^{A(\alpha)} = \int_0^1 e^{A\tau} \frac{dA}{d\alpha} e^{-A\tau} e^A\, d\tau. \tag{12.14}$$

Formula (12.14) gives the rule for differentiating the exponential of an operator in the general case. In the particular case when A is proportional to α, the ordinary rule for differentiating an exponential follows from (12.14).

To calculate the fluctuations of the quantities $\mathcal{P}_k$, we differentiate formula (12.8) with respect to $\mathcal{F}_i$ using the rule (12.14).

We obtain

$$\frac{\partial \langle \mathcal{P}_k \rangle}{\partial \mathcal{F}_i} = \langle \mathcal{P}_i \rangle \langle \mathcal{P}_k \rangle - \int_0^1 \langle \mathcal{P}_k \mathcal{P}_i(\tau) \rangle \, d\tau, \tag{12.15}$$

where

$$\mathcal{P}_i(\tau) = e^{-\tau \sum \mathcal{F}_k \mathcal{P}_k} \mathcal{P}_i e^{\tau \sum \mathcal{F}_k \mathcal{P}_k}; \tag{12.16}$$

consequently,

$$\frac{\partial \langle \mathcal{P}_k \rangle}{\partial \mathcal{F}_i} = - \int_0^1 \langle (\mathcal{P}_k - \langle \mathcal{P}_k \rangle)(\mathcal{P}_i(\tau) - \langle \mathcal{P}_i \rangle) \rangle \, d\tau. \tag{12.17}$$

It is convenient to introduce a more compact notation for the integrand in the right-hand side of (12.17), since we shall encounter it often:

$$(\mathcal{P}_k, \mathcal{P}_i) = \int_0^1 \langle (\mathcal{P}_k - \langle \mathcal{P}_k \rangle)(\mathcal{P}_i(\tau) - \langle \mathcal{P}_i \rangle) \rangle \, d\tau, \tag{12.18}$$

Therefore, (12.17) can be rewritten in the form

$$(\mathcal{P}_k, \mathcal{P}_i) = - \frac{\partial \langle \mathcal{P}_k \rangle}{\partial \mathcal{F}_i} = - \frac{\partial \langle \mathcal{P}_i \rangle}{\partial \mathcal{F}_k} = \frac{\partial^2 \Phi}{\partial \mathcal{F}_i \partial \mathcal{F}_k}. \tag{12.19}$$

Consequently, if the quantities $\mathcal{P}_i$ $(i = 0, 1, \ldots, n)$ do not commute, double differentiation of Φ with respect to the parameters $\mathcal{F}_i$ and $\mathcal{F}_k$, does not give simply the joint fluctuations of $\mathcal{P}_i$ and $\mathcal{P}_k$, but the average over τ of the fluctuations of $\mathcal{P}_k$ and $\mathcal{P}_i(\tau)$.

Only in the special case when the operators $\mathcal{P}_i$ commute or when we can neglect their noncommutativity do we obtain formula (12.9) for the fluctuations. In this case, we can construct the macroscopic distribution function (6.11) introduced in §6.

§13. Thermodynamic Equivalence of the Gibbsian Statistical Ensembles

All the ensembles of statistical mechanics are defined, as was shown in §§3 and 9, by specifying the external conditions in which the systems from which the ensembles are composed are found. For example, the microcanonical ensemble is defined by fixing the energy, the particle number and the volume, the canonical ensemble is defined by fixing the particle number and the volume and by contact with a thermostat, the grand canonical ensemble is defined by fixing the volume and by contact with a thermostat and a particle reservoir, and the isobaric-isothermal ensemble is defined by fixing the particle number and the pressure and by contact with a thermostat.

The application of statistical ensembles to concrete problems is usually not confined to those conditions for which they are defined. In choosing the ensemble, we are guided by calculational convenience and not by the conditions in which the system is found

To justify the replacement of one ensemble by another, it is usually pointed out that the different ensembles differ little from one another, since fluctuations of the quantities which are not specified for them are small. Indeed, as we saw in §12, fluctuations of the energy in the canonical ensemble and of the particle number in the grand canonical ensemble are small. These physical considerations, of course, cannot be regarded as a proof of the equivalence of the statistical ensembles in the thermodynamic sense. To prove this equivalence, it is necessary to show that the different ensembles can replace each other in such a way that the thermodynamic functions calculated by means of them differ little among themselves, and coincide in the thermodynamic limit $V \to$ $N/V = \text{const}$.

The question of the thermodynamic equivalence of the statistical ensembles was considered in [17], where it was shown, for example, that if only the largest term in the partition function of the grand canonical ensemble is retained, the thermodynamic functions obtained in this approximation coincide in the thermodynamic limit with the thermodynamic functions calculated on the basis of a canonical ensemble with the particle number equal to the average particle number in the grand ensemble. However, this is still not a complete proof of the thermodynamic equivalence of the ensembles, since the method of separating out the largest term in the partition functions does not enable us to estimate the discarded terms.

In this section, following [23], we shall give a proof of the thermodynamic equivalence of the statistical ensembles, using the method of steepest descents.

13.1. Thermodynamic Equivalence of the Canonical and Microcanonical Ensembles

Let the system be situated in a thermostat and let it be described by the canonical distribution (9.10). For the system, we shall find the approximate microcanonical distribution by means of which it would be possible to calculate all the thermodynamic functions.

The partition function $Q(\theta, V, N)$ is connected with the statistical weight $\Omega(E, N, V)$ by the relation

$$Q(\theta, V, N)=\sum_E e^{-E/\theta}\Omega(E, N, V), \tag{13.1}$$

where the summation is performed over all the permissible values of the energy, and $\Omega(E, N, V)$ is the number of states for systems with energy E, particle number N, and volume V. If the summation is performed over energy layers ΔE, as we shall assume below, then $\Omega(E, N, V)$ is the number of states in such a layer.

Using the theorem (9.27) on the inversion of the partition function, we obtain for the statistical weight the expression

$$\begin{aligned}\Omega(E, N, V)&=\Gamma(E+\Delta E, N, V)-\Gamma(E, N, V)\\ &=\frac{1}{2\pi i}\int_{a-i\infty}^{a+i\infty} e^{\lambda E}Q(\lambda^{-1}, V, N)\frac{e^{\lambda\,\Delta E}-1}{\lambda}\,d\lambda;\end{aligned} \tag{13.2}$$

$Q(\lambda^{-1}, V, N)$ corresponds to $Q(\lambda)$ in the previous notation, and a is a real positive number. Thus, (13.2) gives the inversion of the partition function (13.1). It is convenient to rewrite formula (13.2) in the form

$$\Omega(E, N, V)=\frac{1}{2\pi i}\int_{a-i\infty}^{a+i\infty} e^{N\chi(\lambda)}\frac{e^{\lambda\,\Delta E}-1}{\lambda}\,d\lambda, \tag{13.3}$$

where

$$\chi(\lambda)=\frac{E}{N}\lambda-\lambda\frac{F(\lambda^{-1}, V, N)}{N}. \tag{13.3a}$$

Here, $F(\theta, V, N)$ is the free energy. In writing the exponential in formula (13.3) in the form shown, we are explicitly taking into account the large magnitude of the factor $\exp(N\chi)$, since χ is finite as $N \to \infty$.

We assume that, for our system, the thermodynamic limit

$$\lim_{N\to\infty} \frac{F}{N} \qquad (V/N = \text{const}).$$

exists. Then, as $N \to \infty$, the function (13.3a) tends to a finite limit; therefore, for large N, the integral (13.3) can be evaluated by the method of steepest descents, if we assume that $\chi(\lambda)$ is an analytic function for $\operatorname{Re}\lambda > A$. Below, we shall see that the necessary conditions for the applicability of the method of steepest descents are indeed fulfilled.

As the saddle point, we choose the real positive root λ_1 of the equation

$$\frac{\partial\chi}{\partial\lambda}(\lambda_1) = 0 \tag{13.4}$$

and put $a = \lambda_1 = 1/\theta$ in (13.3). Then, repeating the same calculations as in §3.3, we obtain

$$\Omega(E, N, V) = e^{N\chi(\lambda_1)} \frac{e^{\lambda_1 \Delta E} - 1}{\lambda_1 \sqrt{2\pi N \chi''(\lambda_1)}} = Q(\theta, V, N)\, e^{E/\theta} \frac{e^{\Delta E/\theta} - 1}{\sqrt{2\pi C_V}}, \tag{13.5}$$

where

$$C_V = \frac{\partial \bar{E}}{\partial \theta}, \qquad \bar{E} = -\theta^2 \frac{\partial}{\partial\theta}\left(\frac{F(\theta, V, N)}{\theta}\right). \tag{13.5a}$$

The condition (13.4) for the saddle point takes the form

$$E = \bar{E} = \langle H \rangle = -\theta^2 \frac{\partial}{\partial\theta}\left(\frac{F(\theta, V, N)}{\theta}\right), \tag{13.6}$$

i.e., E must be chosen to be equal to the average energy (11.17) in the canonical ensemble.

For the method of steepest descents to applicable, it is necessary that there be a maximum at the point λ_1 as we move parallel to the imaginary axis, i.e., that $\chi''(\lambda_1)$ be positive, or that

$$C_V > 0, \tag{13.7}$$

which is one of the conditions for thermodynamic stability of the system; this condition, by virtue of the relation (12.2), is fulfilled for ordinary systems

$$\langle (H - \langle H \rangle)^2 \rangle = \theta^2 C_V > 0. \tag{13.8}$$

We shall construct an approximate microcanonical distribution

$$w(E_k) = \begin{cases} \Omega^{-1}(E, N, V) & \text{for} \quad E \leqslant E_k \leqslant E + \Delta E, \\ 0 & \text{outside this interval,} \end{cases} \tag{13.9}$$

where $\Omega(E, N, V)$ is given by formula (13.5), $E = \overline{E}$, and ΔE is equal to the average energy fluctuations given by (13.8). The relation (13.5) can be rewritten in the form

$$\Omega(E, N, V) = e^{E/\theta} Q(\theta, N, V) \frac{e^{\sqrt{C_V}}}{\sqrt{2\pi C_V}}. \tag{13.10}$$

The expression (13.10), where θ is determined from Eq. (13.6), gives the statistical weight as a function of E, N, and V for the required approximate microcanonical distribution (13.9). The entropy of the approximate microcanonical ensemble is equal to

$$\ln \Omega(E, N, V) = \frac{E - F(\theta, V, N)}{\theta} + \frac{1}{\theta} \sqrt{\langle (H - \langle H \rangle)^2 \rangle} = S(\theta, V, N) + \sqrt{C_V}, \tag{13.11}$$

where we have kept only the principal term in N, and $S(\theta, V, N)$ is the entropy of the canonical ensemble.

Thus, the difference in the entropies calculated by means of the approximate microcanonical distribution (13.9) and the canonical distribution is equal to $(C_V)^{1/2}$. If the fluctuations are normal, i.e., $(C_V)^{1/2} \sim N^{1/2}$, then the last term in (13.11) gives a correction which disappears in the limit

$$\lim_{N \to \infty} \frac{\ln \Omega(E, N, V)}{N} \qquad \left(\frac{V}{N} = \text{const}\right).$$

The temperature θ is connected with the statistical weight by the relation

$$\frac{1}{\theta} = \frac{\partial}{\partial E} \ln \Omega(E, N, V) - \frac{1}{C_V} \left(\frac{\partial \sqrt{C_V}}{\partial \theta}\right)_{V, N}, \tag{13.12}$$

which is obtained by substituting F from (13.11) into (13.6). It follows from (13.12) that, in the thermodynamic limit, the inverse temperature is equal to the energy derivative of the entropy of the approximate microcanonical ensemble.

The relations (13.11) and (13.6) enable us to calculate all the thermodynamic quantities for the approximate microcanonical distribution, if these are known for the canonical distribution. For example, for the pressure and chemical potential we obtain

$$\left.\begin{aligned} \frac{p}{\theta} &= \frac{\partial}{\partial V}\ln\Omega(E, N, V) - \left(\frac{\partial\sqrt{C_V}}{\partial V}\right)_{\theta, N} + \frac{1}{C_V}\left(\frac{\partial\sqrt{C_V}}{\partial\theta}\right)_{V, N}\left(\frac{\partial\bar{E}}{\partial V}\right)_{\theta, N}, \\ \frac{\mu}{\theta} &= -\frac{\partial}{\partial N}\ln\Omega(E, N, V) + \left(\frac{\partial\sqrt{C_V}}{\partial N}\right)_{\theta, V} - \frac{1}{C_V}\left(\frac{\partial\sqrt{C_V}}{\partial\theta}\right)_{V, N}\left(\frac{\partial\bar{E}}{\partial N}\right)_{V, \theta}. \end{aligned}\right\} \quad (13.13)$$

The second and third terms in the right-hand side of Eqs. (13.13) represent the difference between the pressures and chemical potentials (divided by θ) calculated by means of the canonical distribution and the microcanonical distribution (13.9). These differences are small – of order $1/N^{1/2}$ – and vanish in the limit $N \to \infty$.

We must turn our attention to a special case. For a system consisting of a liquid in equilibrium with saturated vapor at constant pressure, the temperature is constant and does not depend on the energy provided. In this case, as was noted by Gibbs [24], the heat capacity is infinite. For such a system, our derivation of the equivalence of the canonical and microcanonical distributio is not valid.

We note, however, that systems with such properties can onl be obtained after the thermodynamic limiting process $N \to \infty$, V → (V/N = const) is performed. It is therefore impossible to put $C_V = \infty$ into the formula (13.11); it is necessary to estimate how tends to infinity and perform the thermodynamic limiting process

Since, at the present time, there is no satisfactory theory of condensation, we cannot generalize the theorem on the equival of the statistical ensembles to the case of systems consisting of several phases in equilibrium.

We shall comment on the meaning of the thermodynamic lim when surface energy is important. It is well known that the ther dynamic method can be applied successfully to surface phenomena, and that there exists a statistical thermodynamics of surface phenomena [25]. For the free energy in the thermodynamic limit, we must also take into account, in addition to the volume terms, terms of higher order of smallness, proportional to the surface of the system, i.e., we must have an expression for the free energy exact to order

$$F \cong Vf + Sf_S,$$

where f and f_S are the densities of the volume and surface energies, and S is the surface area of the system. For this to be possible it is necessary that the energy fluctuations be much smaller than the surface energy, i.e., that

$$\theta \sqrt{C_V} \ll Sf_S.$$

Since $C_V \sim N$ and $S \sim N^{2/3}$, we have

$$\lim_{N \to \infty} \frac{\sqrt{C_V}}{S} = \lim_{N \to \infty} \frac{1}{N^{1/6}} = 0,$$

i.e., the energy fluctuations are small compared with the surface energy, and, consequently, we can take the latter into account while neglecting the fluctuations. This estimate shows that a statistical thermodynamics of surface phenomena is possible, despite the fluctuations of the extensive quantities.

13.2. Thermodynamic Equivalence of the Grand Canonical and Canonical Ensembles

Suppose that systems with a variable number of particles in a thermostat are described by the grand canonical distribution (9.38). Their thermodynamic properties are determined by the partition function (9.39) which is connected with the partition function (13.1) of the canonical ensemble by the relation

$$Q(\theta, \mu, V) = \sum_{N \geqslant 0} \lambda^N Q(\theta, V, N) \equiv Q(\lambda), \tag{13.14}$$

where

$$\lambda = e^{\mu/\theta} \tag{13.14a}$$

is the **absolute activity**. Here, as before, we are using the same symbol Q for the partition functions of the grand canonical and canonical ensembles, distinguishing them only by the arguments on which they depend.

Regarding λ as a complex variable, we obtain the analytic continuation of the function $Q(\lambda)$ into the complex plane.

The inverse of formula (13.14) has the form

$$Q(\theta, V, N) = \frac{1}{2\pi i} \oint Q(\lambda) \frac{d\lambda}{\lambda^{N+1}}, \tag{13.15}$$

where the integration contour encloses the point $\lambda = 0$. Indeed, putting (13.14) into (13.15), we see that only one term $N_1 = N$ in the sum is nonzero, and the contour integral is equal to the residue at the point $\lambda = 0$. The method of Darwin and Fowler [16] is based on the application of the inversion formula (13.15).

In place of $Q(\lambda)$, we introduce the function (9.40)

$$\Omega(\lambda) = -\theta \ln Q(\lambda), \tag{13.16}$$

which plays the role of the thermodynamic potential for the grand ensemble ($\Omega = -pV$), and rewrite (13.15) in the form

$$Q(\theta, V, N) = \frac{1}{2\pi i} \oint \exp\left\{ -\frac{Nv\Omega(\lambda)}{\theta V} \right\} \frac{d\lambda}{\lambda^{N+1}} = \frac{1}{2\pi i} \oint e^{N\varphi(\lambda)} \frac{d\lambda}{\lambda}, \tag{13.17}$$

where

$$\varphi(\lambda) = -\frac{v\Omega(\lambda)}{\theta V} - \ln\lambda, \qquad v = \frac{V}{N}. \tag{13.17a}$$

The function $\varphi(\lambda)$ tends to a finite limit as $N \to \infty$, $v = \text{const}$, since, in the thermodynamic limit, the ratio $\Omega(\lambda)/V$ is finite. In addition, $\varphi(\lambda)$ is assumed to be analytic in the region $\operatorname{Re}\lambda > 0$ of complex values of λ. Therefore, the integral (13.17) can be evaluated by the method of steepest descents, as in the preceding subsection. As a result, we obtain

$$Q(\theta, V, N) = \frac{\exp(N\varphi(\lambda_0))}{\lambda_0 \sqrt{2\pi N\varphi''(\lambda_0)}}, \tag{13.18}$$

where λ_0 is the real positive root of the equation

$$\frac{\partial \varphi(\lambda_0)}{\partial \lambda} = 0, \tag{13.19}$$

which, using (13.17a), we can write in the form

$$\frac{1}{v} = -\frac{1}{V\theta}\lambda_0 \frac{\partial \Omega(\lambda_0)}{\partial \lambda}, \tag{13.19a}$$

or

$$N = -\frac{\partial \Omega}{\partial \mu_0} = \overline{N}, \quad \text{where } \mu_0 = \theta \ln \lambda_0. \tag{13.19b}$$

We shall regard expression (13.18) as the partition function of the approximate canonical ensemble with particle number equal to the average particle number in the grand canonical ensemble, in accordance with (13.19b).

For the method of steepest descents to be applicable, it is necessary that there be a maximum at the point λ_0 as we move away from it parallel to the imaginary axis (cf. §3.3), i.e., that the condition

$$\frac{\partial^2 \varphi(\lambda_0)}{\partial \lambda^2} > 0, \tag{13.20}$$

which can be written in the form

$$\frac{\partial \overline{N}}{\partial \mu} > 0 \tag{13.20a}$$

be fulfilled. This condition, generally speaking, is fulfilled, since, taking (12.4) into account, we have

$$\frac{\partial \overline{N}}{\partial \mu} = \frac{1}{\theta}\overline{(N-\overline{N})^2} > 0. \tag{13.21}$$

We write expression (13.18) for the partition function in the form

$$Q(\theta, V, N) = Q(\theta, \mu, V)\frac{\exp(-N\mu/\theta)}{\sqrt{2\pi \frac{\partial \overline{N}}{\partial \mu}\theta}} = \frac{Q(\theta, \mu, V)\exp(-N\mu/\theta)}{\sqrt{2\pi(\overline{N^2} - (\overline{N})^2)}} \quad (\mu = \mu_0). \tag{13.22}$$

If the fluctuations in the number of particles are small, we can confine ourselves to the principal terms in (13.22), and we obtain

$$Q(\theta, V, N) \cong Q(\theta, \mu, V) \exp(-N\mu/\theta), \tag{13.23}$$

since the discarded factor gives no contribution in the thermodynamic limit.

As is well known, in a region of coexistence of phases, the absolute activity λ does not depend on the specific volume, i.e.,

$$\frac{\partial \lambda}{\partial v} = 0, \quad \text{or} \quad \frac{1}{V}\frac{\partial \bar{N}}{\partial \lambda} = \infty, \quad \text{since} \quad v = V/\bar{N}.$$

In this case, we cannot derive the thermodynamic equivalence of the grand canonical and canonical ensembles from formula (13.22). However, it should be noted that we cannot substitute $\frac{1}{V}\frac{\partial \bar{N}}{\partial \lambda} = \infty$, in formula (13.22), since, as we have already noted in the preceding subsection, systems with such properties can be obtained only after the thermodynamic limit has been taken, and, because of the absence of a consistent theory of condensation, we cannot estimate the order of the increase of $\partial \bar{N}/\partial \lambda$.

Using (13.22) and (13.20), we can calculate all the thermodynamic functions. For the free energy, we obtain

$$F(\theta, V, \bar{N}) = \Omega(\theta, \mu, V) + \mu \bar{N} + \frac{\theta}{2} \ln\left(2\pi\theta \frac{\partial \bar{N}}{\partial \mu}\right), \tag{13.24}$$

where μ is given by Eq. (13.19b). Putting (13.24) into (13.19b), for the chemical potential we shall have the expression

$$\mu = \left(\frac{\partial F}{\partial \bar{N}}\right)_{V,\theta} - \frac{\theta}{2}\frac{\partial^2 \bar{N}}{\partial \mu^2}\left(\frac{\partial \bar{N}}{\partial \mu}\right)^{-2}. \tag{13.25}$$

Differentiating (13.24) with respect to V and θ, taking (13.19b) and (13.25) into account, we obtain for the pressure and entropy respectively

$$p = -\left(\frac{\partial \Omega}{\partial V}\right)_{\theta,\mu} = -\left(\frac{\partial F}{\partial V}\right)_{\theta,\bar{N}} - \frac{\theta}{2}\frac{\partial}{\partial \mu}\left\{\frac{\partial \bar{N}}{\partial V}\left(\frac{\partial \bar{N}}{\partial \mu}\right)^{-1}\right\}, \tag{13.26}$$

$$S = -\left(\frac{\partial \Omega}{\partial \theta}\right)_{V,\mu} = -\left(\frac{\partial F}{\partial \theta}\right)_{V,\bar{N}} - \frac{\theta}{2}\left(\frac{\partial^2 \bar{N}}{\partial \mu^2}\right)\left(\frac{\partial \bar{N}}{\partial \theta}\right)\left(\frac{\partial \bar{N}}{\partial \mu}\right)^{-2} + \frac{1}{2}\frac{\partial}{\partial \theta}\left\{\theta \ln\left(2\pi\theta \frac{\partial \bar{N}}{\partial \mu}\right)\right\}. \tag{13.27}$$

In the expressions (13.25)-(13.27), the first terms in the right-hand sides make the main contribution, and the others are negligibly small in the thermodynamic limit. Neglecting these small terms, we obtain the well-known relations (11.23), (11.20a), and (11.18), i.e., the chemical potential, pressure, and entropy calculated using the partition function (13.14) of the grand ensemble are equal to the corresponding quantities calculated using the partition function (13.23). Thus, the grand canonical and canonical ensembles are thermodynamically equivalent.

§ 14. Passage to the Classical Limit of Quantum Statistics

At sufficiently high temperatures and not too large densities, when we can neglect quantum effects, a Gibbsian quantum-mechanical ensemble goes over into the classical ensemble, and the partition function (9.11) of the quantum ensemble becomes the partition function (3.8) of the classical ensemble. This problem was studied by Wigner [12], Uhlenbeck and Gropper [26], and Kirkwood [27]. We shall follow the latter work.

14.1. Passage to the Limit of Partition Functions

For simplicity, we confine ourselves to a system of N monatomic molecules of mass m in volume V, interacting through the potential

$$v(\boldsymbol{x}_1, \ldots, \boldsymbol{x}_N) = \frac{1}{2} \sum_{i \neq j} \phi(\boldsymbol{x}_i - \boldsymbol{x}_j).$$

In the case of the canonical ensemble, the partition function for the system has the form

$$Q = \sum_k \int \Psi_k^*(\boldsymbol{x}_1, \ldots, \boldsymbol{x}_N)\, e^{-\beta H}\, \Psi_k(\boldsymbol{x}_1, \ldots, \boldsymbol{x}_N)\, d\boldsymbol{x}_1 \ldots d\boldsymbol{x}_N, \tag{14.1}$$

where

$$H = -\sum_j \frac{\hbar^2}{2m} \nabla_j^2 + v(\boldsymbol{x}_1, \ldots, \boldsymbol{x}_N). \tag{14.2}$$

Ψ_k is a complete set of functions of the required symmetry. Here and below, we shall not write out explicitly the arguments θ, V and N on which Q depends.

Since Q does not depend on the choice of functions Ψ_k, for the latter we can choose any complete orthonormal set, symmetric in the case of Bose systems and antisymmetric in the case of Fermi systems. For the Ψ_k, we choose a symmetrized or antisymmetrized product of plane waves, normalized in the volume V:

$$\begin{gathered}\Psi_p = \frac{1}{\sqrt{N!}} \sum_{\mathscr{P}} (\pm 1)^{\mathscr{P}} \mathscr{P} \varphi_{p_1}(\boldsymbol{x}_1) \ldots \varphi_{p_N}(\boldsymbol{x}_N), \\ \varphi_{p_j}(\boldsymbol{x}_j) = V^{-1/2} e^{\frac{i}{\hbar}(\boldsymbol{p}_j \cdot \boldsymbol{x}_j)}, \qquad p = (\boldsymbol{p}_1, \ldots, \boldsymbol{p}_N).\end{gathered} \tag{14.3}$$

The plus sign corresponds to Bose statistics, and the minus sign to Fermi statistics (for simplicity, we do not consider the spin of the particles). The operator $\mathscr{P}$ denotes permutation of the quantities $\mathbf{x}_1, \ldots, \mathbf{x}_N$, or, equivalently, of the quantities $\mathbf{p}_1, \ldots, \mathbf{p}_N$; $(\pm 1)^{\mathscr{P}} = 1$ for a Bose gas, and +1 or −1, depending on the parity of the permutation, for a Fermi gas. The factor $1/\sqrt{N!}$ ensures the normalization to unity of the functions Ψ_p. In fact,

$$\int \Psi_p^* \Psi_p \, d\boldsymbol{x}_1 \ldots d\boldsymbol{x}_N =$$

$$= \frac{1}{V^N N!} \sum_{\mathscr{P}, \mathscr{P}'} (\pm 1)^{\mathscr{P}+\mathscr{P}'} \int \exp\left\{-\frac{i\mathscr{P}}{\hbar} \sum (\boldsymbol{p}_j \cdot \boldsymbol{x}_j)\right\} \exp\left\{\frac{i\mathscr{P}'}{\hbar} \sum (\boldsymbol{p}_j \cdot \boldsymbol{x}_j)\right\} \times$$

$$\times \, d\boldsymbol{x}_1 \ldots d\boldsymbol{x}_N = \frac{1}{V^N N!} \sum_{\mathscr{P}} \int d\boldsymbol{x}_1 \ldots d\boldsymbol{x}_N = \frac{1}{N!} \sum_{\mathscr{P}} 1 = 1,$$

since for $\mathscr{P} \neq \mathscr{P}'$, the integral is equal to zero, and $\sum_{\mathscr{P}} 1$ is equal to the number of permutations of N elements.

We choose the single-particle wave functions $\varphi_{p_j}(\boldsymbol{x}_j)$ in such a way that they possess the property of periodicity in a cube of side $L = V^{1/3}$, i.e.,

$$\varphi_{p_j}(x_j^\alpha + L) = \varphi_{p_j}(x_j^\alpha),$$

The quantum numbers p_j can then take only the values

$$p_j^\alpha = \frac{2\pi\hbar}{L} n_{p_j^\alpha},$$

where $n_{p_j^\alpha}$ are integers 0, ±1, ±2, ... Consequently, in the element $V\, d\boldsymbol{p}_j = V\, dp_j^x\, dp_j^y\, dp_j^z$ of phase volume of each particle, there are $V\, d\boldsymbol{p}_j/(2\pi\hbar)^3$ quantum states.

The summation over the quantum states p_j^α in (14.1) can be replaced by integration over the momenta, since there is a large number of particles and the spectrum is practically continuous:

$$\sum_{p_1, \ldots, p_N} \to \frac{1}{N!} \frac{V^N}{(2\pi\hbar)^{3N}} \int \ldots \int d\boldsymbol{p}_1 \ldots d\boldsymbol{p}_N, \tag{14.4}$$

where the sum over the momenta denotes summation over all the different states. The factor 1/N! allows for the fact that permutations of the particles do not change the state; cf. (1.5b).† In transforming from the sums to the integrals, we are excluding the possible case of Bose-gas degeneracy, when there may be a macroscopically large number of particles in the ground state. This case must be considered separately.

Using (14.3) and (14.4), we write the partition function (14.1) in the form

$$Q = \frac{1}{(N!)^2} \frac{1}{(2\pi\hbar)^{3N}} \int \ldots \int \sum_{\mathscr{P}, \mathscr{P}'} (\pm 1)^{\mathscr{P}+\mathscr{P}'} \exp\left\{ -\frac{i\mathscr{P}}{\hbar} \sum_j (\boldsymbol{p}_j \cdot \boldsymbol{x}_j) \right\} \times$$

$$\times e^{-\beta H} \exp\left\{ \frac{i\mathscr{P}'}{\hbar} \sum_j (\boldsymbol{p}_j \cdot \boldsymbol{x}_j) \right\} dx\, dp, \tag{14.5}$$

where

$$dx\, dp = d\boldsymbol{x}_1 \ldots d\boldsymbol{x}_N\, d\boldsymbol{p}_1 \ldots d\boldsymbol{p}_N.$$

Since the integral in (14.5) is not changed by permutation of the integration variables, which reduces simply to a change in notation, the double sum over the permutations can be replaced by a single sum. Taking into account that

$$\sum_{\mathscr{P}, \mathscr{P}'} \to N! \sum_{\mathscr{P}},$$

we obtain

$$Q = \int \ldots \int \sum_{\mathscr{P}} (\pm 1)^{\mathscr{P}} \exp\left\{ -\frac{i\mathscr{P}}{\hbar} \sum_j (\boldsymbol{p}_j \cdot \boldsymbol{x}_j) \right\} e^{-\beta H} \exp\left\{ \frac{i}{\hbar} \sum_j (\boldsymbol{p}_j \cdot \boldsymbol{x}_j) \right\} d\Gamma, \tag{14.6}$$

†Initially, this factor was erroneously omitted by Kirkwood [27], although he noticed this and corrected it [27a].

where

$$d\Gamma = \frac{dp\, dx}{N!\,(2\pi\hbar)^{3N}}\,. \tag{14.7}$$

Now, the partition function is represented in the form of an integral over the whole phase space of N particles, similar to the partition function (3.8) of classical statistical mechanics, the element of integration (14.7) having precisely the same form (1.5b) as was applied earlier in classical statistical mechanics without a rigorous proof. To calculate the integral (14.6), we must give the explicit form of the result of the action of the operator $e^{-\beta H}$ on the function $\exp\left\{\frac{i}{\hbar}\sum_j (\boldsymbol{p}_j\cdot\boldsymbol{x}_j)\right\}$, i.e., we must find the function

$$u = e^{-\beta H} e^{\frac{i}{\hbar}(px)}, \qquad (px) = \sum_j (\boldsymbol{p}_j\cdot\boldsymbol{x}_j). \tag{14.8}$$

Below, we shall calculate functions of a somewhat more general type:

$$u(\mathcal{P}) = e^{-\beta H} e^{\frac{i\mathcal{P}}{\hbar}(px)}, \tag{14.9}$$

where $\mathcal{P}$ is the permutation operator of the particle coordinates. For $\mathcal{P} = 1$, (14.9) coincides with (14.8). Below, at the end of this section, we shall need these functions to calculate the statistical operators in the semiclassical approximation, i.e., when the quantum effects can be regarded as small corrections to the classical effects. The functions (14.9) satisfy the equation

$$\frac{\partial u}{\partial \beta} = -Hu \qquad \left(\beta = \frac{1}{\theta}\right) \tag{14.10}$$

with the initial condition

$$u\,|_{\beta=0} = e^{\frac{i\mathcal{P}}{\hbar}(px)}. \tag{14.10a}$$

Equation (14.10) is called **Bloch's equation** and plays an important role in quantum statistics. It is applied not only with the initial condition (10.10a), but also with other initial conditions. For example, the operator

$$U = e^{-\beta H}$$

also satisfies Eq. (14.10), but with the initial condition

$$U\,|_{\beta=0}=1.$$

If we make the replacement $\beta \to it/\hbar$ in Bloch's equation, it coincides in external appearance with the Schrödinger equation. This formal analogy is convenient for carrying over the methods developed in quantum mechanics and quantum field theory into quantum statistics. In particular, following this analogy, we make use below of the analog of the Wentzel–Kramers–Brillouin method, well known in quantum mechanics, to solve Eq. (14.10) by the method of expanding in powers of Planck's constant $\hbar$, i.e., by the semiclassical approximation.

In Eq. (14.10), we replace the unknown function u by w:

$$u(\mathscr{P})=e^{-\beta H(p,\,x)}\,e^{\frac{i\mathscr{P}}{\hbar}(px)}\,w(p,\,x,\,\beta), \tag{14.11}$$

where $w(p, x, \beta)$ is a function of the variables p, x, β, and $H(p, x)$ is a function of p and x, and not an operator. Then, Eq. (14.10) is brought to the form

$$\frac{\partial w}{\partial \beta}=e^{\beta H(p,\,x)}\left\{\frac{\hbar^2}{2m}\nabla^2\left(e^{-\beta H(p,\,x)}w\right)+\frac{i\hbar}{m}\mathscr{P}(p\nabla)\,e^{-\beta H(p,\,x)}w\right\}, \tag{14.12}$$

where

$$\nabla=\left\{\frac{\partial}{\partial \boldsymbol{x}_1}\cdots\frac{\partial}{\partial \boldsymbol{x}_N}\right\},$$

with the initial condition

$$w\,|_{\beta=0}=1, \tag{14.12a}$$

or, since

$$H(p,\,x)=\sum_i \frac{p_i^2}{2m}+v(\boldsymbol{x}_1,\,\ldots,\,\boldsymbol{x}_N)=T(p)+v(x),$$

to the form

$$\frac{\partial w}{\partial \beta}=e^{\beta v(x)}\left\{\frac{i\hbar}{m}\mathscr{P}(p\nabla)\,e^{-\beta v(x)}w+\frac{\hbar^2}{2m}\nabla^2\left(e^{-\beta v(x)}w\right)\right\}. \tag{14.13}$$

In this equation, the small parameter $\hbar$ appears in the right-hand side. In fact, as we shall see below, the role of the small parameter is played not by ħ itself, but by the dimensionless ratio of the de Broglie wavelength corresponding to the average thermal velocity to the average distance between the molecules.

The equation (14.13) can be solved by expanding in powers of ħ:

$$w = \sum_{n \geqslant 0} \hbar^n w_n. \tag{14.14}$$

For this, it is convenient to write it in the form of an integral equation

$$w = 1 + \frac{i\hbar}{m} \int_0^\beta e^{v(x)\tau} [\mathcal{P}(p\nabla) e^{-v(x)\tau} w]\, d\tau + \frac{\hbar^2}{2m} \int_0^\beta e^{v(x)\tau} \nabla^2 (e^{-v(x)\tau} w)\, d\tau,$$

whence, by the method of successive approximations, it is easy to find the coefficients w_0, w_1, ... :

$$w_0 = 1;$$

$$w_1 = \frac{i}{m} \int_0^\beta e^{v(x)\tau} \mathcal{P}(p\nabla) e^{-v(x)\tau} d\tau = \frac{-i\beta^2}{2m} \mathcal{P}(p\nabla)\, v(x);$$

$$w_2 = \frac{1}{2m^2} \int_0^\beta e^{v(x)\tau} \mathcal{P}(p\nabla) e^{-v(x)\tau} \mathcal{P}(p\nabla)\, v(x)\, \tau^2 d\tau + \frac{1}{2m} \int_0^\beta e^{v(x)\tau} \nabla^2 e^{-v(x)\tau} d\tau =$$

$$= -\frac{1}{2m} \left\{ \frac{\beta^2}{2} \nabla^2 v - \frac{\beta^3}{3} (\nabla v)^2 - \frac{\beta^3}{3m} (\mathcal{P}(p\nabla))^2 v + \frac{\beta^4}{4m} (\mathcal{P}(p\nabla)\, v)^2 \right\} \tag{14.15}$$

and so on.

Using the expressions obtained and putting $\mathcal{P} = 1$, we write the partition function (14.6) in the form

$$Q = \int e^{-\beta H(p,x)} [1 + w_1 \hbar + w_2 \hbar^2 + \dots]\, d\Gamma + \sum_{\mathcal{P} \neq 1} (\pm 1)^{\mathcal{P}} \int \left[\mathcal{P} e^{-\frac{i}{\hbar}(px)}\right] e^{\frac{i}{\hbar}(px)} e^{-\beta H(p,x)} \times [1 + w_1 \hbar + w_2 \hbar^2 + \dots]\, d\Gamma, \tag{14.16}$$

where $x = (\mathbf{x}_1, \dots, \mathbf{x}_N)$, $p = (\mathbf{p}_1, \dots, \mathbf{p}_N)$ and the w_i are taken for $\mathcal{P} = 1$. The first integral, which corresponds to the identical permutation, has been written out separately.

Performing the permutations of the particle coordinates and confining ourselves to pair permutations only, we transform (14.16) to the form

$$Q = \int e^{-\beta H(p,x)} [1 + w_1\hbar + w_2\hbar^2 + \ldots]\, d\Gamma \pm$$

$$\pm \sum_{k \neq l} \int e^{-\beta H(p,x)} e^{\frac{i}{\hbar}(p_k - p_l,\, x_k - x_l)} [1 + w_1\hbar + w_2\hbar^2 + \ldots]\, d\Gamma. \qquad (14.17)$$

In formula (14.17), the integration over $p_1, \ldots, p_N$ is easily performed, if we make use of the relation

$$\int_{-\infty}^{\infty} e^{-\alpha x^2 + i\gamma x} x\, dx = \frac{1}{i} \frac{\partial}{\partial \gamma} \int_{-\infty}^{\infty} e^{-\alpha x^2 + i\gamma x}\, dx = \frac{1}{i} \sqrt{\frac{\pi}{\alpha}} \frac{\partial}{\partial \gamma} e^{-\frac{\gamma^2}{4\alpha}}.$$

Finally, for the partition function, we obtain the expression

$$Q = \frac{(2\pi m\theta)^{3N/2}}{(2\pi\hbar)^{3N} N!} \int e^{-\beta v} \left\{ 1 - \frac{\hbar^2\beta^2}{12m} \sum_i \left(\nabla_i^2 v - \frac{\beta}{2} (\nabla_i v)^2 \right) \pm \right.$$

$$\left. \pm \sum_{i \neq k} \exp\left\{ -\frac{m |\boldsymbol{x}_{ik}|^2}{\beta\hbar^2} \right\} \left[1 + \frac{\beta}{2} \boldsymbol{x}_{ik} \cdot (\nabla_i v - \nabla_k v) + \ldots \right] \right\} d\boldsymbol{x}_1 \ldots d\boldsymbol{x}_N, \qquad (14.18)$$

$$\boldsymbol{x}_{ik} = \boldsymbol{x}_i - \boldsymbol{x}_k.$$

The first two terms in formula (14.18), which correspond to the identical permutation, were first obtained by Wigner [12], and the other two terms, which take into account the pair permutations, were obtained by Uhlenbeck and Gropper [26].

If in (14.18) we confine ourselves to the first term only, the partition function goes over into the partition function (3.8) of classical statistical mechanics:

$$Q_{\text{cl}} = \frac{(2\pi m\theta)^{3N/2}}{(2\pi\hbar)^{3N} N!} \int e^{-\beta v(\boldsymbol{x}_1 \ldots \boldsymbol{x}_N)} d\boldsymbol{x}_1 \ldots d\boldsymbol{x}_N = \int e^{-\beta H(p,x)} d\Gamma. \qquad (14.19)$$

Thus, if we regard classical statistical mechanics as the limiting case of quantum statistical mechanics, we obtain the correct expression for the partition function (3.8), which corresponds to the normalization (1.5a) of the classical distribution functions. This definition is a good one in that it does not lead to the Gibbs paradox.

The second term in (14.18) gives the quantum corrections associated with the interaction, but without allowance for the effects of exchange. The third term is due to quantum exchange and does not vanish even in the absence of interaction, i.e., for an ideal quantum gas. The last term is associated with both the exchange and the interaction.

The terms associated with exchange contain the exponential factor

$$\exp\left\{-\frac{|\boldsymbol{x}_{ik}|^2 m}{\beta\hbar^2}\right\},$$

and, therefore, they are small at densities that are not too large, when

$$\overline{|\boldsymbol{x}^2|} \gg \frac{\beta\hbar^2}{m} = \frac{\hbar^2}{m\theta}, \tag{14.20}$$

where $|\mathbf{x}^2|$ is the mean square distance between the particles, i.e., when the average distance between the particles is much greater than a distance of the order of the de Broglie wavelength $\hbar/(m\theta)^{1/2}$ corresponding to the particle energy θ.

At low temperatures or high densities, when the condition (14.20) is not fulfilled, the classical and quantum forms may differ greatly, and we have the onset of the phenomenon of degeneracy of quantum gases.

14.2. Passage to the Limit of Equilibrium

Statistical Operators

It is not difficult to obtain semiclassical expansions, similar to (14.16), for equilibrium density matrices also.

By means of the plane waves (14.3), we can write the density matrix in the coordinate x-representation in the form

$$\rho(\boldsymbol{x}_1, \ldots, \boldsymbol{x}_N, \boldsymbol{x}'_1, \ldots, \boldsymbol{x}'_N) = \\ = \sum_p \Psi^*_p(\boldsymbol{x}'_1, \ldots, \boldsymbol{x}'_N)\, e^{-(H-F)/\theta}\, \Psi_p(\boldsymbol{x}_1, \ldots, \boldsymbol{x}_N). \tag{14.21}$$

It is especially simple to find the density matrix in the mixed coordinate-momentum representation:

$$\rho(\boldsymbol{x}_1, \ldots, \boldsymbol{x}_N, \boldsymbol{p}_1, \ldots, \boldsymbol{p}_N) = \frac{1}{(2\pi\hbar)^{3N}} \int \rho(\boldsymbol{x}_1, \ldots, \boldsymbol{x}_N, \boldsymbol{x}'_1, \ldots, \boldsymbol{x}'_N) \times \exp\left\{-\frac{1}{i\hbar}\sum_j (\boldsymbol{p}_j \cdot \boldsymbol{x}'_j)\right\} d\boldsymbol{x}'_1 \ldots d\boldsymbol{x}'_N. \qquad (14.21a)$$

Substituting (14.21) into this, and taking into account that

$$\frac{1}{(2\pi\hbar)^{3N}} \int \exp\left\{\frac{i}{\hbar}\sum_j (\boldsymbol{x}'_j, \boldsymbol{p}'_j - \mathscr{P}\boldsymbol{p}_j)\right\} d\boldsymbol{x}'_1 \ldots d\boldsymbol{x}'_N = \delta(p' - \mathscr{P}p),$$

we obtain

$$\rho(\boldsymbol{x}_1, \ldots, \boldsymbol{x}_N, \boldsymbol{p}_1, \ldots, \boldsymbol{p}_N) = \frac{1}{N!\,(2\pi\hbar)^{3N}} \sum_{\mathscr{P}} (\pm 1)^{\mathscr{P}} \int \delta(p' - \mathscr{P}p)\, e^{-(H-F)/\theta}\, e^{\frac{i}{\hbar}(p'x)}\, d\boldsymbol{p}'_1 \ldots d\boldsymbol{p}'_N,$$

or

$$\rho(\boldsymbol{x}_1, \ldots, \boldsymbol{x}_N, \boldsymbol{p}_1, \ldots, \boldsymbol{p}_N) = \frac{1}{N!\,(2\pi\hbar)^{3N}} \sum_{\mathscr{P}} (\pm 1)^{\mathscr{P}} e^{-\frac{H-F}{\theta}} e^{\frac{i\mathscr{P}}{\hbar}(px)}. \qquad (14.22)$$

Thus, to calculate the density matrix in the mixed representation, we must find the function (14.9); we have already studied the calculation of this function, in the preceding section.

Using the relations (14.9) and (14.11), we obtain

$$\rho(\boldsymbol{x}_1, \ldots, \boldsymbol{x}_N, \boldsymbol{p}_1, \ldots, \boldsymbol{p}_N) = \frac{1}{N!\,(2\pi\hbar)^{3N}} e^{-\frac{H(p,x)-F}{\theta}} \sum_{\mathscr{P}} (\pm 1)^{\mathscr{P}} e^{\frac{i\mathscr{P}}{\hbar}(px)} (1 + w_1\hbar + w_2\hbar^2 + \ldots), \qquad (14.22a)$$

where w_1 and w_2 are given by the formulas (14.15). Retaining only the first term in (14.22a), we obtain the classical distribution function, exact within an unimportant phase factor:

$$\rho(\boldsymbol{x}_1, \ldots, \boldsymbol{x}_N, \boldsymbol{p}_1, \ldots, \boldsymbol{p}_N) = \frac{1}{N!\,(2\pi\hbar)^{3N}} e^{-\frac{H(p,x)-F}{\theta}} e^{\frac{i(px)}{\hbar}}. \qquad (14.23)$$

In order to avoid the appearance of such phase factors, Wigner [12] proposed a somewhat different definition for an operator in the mixed representation. Wigner introduced the function

$$f(\boldsymbol{x}_1, \ldots, \boldsymbol{x}_N, \boldsymbol{p}_1, \ldots, \boldsymbol{p}_N) = \\ = \frac{1}{(2\pi\hbar)^{3N}} \int \rho\left(\boldsymbol{x}_1 + \frac{\xi_1}{2}, \ldots, \boldsymbol{x}_N + \frac{\xi_N}{2}, \quad \boldsymbol{x}_1 - \frac{\xi_1}{2}, \ldots, \boldsymbol{x}_N - \frac{\xi_N}{2}\right) \times \\ \times \exp\left\{\frac{i}{\hbar}\sum_j (\boldsymbol{p}_j \cdot \xi_j)\right\} d\xi_1 \ldots d\xi_N. \qquad (14.24)$$

Integrals of this function over all p and all x have the form of diagonal elements of the density matrix in the x- and p-representations respectively:

$$\int f(x, p)\, dp = \rho(x, x), \qquad \int f(x, p)\, dx = \rho(p, p), \qquad (14.25)$$

where $x = (\boldsymbol{x}_1, \ldots, \boldsymbol{x}_N)$, and $p = (\boldsymbol{p}_1, \ldots, \boldsymbol{p}_N)$. We obtain the relations (14.25) by making a change of variables $\boldsymbol{x}_i - (\xi_i/2) = \boldsymbol{x}_i'$, $\boldsymbol{x}_i + (\xi_i/2) = \boldsymbol{x}_i''$.

The function f(x, p), of course, cannot be regarded as the distribution function of the coordinates and momenta. Only its integrals give distribution functions of the coordinates and momenta; the function itself can be negative and does not have the meaning of a probability density. From the Hermiticity of the density matrix, it follows only that f(x, p) is real.

It is possible to treat the limiting transition to classical statistics by means of the density matrix f(x, p) in the Wigner mixed representation, as Wigner himself has done [12].

Thus, we have seen that the quantum-mechanical canonical ensemble goes over into the classical one as $\hbar \to 0$. It would also be possible to examine the more general question of the transition from quantum Poisson brackets to classical Poisson brackets and from density matrices to distribution functions when the system is described by an equilibrium Gibbsian ensemble. This question was considered in the paper [28], where it was shown that quantum Poisson brackets go over to the classical ones as $\hbar \to 0$ but that the limiting process must be carried out for density matrices with "classical symmetry," i.e., density matrices that are symmetric only with respect to simultaneous permutations of both groups of variables $\mathbf{x}_1, \ldots, \mathbf{x}_N$ and $\mathbf{x}'_1, \ldots, \mathbf{x}'_N$, since the quantum symmetry property (which conerns permutations of only one group of variables) has no classical analog.

In essence, this question is connected with the passage to the classical limit of quantum mechanics, and we shall not study it here.

Chapter III

Irreversible Processes Induced by Mechanical Perturbations

§ 15. Response of a System to External Mechanical Perturbations

Until now, we have considered only equilibrium processes. We now turn to the study of nonequilibrium, irreversible processes.

One of the principal problems of the theory of irreversible processes is the study of the effects on statistical ensembles of different perturbations disturbing the equilibrium. The effect of change of the external parameters has already been considered, in essence, in Chapters I and II, which were devoted to equilibrium statistical mechanics, in the derivation of the thermodynamic equalities in §§5 and 11; there, however, it was assumed that the change in the parameters was infinitely slow and quasi-static, such that the system could be assumed to be in statistical equilibrium at each moment of time. In the theory of irreversible processes, we also consider change of the external parameters, but this change may no longer be infinitely slow.

A nonequilibrium ensemble can arise, for example, when certain external processes begin to affect an equilibrium ensemble (describable, consequently, by one of the Gibbsian distributions), leading to changes of the parameters used to specify the ensemble (volume, particle number, temperature, chemical potential, and so on). The cause of these perturbations may be work performed on the system by changing its volume, or interaction with other ensembles (possessing a different temperature or chemical potential), or, finally, the switching on of external fields acting directly on the particles of the system. This last case, of irreversible processes induced by mechanical perturbations, will be considered in this chapter. The mechanism of the appearance of irreversibility is most easily elucidated in this case.

A change in the external parameters, generally speaking, does not have a direct effect on the distribution function (or statistical operator), but an indirect one; it creates a statistically nonequilibrium state, which then tends to an equilibrium state, if there are no effects impeding this. Only in the case when the perturbation is induced by external fields does the change affect the distribution function directly, and this is a reason for the relative simplicity of the study of such perturbations.

We shall make a few comments on terminology. In macroscopic thermodynamics, work performed on the system, for example, by changing the volume by a movable piston (mechanical contact), is called mechanical action on the system. A perturbation induced by contact of the system with other thermodynamic systems with a different temperature (thermal contact) is said to be a thermal effect. Contact with a reservoir with which matter can be exchanged (material contact) is of the same type. In nonequilibrium statistical thermodynamics, a somewhat different terminology is used. By *mechanical perturbations*, we mean only those perturbations which represent the action of external fields and which can be described completely by adding the corresponding energy of interaction of the system with the field to the Hamiltonian. Perturbations which, generally speaking, cannot be represented in this way are called, in the terminology of Kubo [1-4], thermal perturbations. Below, we shall use this terminology. Work performed on the system by change of its volume (or of other parameters not conjugate to real external fields) belong to thermal perturbations also.

We remark that it is sometimes possible to represent thermal perturbations formally as the result of certain mechanical perturbations, if we introduce the appropriate fictitious fields [5-9]. For example, diffusional flow can be regarded as the consequence of switching on an auxiliary gravitational or centrifugal field [5], and viscous flow as a consequence of motion of the walls of the vessel [5]. Depending on the nature of this motion, we can obtain both the shear viscosity [5] and the bulk viscosity [6]. To take into account nonuniformity of temperature, we can introduce an auxiliary gravitational field [7], since according to the general theory of relativity, in a gravitational field the temperature is nonuniform even in a state of statistical equilibrium.

It should be noted that to replace thermal perturbations by mechanical ones is simply a convention. In reality, the motion of the walls cannot be transmitted instantaneously by the molecules, and its influence is indirect. It creates a nonequilibrium state, which propagates with the speed of sound and cannot be represented exactly as the effect of an external field. But in the case of sufficiently slow changes of the parameters, these analogies, if not taken too literally, lead to reasonable results and are very useful. They are based on the fact that mechanical and thermal perturbations can induce the same transport processes.

We turn now to discuss the theory of the linear response of statistical systems to mechanical perturbations.

15.1. Linear Response of a System (Case of Classical Statistics)

The response of a statistical ensemble to an external time-dependent perturbations is easily investigated in both the classical and the quantum case [1-14]. It is closely connected with the theory of retarded Green functions [12-15]. We shall discuss this problem first on the basis of classical statistical mechanics (we shall consider the quantum case in the next subsection).

We shall consider a statistical ensemble of systems with a Hamiltonian $H(p, q)$ that does not depend explicitly on time; q and p are the aggregates of coordinates and momenta of all the particles, and the Hamiltonian includes all possible interactions between the particles. We shall study the response of the ensemble to the switching on of an external time-dependent perturbation $H_t^1(p, q)$.

The dynamic variable $H_t^1(p, q)$ represents the energy of interaction of the system with the external field. The subscript argument t denotes only the explicit time dependence of the perturbation. In addition, there may be an implicit time dependence throu the coordinates and momenta q(t), p(t) of the particles, in accordance with the equations of motion.

The total Hamiltonian describing the system and its interaction with the field is

$$H(p, q) + H_t^1(p, q). \tag{15.1}$$

The Hamiltonian of the external field responsible for the interaction $H^1(p, q)$ is not included here, since the external field is assu fixed.

We assume that the external perturbation was absent at t = − i.e.,

$$H_t^1(p, q)\big|_{t=-\infty} = 0. \tag{15.1a}$$

The perturbation $H_t^1(p, q)$ can often be represented in the for of a sum

$$H_t^1(p, q) = -\sum_j B_j(p, q)\, F_j(t), \tag{15.2}$$

where $F_j(t)$ are the external driving forces, which are functions of time and independent of the coordinates and momenta of the pa ticles, and $B_j(p, q)$ are the dynamic variables conjugate to the fie $F_j(t)$ and are not explicitly time dependent.

For adiabatic switching on of a periodic perturbation,

$$H_t^1(p, q) = \sum_\omega e^{\varepsilon t - i\omega t} B_\omega(p, q) \qquad (\varepsilon > 0), \tag{15.2a}$$

where ε is an infinitesimally small positive quantity and B_ω = B since the energy H_t^1 is real. For instantaneous switching on of a periodic perturbation,

$$H_t^1(p, q) = \begin{cases} 0 & \text{for} \quad t < t_0, \\ \sum\limits_\omega e^{-i\omega t} B_\omega(p, q) & \text{for} \quad t \geqslant t_0. \end{cases} \tag{15.3}$$

The distribution function satisfies Liouville's equation (2.11)

$$\frac{\partial f}{\partial t} = \{H + H_t^1, f\}, \tag{15.4}$$

where $\{\ldots\}$ is the classical Poisson bracket (2.10). In addition, f satisfies the initial condition

$$f(t)|_{t=-\infty} = f_0 = Q^{-1}(\theta, V, N)e^{-H/\theta}, \quad Q(\theta, V, N) = \int e^{-H/\theta}\,d\Gamma, \tag{15.5}$$

which means that at t $= -\infty$, the system was in a state of statistical equilibrium and was described by the canonical distribution (3.7). In the initial condition (15.5), we can also take for f_0 the grand canonical distribution (3.30):

$$f(t)|_{t=-\infty} = f_0 = Q^{-1}(\theta, \mu, V)e^{-(H-\mu N)/\theta},$$
$$Q(\theta, \mu, V) = \sum_N \int e^{-(H-\mu N)/\theta}\,d\Gamma_N. \tag{15.5a}$$

Liouville's equation (15.4) and the initial condition (15.5) or (15.5a) completely determine the distribution function f(t) for any time t.

This formulation of the problem of mechanical perturbations of a statistical system is far from being as self-evident as it appears to be at first sight. The initial condition (15.5) means that the system was in statistical equilibrium with a thermostat in the remote past, while the use of Liouville's equation (15.4) shows that the influence of the thermostat is later not taken into account, since the Hamiltonian (15.1) refers only to the system itself. It is assumed that the external perturbation H_t^1 is adiabatically switched on at t = $-\infty$, the thermostat moves away, as it were, and the system then develops as an isolated system, in a manner appropriate to its Hamiltonian and the external perturbation. This interpretation is not entirely satisfactory. The moving away of the thermostat is not a real physical operation. If the external perturbation is large, then, after the thermostat has moved away and after a sufficiently long interval of time, it may no longer be important that the system was in statistical equilibrium with it at t $= -\infty$, and the system may be in a strongly nonequilibrium state. In this case, it is not effective to use the equilibrium state as the zeroth approximation.

In another possible interpretation (Bernard and Callen [11]), it is assumed that the thermostat does not move away, but that in the Hamiltonian of the system there is an additional small term, describing its interaction with the thermostat. This interaction is not written out explicitly, but is assumed that it causes incoherent transitions between the states of the system, owing to which the system forgets the details of the initial state after a sufficiently large interval of time. This interpretation also cannot be considered completely satisfactory, since the influence of the thermostat is essentially not taken into account at all in it. The question of how the influence of the thermostat can be taken into account will be considered below, in Chapter IV and in Appendices II and III.†

The above criticisms apply mainly to strongly nonequilibrium states. For weakly nonequilibrium states, when $f(t)$ differs little from f_0, the above method of treating mechanical perturbations gives rise to no objections, and it is this case that we shall consider below.

We now examine the solution of Eq. (15.4) with the initial condition (15.5) or (15.5a). We go over from the function f to f_1 by means of the transformation

$$f_1 = e^{-itL} f, \tag{15.6}$$

where L is the Liouville operator, i.e., the differential operator (2.14a), (2.16):

$$iLf = \{H, f\}; \tag{15.7}$$

e^{-itL} is the evolution operator, which acting on an arbitrary function of the coordinates and momenta q(0), p(0), transforms it to a function of q(t), p(t), which are solutions of Hamilton's equations with the initial conditions $p(t)|_{t=0} = p(0)$, $q(t)|_{t=0} = q(0)$ [cf. the remarks on formula (2.21a)].

After the transformation (15.6), Liouville's equation takes the form

$$\frac{\partial f_1}{\partial t} = e^{-itL}\{H_t^1, e^{itL} f_1\}.$$

†A third possible interpretation of the response theory is that an infinitesimally small term, causing the system to tend to equilibrium, is introduced not into the Hamiltonian, but into Liouville's equation

$$\frac{\partial f}{\partial t} - \left\{H + H_t^1 f\right\} = -\varepsilon(f - f_0)$$

and $\varepsilon \to +0$ after the thermodynamic limiting process. If we denote $\varepsilon = 1/\tau$, then the right-hand side of this equation has the form of the usual relaxation term $(f_0 - f)/\tau$.

Taking into account that the evolution operator e^{-itL} acts on both H_t^1 and $e^{itL} f_1$ we write this equation in the form

$$\frac{\partial f_1}{\partial t} = \{H_t^1(t), f_1\}, \tag{15.8}$$

where

$$H_t^1(t) = e^{-it'L} H_t^1 \tag{15.9}$$

is the analog in classical statistical mechanics of the Heisenberg picture in quantum mechanics. Equation (15.8) must be supplemented by the initial condition

$$f_1(t)|_{t=-\infty} = f_0, \tag{15.10}$$

which follows from (15.5) and (15.5a), since $e^{-itL} f_0 = f_0$, as f_0 is an integral of Liouville's equation. The right-hand side of Eq. (15.8) contains only the perturbation energy $H_t^1(t)$, and (15.8) is therefore convenient for studying the behavior of the system for small perturbations.

It is convenient to write Eq. (15.8) and the initial condition (15.10) in the form of a single integral equation

$$f_1(t) = f_0 + \int_{-\infty}^{t} \{H_{t'}^1(t'), f_1(t')\}\, dt', \tag{15.11}$$

or

$$f(t) = f_0 + \int_{-\infty}^{t} e^{i(t-t')L} \{H_{t'}^1, f(t')\}\, dt'. \tag{15.11a}$$

If the perturbation H_t^1 is small, the solution of Eq. (15.11a) can be obtained by iteration, taking f_0 as the zeroth approximation. In the first approximation, we find

$$f(t) = f_0 + \int_{-\infty}^{t} \{H_{t'}^1(t'-t), f_0\}\, dt', \tag{15.12}$$

where the Poisson bracket is equal to

$$\{H_t^1, f_0\} = \frac{\partial H_t^1}{\partial q}\frac{\partial f_0}{\partial p} - \frac{\partial H_t^1}{\partial p}\frac{\partial f_0}{\partial q}.$$

In the right-hand side, summation over all the particles is to be understood. Noting that

$$\{H_t^1, f_0\} = -\beta\{H_t^1, H\} f_0, \qquad \beta = \frac{1}{\theta}, \tag{15.13}$$

we write (15.12) in the form

$$f(t) = f_0\left(1 - \beta \int_{-\infty}^{t} \{H_{t'}^1(t'-t), H\}\, dt'\right). \tag{15.14}$$

Using (15.12) or (15.14), we can calculate the average value of any dynamic variable A(p, q) in the linear approximation in H_t^1:

$$\langle A\rangle = \int A(p, q) f(p, q, t)\, d\Gamma, \quad \text{where } d\Gamma = \frac{dp\, dq}{N!\,(2\pi\hbar)^{3N}}. \tag{15.15}$$

Substituting (15.12) in (15.15) and integrating by parts, we obtain

$$\langle A\rangle = \langle A\rangle_0 + \int_{-\infty}^{t} \langle\{A, H_{t'}^1(t'-t)\}\rangle_0\, dt', \tag{15.16}$$

where

$$\langle\ldots\rangle_0 = \int \ldots f_0\, d\Gamma$$

denotes averaging with the equilibrium distribution function.

Formula (15.16) gives the formal solution of the problem of the response of a classical statistical system to a mechanical perturbation. But the actual calculation of the right-hand side in (15.16 is far from simple, since it contains averages of dynamic variables at different times and requires that we solve a dynamic problem. In a number of cases, this can be done for systems with a small parameter.

Expression (15.16) describes the retarded response of the average values of A to a variable perturbation H_t^1. This response has a causal character, since only the effect of perturbations at $t' \leq t$, i.e., which took place at past moments of time, are found to be important.

In formula (15.16), it is convenient to extend the integration formally to $+\infty$ by introducing the discontinuous function $\theta(t-t')$:

$$\theta(t) = \begin{cases} 1 & \text{for} \quad t > 0, \\ 0 & \text{for} \quad t < 0. \end{cases} \tag{15.17}$$

Then

$$\langle A\rangle = \langle A\rangle_0 + \int_{-\infty}^{\infty} \langle\langle A H^1_{t'}(t'-t)\rangle\rangle\, dt', \tag{15.18}$$

where

$$\langle\langle AB(t'-t)\rangle\rangle = \theta(t-t')\langle\{A,\ B(t'-t)\}\rangle_0 = \theta(t-t')\langle\{A(t),\ B(t')\}\rangle_0 \tag{15.19}$$

is a **retarded two-time Green function** of classical statistical mechanics.

The latter equality in (15.19) is connected with the fact that the average values $\langle A(t)B(t')\rangle_0$ of a product of dynamic variables (or of the corresponding Poisson bracket) over a state of statistical equilibrium depends only on the difference of the time arguments. The fact that the average of a product of dynamic variables, taken at different times, depends only on the difference of the times is the stationary condition, well-known from the theory of random stationary processes [16, 17]. It means that, in the stationary case, time correlation functions cannot depend on the choice of origin for the time, i.e.,

$$\langle A(t+\tau)\, B(t'+\tau)\rangle = \langle A(t)\, B(t')\rangle,$$

whence, putting $\tau = -t$, we obtain

$$\langle A(t)\, B(t')\rangle = \langle AB(t'-t)\rangle,$$

where the averaging is performed over a stationary state.

This property for averaging over an equilibrium state is also easily proved directly. By definition, we have

$$\langle A(t_1)\, B(t_2)\rangle_0 =$$

$$= Q^{-1} \int e^{-\beta H(p,\, q)}\, A(p(t_1),\ q(t_1))\, B(p(t_2),\ q(t_2))\, \frac{dp(0)\, dq(0)}{N!\, h^{3N}},$$

where

$$p(t) = e^{-itL}\, p(0), \qquad q(t) = e^{-itL}\, q(0).$$

Using Liouville's theorem (2.2)

$$dp(0)\, dq(0) = dp(t_1)\, dq(t_1),$$

we transform the integral in the right-hand side of the form

$$Q^{-1} \int e^{-\beta H(p,\, q)}\, A(p(t_1),\ q(t_1))\, B(p(t_2),\ q(t_2))\, \frac{dp(t_1)\, dq(t_1)}{N!\, h^{3N}}.$$

Introducing the new integration variables $p' = p(t_1)$, $q' = q(t_1)$, we obtain

$$\langle A(t_1) B(t_2)\rangle_0 =$$

$$= Q^{-1} \int e^{-\beta H(p', q')} A(p', q') B(p'(t_2 - t_1), q'(t_2 - t_1)) \frac{dp'\, dq'}{N!\, h^{3N}} =$$

$$= \langle AB(t_2 - t_1)\rangle_0,$$

since

$$p(t_2) = e^{-i(t_2 - t_1)L} p(t_1) = e^{-i(t_2 - t_1)L} p',$$

$$q(t_2) = e^{-i(t_2 - t_1)L} q(t_1) = e^{-i(t_2 - t_1)L} q',$$

and $H(p, q)$ is an integral of motion. Thus, the required property is proved.

Thus, we have seen that the influence of an external mechanical perturbation on the average value of a dynamic variable is described by the retarded Green function coupling this variable with the perturbation.

The retarded Green functions (15.19), which were introduced by Bogolyubov and Tyablikov [15] for the case of quantum statistics, are very convenient for application in the statistical mechanics of equilibrium and nonequilibrium systems, owing to their transparent physical meaning and their simple analytic properties [12-15] (cf., further, §16). They are also useful in classical statistical mechanics [18-20].

The physical meaning of the retarded Green functions can be elucidated by considering the effect of an instantaneous δ-function perturbation

$$H_t^1 = B\,\delta(t - t_1) \tag{15.20}$$

on the average value of a dynamic quantity A. Substituting (15.20) in (15.18), we obtain

$$\langle A\rangle = \langle A\rangle_0 + \langle\langle AB(t_1 - t)\rangle\rangle = \langle A\rangle_0 + \langle\langle A(t) B(t_1)\rangle\rangle. \tag{15.21}$$

Consequently, the retarded Green function is equal to the change of the average value of A by the time t due to the instantaneous switching on of the δ-function perturbation at time t_1.

By means of (15.14), we can obtain another form for the relation (15.18):

$$\langle A\rangle = \langle A\rangle_0 - \frac{1}{\theta}\int_{-\infty}^{t} \langle A\dot{H}^1_{t'}(t'-t)\rangle_0\, dt' =$$

$$= \langle A\rangle_0 + \frac{1}{\theta}\int_{-\infty}^{t} \langle \dot{A}(t-t')\, H^1_{t'}\rangle_0\, dt' \qquad (\theta = \beta^{-1}), \qquad (15.22)$$

where

$$\dot{A} = \{A,\ H\}. \qquad (15.22a)$$

The latter equality in (15.22) follows from the stationary condition

$$\langle A\dot{H}^1_{t'}(t'-t)\rangle_0 = -\langle \dot{A}(t-t')\, H^1_{t'}\rangle_0. \qquad (15.23)$$

In fact, the average value of a product of dynamic variables over a state of statistical equilibrium depends only on the time difference:

$$\langle AH^1_{t'}(t'-t)\rangle_0 = \langle A(t-t')\, H^1_{t'}\rangle_0,$$

whence, by differentiating with respect to t, we obtain the relation (15.23).

Thus, the change of the average value of A under the influence of a perturbation H^1_t is determined in the linear approximation by the time correlation function coupling A with $\dot{H}^1_t$ or $\dot{A}$ with H^1_t.

If the external perturbation has the form (15.2), we can write formulas (15.18) and (15.22) in the form

$$\langle A\rangle = \langle A\rangle_0 - \sum_j \int_{-\infty}^{\infty} \langle\langle A(t)\, B_j(t')\rangle\rangle\, F_j(t')\, dt', \qquad (15.24)$$

$$\langle A\rangle = \langle A\rangle_0 + \sum_j \frac{1}{\theta}\int_{-\infty}^{t} \langle A(t)\, \dot{B}_j(t')\rangle_0\, F_j(t')\, dt'. \qquad (15.24a)$$

These relations for the linear response of a system are called the *Kubo formulas*.

Kubo made a detailed study of the response of classical and quantum systems to the switching on of an external perturbation, starting from Liouville's equation and the equilibrium condition at $t = -\infty$ [1, 3, 4]. However, relations similar to the Kubo formulas had been obtained earlier, for a particular case, by Kirkwood [21], who expressed the coefficient of friction of a Brownian particle for the classical case in terms of the correlation function of the forces acting on it† (cf. §26), and by Callen and Welton [22], who proved a generalized Nyquist theorem on the relation between the susceptibilities (or kinetic coefficients) in linear dissipative processes and the equilibrium fluctuations; we shall study this theorem in §17. The great merit of Kubo's work lies in the fact that he gave the most general proof of these formulas, applied them widely to the theory of linear dissipative processes, and drew the attention of physicists to them.

A remarkable feature of the Kubo relations is that they express nonequilibrium properties in the form of averages over a state of statistical equilibrium and have an extremely general character.

In the case of adiabatic switching on of the periodic perturbation (15.2a), the formula (15.18) takes the form

$$\langle A\rangle = \langle A\rangle_0 + \sum_{\omega} e^{\varepsilon t - i\omega t}\,\langle\langle AB_{\omega}\rangle\rangle_{\omega}, \tag{15.25}$$

where

$$\langle\langle AB_{\omega}\rangle\rangle_{\omega} = \int_{-\infty}^{\infty} e^{\varepsilon t - i\omega t}\,\langle\langle AB_{\omega}(t)\rangle\rangle\, dt \tag{15.25a}$$

is the Fourier component of the retarded Green function.‡

If the external perturbation contains only one harmonic of frequency ω, i.e.,

$$H_t^1 = -\, F\cos\omega t\, e^{\varepsilon t} B, \tag{15.26}$$

†An expression for the coefficient of friction of a Brownian particle in terms of the correlation function of the forces acting upon it was first obtained by Uhlenbeck and Ornstein [154], and later by Krylov and Bogolyubov [155].

‡The normalization of the Fourier component (15.25a) of the Green function differs by a factor 2π from that used in [12]. This is convenient since, with this normalization, factors of $1/2\pi$ are eliminated in many formulas.

where F is the amplitude of the periodic force and is independent of the coordinates and momenta, and B is the dynamic variable, then the linear response has the form

$$\langle A\rangle = \langle A\rangle_0 + \mathrm{Re}\,\{\varkappa(\omega)\, Fe^{-i\omega t+\varepsilon t}\}, \tag{15.27}$$

where Re denotes the real part of the expression, and $\varkappa(\omega)$ is the complex generalized susceptibility, equal to

$$\varkappa(\omega) = -\langle\langle AB\rangle\rangle_\omega, \quad \text{where } \langle\langle AB\rangle\rangle_\omega = \int_{-\infty}^{\infty} e^{-i\omega t+\varepsilon t}\langle\langle AB(t)\rangle\rangle\, dt. \tag{15.28}$$

This is the Kubo formula for the susceptibility. The relations (15.27) and (15.28) show that the Fourier components of the retarded Green function have the meaning of the generalized complex susceptibility describing the effect of the perturbation (15.26) on the average value of A.

We remark that in calculating the complex susceptibility (15.28) we must perform two limiting processes: the usual limiting process of statistical mechanics $V \to \infty$ $(V/N = \text{const})$ in the calculation statistical averages, and the limiting process $\varepsilon \to 0$. The results depends on the order in which these limiting processes are performed. The correct order is that with $V \to \infty$ first, and then $\varepsilon \to 0$.†

Just as it is always assumed in equilibrium statistical thermodynamics that the thermodynamic limit $V \to \infty$ $(V/N = \text{const})$ exists, in nonequilibrium statistical thermodynamics it is assumed, in addition, that for the expressions (15.28) the limit exists in which $V \to \infty$ first, and then $\varepsilon \to 0$. We shall call this the (V, ε)-limit. However, whereas the existence of the thermodynamic limit is proved, with certain restrictions on the interaction potential (cf. the references [13–17e] to Chapter I), there is still no rigorous mathematical proof of the existence of the (V, ε)-limit. Later, in §§15.4 and 17.4 of this chapter, we shall return to the discussion of the order of the limiting processes.

---

†It is shown in Appendix I, using the example of the quantum-mechanical scattering problem (where the role of V is played by the volume in which the eigenfunctions are normalized), that this order of the limits corresponds to the exclusion of the advanced solutions of the Schrödinger equation.

15.2. Linear Response of a System (Case of Quantum Statistics)

We shall consider the response of a quantum statistical ensemble of systems, with a Hamiltonian H independent of time, to the switching on of a time-dependent external perturbation H_t^1.

The total Hamiltonian of the system, including the external perturbation, is

$$H + H_t^1, \tag{15.29}$$

where H_t^1 is the operator of the interaction of the system of particles with the external field. The Hamiltonian of the external field with which the particles interact is not included in (15.29), since the field is regarded as fixed.

We assume, as before, that at $t = -\infty$ the external perturbation was absent, i.e.,

$$H_t^1 \big|_{t=-\infty} = 0. \tag{15.29a}$$

The perturbation H_t^1 can often be represented in the form

$$H_t^1 = -\sum_j B_j F_j(t), \tag{15.30}$$

where $F_j(t)$ are the external driving forces, which are functions of time and are C-numbers,† and B_j are the operators conjugate to the fields $F_j(t)$ and do not depend explicitly on time. The expression (15.30) is analogous to the classical expression (15.2).

If a periodic perturbation is switched on adiabatically, then

$$H_t^1 = -\sum_\omega e^{\varepsilon t - i\omega t} B_\omega \qquad (\varepsilon > 0), \tag{15.30a}$$

where ε is an infinitesimally small positive quantity, and B_ω is a quantum-mechanical operator that does not depend explicitly on time. It follows from the Hermiticity of (15.30a) that $B_\omega^+ = B_{-\omega}$.

The statistical operator ρ satisfies the quantum Liouville equation (8.6)

$$i\hbar \frac{\partial \rho}{\partial t} = \left[H + H_t^1, \rho\right] \tag{15.31}$$

†In the terminology customary in quantum mechanics, quantities which do not have an operator structure are called C-numbers.

and the initial condition

$$\rho|_{t=-\infty} = \rho_0 = Q^{-1}(\theta, V, N)\, e^{-H/\theta}, \tag{15.32}$$

which means that, at t $= -\infty$, the system was in a state of statistical equilibrium and was described by the canonical ensemble (9.14). As the initial condition, we can also apply the grand canonical ensemble (9.42):

$$\rho|_{t=-\infty} = \rho_0 = Q^{-1}(\theta, \mu, V)\, e^{-(H-\mu N)/\theta}. \tag{15.32a}$$

We go over from the statistical operator ρ to ρ_1, by means of the canonical transformation

$$\rho_1 = e^{iHt/\hbar}\rho\, e^{-iHt/\hbar}. \tag{15.33}$$

Then the quantum Liouville equation is transformed into

$$i\hbar\frac{\partial\rho_1}{\partial t} = \left[H_t^1(t), \rho_1\right] \tag{15.34}$$

with the initial condition

$$\rho_1|_{t=-\infty} = \rho_0, \tag{15.35}$$

where

$$H_t^1(t) = e^{iHt/\hbar} H_t^1 e^{-iHt/\hbar} \tag{15.36}$$

is the perturbation operator in the Heisenberg picture with Hamiltonian H; with respect to the total Hamiltonian (15.29), formula (15.36) gives the interaction picture.

Equation (15.34) and the initial condition (15.35) can be written in the form of a single integral equation

$$\rho_1(t) = \rho_0 + \int_{-\infty}^{t} \frac{1}{i\hbar}\left[H_{t'}^1(t'), \rho_1(t')\right] dt', \tag{15.37}$$

or

$$\rho(t) = \rho_0 + \int_{-\infty}^{t} e^{-iH(t-t')/\hbar}\frac{1}{i\hbar}\left[H_{t'}^1, \rho(t')\right] e^{iH(t-t')/\hbar}\, dt'; \tag{15.37a}$$

these equations are analogous to the classical equations (15.11) and (15.11a).

If the perturbation H_t^1 is small, the solution of Eq. (15.37) can be obtained by iteration, using ρ_0 as the zeroth approximation. In the first approximation, we shall have

$$\rho = \rho_0 + \int_{-\infty}^{t} \frac{1}{i\hbar} \left[H_{t'}^1 (t' - t),\ \rho_0 \right] dt'. \tag{15.38}$$

Up to this point, all the relations are valid when ρ_0 is not only the canonical or grand canonical distribution, but any equilibrium distribution, for example, the microcanonical distribution, since nowhere have we used the explicit form of ρ_0. We assume now that ρ_0 is the canonical distribution (15.32).

Using the identity, valid for any operator A,

$$[A,\ e^{-\beta H}] = -\, e^{-\beta H} \int_0^{\beta} e^{\lambda H} [A,\ H]\, e^{-\lambda H}\, d\lambda, \tag{15.39}$$

which is usually called the Kubo identity and which we shall prove somewhat further on [cf. formula (15.42) and after], we obtain

$$\rho = \rho_0 \left(1 - \int_0^{\beta} \int_{-\infty}^{t} e^{\lambda H}\, \dot{H}_{t'}^1 (t' - t)\, e^{-\lambda H}\, d\lambda\, dt' \right), \tag{15.40}$$

where

$$\dot{H}_{t'}^1 (t' - t) = \frac{1}{i\hbar} \left[H_{t'}^1 (t' - t),\ H \right]. \tag{15.41}$$

In the case when ρ_0 is the grand canonical distribution (15.32a), formula (15.40) remains valid, but in it we must put

$$\dot{H}_{t'}^1 (t' - t) = \frac{1}{i\hbar} \left[H_{t'}^1 (t' - t),\ H - \mu N \right]. \tag{15.41a}$$

If H_t^1 commutes with N as is often the case, then (15.41a) coincides with (15.41). If ρ_0 is the microcanonical distribution, formula (15.40) is no longer valid.

We shall now derive the Kubo identity (15.39). We put

$$[A,\ e^{-\beta H}] = e^{-\beta H}\, S(\beta), \tag{15.42}$$

where $S(\beta)$ is an operator to be determined. Differentiating (15.42) with respect to β, we obtain a differential equation for $S(\beta)$:

$$\frac{\partial S}{\partial \beta} = - e^{\beta H} [A, H] e^{-\beta H}$$

with the initial condition $S|_{\beta=0} = 0$. Integrating this, taking the initial condition into account, we obtain Kubo's identity (15.39).

Formulas (15.37a) and (15.40) enable us to calculate, in the linear approximation in H_t^1, the average value of any observable quantity represented by the operator A:

$$\langle A \rangle = \mathrm{Tr}\,(\rho A). \tag{15.43}$$

Substituting (15.38) in (15.43) and using the invariance of the trace with respect to a cyclic permutation of the operators, we obtain

$$\langle A \rangle = \langle A \rangle_0 + \int_{-\infty}^{t} \frac{1}{i\hbar} \langle [A(t), H_{t'}^1(t')] \rangle_0 \, dt', \tag{15.44}$$

where

$$A(t) = e^{iHt/\hbar} A e^{-iHt/\hbar} \tag{15.45}$$

is the operator A in the Heisenberg picture, and

$$\langle \ldots \rangle_0 = \mathrm{Tr}\,(\rho_0 \ldots) \tag{15.46}$$

represents averaging with the equilibrium statistical operator (15.32) or (15.32a).

Equation (15.44) describes the retarded response of the average values of an operator A to the switching on of a perturbation H_t^1, for a quantum-statistical ensemble. It has exactly the same form as (15.16) in classical statistical mechanics, except that the classical Poisson bracket is replaced by a quantum Poisson bracket and the classical averaging by quantum averaging.

Extending the time integration in (15.44) to $+\infty$ by introducing the discontinuous function $\theta(t - t')$ (15.17), Eq. (15.44) can be written conveniently in the form

$$\langle A \rangle = \langle A \rangle_0 + \int_{-\infty}^{\infty} \langle\langle A(t) H_{t'}^1(t') \rangle\rangle \, dt', \tag{15.47}$$

where

$$\langle\langle A(t) B(t') \rangle\rangle = \theta(t - t') \frac{1}{i\hbar} \langle [A(t), B(t')] \rangle_0 \tag{15.48}$$

is the retarded two-time Green function in quantum statistical mechanics, introduced by Bogolyubov and Tyablikov [12-15].

Formulas (15.47) and (15.48) are analogous to formulas (15.18) and (15.19) of classical statistical mechanics. Thus, the effect of external perturbations on the average values of observable quantities in quantum, as in classical statistics is described by retarded quantum Green functions coupling the observable quantity with the perturbation.

The Green functions (15.48) depend on the difference $t - t'$ of the time arguments, as do the time correlation functions

$$\langle A(t)\, B(t')\rangle_0 = \langle A(t-t')\, B\rangle_0 = \langle AB\,(t'-t)\rangle_0, \tag{15.48a}$$

inasmuch as the averaging is performed over an equilibrium ensemble. It is easy to see this directly, by using a cyclic permutation of the operators in the trace. In fact,

$$\langle A(t)\, B(t')\rangle = Q^{-1}\,\mathrm{Tr}\left\{e^{-\beta H} e^{\frac{iHt}{\hbar}} A e^{\frac{-iH(t-t')}{\hbar}} B e^{\frac{-iHt'}{\hbar}}\right\} =$$

$$= Q^{-1}\,\mathrm{Tr}\left\{A e^{\frac{iH(t'-t)}{\hbar}} B e^{\frac{-iH(t'-t)}{\hbar}} e^{-\beta H}\right\} = \langle AB\,(t'-t)\rangle,$$

as we wished to prove. This same relation is valid for averaging over the grand ensemble; we need only replace $H \to H - \mu N$ in the proof.

The physical meaning of the retarded quantum Green functions is the same as for the classical ones. An instantaneous δ-function perturbation of the type (15.20) influences the average value of an observable quantity A through the Green function

$$\langle A\rangle = \langle A\rangle_0 + \langle\langle A(t)\, B(t_1)\rangle\rangle. \tag{15.49}$$

Just as in classical statistics, the effect of a perturbation on average values can also be expressed in terms of time correlation functions. For this, we make use of the expression (15.40) for the perturbed statistical operator. Then,

$$\langle A\rangle = \langle A\rangle_0 - \int_0^{\beta}\int_{-\infty}^{t} \langle e^{\lambda H}\, \dot{H}^1_{t'}(t')\, e^{-\lambda H}\, A(t)\rangle_0\, d\lambda\, dt'$$

$$= \langle A\rangle_0 + \int_0^{\beta}\int_{-\infty}^{t} \langle e^{\lambda H}\, H^1_{t'}(t')\, e^{-\lambda H}\, \dot{A}(t)\rangle_0\, d\lambda\, dt', \tag{15.50}$$

where the stationarity condition (15.23) has been used, as in the derivation of (15.22). Formula (15.50) can also be written in the form

$$\begin{aligned}\langle A\rangle &= \langle A\rangle_0 - \int_0^\beta \int_{-\infty}^t \langle \dot{H}^1_{t'}(t' - i\hbar\lambda)\, A(t)\rangle_0 \, d\lambda \, dt' \\ &= \langle A\rangle_0 + \int_0^\beta \int_{-\infty}^t \langle H^1_{t'}(t' - i\hbar\lambda)\, \dot{A}(t)\rangle_0 \, d\lambda \, dt'. \end{aligned} \tag{15.51}$$

Comparing (15.51) with the classical expression (15.22) for the linear response, we note that (15.51) goes over into (15.22) if we put, formally, $\hbar = 0$ and replace the quantum averaging by classical averaging. This simple rule can be applied to obtain the classical formulas from the quantum formulas.

The formulas (15.47) and (15.51) give expressions for the linear response of a quantum statistical ensemble to mechanical perturbations, in terms of Green functions or quantum time correlation functions. For an external perturbation in the form (15.30), these formulas can be written in a form

$$\begin{aligned}\langle A\rangle &= \langle A\rangle_0 - \sum_j \int_{-\infty}^{\infty} \langle\langle A(t)\, B_j(t')\rangle\rangle\, F_j(t')\, dt', \\ \langle A\rangle &= \langle A_0\rangle + \sum_j \int_{-\infty}^{t} \int_0^\beta \langle e^{\lambda H}\, \dot{B}_j(t')\, e^{-\lambda H}\, A(t)\rangle_0\, F_j(t')\, dt'\, d\lambda, \end{aligned} \tag{15.51a}$$

similar to formulas (15.24) and (15.24a) of classical statistical mechanics. These are the **Kubo formulas** for the linear response of a quantum system.

The formulas (15.51a) for the linear response of a system are sometimes also represented in the form

$$\langle A\rangle = \langle A\rangle_0 + \sum_j \int_{-\infty}^{t} \varphi_{AB_j}(t - t')\, F_j(t')\, dt',$$

where

$$\varphi_{AB_j}(t - t') = \int_0^\beta \langle e^{\lambda H}\, \dot{B}_j(t')\, e^{-\lambda H}\, A(t)\rangle_0\, d\lambda \equiv \beta\,(\dot{B}_j(t'),\ A(t))$$

is the **response function**, or **after-effect function**, describing the effect of the perturbation B_j on the average value of A. It is sometimes called the response. In the classical case, it goes over into the time correlation function

$$\varphi_{AB_j}(t-t') = \beta \langle \dot{B}_j(t') A(t) \rangle_0$$

The response function differs from the Green function only by the discontinuous factor $\theta(t-t')$. Indeed, from a comparison with (15.51a),

$$\langle\langle A(t) B(t') \rangle\rangle = -\theta(t-t')\, \varphi_{AB}(t-t').$$

Sometimes, in addition to the response function, the relaxation function

$$\Phi_{AB}(t) = \int_t^{\infty} \varphi_{AB}(t')\, e^{-\varepsilon t'}\, dt'$$

is introduced.

It is possible to develop the theory of linear response using these functions, as Kubo does (cf. [1, 3, 4]), but since they are simply related to the retarded Green functions, which have a very simple physical meaning, it is evidently simpler to construct a theory of linear response on the basis of the retarded Green functions, and this is the path we shall follow in this book.

15.3. Nonlinear Response of a System

The nonlinear response of a statistical system to external mechanical perturbations can be studied by the same method as was used in §§15.1 and 15.2 to treat a linear response [1, 11]. In this subsection, we shall consider only the quantum case, since the classical case can be treated analogously by replacing the quantum Poisson brackets by classical Poisson brackets and the quantum averaging by classical averaging.

We again start from the quantum Liouville equation (15.31) and the initial conditions (15.32) or (15.32a) and transform Liouville's equation to the integral form (15.37). Iterating Eq. (15.37a), we obtain a perturbation theory series for the statistical operator [1]

$$\rho(t) = \rho_0 + \sum_{n=1}^{\infty} \frac{1}{(i\hbar)^n} \int_{-\infty}^{t} dt_1 \int_{-\infty}^{t_1} dt_2 \ldots \int_{-\infty}^{t_{n-1}} dt_n\, e^{-iHt/\hbar} \left[H^1_{t_1}(t_1)\left[H^1_{t_2}(t_2) \ldots\right.\right.$$

$$\left.\left.\ldots \left[H^1_{t_n}(t_n),\ \rho_0\right] \ldots\right] e^{iHt/\hbar} \qquad (15.52)$$

and for the average value of the operator A

$$\langle A\rangle = \langle A\rangle_c + \sum_{n=1}^{\infty} \frac{1}{(i\hbar)^n} \int_{-\infty}^{t} \int_{-\infty}^{t_1} \cdots \int_{-\infty}^{t_{n-1}} \mathrm{Tr}\,\{A(t)\left[H^1_{t_1}(t_1)\left[H^1_{t_2}(t_2)\ \ldots\right.\right.$$

$$\ldots \left[H^1_{t_n}(t_n),\ \rho_0\right]\ldots]\}\, dt_1 \ldots dt_n. \qquad (15.52a)$$

The series (15.52a) describes the nonlinear response of a statistical system to the switching on of a perturbation H_t^1.

It is also easy to obtain a more compact formula for the nonlinear response of a system, if we start from the equation of motion for the evolution operator U(t)

$$i\hbar \frac{\partial U(t)}{\partial t} = (H + H_t^1)\, U(t). \qquad (15.53)$$

For $H_t^1 = 0$, the solution of Eq. (15.53) has the form

$$U(t) = e^{-iHt/\hbar},$$

since, in this case, the Heisenberg picture (8.17a) has the form (8.17). Consequently, Eq. (15.53) must be supplemented by the initial condition

$$e^{iHt/\hbar} U(t)\,|_{t=-\infty} = 1. \qquad (15.53a)$$

It is easy to see that, if U(t) satisfies Eq. (15.53) and the initial condition (15.53a), then

$$\rho(t) = U(t)\,\rho_0\, U^+(t) \qquad (15.53b)$$

satisfies Liouville's equation (15.31) and the initial condition (15.32) or (15.32a). Indeed, differentiating (15.53b) and taking (15.53) into account, we obtain (15.31). In addition, according to (15.53a), as $t \to -\infty$, $\rho(t)$ tends to

$$e^{-iHt/\hbar} \rho_0\, e^{iHt/\hbar} = \rho_0,$$

i.e., $\rho(t)$ satisfies the initial condition (15.32) or (15.32a), as we wished to prove.

It is convenient to multiply Eq. (15.53) by $e^{iHt/\hbar}$ on the left and transform it to

$$i\hbar \frac{\partial}{\partial t} \left(e^{iHt/\hbar}\, U(t)\right) = H_t^1(t)\, e^{iHt/\hbar}\, U(t), \qquad (15.54)$$

where

$$H_t^1(t) = e^{iHt/\hbar} H_t^1 e^{-iHt/\hbar} \tag{15.55}$$

is the perturbation energy operator in the Heisenberg picture. Integrating Eq. (15.54) over t from $-\infty$ to t, taking the initial condition (15.53a) into account, we obtain an integral equation for U(t)

$$U(t) = e^{-iHt/\hbar}\left\{1 + \frac{1}{i\hbar}\int_{-\infty}^{t} H_{t_1}^1(t_1)\, e^{iHt_1/\hbar}\, U(t_1)\, dt_1\right\}. \tag{15.56}$$

It is convenient to go over to the operator

$$U_1(t) = e^{iHt/\hbar}\, U(t),$$

which satisfies the simpler integral equation

$$U_1(t) = 1 + \frac{1}{i\hbar}\int_{-\infty}^{t} H_{t_1}^1(t_1)\, U_1(t_1)\, dt_1$$

with the initial condition

$$U_1(t)\,|_{t=-\infty} = 1.$$

Solving this integral equation by iteration, we obtain a perturbation theory series for U(t):

$$U(t) = e^{-\frac{iHt}{\hbar}} \sum_{n=0}^{\infty}\left(\frac{1}{i\hbar}\right)^n \int_{-\infty}^{t} dt_1 \int_{-\infty}^{t_1} dt_2 \ldots \int_{-\infty}^{t_{n-1}} dt_n\, H_{t_1}^1(t_1)\, H_{t_2}^1(t_2) \ldots H_{t_n}^1(t_n). \tag{15.56a}$$

The operator U(t) can be written in a more compact form by using the time-ordering operator P, which, acting on any product of time-dependent operators, places them in chronological order of decreasing times, i.e.,

$$P[A(t_1)\, B(t_2) \ldots L(t_n)] = A(t_1)\, B(t_2) \ldots L(t_n),$$

where $t_1 > t_2 > \ldots > t_n$; A, B, ..., L are arbitrary time-dependent operators, for example, operators in the Heisenberg or interaction pictures. Using P, we can represent the n-th term of the series (15.56a) in the form of a multiple integral with the same upper limits of integration:

$$\frac{1}{(i\hbar)^n}\int_{-\infty}^{t} dt_1 \int_{-\infty}^{t_1} dt_2 \ldots \int_{-\infty}^{t_{n-1}} dt_n\, H^1_{t_1}(t_1)\, H^1_{t_2}(t_2) \ldots H^1_{t_n}(t_n) =$$

$$= \frac{1}{n!}\frac{1}{(i\hbar)^n}\int_{-\infty}^{t} dt_1 \int_{-\infty}^{t} dt_2 \ldots \int_{-\infty}^{t} dt_n\, P\left[H^1_{t_1}(t_1)\, H^1_{t_2}(t_2) \ldots H^1_{t_n}(t_n)\right],$$

since the integral in the right-hand side of this equality is symmetric with respect to $t_1, \ldots, t_n$, and the number of possible permutations of the time arguments is equal to n!. Consequently, the evolution operator (15.56a) can be represented in the form of an ordered P-exponential

$$U(t) = e^{-iHt/\hbar}\, P \exp\left\{\frac{1}{i\hbar}\int_{-\infty}^{t} H^1_{t_1}(t_1)\, dt_1\right\}, \tag{15.56b}$$

as is usually done in quantum field theory [23-26].

The formula (15.53b) can be written in the explicit form

$$\rho(t) = U(t)\, e^{-\beta(H-F)}\, U^+(t), \tag{15.57}$$

or

$$\rho(t) = U(t)\, e^{-\beta(H-\mu N-\Omega)}\, U^+(t) \tag{15.57a}$$

for the canonical and grand canonical ensembles, respectively.

We shall obtain one more formula for the statistical operators (15.57) and (15.57a). We note that for $f(A)$ – an arbitrary function of the operator A – the relation

$$Uf(A)\,U^+ = f(UAU^+), \tag{15.58}$$

holds, where U is the operator of an arbitrary unitary transformation ($U^+U = UU^+ = 1$), for example, (15.56). The relation (15.58) is easily proved by expanding $f(A)$ in a Taylor series and taking into account that

$$UA^nU^+ = UAU^+UA \ldots UAU^+ = (UAU^+)^n,$$

since $U^+U = 1$. Using (15.58), we obtain

$$\rho(t) = \exp\{-\beta[U(t)\, HU^+(t) - F]\} \tag{15.59}$$

or

$$\rho(t) = \exp\{-\beta[U(t)(H-\mu N)U^{+}(t)-\Omega]\} \tag{15.59a}$$

in place of (15.57), and (15.57a). Together with (15.56b), these formulas give a compact form of the perturbation theory series (15.52) for the statistical operator.

We remark, however, that the theory of nonlinear response to mechanical perturbations, described above, is less well-founded than the theory of linear response. We took contact with a thermostat as the initial condition at $t = -\infty$, and then studied the evolution of the system as if it were isolated from all external influences apart from the force field. In fact, in almost all real cases, the system receiving energy from the external field can pass it on to its surroundings. This is especially obvious if we think of the system as a distinct part of a large system.

Even if we assume that at the initial moment the system was in equilibrium with a thermostat, this equilibrium is disturbed as a result of the mechanical perturbations, and thermal perturbations, which cannot be described by an external field, arise. Only in the linear approximation are mechanical and thermal perturbations additive.

A system of a large number of particles, receiving energy from an external field, can distribute it between its particles, for example, in the form of Joule heating, and is characterized by a time-dependent temperature. A number of experiments on the behavior of magnetic materials placed in a constant magnetic field with a superimposed alternating magnetic field can be interpreted if a time-dependent spin temperature is introduced [27]. On the other hand, the theory of nonlinear response developed above contains only the equilibrium values of the parameters $\beta = 1/kT$ and μ, which entered because of the initial condition of statistical equilibrium at $t = -\infty$. In its usual form, this theory is not very well adapted for the introduction of the concept of a time-dependent temperature. This dependence is usually obtained from the energy balance condition and from the work done by the external field on the system, by taking the effect of the field to be adiabatic [27], but this procedure is somewhat artificial. In the papers [28], it is shown that, for spin systems, the time dependence of the spin temperature can be treated on the basis of the theory of nonlinear response, by summing those terms in the per-

turbation theory series (15.52a) that are important at large times. The analogous procedure in nonlinear mechanics is the selection and summation of the secular terms. In Chapter IV it will be shown how one can find solutions of Liouville's equation with time-dependent parameters.†

We remark that, in the theory of nonlinear response, there is not always a one-to-one correspondence between the external perturbation H_t^1 and $\langle A \rangle - \langle A_0 \rangle$ – the resulting response of the system. This is most easily illustrated by invoking an analogy with the theory of nonlinear automatic systems [29]. In the language of this theory, the perturbation H_t^1 can be called the input signal, and $\langle A \rangle - \langle A_0 \rangle$, the output signal. It is well known that, for nonlinear automatic systems with feedback, there may be no one-to-one relation between the input and output signals. Such "autonomous" systems are possible not only in cybernetics, but also in statistical mechanics, since here also, a feedback mechanism is possible. For example, a self-oscillating regime is possible in turbulent flow,‡ in nonlinear acoustics in the thermal generation of sound [30], and in a quantum generator [31, 32]. A self-oscillating regime is also possible in chemical kinetics (chemical oscillations) where a feedback mechanism is created either by chemical autocatalysis (kinetic oscillations) or by heat which is liberated in the reaction and increases its rate (thermo-kinetic oscillations) [33]. The biological rhythms in a living organism may also be connected with periodic chemical processes [33].

In autonomous systems, small perturbations increase up to a certain finite magnitude, independent of the initial conditions. An analogous situation is also found in nonlinear mechanics [34-36], where an oscillation tends to a limiting cycle, irrespective of the initial conditions.

We shall explain in somewhat more detail the analogy between the theory of nonlinear response in statistical mechanics and the theory of nonlinear automatic systems.

In the theory of linear response, the connection between the driving force F(t) (the input signal) and the response of the system $\Delta A = \langle A \rangle - \langle A \rangle_0$ (the output signal) is given by the linear integral relation (15.51a)

$$\Delta A(t) = \int_{-\infty}^{t} L(t - t') F(t') \, dt', \qquad (15.60)$$

†The theory of nonlinear response has been considered in the papers of Miyake and Kubo [119], Keldysh [120], Muroyama [121], and Tani [122]. Recently, Kalashnikov [123] applied the method of the nonequilibrium statistical operator (described in Chapter IV of this book) to the problem of nonlinear response and used nonequilibrium Green functions.

‡A feedback mechanism was formally introduced by Landau in the theory of turbulence to describe the establishment of a limit on the growth of turbulent pulsations in hydrodynamically unstable flows [86].

where

$$L(t-t') = -\langle\langle A(t)\, B(t')\rangle\rangle$$

is a retarded Green function. In the nonlinear theory of automatic systems with feedback, the input signal F(t) and the output signal ΔA are connected by a nonlinear integral relation [29]

$$\cdot\, \Delta A(t) = \int_{-\infty}^{\infty} L(t,\ t')\, f\left[F(t') - \int_{-\infty}^{\infty} K(t',\ t'')\, \Delta A(t'')\, dt''\right] dt', \qquad (15.61)$$

where the function L(t, t') determines the response ΔA of the system, K(t', t"), is the feedback response (the back effect of the response of the system on the perturbation F), and $f[\ldots]$ is a noninertial nonlinear transformation in a straight chain. In the linear theory, without feedback,

$$f[F(t)] = F(t), \qquad L(t,\ t') = L(t-t'), \qquad K(t',\ t'') = 0,$$

and (15.61) goes over into (15.60). In the theory of nonlinear response of a statistical ensemble without feedback, K(t', t") = 0, and the transformation f, as can be seen from (15.52a), has a nonlinear and retarded character (each power of F gives one integral of the retarded type).

For positive feedback, the nonlinear integral equation (15.61), as is well known [29], determines the regime of automatic control. Analogous processes are also possible in nonequilibrium statistical mechanics when an unstable state is excited and generation arises. We can take (15.61) as a model for such a process.

For a nonlinear system with feedback, it is impossible to obtain relations between the input and output signals in explicit form. In such a system, small perturbations grow up to a certain value, and then the system fluctuates about this value. In this case, the statistical characteristics of the transformed signal are studied in terms of the given characteristics of the input signal. For unstable statistical systems, an analogous formulation of the problem is necessary. The above ideas on the analogy between the theory of automatic control and the theory of the nonlinear response of a system are given in order to indicate the limitations of ordinary nonlinear-response theory and to indicate a possible general point of view in the theory of active media; it does not pretend to more than this.

The usual theory of nonlinear response described in this subsection has, nevertheless, its own region of applicability, namely, when the thermal perturbations arising as a result of the mechanical ones can be neglected, and when the medium is passive, i.e., there is no feedback and generation in it is impossible. These cas can be realized, for example, in the theory of magnetic resonance [27, 37], in nonlinear optics [38], and in nonlinear acoustics [30].

15.4. Effect of an Alternating Electric Field. Electrical Conductivity

In subsections 15.1-15.3, we examined the linear and nonlinear response of a system to a perturbation H_t^1, without specifying its nature precisely and assuming only that it is the result of the action of a real external field. This perturbation may be induced by an alternating electric, magnetic (or, in general, electromagnetic) field, and also by a gravitational field.

We shall consider the effect on a statistical ensemble of switching on a spatially uniform alternating external electric field, periodic in time:

$$\boldsymbol{E}^0(t) = \boldsymbol{E}^0 \cos \omega t \, e^{\varepsilon t} = \mathrm{Re}\{e^{-i\omega t + \varepsilon t} \boldsymbol{E}^0\}. \tag{15.62}$$

The electric field $\mathbf{E}$ in the medium containing the charges is not the same as the external field, because of the strong screening effect of the charges, associated with the Coulomb interaction between them. In the derivation of the formulas of the theory of linear response in an electric field, sufficient attention has not always been given to the difference between the external field and the mean field [1, 4, 5, 11, 12, 39]. For a detailed discussion of this question, see [3, 40, 41].

The electrical conductivity is defined as the coefficient of proportionality between the current density and the mean field in the medium, or, if there is dispersion, as the coefficient between their space and time Fourier components (cf. §18). To determine the conductivity, we need to know how the external field $\mathbf{E}^0$ acting on the charges is related to the mean field. We must distinguish two cases:

1) the Coulomb interaction between the charges is taken into account by introducing a screening field, as is often done in the electron theory of metals [42, 43]. In this case, it is not necessary to take account of the screening effect a second time, and the external field is equal to the mean electric field in the medium $\mathbf{E}^0 = \mathbf{E}$. Most of the authors who have studied the linear response of a system to an external electric field have considered precisely this case [1, 3, 5, 11, 12, 39], although this is not always stated.

In this case, to describe electrons interacting with a lattice, we apply the Fröhlich model [44-46], in which the Coulomb interaction is taken into account only indirectly, by a modification of the matrix elements of the interaction of the electrons with the lattice.†

2) The Coulomb interaction between the electrons is taken into account explicitly. To describe the interaction of the electrons with the lattice, we must apply a model in which the matrix elements of the interaction do not include the Coulomb interaction between the electrons, which is taken into account separately.‡

If this model is used, it is necessary to take into account the screening effect and the dielectric polarization of the medium [4, 40, 41]. In this case, the acting external field $\mathbf{E}^0$ is equal to the induction, $\mathbf{E}^0 = \mathbf{D}$.

We shall examine the response of a system to an external electric field. The operator corresponding to the perturbation due to the field (15.62) is

$$H_t^1 = -\sum_j e_j (\boldsymbol{E}^0 \cdot \boldsymbol{x}_j) \cos \omega t \, e^{\varepsilon t} = -(\boldsymbol{E} \cdot \boldsymbol{P}) \cos \omega t \, e^{\varepsilon t}, \tag{15.63}$$

†In the Fröhlich model, the expression

$$H_{\text{int}} = \frac{1}{\sqrt{V}} \sum_{\substack{k_1 k_2 q \\ k_1 - k_2 = q}} v_q \left(\frac{\hbar}{2\omega_q}\right)^{1/2} (b_q + b_{-q}^+)\, a_{k_1\sigma}^+ a_{k_2\sigma},$$

is used for the Hamiltonian of the electron–phonon interaction in the second-quantized representation, where ω_q is the phonon frequency, and $v_q \sim q$ is the screened energy of interaction of the electrons with a lattice potential deformed as a result of the motion of the atoms (the proportionality of v_q to the phonon wave number q arises as a result of the screening of the Coulomb interaction between the electrons [44-46]. The quantity

$$\left(\frac{\hbar}{2\omega_q}\right)^{1/2} (b_q + b_{-q}^+)$$

corresponds to the generalized coordinate of the normal lattice vibrations.

‡For the Hamiltonian of the electron–phonon interaction in this case, we must take an expression similar to H_{int} in the Fröhlich model, but with $v_q = v_q^i$ – the "bare" electron–phonon interaction, where $v_q^i \sim 1/q$ and does not include the effects of the Coulomb screening. If we take the Coulomb interaction into account through the screening effect, we can exclude it from the Hamiltonian by replacing v_q^i by $v_q \sim q$ (cf. [44, 46]). We then arrive at the Fröhlich model.

where e_j is the charge of a particle, $\mathbf{x}_j$ is its position vector, and

$$\boldsymbol{P} = \sum_j e_j \boldsymbol{x}_j \tag{15.64}$$

is the polarization vector, regarded as a quantum-mechanical operator, or, in the classical case, as a dynamic variable. In accordance with (15.47), under the influence of the perturbation (15.63) an electric current

$$\langle J_\alpha \rangle = \int_{-\infty}^{\infty} \langle\langle J_\alpha (t)\, H_{t'}^1 (t') \rangle\rangle\, dt' . \tag{15.65}$$

arises in the system. There is no constant term in this formula, since the mean current is equal to zero, $\langle J_\alpha \rangle_0 = 0$, in statistical equilibrium. In formula (15.65),

$$H_t^1 (t) = - \boldsymbol{E}^0 \cdot \boldsymbol{P} (t) \cos \omega t\, e^{\varepsilon t}, \qquad J_\alpha (t) = \sum_j e_j \dot{x}_{j\alpha} (t) = \dot{P}_\alpha (t). \tag{15.65a}$$

J_α is the electric current operator, and $\dot{x}_{j\alpha}$ is the α-component of the velocity operator of the j-th particles. In the second-quantized representation, the electric current operator has the form

$$J_\alpha = \frac{e}{m} \sum_{p,\, \sigma} p_\alpha a_{p\sigma}^+ a_{p\sigma}, \tag{15.65b}$$

where $a_{p\sigma}^+$ and $a_{p\sigma}$ are operators which respectively create and destroy particles in a state with momentum p and spin σ (we assume that the particles have the same charge, equal to e).

Taking (15.65a) into account, we write the expression (15.65) in the form

$$\langle J_\alpha \rangle = - \sum_\beta \int_{-\infty}^{\infty} \langle\langle J_\alpha (t)\, P_\beta (t') \rangle\rangle\, E_\beta^0 \cos \omega t'\, e^{\varepsilon t'}\, dt' . \tag{15.65c}$$

Up to this point, all the arguments are valid both for the first case, when $\mathbf{E}^0 = \mathbf{E}$, and for the second case, when $\mathbf{E}^0 = \mathbf{D}$. We shall now examine the first case, when the acting field is equal to the mean field. Formula (15.65c) can then be rewritten in the form

$$\langle J_\alpha \rangle = \sum_\beta \mathrm{Re} \{ \sigma_{\alpha\beta} (\omega)\, e^{-i\omega t + \varepsilon t} \} E_\beta, \tag{15.66}$$

where

$$\sigma_{\alpha\beta}(\omega) = -\int_{-\infty}^{\infty} e^{-i\omega t+\varepsilon t} \langle\langle J_\alpha P_\beta(t)\rangle\rangle\, dt \tag{15.66a}$$

is the tensor of the electrical conductivity in a periodic field. The limit $\varepsilon \to 0$ is taken after the thermodynamic limit $V \to \infty$ ($V/N =$ const).

We now consider the second case, when it is necessary to take into account the dielectric polarization of the medium,† and the external field $\mathbf{E}^0$ inside the medium corresponds not to the mean field $\mathbf{E}$, but to the induction $\mathbf{D}$:

$$\boldsymbol{E}^0 = \boldsymbol{D} = \varepsilon(\omega)\cdot \boldsymbol{E}, \tag{15.67}$$

where $\boldsymbol{\varepsilon}(\omega)$ is the dielectric permittivity tensor, for which we shall obtain an explicit expression below. We represent formula (15.65) in the form

$$\langle J_\alpha\rangle = \sum_\beta \mathrm{Re}\{\varkappa_{\alpha\beta}(\omega)\, e^{-i\omega t+\varepsilon t}\}\, D_\beta, \tag{15.68}$$

where

$$\varkappa_{\alpha\beta}(\omega) = -\int_{-\infty}^{\infty} e^{-i\omega t+\varepsilon t} \langle\langle J_\alpha P_\beta(t)\rangle\rangle\, dt \tag{15.68a}$$

is the tensor of the electric susceptibility in a periodic field. The expressions (15.66a) and (15.68a) are the same in appearance the only difference being in the meaning of the averaging operation $\langle\ldots\rangle$, that is, in how the Coulomb interaction is taken into account in the Hamiltonian with which the averaging is performed — as a self-consistent screening field, or explicitly. In the first case, the Hamiltonian does not contain the Coulomb interaction, but the matrix elements of the electron–phonon interaction are modified to allow for the Coulomb interaction (cf. footnote † on p. 168). In the second case, the Hamiltonian contains both a term for the interaction with the phonons, and the direct Coulomb interaction (cf. footnote ‡ on p. 168).

†In the case of a nonuniform field (cf. §18), we must take into account that only its longitudinal part, i.e., the part parallel to the wave vector, is screened.

Using (15.67), we can rewrite the formula (15.68) in the form

$$\langle J_\alpha \rangle = \sum_\beta \mathrm{Re}\,\{\sigma_{\alpha\beta}(\omega)\, e^{-i\omega t + \varepsilon t} E_\beta\} = \mathrm{Re}\,\{\sigma(\omega)\cdot \boldsymbol{E}\, e^{-i\omega t+\varepsilon t}\}_\alpha, \qquad (15.69)$$

here

$$\sigma_{\alpha\beta}(\omega) = \sum_\gamma \varkappa_{\alpha\gamma}(\omega)\, \varepsilon_{\gamma\beta}(\omega),$$
$$\sigma(\omega) = \varkappa(\omega)\cdot\varepsilon(\omega) \qquad (15.70)$$

s the electrical conductivity tensor in a periodic field.

The dielectric permittivity tensor $\varepsilon(\omega)$ can be expressed in erms of the susceptibility tensor $\varkappa(\omega)$ by making use of the re- ation†

$$\boldsymbol{D} = \boldsymbol{E} + 4\pi\langle \boldsymbol{P}\rangle, \qquad (15.71)$$

hich, in our case of the periodic field (15.62), in complex nota- on has the form

$$\boldsymbol{D} = \boldsymbol{E} - \frac{4\pi}{i\omega}\langle \boldsymbol{J}\rangle,$$

ince

$$\langle \boldsymbol{J}\rangle = \langle \dot{\boldsymbol{P}}\rangle$$

ccording to (15.65a). Using this relation and (15.67)-(15.70), we btain for $\omega \neq 0$

$$\varepsilon^{-1}(\omega) = 1 + \frac{4\pi\varkappa(\omega)}{i\omega}, \qquad \sigma(\omega) = \varkappa(\omega)\left(1 + \frac{4\pi\varkappa(\omega)}{i\omega}\right)^{-1}. \qquad (15.72)$$

hus, the electric susceptibility tensor (15.68a) enables us to de- ermine the dielectric permittivity tensor $\varepsilon(\omega)$ and the electrical onductivity tensor $\sigma(\omega)$ in a periodic field.

Thus, the adiabatic switching on of an electric field leads to e appearance of an electric current (15.66) in a system with nite electrical conductivity, i.e., to the appearance of an ir- eversible process. Generally speaking, the electrical conduc- vity also remains finite for a static electrical field, when $\omega = 0$.

---

n the case of a nonuniform field (cf. §18), the relation (15.71) is valid only for the ongitudinal components of the fields, since only they are screened.

In this case, from (15.66a) in the limit $\omega \to 0$, we find the static conductivity

$$\sigma_{\alpha\beta}(0) = -\lim_{\varepsilon\to 0}\int_{-\infty}^{\infty} e^{\varepsilon t}\langle\langle J_\alpha P_\beta(t)\rangle\rangle\, dt. \qquad (15.73)$$

It is also possible to obtain another expression for the tensor $\varkappa_{\alpha\beta}(\omega)$, if we write (15.68a) in the form

$$\varkappa_{\alpha\beta}(\omega) = -\frac{1}{i\hbar}\int_{-\infty}^{0} e^{-i\omega t+\varepsilon t}\,\mathrm{Tr}\,\{[P_\beta(t),\ \rho_0]\,J_\alpha\}\,dt$$

and use the Kubo identity (15.39), according to which

$$[P_\beta(t),\ \rho_0] = -i\hbar\rho_0\int_0^\beta e^{\lambda H}\dot{P}_\beta(t)\,e^{-\lambda H}\,d\lambda. \qquad (15.74)$$

We then obtain

$$\varkappa_{\alpha\beta}(\omega) = \int_0^\beta\int_0^\infty e^{i\omega t-\varepsilon t}\langle e^{\lambda H}J_\beta e^{-\lambda H}J_\alpha(t)\rangle\,d\lambda\,dt =$$

$$= \int_0^\beta\int_0^\infty e^{i\omega t-\varepsilon t}\langle J_\beta J_\alpha(t+i\hbar\lambda)\rangle\,d\lambda\,dt \equiv \beta\int_0^\infty e^{i\omega t-\varepsilon t}(J_\beta,\ J_\alpha(t))\,dt. \qquad (15.75)$$

The formula (15.66a) is also brought to the same form:

$$\sigma_{\alpha\beta}(\omega) = \int_0^\beta\int_0^\infty e^{i\omega t-\varepsilon t}\langle J_\beta J_\alpha(t+i\hbar\lambda)\rangle\,d\lambda\,dt, \qquad (15.76)$$

which, in the classical limit $\hbar \to 0$, has the form

$$\sigma_{\alpha\beta}(\omega) = \beta\int_0^\infty e^{i\omega t-\varepsilon t}\langle J_\beta J_\alpha(t)\rangle\,dt. \qquad (15.77)$$

This formula also follows directly from the classical formula (15.22) for the linear response, if (15.63) is taken into account.

In the case of a static field, the formula (15.76) becomes

$$\sigma_{\alpha\beta}(0) = \lim_{\varepsilon\to 0}\int_0^\beta\int_0^\infty e^{-\varepsilon t}\langle J_\beta J_\alpha(t+i\hbar\lambda)\rangle\,d\lambda\,dt. \qquad (15.78)$$

Expressions for the kinetic coefficients of the type (15.76)-(15.78), and the equivalent expressions (15.73), are usually called Kubo formulas, although they were known earlier to other authors (cf. the comments on pp. 151-152).

The order of the limits $\varepsilon \to 0$ and $V \to \infty$ in the Kubo formulas is very important, since there is no uniformity property with respect to these two limits and the result depends on the order in which they are taken. The limiting process in which $V \to \infty$ first (with $V/N = \text{const}$), and then $\varepsilon \to 0$, i.e., the (V, ε)-limit, corresponds to the imposition of a causality condition on the solution of the Liouville equation. It implies the exclusion of the advanced solutions, as is clear from an examination of the boundary conditions of formal scattering theory (cf. Gell-Mann and Goldberger [47]† and Appendix I). Only this order of the limits can give a finite value for the kinetic coefficients (15.66a) and (15.68a).

The other order of the limits, in which first $\varepsilon \to 0$ at finite volume, and then $V \to \infty$ (with $V/N = \text{const}$), is not suitable, since it leads as $\omega \to 0$ to meaningless expressions for the kinetic coefficients. We shall return to this question in §17.

Using the Kubo formulas (15.78), we can calculate the static electrical conductivity [48-50], and by means of (15.76) we can calculate the nonstatic electrical conductivity [41, 49, 51-56, 103] without using a kinetic equation; this, however, is not a simple problem, since it requires the calculation of time correlation functions.

15.5. Effect of an Alternating Magnetic Field. Magnetic Susceptibility

To conclude this section, we shall examine the effect on a statistical ensemble of switching on a spatially uniform alternating periodic external magnetic field $\mathbf{H}(t)$ with frequency ω:

$$\boldsymbol{H}(t) = \boldsymbol{H} \cos \omega t \, e^{\varepsilon t} = \operatorname{Re} \{ e^{-i\omega t + \varepsilon t} \boldsymbol{H} \}. \tag{15.79}$$

This problem is analogous to the problem of the effect of an electric field, treated above in §15.4; we shall therefore treat it more briefly.

†This paper, although devoted to formal scattering theory, enables us to understand the nature of the origin of irreversibility.

The operator corresponding to the perturbation due to the fi (15.79) is

$$H_t^1 = -(\boldsymbol{M} \cdot \boldsymbol{H}(t)) = -(\boldsymbol{MH}) \cos \omega t \, e^{\varepsilon t}, \quad (15.8$$

where **M** is the operator of the total magnetic moment of the sys tem. According to (15.47), under the influence of the perturba- tion (15.80) the magnetic moment changes with time according tc the formula

$$\langle M_\alpha \rangle = \langle M_\alpha \rangle_0 + \int_{-\infty}^{\infty} \langle\langle M_\alpha (t) H_{t'}^1 (t') \rangle\rangle \, dt' \quad (15.8$$

where $\langle M_\alpha \rangle_0$ is the mean component of the magnetic moment along the α-axis in the state of statistical equilibrium. If a mag netic field is present in the equilibrium state, then $\langle M_\alpha \rangle_0 \neq 0$. We write the formula (15.81) in the form

$$\langle M_\alpha \rangle = \langle M_\alpha \rangle_0 + \sum_\beta \operatorname{Re} \{\chi_{\alpha\beta} (\omega) \, e^{-i\omega t + \varepsilon t}\} \, H_\beta, \quad (15.8$$

where

$$\chi_{\alpha\beta} (\omega) = - \int_{-\infty}^{\infty} e^{-i\omega' + \varepsilon t} \langle\langle M_\alpha M_\beta (t) \rangle\rangle \, dt \quad (15.8$$

is the magnetic susceptibility tensor in a periodic magnetic fielc The magnetic susceptibility tensor (15.83) can also be written, using Kubo's identity (15.39), in the form

$$\chi_{\alpha\beta} (\omega) = \int_0^\beta \int_0^\infty e^{i\omega t - \varepsilon t} \langle \dot{M}_\beta M_\alpha (t + i\hbar\lambda) \rangle \, d\lambda \, dt, \quad (15.8$$

which is analogous to formula (15.75). For the application of the formulas (15.83) and (15.84) in the theory of magnetic resonance see the papers [57-59] and the monographs [13, 14, 27, 37].

§ 16. Two-Time Green Functions

The response of a quantum system to an external mechanica perturbation can be expressed, as was shown in the preceding section, in terms of the retarded two-time Green functions (15.4 and the response of a classical system in terms of (15.19). We shall now examine the fundamental properties of quantum two-ti Green functions and discuss the relations between Green functio of different types, their spectral representations, dispersion relations and symmetry properties. This will enable us later, in §17, to derive very easily the fluctuation-dissipation theorems and all the properties of the kinetic coefficients. The properties of the Green functions (especially the retarded ones) will be used widely later. We shall confine ourselves to the case of quantum Green functions, since the classical case is easily obtained analogously, or by the limiting process $\hbar \to 0$.

There are very many papers in which Green functions are applied to different problems in statistical mechanics, both for the equilibrium and for the nonequilibrium case (cf. [12-15, 60-64], where a detailed survey of the literature can be found).

Green functions of different types have been applied: the differences lie in the nature of the averaging performed in them, in the arguments on which they depend, and in their analytic properties. If the averaging is taken over the ground state, we have the *field Green functions* that are usually applied in quantum field theory; if the averaging is performed over a statistical ensemble, we have *thermodynamic Green functions*. If the Green functions depend on the time variables, they are called *time Green functions*; if they depend on the temperature as an explicit argument, they are called *temperature* or *Matsubara Green functions*, since they were first introduced by Matsubara [65].

The different Green functions have their advantages and deficiencies. Causal Green functions have a more complicated analytic structure, but are intimately connected with perturbation theory. Retarded Green functions have a simple analytic structure and are related simply to the kinetic coefficients, but connected more indirectly with perturbation theory. In the theory of irreversible processes, retarded Green functions are evidently the most convenient, and so we shall give them special attention.

In this section, we shall examine the spectral representations, dispersion relations, sum rules, symmetry properties, and certain other properties of Green functions and correlation functions.

16.1. Retarded, Advanced, and Causal Green Functions

Green functions in statistical mechanics are a convenient generalization of the concept of correlation functions. Like the latter, they are intimately connected with the calculation of observable quantities, but have advantages in the construction and solution of the equations determining them.

In statistical mechanics, as in quantum field theory, we can consider retarded $G_r(t, t')$, advanced $G_a(t, t')$, and causal $G_c(t, t')$ Green functions:

$$\begin{aligned} G_r(t, t') &= \langle\langle A(t) B(t')\rangle\rangle_r = \frac{1}{i\hbar}\theta(t-t')\langle [A(t), B(t')]\rangle, \\ G_a(t, t') &= \langle\langle A(t) B(t')\rangle\rangle_a = -\frac{1}{i\hbar}\theta(t'-t)\langle [A(t), B(t')]\rangle, \\ G_c(t, t') &= \langle\langle A(t) B(t')\rangle\rangle_c = \frac{1}{i\hbar}\langle TA(t) B(t')\rangle. \end{aligned} \tag{16.1}$$

Here $\langle ... \rangle$ is an average over the grand canonical ensemble (9.42). The subscript 0 on the brackets, denoting equilibrium averages, will be omitted, since in this section we shall study averaging only over an equilibrium ensemble:

$$\langle \ldots \rangle = Q^{-1}\,\mathrm{Tr}\,(e^{-\mathscr{H}/\theta}\ldots), \qquad Q = \mathrm{Tr}\; e^{-\mathscr{H}/\theta} = e^{-\Omega/\theta}; \tag{16.2}$$

and Ω is the thermodynamic potential (9.40) in the variables θ, μ, and V. The operator $\mathscr{H}$ includes a term with the chemical potential μ:

$$\mathscr{H} = H - \mu N. \tag{16.3}$$

The time arguments in the operators A(t) and B(t') denote the Heisenberg picture:

$$A(t) = e^{i\mathscr{H}t/\hbar} A e^{-i\mathscr{H}t/\hbar}. \tag{16.4}$$

The symbol T denotes a chronological product of operators†:

$$TA(t) B(t') = \theta(t-t') A(t) B(t') + \eta\theta(t'-t) B(t') A(t), \tag{16.5}$$

where θ (t) is the discontinuous function (15.17). Finally, [A, B] is a commutator or anticommutator, according to the sign of η:

$$[A,B] = AB - \eta BA, \tag{16.6}$$

i.e., for $\eta = 1$, it is a commutator, and for $\eta = -1$, an anticommutator.

†In §15, we have used the particular case of this operation with $\eta = 1$ to write the ordered P-exponential (15.56b).

The sign of η in formulas (16.5) and (16.7) is chosen to be plus r minus from considerations of convenience in the problem. If . and B are Bose operators, the plus sign is usually chosen, and : Fermi operators, the minus sign (for this choice of η, we have $\text{\textbackslash}, A^+] = 1$), although the other choice of the sign of η is also pos-ible. We have already introduced Green functions $G_r(t, t')$ with $= 1$ [cf. formula (15.48)]. In general, A and B are neither Fermi or Bose operators, since a product of operators may satisfy more omplicated commutation relations than do the original operators. ı quantum field theory, the sign of η in a T-product is determined y the parity of the permutation of the Fermi operators required ɔ bring the product to chronological order.

We note that, because of the discontinuous factor $\theta(t - t')$, reen functions are not defined at the point $t = t'$ at which the time rguments coincide. This indeterminacy is well known in quantum ield theory [23-26].

It follows from the definitions (16.1)-(16.6) that the Green ınctions applied in statistics differ from the field Green functions nly in the method of averaging. Instead of averaging over the ɔwest vacuum state of the system, the averaging is performed ver the grand canonical ensemble (16.2). Consequently, the reen functions introduced depend both on time and on temper-ture. It is obvious that, as the temperature tends to zero, the reen functions (16.1) go over to the usual field Green functions, ı which the averaging is performed over the lowest energy state. /e observe that, in contrast to quantum field theory, where the acuum averages are infinite and are discarded as having no physi-al meaning, in statistical mechanics averages over the ground tate of the system in the thermodynamic limit (cf. §3.2) give ob-ervable quantities.

It is not by chance that the grand ensemble is applied. It is ery convenient in that, in working with it, it is not necessary ɔ take into account the supplementary constraint that the total umber of particles be constant, as we had to do for the canonical nsemble (9.14), and the occupation numbers of the different states re independent.

The Green functions (16.1) for the case of statistical equi-ibrium depend only on $t - t'$, as follows from the permissibility f cyclic permutations in the trace [cf. (15.48a)].

For many problems in statistical mechanics, we need not have recourse to many-time functions, but can confine ourselves to two-time functions. The latter are very convenient, since for them we can use simple spectral expansions, which make it very much easier to solve the equations for the Green functions. On the other hand, they contain a sufficiently large amount of information on the equilibrium and nonequilibrium properties of many-particle systems. Of the two-time Green functions, those most convenient to use are the retarded and advanced Green functions G_r and G_a, since their Fourier components can be analytically continued into the complex energy plane.

We shall obtain a system of equations for the Green functions (16.1). The operators A(t) and B(t) satisfy equations of motion of the form

$$i\hbar \frac{dA}{dt} = A\mathcal{H} - \mathcal{H}A, \tag{16.7}$$

where A is an operator in the Heisenberg picture (16.4). The right hand side of Eq. (16.7) can be written explicitly using the explicit form of the Hamiltonian and the commutation relations for the operators. Differentiating the Green functions (16.1) with respect to t, we obtain the equation

$$i\hbar \frac{dG(t-t')}{dt} = i\hbar \frac{d}{dt} \langle\langle A(t) B(t')\rangle\rangle =$$
$$= \frac{d\theta(t-t')}{dt} \langle [A(t),\ B(t')] \rangle + \langle\langle i\hbar \frac{dA(t)}{dt} B(t') \rangle\rangle,$$

which is the same for all three Green functions G_r, G_a, and G_c, since

$$\frac{d}{dt}\theta(-t) = -\frac{d}{dt}\theta(t).$$

Therefore, we shall write simply G and $\langle\langle\ldots\rangle\rangle$ without the subscripts indicating the type of Green function. Taking into account the relation between the discontinuous function $\theta(t)$ and the δ-function of t,

$$\theta(t) = \int_{-\infty}^{t} \delta(t')\, dt', \tag{16.8}$$

and the equation of motion (16.7) for the operator A(t), we write the equation for the Green function in the form

$$i\hbar \frac{dG(t-t')}{dt} = \delta(t-t')\langle[A(0),\ B(0)]\rangle + \langle\langle(A(t)\mathcal{H} - \mathcal{H}A(t))\,B(t')\rangle\rangle, \tag{16.9}$$

where we have taken into account that, in the case of statistical equilibrium, $\langle A(t)B(t')\rangle$ depends only on the difference $t-t'$, and, consequently,

$$\langle A(t)\,B(t)\rangle = \langle A(0)\,B(0)\rangle, \qquad A(0) = A,\ B(0) = B.$$

In the right-hand side of Eq. (16.9), two-time Green functions appear, which, generally speaking, are of higher order than the original ones. For these, we can also set up equations of the type (16.9) and obtain a chain of coupled equations for the Green functions.

Chains of equations of the type (16.9) are simply equations of motion for the Green functions. Alone, these equations are still insufficient as is clear from the fact that they are the same for all the Green functions G_r, G_a, and G_c, if these are constructed from the same operators A and B. They must be supplemented by boundary conditions. This will be done below in §16.3, by means of spectral theorems.

The equations (16.9) are exact, and the solution of this chain is, therefore, an exceedingly complicated problem. Sometimes, by means of some approximation technique, a chain of equations of the type (16.9) can be decoupled, i.e., transformed into a finite system of equations, and solved. For example of such decouplings, see [12-15, 63, 64, 66, 124-137, 145-153]. If the system contains a small parameter, for example, a small interaction or small density, such decouplings can be justified.

16.2. Spectral Representations for the Time Correlation Functions

To solve the equations (16.9) for the Green function, it is important to have spectral representations for them, which supplement the system of equations with the necessary boundary conditions. We shall obtain spectral representations for the Green functions (16.1) in the following subsection; here, we obtain them for the corresponding correlation functions

$$F_{BA}(t'-t)=\langle B(t')A(t)\rangle, \quad F_{AB}(t-t')=\langle A(t)B(t')\rangle. \tag{16.10}$$

Let C_ν and E_ν be the eigenfunctions and eigenvalues of the Hamiltonian (16.3):

$$\mathscr{H}C_\nu = E_\nu C_\nu. \tag{16.11}$$

We write in explicit form the statistical averaging operation for the time correlation functions (16.10):

$$\langle B(t')A(t)\rangle = Q^{-1}\sum_\nu \left(C_\nu^* B(t')A(t)C_\nu\right)e^{-\frac{E_\nu}{\theta}}, \tag{16.12}$$

where $(C_\nu^* B(t')A(t)C_\nu)$ are the diagonal matrix elements of $B(t')A(t)$. We shall make use of a common device in the theory of dispersion relations, based on the completeness of the set of functions C_ν [23, 25, 67], and represent (16.12) in the form

$$\langle B(t')A(t)\rangle = Q^{-1}\sum_{\nu,\mu}\left(C_\nu^* B(t')C_\mu\right)\left(C_\mu^* A(t)C_\nu\right)e^{-\frac{E_\nu}{\theta}} =$$

$$= Q^{-1}\sum_{\nu,\mu}\left(C_\nu^* B(0)C_\mu\right)\left(C_\mu^* A(0)C_\nu\right)e^{-E_\nu/\theta}\exp\left\{\frac{i}{\hbar}(E_\mu - E_\nu)(t-t')\right\}, \tag{16.13}$$

since

$$e^{-i}\quad C_\nu = e^{-\frac{iE_\nu t}{\hbar}}C_\nu, \quad C_\mu^* e^{\frac{i\mathscr{H}t}{\hbar}} = C_\mu^* e^{\frac{iE_\mu t}{\hbar}}$$

On the other hand,

$$\langle A(t)B(t')\rangle = Q^{-1}\sum_{\nu,\mu}\left(C_\nu^* A(0)C_\mu\right)\left(C_\mu^* B(0)C_\nu\right)e^{-E_\nu/\theta}\exp\left\{\frac{i}{\hbar}(E_\mu - E_\nu)(t'-t)\right\}. \tag{16.14}$$

Interchanging the summation indices μ and ν in the latter equality and comparing (16.13) and (16.14), we observe that they can be represented in the form

$$\langle B(t')A(t)\rangle = \frac{1}{2\pi}\int_{-\infty}^{\infty} J_{BA}(\omega)e^{i\omega(t'-t)}d\omega,$$

$$\langle A(t)B(t')\rangle = \frac{1}{2\pi}\int_{-\infty}^{\infty} J_{BA}(\omega)e^{\frac{\hbar\omega}{\theta}}e^{i\omega(t'-t)}d\omega, \tag{16.15}$$

where we have introduced the notation

$$J_{BA}(\omega) = 2\pi Q^{-1} \sum_{\nu,\mu} (C_\mu^* B(0) C_\nu)(C_\nu^* A(0) C_\mu) e^{-\frac{E_\mu}{\theta}} \delta\left(\frac{E_\mu - E_\nu}{\hbar} - \omega\right). \tag{16.16}$$

The relations (16.15) are the required spectral representations of the time correlation functions†: $J_{BA}(\omega)$ is the spectral intensity of the function $\langle B(t') A(t) \rangle$.

Comparing the first of the relations (16.15) with the second, we obtain an important property of the spectral intensity

$$J_{AB}(-\omega) = J_{BA}(\omega) e^{\beta\hbar\omega}. \tag{16.16a}$$

For the systems studied in statistical mechanics, the spectrum of the E_μ is practically continuous because of the large dimensions of the system and, therefore, the summation over the states in (16.16) is, in essence, an integration, which can "remove" the δ-function. Consequently, in general, the spectral intensity does not have a δ-function form; only in the particular cases of "ideal" systems in which the elementary excitations do not attenuate can it be of δ-function form.

It is also possible to derive the relations (16.15) without using the eigenfunctions of the operator $\mathcal{H}$. For this, it is sufficient to note that $\langle B(t') A(t) \rangle$ depends only on the difference $t - t'$. Thus, the first of the relations (16.15) is simply the definition of the Fourier component of the time correlation function, it being assumed that such an expansion is possible.‡ The second relation of (16.15) can be obtained from the first by the replacement $t - t' \rightarrow$

†Spectral representations for the time correlation functions and Green functions of statistical mechanics were first applied in a paper by Callen and Welton [22] on the theory of fluctuations and noise, and were later widely used by many authors [1, 3, 4, 60-64, 68].

‡Khinchin [16] proved that, for a continuous stationary random process, the correlation functions can be represented in the form of Fourier–Stieltjes integrals:

$$\langle (A - \langle A \rangle)(A(t) - \langle A \rangle) \rangle = \int_{-\infty}^{\infty} \cos \omega t \, dF(\omega)$$

(A are dynamic variables of classical mechanics, $A^* = A$), where $F(\omega)$ is a non-decreasing function with bounded variation, called the spectrum of the process. In place of the Fourier–Stieltjes integral, we can simply use the Fourier integral, if we assume that $dF(\omega)/d\omega = J_{AA}(\omega)$ can be a generalized function.

$t - t' + (i\hbar/\theta)$, since

$$\left\langle B(0)\, A\left(t + \frac{i\hbar}{\theta}\right)\right\rangle = \langle A(t)\, B(0)\rangle, \tag{16.17}$$

as can be easily verified directly by performing a cyclic permutation of the operators in the trace. Indeed, after the permutation which moves the operator B into the first position, we obtain

$$\begin{aligned}\langle A(t)\, B(0)\rangle &= Q^{-1}\,\mathrm{Tr}\,\{e^{-\beta\mathcal{H}} e^{i\mathcal{H}t/\hbar} A e^{-i\mathcal{H}t/\hbar} B\} \\ &= Q^{-1}\,\mathrm{Tr}\left\{B e^{-\left(\beta - \frac{it}{\hbar}\right)\mathcal{H}} A e^{\left(\beta - \frac{it}{\hbar}\right)\mathcal{H}} e^{-\beta\mathcal{H}}\right\},\end{aligned}$$

i.e., the relation (16.17).

For all real systems, there is attenuation of the correlations in time, i.e.,

$$\lim_{|t-t'|\to\infty} \langle A(t)\, B(t')\rangle = \langle A\rangle \langle B\rangle$$

(A and B are not integrals of motion); consequently, if $\langle A\rangle = 0$, or $\langle B\rangle = 0$, then

$$\lim_{|t-t'|\to\infty} \langle A(t)\, B(t')\rangle = 0.$$

If we introduce the new operators $A(t) - \langle A\rangle$, and $B(t) - \langle B\rangle$ the correlations for these will attenuate in time. Then, the expansion (16.15) can be written in the form

$$\begin{aligned}\langle (B(t') - \langle B\rangle)(A(t) - \langle A\rangle)\rangle &= \langle B(t')\, A(t)\rangle - \langle B\rangle \langle A\rangle = \\ &= \frac{1}{2\pi}\int_{-\infty}^{\infty} J_{BA}(\omega)\, e^{i\omega(t'-t)} d\omega.\end{aligned} \tag{16.15a}$$

Below, we shall always assume that, if $\langle B(t') A(t)\rangle$ tends to a finite limit as $|t - t'| \to \infty$, its expansion as a Fourier integral has the meaning of (16.15a).

If $J_{BA}(\omega)$ is a continuous function, then, according to the Riemann–Lebesgue lemma,

$$\lim_{|t-t'|\to\infty} \langle (B(t') - \langle B\rangle)(A(t) - \langle A\rangle)\rangle = 0.$$

If $J_{BA}(\omega)$ has δ-function singularities at certain frequencies, the correlation functions (16.15a) will not tend to zero as $|t-t'|\to\infty$, but will oscillate. If these frequencies are incommensurable, the correlation functions are almost-periodic.

The time average of the correlation functions is equal to zero:

$$\lim_{T\to\infty}\frac{1}{2T}\int_{-T}^{T}\frac{1}{2\pi}\int_{-\infty}^{\infty}J_{BA}(\omega)\,e^{-i\omega t}\,dt\,d\omega=$$

$$=\lim_{T\to\infty}\frac{1}{2T}\int_{-\infty}^{\infty}J_{BA}(\omega)\,\delta(\omega)\,d\omega=\lim_{T\to\infty}\frac{J_{BA}(0)}{2T}=0,$$

if the spectral intensity is finite at $\omega=0$, as is characteristic for the spectrum of a random ergodic process.† Below, we shall assume that $J_{BA}(\omega)$ does not have a δ-function form at $\omega=0$. This is essential for the unique determination of the Green function from Eqs. (16.9) [70, 138-142].

The time correlation functions for any finite system, i.e., before the limiting process $V\to\infty$, are almost-periodic functions of $t-t'$, as was proved by Bocchieri and Loinger [71] and by Percival [71a]. This quantum-mechanical theorem is analogous to Poincaré's classical recurrence theorem.‡ Attenuation of the correlation functions for real systems can be obtained only after the limiting process $V\to\infty$ ($V/N=$ const), which excludes, as it were, long Poincaré cycles.

If A and B is an integral of motion, the time correlation function does not depend at all on time. Indeed, let A be an integral of motion. Then

$$A(t)=e^{iHt/\hbar}Ae^{-iHt/\hbar}=A(0)$$

and, consequently,

$$\langle A(t)\,B(t')\rangle=\langle A(t-t')\,B(0)\rangle=\langle A(0)\,B(0)\rangle.$$

---

†The relation between the spectrum of a stationary random process and the ergodicity property is established by the Wiener–Khinchin theorem [16, 69].

‡According to Poincare's recurrence theorem, any isolated mechanical system in a finite volume will repeatedly approach arbitrarily close to its initial state. For a very elegant and simple proof of this theorem and its exact formulation, see the book by Kac [72].

We shall prove one important property of the spectral intensity, which we shall need later. We shall show that the spectral intensity $J_{A^+A}(\omega)$ of a time correlation function constructed from the conjugate operators A^+ and A is nonnegative, i.e., in the spectral expansion

$$\langle A^+(t')\,A(t)\rangle = \frac{1}{2\pi}\int_{-\infty}^{\infty} J_{A^+A}(\omega)\,e^{i\omega(t'-t)}\,d\omega \tag{16.18}$$

we have, always,

$$J_{A^+A}(\omega) \geqslant 0. \tag{16.18a}$$

It follows from the expression (16.16) for the spectral intensity that

$$J_{A^+A}(\omega) = 2\pi Q^{-1}\sum_{\nu,\,\mu}(C^*_\mu A^+(0)\,C_\nu)(C^*_\nu A(0)\,C_\mu)\,e^{-E_\mu/\theta}\delta\left(\frac{E_\mu - E_\nu}{\hbar} - \omega\right) \geqslant 0,$$

since all the terms in the summation are nonnegative. Thus, the spectral intensity $J_{A^+A}(\omega)$ cannot be negative. It follows directly from what we have proved that $J_{AA^+}(\omega)$ is also nonnegative.

16.3. Spectral Representations and Dispersion Relations for the Green Functions

We now study the spectral representations for the Green functions, treating the retarded and advanced ones first. They are easily obtained by means of the spectral representations (16.15) for the time correlation functions. Indeed, let $G_r(\omega)$ be the Fourier component of the Green function $G_r(t - t')$:

$$G_r(t-t') = \frac{1}{2\pi}\int_{-\infty}^{\infty} G_r(\omega)\,e^{-i\omega(t-t')}\,d\omega,$$

$$G_r(\omega) = \int_{-\infty}^{\infty} G_r(t)\,e^{i\omega t}\,dt. \tag{16.19}$$

We are using the same notation for the Fourier components of Green functions as for the Green functions themselves, distinguishing them only by their arguments. Sometimes, we shall also use the notation $\langle\langle A\,|\,B\rangle\rangle_\omega$.

$$\langle\langle A(t)\,B(t')\rangle\rangle = \frac{1}{2\pi}\int_{-\infty}^{\infty}\langle\langle A|B\rangle\rangle_\omega\,e^{-i\omega(t-t')}\,d\omega. \tag{16.19a}$$

Substituting the expression (16.1) for $G_r(t)$ into the second of the equalities (16.19), we obtain

$$G_r(\omega) = \frac{1}{i\hbar} \int_{-\infty}^{\infty} \theta(t - t') \{\langle A(t) B(t')\rangle - \eta \langle B(t') A(t)\rangle\} e^{i\omega(t-t')} dt.$$

In the integrand here, there are, in addition to the discontinuous factor, time correlation functions. Using the representation (16.15) for them, we shall have

$$G_r(\omega) = \frac{1}{2\pi} \int_{-\infty}^{\infty} d\omega' J_{BA}(\omega') (e^{\hbar\omega'/\theta} - \eta) \frac{1}{i\hbar} \int_{-\infty}^{\infty} dt\, e^{i(\omega-\omega')t} \theta(t). \tag{16.20}$$

We represent the discontinuous function $\theta(t)$ in the form

$$\theta(t) = \int_{-\infty}^{t} e^{\varepsilon t} \delta(t)\, dt \qquad (\varepsilon \to 0,\ \varepsilon > 0), \tag{16.21}$$

or, since

$$\delta(t) = \frac{1}{2\pi} \int_{-\infty}^{\infty} e^{-ixt}\, dx, \tag{16.22}$$

in another very convenient form:

$$\theta(t) = \frac{i}{2\pi} \int_{-\infty}^{\infty} \frac{e^{-ixt}}{x + i\varepsilon}\, dx. \tag{16.23}$$

It is easily verified that the function defined in this way does indeed possess the properties of the discontinuous θ-function. We

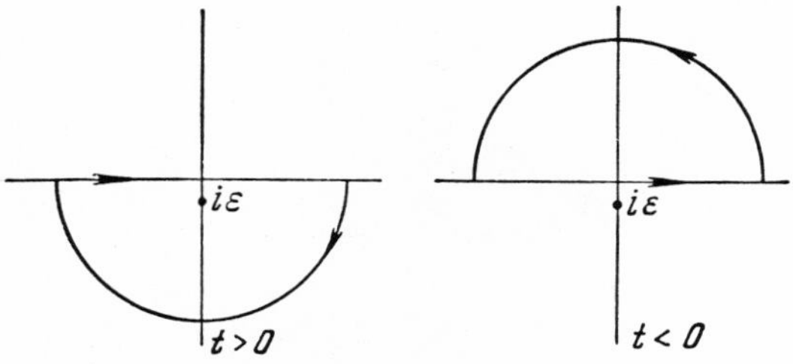

Fig. 1

shall regard x as a complex variable and assume that the integral is taken over the contour depicted in Fig. 1. The integrand has a pole in the lower half-plane. For $t > 0$, the contour must be closed in the lower half-plane, and the integral (16.23) is equal to unity. For $t < 0$, we must close to the contour in the upper half-plane, and the integral (16.23) is equal to zero.

The discontinuous θ-function can also be represented by means of the contour integral

$$\theta(t) = \frac{i}{2\pi} \int\limits_{C_r} e^{-itx} \frac{dx}{x} = \begin{cases} 0 & \text{for} \quad t<0, \\ 1 & \text{for} \quad t>0, \end{cases}$$

where the integration is performed over the contour C_r depicted in Fig. 2.

Using (16.23) and (16.22), we obtain

$$\frac{1}{2\pi i} \int\limits_{-\infty}^{\infty} e^{i(\omega-\omega')t}\, \theta(t)\, dt = \frac{1}{2\pi} \int\limits_{-\infty}^{\infty} \frac{\delta(\omega-\omega'-x)}{x+i\varepsilon}\, dx = \frac{1}{2\pi} \, \frac{1}{\omega-\omega'+i\varepsilon}.$$

Consequently, $G_r(\omega)$, the Fourier component of the retarded Green function $G_r(t)$, is equal to

$$G_r(\omega) = \frac{1}{2\pi\hbar} \int\limits_{-\infty}^{\infty} (e^{\hbar\omega'/\theta} - \eta)\, J_{BA}(\omega') \frac{d\omega'}{\omega-\omega'+i\varepsilon}. \tag{16.24}$$

Repeating the same calculations for $G_a(\omega)$, the Fourier component of the advanced Green function $G_a(t)$, we obtain

$$G_a(\omega) = \frac{1}{2\pi\hbar} \int\limits_{-\infty}^{\infty} (e^{\hbar\omega'/\theta} - \eta)\, J_{BA}(\omega') \frac{d\omega'}{\omega-\omega'-i\varepsilon}. \tag{16.24a}$$

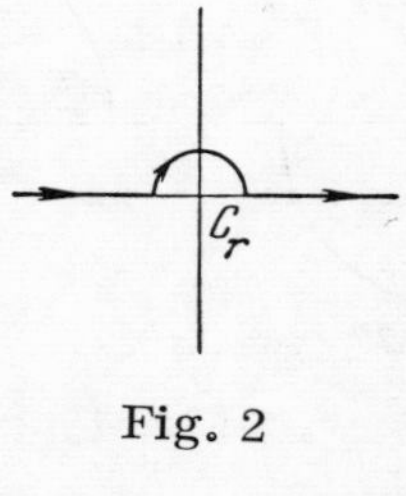

Fig. 2

We write formulas (16.24) and (16.24a) as one formula

$$G_{r,a}(\omega)=\frac{1}{2\pi\hbar}\int_{-\infty}^{\infty}(e^{\hbar\omega'/\theta}-\eta)\,J_{BA}(\omega')\frac{d\omega'}{\omega-\omega'\pm i\varepsilon}, \tag{16.24b}$$

where the plus corresponds to the subscript r, and the minus to the subscript a.

Thus far, we have treated ω as a real quantity. The function (16.24b) can be analytically continued into the complex ω-plane. Indeed, assuming that ω is complex, we obtain

$$\frac{1}{2\pi\hbar}\int_{-\infty}^{\infty}(e^{\hbar\omega'/\theta}-\eta)\,J_{BA}(\omega')\frac{d\omega'}{\omega-\omega'}=\begin{cases}G_r(\omega) & \text{for}\quad \operatorname{Im}\omega>0,\\ G_a(\omega) & \text{for}\quad \operatorname{Im}\omega<0.\end{cases} \tag{16.25}$$

Therefore, following [15], we can treat the function $G_{r,a}(\omega)$ as a single analytic function in the complex plane, with singularities on the real axis. Below, we shall omit the subscripts r,a and write simply $G(\omega)$, regarding ω as complex. We then write the relation (16.25) in the form

$$G(\omega)=\langle\langle A\,|\,B\rangle\rangle_\omega=\frac{1}{2\pi\hbar}\int_{-\infty}^{\infty}(e^{\hbar\omega'/\theta}-\eta)\,J_{BA}(\omega')\frac{d\omega'}{\omega-\omega'}. \tag{16.26}$$

The analyticity of $G(\omega)$ follows from a theorem proved by Bogolyubov and Parasyuk in the theory of dispersion relations [73]. According to this theorem, for $G(\omega)$ to be analytic, it is necessary and sufficient that $G_r(t)$ [or $G_a(t)$] be generalized functions in the Sobolev–Schwartz sense; this is not too stringent a requirement.

First, we shall examine the analytic properties of $G_r(\omega)$. According to (16.19), we have

$$G_r(\omega)=\int_{-\infty}^{\infty}G_r(t)\,e^{i\omega t}\,dt, \tag{16.27}$$

with $G_r(t)=0$ for $t<0$.

We shall show that the function $G_r(\omega)$ can be analytically continued into the region of complex ω in the upper half-plane. Let ω have a positive and nonzero imaginary part γ:

$$\omega=\operatorname{Re}\omega+i\operatorname{Im}\omega=\alpha+i\gamma,\qquad \gamma>0.$$

Then we shall have

$$G_r(\alpha+i\gamma)=\int_{0}^{\infty}G_r(t)\,e^{i\alpha t-\gamma t}\,dt. \tag{16.28}$$

The factor $e^{-\gamma t}$ plays the role of a cut-off factor which ensures the convergence of the integral (16.28) and of its derivatives with respect to ω under sufficiently general assumptions about the function $G_r(t)$ [73]. Consequently, the function $G_r(\omega)$ can be analytically continued into the upper half-plane.

It can be shown analogously that the function $G_a(\omega)$ can be analytically continued into the lower half-plane:

$$\omega = \alpha + i\gamma, \qquad \gamma < 0.$$

If we make a cut along the real axis, the function

$$G(\omega) = \begin{cases} G_r(\omega) & \text{for} \quad \operatorname{Im} \omega > 0, \\ G_a(\omega) & \text{for} \quad \operatorname{Im} \omega < 0 \end{cases} \tag{16.29}$$

can be regarded as a single analytic function, consisting of two branches, one of which is defined in the upper and the other in the lower half-plane of complex values of ω.

If $G(\omega)$ is known, we can also find the spectral intensity $J_{BA}(\omega)$ from the relation

$$G(\omega + i\varepsilon) - G(\omega - i\varepsilon) = \frac{1}{i\hbar}\left(e^{\hbar\omega/\theta} - \eta\right) J_{BA}(\omega), \tag{16.30}$$

where ω is real and $\varepsilon \to 0$. In fact, taking the difference of the expressions (16.26) for real ω

$$G(\omega + i\varepsilon) - G(\omega - i\varepsilon) = \frac{1}{2\pi\hbar}\int_{-\infty}^{\infty}\left(e^{\hbar\omega'/\theta} - \eta\right) J_{BA}(\omega')\left\{\frac{1}{\omega - \omega' + i\varepsilon} - \frac{1}{\omega - \omega' - i\varepsilon}\right\} d\omega'$$

and making use of the representation

$$\delta(\omega - \omega') = \lim_{\varepsilon \to 0} \frac{1}{2\pi i}\left\{\frac{1}{\omega - \omega' - i\varepsilon} - \frac{1}{\omega - \omega' + i\varepsilon}\right\}, \tag{16.31}$$

of the δ-function, we arrive at the relation (16.30) which plays an important role in the applications of Green functions.

Thus, if we could somehow succeed in decoupling the chain of equations (16.9) for the Green functions and find the Green function $G(\omega)$, we could, by means of (16.30), construct the spectral intensity $J_{BA}(\omega)$ and find expressions for the time correlation functions (16.15).

We shall give some further simple, but important, relations between the Green functions and the spectral intensities of the correlation functions.

In the integrand in formulas (16.24) and (16.24a), we shall make use of the symbolic identity

$$\frac{1}{\omega - \omega' \pm i\varepsilon} = \mathcal{P}\frac{1}{\omega - \omega'} \mp i\pi\delta(\omega - \omega'), \tag{16.32}$$

where $\varepsilon \to 0$, $\varepsilon > 0$, and $\mathcal{P}$ denotes that the integral must be taken as a principal-value integral. Here, $\omega - \omega'$ is regarded as a real variable. We then obtain†

$$\begin{aligned} G_r(\omega) &= \frac{1}{2\pi\hbar}\mathcal{P}\int_{-\infty}^{\infty}(e^{\hbar\omega'/\theta} - \eta)\,J_{BA}(\omega')\frac{d\omega'}{\omega - \omega'} - \frac{i}{2\hbar}(e^{\hbar\omega/\theta} - \eta)\,J_{BA}(\omega), \\ G_a(\omega) &= \frac{1}{2\pi\hbar}\mathcal{P}\int_{-\infty}^{\infty}(e^{\hbar\omega'/\theta} - \eta)\,J_{BA}(\omega')\frac{d\omega'}{\omega - \omega'} + \frac{i}{2\hbar}(e^{\hbar\omega/\theta} - \eta)\,J_{BA}(\omega), \end{aligned} \tag{16.33}$$

whence follows the connection between the real and imaginary parts of the Green functions [($J_{BA}(\omega)$ is assumed real]:

$$\begin{aligned} \operatorname{Re} G_r(\omega) &= \frac{1}{\pi}\mathcal{P}\int_{-\infty}^{\infty}\frac{\operatorname{Im} G_r(\omega')}{\omega' - \omega}\,d\omega', \\ \operatorname{Re} G_a(\omega) &= -\frac{1}{\pi}\mathcal{P}\int^{\infty}\frac{\operatorname{Im} G_a(\omega')}{\omega' - \omega}\,d\omega'. \end{aligned} \tag{16.34}$$

The relations (16.34) are called **dispersion relations**. If the spectral intensity is real, they follow directly from (16.33), since then

$$\begin{aligned} \operatorname{Im} G_r(\omega) &= -\frac{1}{2\hbar}(e^{\hbar\omega/\theta} - \eta)\,J_{BA}(\omega), \\ \operatorname{Im} G_a(\omega) &= \frac{1}{2\hbar}(e^{\hbar\omega/\theta} - \eta)\,J_{BA}(\omega), \end{aligned} \tag{16.34a}$$

†The relations (16.33) express the well-known properties of the limit values of the Cauchy integral as the point ω moves on to the integration contour; these were first established by Sokhotskii in 1873, and then by Plemel in 1908 (cf. the textbook by Lavrent'ev and Shabat [74] on the theory of functions of a complex variable).

although (16.34) are valid not only in this case. The dispersion relations (16.34) follow from the analyticity of $G_r(\omega)$ in the upper, and of $G_a(\omega)$ in the lower, half-plane of complex values of ω [cf. (16.28)]. Indeed, if $G_r(\omega)$ is analytic in the upper half-plane, and $G_a(\omega)$ in the lower, they can be represented in the form of Cauchy integrals:

$$G_r(\omega) = \frac{1}{2\pi i} \int_{i\delta-\infty}^{i\delta+\infty} \frac{G_r(\omega')}{\omega' - \omega} d\omega' \qquad (\operatorname{Im}\omega > \delta > 0),$$
$$G_a(\omega) = -\frac{1}{2\pi i} \int_{-i\delta-\infty}^{-i\delta+\infty} \frac{G_a(\omega')}{\omega' - \omega} d\omega' \quad (\operatorname{Im}\omega < -\delta < 0). \tag{16.35}$$

It is assumed that, for any positive δ,

$$|G_{r,a}(\omega)| \leqslant \frac{A(\delta)}{|\omega|} \quad \text{for } \operatorname{Im}\omega > \delta \ (\text{or } \operatorname{Im}\omega < -\delta),$$

and, therefore, the integral over a closed contour of radius R in the upper (or lower) half-plane tends to zero as $R \to \infty$.

We shall displace the integration path on to the real axis by making $\delta \to 0$. Then

$$G_{r,a}(\omega) = \pm \frac{1}{2\pi i} \int_{-\infty}^{\infty} \frac{G_{r,a}(\omega')}{\omega' - \omega} d\omega'. \tag{16.36}$$

We now displace the point ω on to the real axis, putting $\omega \to \omega \pm i\varepsilon$ ($\varepsilon \to 0$), and rewrite (16.36) in the form

$$G_{r,a}(\omega) = \pm \frac{1}{2\pi i} \int_{-\infty}^{\infty} \frac{G_{r,a}(\omega')}{\omega' - \omega \mp i\varepsilon} d\omega'. \tag{16.36a}$$

If the symbolic identity (16.32) is taken into account, the dispersion relations (16.34) follow from (16.36a).

Consequently, to derive the dispersion relations, we do not require a detailed knowledge of $G_r(t - t')$ [or $G_a(t - t')$]; it is sufficient to know that it is a retarded (or advanced) function, i.e., that it is equal to zero for $t < t'$ (or for $t > t'$) and that its Fourier components fall off sufficiently rapidly as $\omega \to \infty$. The spectral representations (16.26) and the dispersion relations (16.34) for the

Green functions lead directly to spectral representations and dispersion relations for the kinetic coefficients, since the latter can be expressed in terms of retarded Green functions [cf. (15.56a)]. We shall examine these questions in the following section, which is devoted to the fluctuation-dissipation theorems.

We have examined above the spectral representations for the retarded and advanced Green functions, which are applied in the theory of irreversible processes. However, the causal Green functions are also fairly frequently applied in statistical physics, and a diagram technique has been developed for these [60-64, 75]. For the spectral representations for causal Green functions, see in [76, 60, 61].

16.4. Sum Rules

From the existence of the spectral expansions for the Green functions, certain simple identities follow; these are the *sum rules*, which find application in the theory of nonequilibrium processes, for example, in the theories of electrical conductivity and magnetism [1, 3, 4].

For the retarded Green functions, we have, by definition,

$$\langle\langle A \mid B\rangle\rangle_\omega^r = \int_{-\infty}^{\infty} \langle\langle AB(t)\rangle\rangle^r e^{-i\omega t}\, dt. \tag{16.37}$$

Integrating this relation over all ω, we obtain

$$\int_{-\infty}^{\infty} \langle\langle A \mid B\rangle\rangle_\omega^r \, d\omega = \int_{-\infty}^{\infty} d\omega \int_{-\infty}^{0} \frac{1}{i\hbar} \langle [A,\ B(t)] \rangle e^{-i\omega t}\, dt$$

$$= \int_{-\infty}^{0} \frac{2\pi}{i\hbar} \langle [A,\ B(t)] \rangle \delta(t)\, dt.$$

Consequently, for the retarded Green functions, the identity

$$\int_{-\infty}^{\infty} \langle\langle A \mid B\rangle\rangle_\omega^r \, d\omega = \frac{\pi}{i\hbar} \langle [A(0),\ B(0)] \rangle, \tag{16.38}$$

holds; this is called a sum rule. The name is connected with the fact that such relations were first obtained in matrix form and contained sums over quantum states.

For the advanced Green functions, an analogous relation is valid, but with the opposite sign in the right-hand side:

$$\int_{-\infty}^{\infty} \langle\langle A \mid B \rangle\rangle_{\omega}^{a} \, d\omega = -\frac{\pi}{i\hbar} \langle [A(0), \; B(0)] \rangle, \tag{16.38a}$$

To obtain sum rules of another type, we integrate (16.37) by parts, putting $\langle [A, B(t)] \rangle \mid_{t=-\infty} = 0$. We obtain

$$\langle\langle A \mid B \rangle\rangle_{\omega}^{r} = \frac{1}{\hbar\omega} \langle [A(0), \; B(0)] \rangle - \frac{1}{\hbar\omega} \int_{-\infty}^{0} \left\langle \left[A, \; \frac{dB(t)}{dt} \right] \right\rangle e^{-i\omega t} \, dt,$$

and, consequently,

$$\int_{-\infty}^{\infty} \{ \hbar\omega \langle\langle A \mid B \rangle\rangle_{\omega}^{r} - \langle [A(0), \; B(0)] \rangle \} \, d\omega =$$

$$= -2\pi \int_{-\infty}^{0} \left\langle \left[A, \; \frac{dB(t)}{dt} \right] \right\rangle \delta(t) \, dt = -\pi \left\langle \left[A, \; \frac{dB(t)}{dt} \right] \right\rangle_{t=0}, \tag{16.39}$$

and, analogously, for the advanced Green functions,

$$\int_{-\infty}^{\infty} \{ \hbar\omega \langle\langle A \mid B \rangle\rangle_{\omega}^{a} - \langle [A(0), \; B(0)] \rangle \} \, d\omega = \pi \left\langle \left[A, \; \frac{dB(t)}{dt} \right] \right\rangle_{t=0}. \tag{16.39a}$$

The relations (16.39) and (16.39a) are sum rules of the second kind.

It follows from the convergence of the integrals in the left-hand sides of (16.39) and (16.39a) that, as $\omega \to \infty$, the retarded and advanced Green functions behave like $1/\omega$:

$$\langle\langle A \mid B \rangle\rangle_{\omega}^{r,\,a} \approx \langle [A(0), \; B(0)] \rangle \frac{1}{\hbar\omega}, \tag{16.40}$$

provided that the average commutator (or anticommutator) in the right-hand side of relation (16.40) is nonzero; otherwise, they fall off more rapidly with ω.

Continuing the integration by parts in (16.37) up to terms of the n-th order, we obtain

$$\langle\langle A \mid B \rangle\rangle_{\omega}^{r} = \frac{1}{\hbar\omega} \left\{ \langle [A(0), \; B(0)] \rangle + \sum_{k=1}^{n} \left\langle \left[A, \; \frac{d^{k}B(t)}{dt^{k}} \right] \right\rangle_{t=0} \frac{1}{(i\omega)^{k}} \right\} -$$

$$- \frac{i}{(i\omega)^{n+1}} \int_{-\infty}^{0} \left\langle \left[A, \; \frac{d^{n+1}B(t)}{dt^{n+1}} \right] \right\rangle e^{-i\omega t} \, dt, \tag{16.41}$$

whence follows the generalized sum rule of the second kind:

$$\int_{-\infty}^{\infty}\left\{\langle\langle A|B\rangle\rangle_{\omega}^{r}-\frac{1}{\hbar\omega}\left\{\langle[A, B]\rangle-\sum_{k=1}^{n}\left\langle\left[A, \frac{d^{k}B(t)}{dt^{k}}\right]\right\rangle_{t=0}\frac{1}{(i\omega)^{k}}\right\}\right\}\omega^{n+1}\,d\omega=$$

$$=-\frac{\pi}{i^{n}}\left\langle\left[A, \frac{d^{n+1}B(t)}{dt^{n+1}}\right]\right\rangle_{t=0}. \tag{16.41a}$$

The analogous relations for the advanced Green functions differ from (16.41a) only in the sign in the right-hand side of the equality.

From (16.41a), we obtain, as $n \to \infty$, an asymptotic expansion in powers of $1/\omega$, which is the same for the retarded and advanced Green functions. We obtain this same expansion, if in the equation (16.9) for the Green function in the Fourier representation

$$\langle\langle A \mid B\rangle\rangle_{\omega}=\frac{1}{\hbar\omega}\langle[A, B]\rangle-\frac{i}{\omega}\left\langle\left\langle A \,\middle|\, \frac{dB}{dt}\right\rangle\right\rangle_{\omega}$$

we perform successive iteration and eliminate the terms with time derivatives by means of the same equations in which $B \to \frac{dB}{dt}$, $\frac{dB}{dt} \to \frac{d^2B}{dt^2}$, etc.

The sum rules studied above can be applied in the theory of irreversible processes, for example, for the electrical conductivity and magnetic susceptibility tensors (cf. §18) which are expressed in terms of retarded Green functions.

16.5. Symmetry of the Green Functions

We shall now examine the symmetry properties of the correlation functions and Green functions (cf. [1, 77]).

From the definitions (16.1) of the retarded and advanced Green functions, it follows that

$$\langle\langle A(t)\, B(t')\rangle\rangle_{r}=\eta\,\langle\langle B(t')\, A(t)\rangle\rangle_{a}, \tag{16.42}$$

i.e., that the retarded Green functions with a commutator ($\eta = 1$) are equal to the advanced Green functions of the same type with the operators interchanged, while retarded functions with an anti-commutator ($\eta = -1$) on interchange of the operators go over into the advanced functions with the opposite sign.

Going over in (16.1) to the Fourier components of the Green functions

$$\langle\langle A(t)\, B(t')\rangle\rangle_{r}=\frac{1}{2\pi}\int_{-\infty}^{\infty}\langle\langle A \mid B\rangle\rangle_{\omega}^{r}\, e^{-i\omega(t-t')}\,d\omega,$$

$$\langle\langle B(t')\,A(t)\rangle\rangle_a = \frac{1}{2\pi}\int_{-\infty}^{\infty}\langle\langle B\,|\,A\rangle\rangle_\omega^a\, e^{-i\omega(t'-t)}\,d\omega,$$

we obtain for them the symmetry condition:

$$\langle\langle A\,|\,B\rangle\rangle_\omega^r = \eta\langle\langle B\,|\,A\rangle\rangle_{-\omega}^a. \tag{16.43}$$

Here, ω is real throughout.

We use the analytic continuation (16.29) of the Green functions into the complex plane and write the symmetry condition (16.43) in the form

$$\langle\langle A\,|\,B\rangle\rangle_\omega = \eta\,\langle\langle B\,|\,A\rangle\rangle_{-\omega}. \tag{16.43a}$$

We obtain one further useful symmetry property by taking the complex conjugate of the expressions (16.1) for the retarded and advanced Green functions:

$$\langle\langle A(t)\,B(t')\rangle\rangle^* = \eta\,\langle\langle A^+(t)\,B^+(t')\rangle\rangle. \tag{16.44}$$

In the particular case of Hermitian operators and commutator Green functions, we obtain

$$\langle\langle A(t)\,B(t')\rangle\rangle^* = \langle\langle A(t)\,B(t')\rangle\rangle \tag{16.44a}$$

with $A = A^+$, $B = B^+$, and $\eta = 1$. Consequently, commutator Green functions of Hermitian operators are real.

In the case when the equations of motion for the operators are invariant with respect to time reversal, i.e., with respect to the replacements

$$t \to -t, \quad t' \to -t', \quad i \to -i,$$

the Green functions have certain further symmetry properties.† For example, the equations of motion of particles in the absence of a magnetic field are symmetric with respect to time reversal.

†More accurately, by the time-reversal operation, we mean the action of the antiunitary operator TK, where T is the operator of the transformation $t \to -t$, and K is the complex-conjugation operator acting on all the state vectors (cf. [143], Chapter 13). In the second-quantized representation, under the time-reversal operation we must reverse the momenta, and change all creation operators into annihilation operators and vice versa. In addition, we must reverse the order of the operators (cf. [144], Chapter 12).

Let the equations of motion for A and B be invariant with respect to time reversal, for which $A \to \varepsilon_A A$ and $B \to \varepsilon_B B$, where $\varepsilon_A, \varepsilon_B = \pm 1$, depending on the polarity of the operators on reversal of the velocities. We shall examine the spectral expansion

$$\langle A(t) B(t') \rangle = \frac{1}{2\pi} \int_{-\infty}^{\infty} J_{AB}(\omega) e^{i\omega(t-t')} d\omega. \tag{16.45}$$

With the replacement $t \to -t$, $t' \to -t'$, $i \to -i$, the left-hand side of this equality is multiplied by $\varepsilon_A \varepsilon_B$, and in the right-hand side, $J_{AB}(\omega)$ goes over into $J^*_{AB}(\omega)$ (as a result of replacing i by $-i$). Consequently, in the case under consideration,

$$\begin{aligned} J_{AB}(\omega) &= J^*_{AB}(\omega) \varepsilon_A \varepsilon_B, \\ J_{AB}(\omega) &= J^*_{AB}(\omega) \quad \text{for} \quad \varepsilon_A \varepsilon_B = 1, \end{aligned} \tag{16.46}$$

i.e., the spectral intensity is real for operators of the same parity.

Comparing (16.45) with the relation conjugate to it

$$\langle B^+(t') A^+(t) \rangle = \frac{1}{2\pi} \int_{-\infty}^{\infty} J_{AB}(\omega) e^{-i\omega(t-t')} d\omega, \tag{16.46a}$$

where we have assumed that the spectral intensity is real, we see that

$$\langle A(t) B(t') \rangle = \langle B^+(t) A^+(t') \rangle, \tag{16.47}$$

and, consequently,

$$\langle\langle A | B \rangle\rangle_\omega = \langle\langle B^+ | A^+ \rangle\rangle_\omega. \tag{16.48}$$

In the case when the magnetic field is nonzero, the spectral intensity of the time correlation functions will no longer be real, but since the equations of motion are invariant with respect to time reversal with simultaneous reversal of the magnetic field direction ($\mathbf{H} \to -\mathbf{H}$), the spectral intensity possesses the symmetry property

$$J_{AB,\,H}(\omega) = J^*_{AB,\,-H}(\omega) \varepsilon_A \varepsilon_B \tag{16.49}$$

in place of (16.46) in the absence of the field. Consequently, the Green functions possess the symmetry property

$$\begin{aligned} \langle\langle B^+(t) A^+(t') \rangle\rangle_H &= \langle\langle A(t) B(t') \rangle\rangle_{-H} \varepsilon_A \varepsilon_B, \\ \langle\langle B^+ | A^+ \rangle\rangle_{\omega,\,H} &= \langle\langle A | B \rangle\rangle_{\omega,\,-H} \varepsilon_A \varepsilon_B \end{aligned} \tag{16.50}$$

in place of (16.47) and (16.48).

In §17, we shall need the symmetry relations studied in this subsection to establish the properties of the complex conductivity and susceptibility.

§ 17. Fluctuation-Dissipation Theorem and Dispersion Relations

Relations between, on the one hand, the kinetic coefficients or susceptibilities that determine the response of a system to an external perturbation and, on the other, the equilibrium fluctuations are called fluctuation-dissipation theorems. The Kubo formulas (15.76) and (15.77) for the electrical conductivity are a particular case of these theorems. The fluctuation-dissipation theorems were first formulated for the general case by Callen and Welton [22] as a generalization of the Nyquist theorem [78] well-known from the theory of electrical noise in linear circuits. The fluctuation-dissipation theorems were further generalized and applied to the thermodynamics of irreversible processes in the papers [79-81, 1, 11].

In §15, expressions were obtained for the kinetic coefficients (or for the generalized susceptibility and conductivity) in terms of retarded Green functions, and the general properties of the latter were investigated in §16. The Callen–Welton fluctuation-dissipation theorem, the dispersion relations, sum rules and symmetry properties for the kinetic coefficients and the generalized susceptibility follow directly from these properties.

17.1. Dispersion Relations, Sum Rules, and Onsager's Reciprocity Relations for the Generalized Susceptibility

Suppose that a system is acted upon by a time-dependent perturbation of the mechanical type, which is switched on adiabatically and is described by adding to the Hamiltonian the term

$$H_t^1 = -\sum_{j=1}^{n} F_j(t)\,\alpha_j, \tag{17.1}$$

$$F_j(t) \sim e^{\varepsilon t} \quad \text{as} \quad t \to -\infty, \ \varepsilon > 0,$$

where α_j are dynamic variables, and $F_j(t)$ is the "force" with which the external field acts on the variable α_j, i.e., its conjugate force; $F_j(t)$ is assumed to be a known function of time.

The perturbation (17.1) can be represented conveniently in the form of a scalar product of n-dimensional vectors:

$$H'_t = -(\boldsymbol{F}(t) \cdot \boldsymbol{\alpha}), \tag{17.1a}$$

where

$$\boldsymbol{\alpha} = (\alpha_1, \ldots, \alpha_n), \qquad \boldsymbol{F}(t) = (F_1(t), \ldots, F_n(t)).$$

Taking the perturbation in the form (17.1), we assume that $F_j^0 = 0$ (or $\langle \alpha_j \rangle_0 = 0$) in a state of statistical equilibrium. If this is not so, and $F_j^0 \neq 0$ in statistical equilibrium, we must take for the perturbation not (17.1), but the deviation of the interaction energy from its equilibrium value:

$$H_t^1 = -\sum_{j=1}^{n} (F_j(t) - F_j^0)\alpha_j. \tag{17.1b}$$

Below, for brevity, we shall write the perturbation in the form (17.1), assuming that, in the case when $F_j^0 \neq 0$, the equilibrium values of this parameter have been subtracted from the $F_j(t)$.

According to (15.47), the response of a system to the perturbation (17.1) is equal to

$$\langle \boldsymbol{\alpha} \rangle = \langle \boldsymbol{\alpha} \rangle_0 + \int_{-\infty}^{t} \varkappa(t-t') \cdot \boldsymbol{F}(t')\, dt', \tag{17.2}$$

where

$$\varkappa(t-t') = -\langle\langle \boldsymbol{\alpha}(t)\, \boldsymbol{\alpha}(t') \rangle\rangle \tag{17.2a}$$

is the generalized reaction matrix with components

$$\varkappa_{ik}(t-t') = -\langle\langle \alpha_i(t)\, \alpha_k(t') \rangle\rangle. \tag{17.2b}$$

The double brackets denote the retarded Green function (15.48) in the quantum case, and (15.19) in the classical case.

Since the retarded Green function is nonzero only for positive arguments, we have always

$$\varkappa(t-t') = 0 \quad \text{for} \quad t < t', \tag{17.3}$$

which expresses the causality principle: the response of a system cannot precede in time the perturbation which gives rise to it. In the phenomenological theory, when there are no explicit expressions (17.2a) for the reaction matrix, the causality principle (17.3) is taken as the basic physical postulate of the theory [82].

We shall expand the functions $\mathbf{F}(t)$ and $\langle \boldsymbol{\alpha} \rangle$ in Fourier integrals:

$$\langle \alpha \rangle = \langle \alpha \rangle_0 + \frac{1}{2\pi} \int_{-\infty}^{\infty} \alpha(\omega) e^{-i\omega t} \, d\omega,$$
$$\boldsymbol{F}(t) = \frac{1}{2\pi} \int_{-\infty}^{\infty} \boldsymbol{F}(\omega) e^{-i\omega t} \, d\omega, \tag{17.4}$$

where $\boldsymbol{\alpha}(\omega)$ and $\mathbf{F}(\omega)$ are the Fourier components

$$\alpha(\omega) = \int_{-\infty}^{\infty} (\langle \alpha \rangle - \langle \alpha \rangle_0) e^{i\omega t} \, dt,$$
$$\boldsymbol{F}(\omega) = \int_{-\infty}^{\infty} \boldsymbol{F}(t) e^{i\omega t} \, dt. \tag{17.4a}$$

Substituting the expansions (17.4) in (17.2), we obtain in place of the integral relation the algebraic linear-response equation

$$\alpha(\omega) = \varkappa(\omega) \cdot \boldsymbol{F}(\omega), \tag{17.5}$$

where

$$\varkappa_{ik}(\omega) = \int_{-\infty}^{\infty} \varkappa_{ik}(t) e^{i\omega t} \, dt = -\langle\langle \alpha_i \mid \alpha_k \rangle\rangle_\omega = \int_0^{\infty} \int_0^{\beta} \langle \dot{\alpha}_k \alpha_i (t + i\hbar\lambda) \rangle e^{i\omega t - \varepsilon t} \, dt \, d\lambda \tag{17.5a}$$

is the generalized susceptibility matrix.† Formula (17.5a) expresses Kubo's fluctuation-dissipation theorem.

---

†Occasionally, for example in [82], in addition to the generalized susceptibility matrix $\varkappa(\omega)$, a generalized admittance matrix

$$Y(\omega) = -i\omega\varkappa(\omega)$$

and its inverse, the generalized impedance matrix

$$Z(\omega) = i\omega^{-1}\varkappa^{-1}(\omega).$$

are introduced. Then the linear-response equation (17.5) takes the form

$$\alpha(\omega) = \frac{1}{-i\omega} Y(\omega) \cdot \boldsymbol{F}(\omega), \text{ where } Y(\omega) = i\omega \langle\langle \alpha \mid \alpha \rangle\rangle_\omega.$$

Since the generalized susceptibility can be expressed in terms of retarded Green functions, all the properties of the latter studied in §16 can be extended to it.

Decomposing $\varkappa(\omega)$ into its real and imaginary parts,

$$\varkappa(\omega)=\varkappa'(\omega)+i\varkappa''(\omega), \tag{17.6}$$

we obtain from (16.34) the **dispersion relations for the generalized susceptibility** or **Kramers – Krönig relations**:

$$\varkappa'(\omega)=\frac{1}{\pi}\mathscr{P}\int_{-\infty}^{\infty}\frac{\varkappa''(u)}{u-\omega}\,du,\qquad \varkappa''(\omega)=-\frac{1}{\pi}\mathscr{P}\int_{-\infty}^{\infty}\frac{\varkappa'(u)}{u-\omega}\,du. \tag{17.7}$$

Similar relations between the real and imaginary parts of the refractive index were first obtained in classical electrodynamics by Kramers [83] and Krönig [84] in 1926-1927.

It follows from the reality of $\boldsymbol{\alpha}$ and $\mathbf{F}(t)$ that

$$\alpha(\omega)=\alpha^*(-\omega),\quad F(\omega)=F^*(-\omega), \tag{17.8}$$

and, therefore,

$$\varkappa(\omega)=\varkappa^*(-\omega), \tag{17.9}$$

whence it follows that the real part of the generalized susceptibility $\varkappa(\omega)$ is even, and the imaginary part odd, in ω:

$$\varkappa'(\omega)=\varkappa'(-\omega),\quad \varkappa''(\omega)=-\varkappa''(-\omega). \tag{17.9a}$$

Using (17.9a), we can write the Kramers–Krönig relations in the form

$$\begin{aligned}\varkappa'(\omega)&=\frac{1}{\pi}\mathscr{P}\int_{0}^{\infty}\varkappa''(u)\frac{2u}{u^2-\omega^2}\,du,\\ \varkappa''(\omega)&=-\frac{1}{\pi}\mathscr{P}\int_{0}^{\infty}\varkappa'(u)\frac{2\omega}{u^2-\omega^2}\,du.\end{aligned} \tag{17.10}$$

From the sum rules (16.38) and (16.39) for the retarded Green functions, there follow sum rules for the generalized susceptibility:

$$\begin{aligned}\frac{1}{\pi}\int_{-\infty}^{\infty}\varkappa_{ik}(\omega)\,d\omega&=\frac{i}{\hbar}\langle[\alpha_i,\ \alpha_k]\rangle,\\ \frac{1}{\pi}\int_{-\infty}^{\infty}\{\hbar\omega\varkappa_{ik}(\omega)+\langle[\alpha_i,\ \alpha_k]\rangle\}\,d\omega&=\left\langle\left[\alpha_i,\ \frac{d\alpha_k}{dt}\right]_{t=0}\right\rangle.\end{aligned} \tag{17.11}$$

It follows from the symmetry property (16.50) of the Green functions that, for the generalized susceptibility,

$$\varkappa_{ik}(\omega, \boldsymbol{H}) = \varkappa_{ki}(\omega, -\boldsymbol{H})\, \varepsilon_i \varepsilon_k, \qquad \varepsilon_i,\ \varepsilon_k = \pm 1, \tag{17.12}$$

and in the absence of a magnetic field,

$$\varkappa_{ik}(\omega) = \varkappa_{ki}(\omega)\, \varepsilon_i \varepsilon_k. \tag{17.12a}$$

Decomposing the susceptibility into symmetric and antisymmetric parts,

$$\varkappa_{ik}^{s} = \frac{1}{2}(\varkappa_{ik} + \varkappa_{ki}), \qquad \varkappa_{ik}^{a} = \frac{1}{2}(\varkappa_{ik} - \varkappa_{ki}), \tag{17.13}$$

we see that $\varkappa^{s}$ is even and $\varkappa^{a}$ is odd with respect to reversal of the magnetic field **H**:

$$\varkappa_{ik}^{s}(\omega, \boldsymbol{H}) = \varkappa_{ik}^{s}(\omega, -\boldsymbol{H}), \quad \varkappa_{ik}^{a}(\omega, \boldsymbol{H}) = -\varkappa_{ki}^{a}(\omega, -\boldsymbol{H}) \quad \text{for } \varepsilon_i \varepsilon_k = 1. \tag{17.14}$$

These symmetry properties (17.12) and (17.14) are called the Onsager reciprocity relations for the generalized susceptibility. They stem, thus, from the theory of the linear response of a system to mechanical perturbations and from the invariance of the equations of motion under time reversal with the replacement $\mathbf{H} \to -\mathbf{H}$. They are also valid for the kinetic coefficients, irrespective of the type of perturbation – mechanical or thermal – giving rise to the irreversible process.

The reciprocity relations for the kinetic coefficients were established in 1931 by Onsager [85], who started from a hypothesis on the nature of the attenuation of the fluctuations in time; namely, he postulated that they obey the same equations as the average values, and used the invariance of the equations of motion of the particles under time reversal and reversal of the magnetic field.

The Onsager reciprocity relations reflect, on the macroscopic level, the invariance of the microscopic equations of motion. They have very great significance for the theory of irreversible processes and, in effect, form the basis of the whole of nonequilibrium thermodynamics, a good account of which can be found in the monograph by de Groot and Mazur [82], and also in the books [86-90]. We shall come back to a discussion of the Onsager reciprocity relations in §17.3 and, for the case of thermal perturbations, in Chapter IV.

17.2. The Callen–Welton Fluctuation-Dissipation Theorem for the Generalized Susceptibility

The fluctuation-dissipation theorems relate the characteristics of a dissipative process [for example, the complex susceptibility $\varkappa_{ik}(\omega)$ or conductivity $\sigma_{\alpha\beta}(\omega)$] to the equilibrium fluctuations in the system. Thus, they express nonequilibrium properties in terms of equilibrium properties.

In §17.1, we studied Kubo's fluctuation theorem (17.5a). The Callen–Welton fluctuation-dissipation theorem is another representation of the Kubo formula. It follows from the theory of linear response and from the spectral representation (16.24) for the retarded Green function. For the complex susceptibility $\varkappa_{ik}(\omega)$, we obtain

$$\varkappa_{ik}(\omega) = -\frac{1}{2\pi\hbar}\int_{-\infty}^{\infty}(e^{\hbar u/\theta}-1)\,J_{\alpha_k\alpha_i}(u)\frac{du}{\omega-u+i\varepsilon} =$$

$$= \frac{i}{2\hbar}(e^{\hbar\omega/\theta}-1)\,J_{\alpha_k\alpha_i}(\omega) - \frac{1}{2\pi\hbar}\,\mathcal{P}\int_{-\infty}^{\infty}(e^{\hbar u/\theta}-1)\,J_{\alpha_k\alpha_i}(u)\,\frac{du}{\omega-u}, \tag{17.15}$$

where

$$J_{\alpha_k\alpha_i}(\omega) = \int_{-\infty}^{\infty}\langle(\alpha_k(t)-\langle\alpha_k\rangle_0)(\alpha_i(t')-\langle\alpha_i\rangle_0)\rangle\,e^{-i\omega(t-t')}\,dt \tag{17.15a}$$

is the Fourier component of the time correlation function coupling α_k and α_i. In the time correlation function in (17.15a), the averaging is performed over an equilibrium ensemble; we shall not denote this by a subscript 0 on the brackets, but use this subscript only for $\langle\alpha_k\rangle_0$.

Formula (17.15) represents the complex susceptibility in terms of the spectral intensity (17.15a) of the equilibrium fluctuations, and expresses the Callen – Welton fluctuation-dissipation theorem. In the classical case, passing to the limit $\hbar \to 0$ in (17.15), we obtain

$$\varkappa_{ik}(\omega) = -\frac{1}{2\pi\theta}\int_{-\infty}^{\infty}J_{\alpha_k\alpha_i}(u)\frac{u\,du}{\omega-u+i\varepsilon}. \tag{17.15b}$$

We express the spectral intensity $J_{\alpha_k\alpha_i}(\omega)$ in terms of $J_{\alpha_i\alpha_k}(-\omega)$

by means of (16.16a):

$$J_{\alpha_k\alpha_i}(\omega) = J_{\alpha_i\alpha_k}(-\omega)\, e^{-\hbar\omega/\theta}. \tag{17.16}$$

We introduce the symmetrized time correlation functions

$$\{\alpha_k(t),\ \alpha_i(t')\} = \frac{1}{2}\left(\langle\alpha_k(t)\,\alpha_i(t')\rangle + \langle\alpha_i(t')\,\alpha_k(t)\rangle\right), \tag{17.17}$$

which are symmetric with respect to i and k with the replacement $t \to t'$, and have the Fourier components

$$J_{\{\alpha_k\alpha_i\}}(\omega) = \frac{1}{2}\left\{J_{\alpha_k\alpha_i}(\omega) + J_{\alpha_i\alpha_k}(-\omega)\right\} = J_{\alpha_k\alpha_i}(\omega)\,\frac{1}{2}\left(1 + e^{\hbar\omega/\theta}\right), \tag{17.17a}$$

as follows from (17.16). Using (17.17a), we write the fluctuation-dissipation theorem (17.15) in the form

$$\varkappa_{ik}(\omega) = -\frac{1}{\pi\hbar}\int_{-\infty}^{\infty} \tanh\frac{\hbar u}{2\theta}\, J_{\{\alpha_k\alpha_i\}}(u)\,\frac{du}{\omega - u + i\varepsilon}. \tag{17.18}$$

Up to this point, we have expressed the generalized susceptibility in terms of time correlation functions. We can also obtain the inverse relations, by expressing the fluctuations in terms of the generalized susceptibility. It was in such a form that the fluctuation-dissipation theorem was first obtained by Callen and Welton.

It follows from formula (17.18) that

$$\varkappa^*_{ik}(\omega) - \varkappa_{ki}(\omega) =$$

$$= -\frac{1}{\pi\hbar}\int_{-\infty}^{\infty} \tanh\frac{\hbar u}{2\theta}\left\{ J^*_{\{\alpha_k\alpha_i\}}(u)\,\frac{1}{\omega - u - i\varepsilon} - J_{\{\alpha_i\alpha_k\}}(u)\,\frac{1}{\omega - u + i\varepsilon}\right\} du.$$

Taking into account that, because α_i and α_k are assumed Hermitian

$$J^*_{\alpha_k\alpha_i}(u) = J_{\alpha_i\alpha_k}(u),$$

we obtain

$$\varkappa^*_{ik}(\omega) - \varkappa_{ki}(\omega) = \frac{2}{i\hbar}\tanh\frac{\hbar\omega}{2\theta}\, J_{\{\alpha_i\alpha_k\}}(\omega), \tag{17.19}$$

i.e.,

$$\operatorname{Im}\varkappa^s_{ik}(\omega) = \frac{1}{\hbar}\tanh\frac{\hbar\omega}{2\theta}\operatorname{Re} J_{\{\alpha_i\alpha_k\}}(\omega), \qquad \operatorname{Re}\varkappa^a_{ik}(\omega) = \frac{1}{\hbar}\tanh\frac{\hbar\omega}{2\theta}\operatorname{Im} J_{\{\alpha_i\alpha_k\}}(\omega), \tag{17.19a}$$

and the inverse relation

$$\{(\alpha_i(t)-\langle\alpha_i\rangle_0),\ (\alpha_k(t')-\langle\alpha_k\rangle_0)\} = \\ = \frac{i\hbar}{4\pi}\int_{-\infty}^{\infty}\left(\varkappa_{ik}^{*}(\omega)-\varkappa_{ki}(\omega)\right)\coth\frac{\hbar\omega}{2\theta}\,e^{i\omega(t-t')}\,d\omega, \qquad (17.19b)$$

whence follows, for t = t',

$$\{(\alpha_i-\langle\alpha_i\rangle_0),\ (\alpha_k-\langle\alpha_k\rangle_0)\} = \frac{i\hbar}{4\pi}\int_{-\infty}^{\infty}\left(\varkappa_{ik}^{*}(\omega)-\varkappa_{ki}(\omega)\right)\coth\frac{\hbar\omega}{2\theta}\,d\omega. \qquad (17.19c)$$

For the particular case of one variable, we obtain

$$\langle(\alpha-\langle\alpha\rangle_0)^2\rangle = \frac{\hbar}{2\pi}\int_{-\infty}^{\infty}\varkappa''(\omega)\coth\frac{\hbar\omega}{2\theta}\,d\omega. \qquad (17.19d)$$

The factor $\coth(\hbar\omega/2\theta)$ in the integrands of the formulas (19b), (19c), and (19d) are proportional to the average energy of an oscillator with frequency ω. Since

$$2\bar{n}+1 = \tanh\frac{\hbar u}{2\theta}, \qquad \bar{n} = \frac{1}{e^{\hbar u/\theta}-1}.$$

The relations (17.19) and (17.19c) were obtained by Callen, Barash and Jackson [81], and formula (17.19d) by Callen and Welton [22]. The relations (17.19c) and (17.19d) have the form of sum rules (cf. §16.4).

17.3. Linear Relations between the Fluxes and Forces. Kinetic Coefficients and Their Properties

The rate of change with time $\dot{\alpha}_j$ of a dynamic variable α_j is called the flux operator (or dynamic variable), and its average value $\langle\dot{\alpha}_j\rangle$ is called a flux. In a state of statistical equilibrium there are no fluxes, since

$$\langle\dot{\alpha}_j\rangle_0 = \frac{d}{dt}\langle\alpha_j\rangle_0 = 0;$$

they are, therefore, characteristic of a nonequilibrium state.

It is convenient to introduce an n-dimensional vector of the flux operators

$$\dot{\alpha} = (\dot{\alpha}_1,\ \ldots,\ \dot{\alpha}_n).$$

By means of the theory of linear response, we find the relations between the fluxes and the forces **F** (t), when the latter are sufficiently small.

According to (15.47), under the influence of the perturbation (17.1), the fluxes

$$\langle \dot{\alpha}_i \rangle = \sum_k \int_{-\infty}^{t} L_{ik}(t-t')\, F_k(t')\, dt' \tag{17.20}$$

or

$$\langle \dot{\alpha} \rangle = \int_{-\infty}^{t} L(t-t') \cdot \boldsymbol{F}(t')\, dt', \tag{17.20a}$$

arise in the system, where

$$L(t-t') = -\langle\langle \dot{\boldsymbol{\alpha}}(t)\, \boldsymbol{\alpha}(t') \rangle\rangle \tag{17.20b}$$

is a tensor with components

$$L_{ik}(t-t') = -\langle\langle \dot{\alpha}_i(t)\, \alpha_k(t') \rangle\rangle = \int_0^{\beta} \langle \dot{\alpha}_k(t'-i\hbar\lambda)\, \dot{\alpha}_i(t) \rangle\, d\lambda. \tag{17.20c}$$

The relations (17.20) and (17.20a) are called linear relations between the forces and fluxes.

The linear relations (17.20) are integral relations, of the retarded type. Expanding the forces $F_j(t)$ and fluxes $\langle \dot{\alpha}_i \rangle$ in Fourier integrals, we obtain algebraic relations between the Fourier components of the forces and of the fluxes:

$$\dot{\alpha}(\omega) = L(\omega) \cdot \boldsymbol{F}(\omega), \tag{17.21}$$

where

$$L_{ik}(\omega) = \int_{-\infty}^{\infty} L_{ik}(t)\, e^{i\omega t}\, dt = -\langle\langle \dot{\alpha}_i \mid \alpha_k \rangle\rangle_{\omega} =$$

$$= \int_0^{\infty} \int_0^{\beta} \langle \dot{\alpha}_k \dot{\alpha}_i(t+i\hbar\lambda) \rangle\, e^{i\omega t - \varepsilon t}\, dt\, d\lambda \tag{17.21a}$$

is the tensor of the kinetic coefficients for periodic

processes with frequency ω, and

$$\dot{\alpha}(\omega) = \int_{-\infty}^{\infty} \langle \dot{\alpha} \rangle e^{i\omega t}\, dt \tag{17.21b}$$

are the Fourier components of the fluxes.

The function $L_{ik}(t - t')$ differs appreciably from zero only in the region $|t - t'| \sim \tau$, where τ is the correlation time, and tends to zero as $|t - t'| \to \infty$. If the forces $F_k(t')$ vary sufficiently slowly with time, so that their change in time τ can be neglected, we can take the slowly varying function $F_k(t')$ at the point $t' = t$ outside the integral in the right-hand side of (17.20). Then the retarded linear relations (17.20) go over into the usual linear relations connecting the instantaneous values of the forces and fluxes:

$$\langle \dot{\alpha}_i \rangle = \sum_k L_{ik} F_k(t), \tag{17.22}$$

where

$$L_{ik} = \int_0^{\infty} e^{-\varepsilon t} L_{ik}(t)\, dt = -\int_0^{\infty} e^{-\varepsilon t} \frac{1}{i\hbar} \langle [\dot{\alpha}_i(t), \alpha_k] \rangle\, dt =$$

$$= \int_0^{\infty}\int_0^{\beta} e^{-\varepsilon t} \langle \dot{\alpha}_k \dot{\alpha}_i(t + i\hbar\lambda) \rangle\, dt\, d\lambda = \int_{-\infty}^{0}\int_0^{\beta} e^{\varepsilon t} \langle \dot{\alpha}_i \dot{\alpha}_k(t + i\hbar\lambda) \rangle\, dt\, d\lambda \qquad (\varepsilon > 0) \tag{17.22a}$$

is the **tensor of the kinetic coefficients** for stationary processes.

The generalized susceptibility $\varkappa_{ik}(\omega)$ (17.5a) introduced in the preceding subsection is connected with the kinetic coefficients $L_{ik}(\omega)$ (17.21a) by the relation

$$L_{ik}(\omega) = \frac{1}{i\hbar} \langle [\alpha_i, \alpha_k] \rangle - i\omega \varkappa_{ik}(\omega), \tag{17.23}$$

which is obtained by integrating (17.21a) by parts. It follows from (17.23) that the imaginary part of the susceptibility is related to the real part of the kinetic coefficient, and vice versa.

Dispersion relations for the kinetic coefficients $L_{ik}(\omega)$ follows from (16.34):

$$L'_{ik}(\omega) = \frac{1}{\pi} \mathscr{P} \int_{-\infty}^{\infty} \frac{L''_{ik}(u)}{u - \omega}\, du,$$

$$L''_{ik}(\omega) = -\frac{1}{\pi}\mathcal{P}\int_{-\infty}^{\infty}\frac{L'_{ik}(u)}{u-\omega}\,du, \tag{17.24}$$

$$L_{ik}(\omega) = L'_{ik}(\omega) + iL''_{ik}(\omega);$$

they have the same form as for $\varkappa_{ik}(\omega)$.

From the reality of $\langle\dot{\alpha}\rangle$ and $\mathbf{F}(t)$, the kinetic coefficients have the properties:

$$L_{ik}(\omega) = \overset{*}{L}_{ik}(-\omega). \tag{17.25}$$

If these are taken into account, (17.24) is transformed to

$$\begin{aligned} L'_{ik}(\omega) &= \frac{1}{\pi}\mathcal{P}\int_0^{\infty} L''_{ik}(u)\,\frac{2u}{u^2-\omega^2}\,du, \\ L''_{ik}(\omega) &= -\frac{1}{\pi}\mathcal{P}\int_0^{\infty} L'_{ik}(u)\,\frac{2\omega}{u^2-\omega^2}\,du. \end{aligned} \tag{17.24a}$$

The sum rules (16.38) and (16.39) give for the kinetic coefficients the relations

$$\begin{aligned} &\frac{1}{\pi}\int_{-\infty}^{\infty} L_{ik}(\omega)\,d\omega = \frac{i}{\hbar}\langle[\dot{\alpha}_i,\ \alpha_k]\rangle, \\ &\frac{1}{\pi}\int_{-\infty}^{\infty}\{\hbar\omega L_{ik}(\omega) + \langle[\dot{\alpha}_i,\ \alpha_k]\rangle\}\,d\omega = \left\langle\left[\frac{d\alpha_i}{dt},\ \frac{d\alpha_k}{dt}\right]\right\rangle_{t=0}. \end{aligned} \tag{17.26}$$

From the symmetry properties (16.50) of the Green functions, which are associated with the invariance of the equation of motion under time reversal when the direction of the magnetic field is reversed, symmetry properties follow for the frequency-dependent kinetic coefficients

$$L_{ik}(\omega,\ \boldsymbol{H}) = L_{ki}(\omega,\ -\boldsymbol{H})\,\varepsilon_i\varepsilon_k \tag{17.27}$$

and for the stationary kinetic coefficients ($\omega = 0$)

$$L_{ik}(\boldsymbol{H}) = L_{ki}(-\boldsymbol{H})\,\varepsilon_i\varepsilon_k, \qquad (L_{ik}(\boldsymbol{H}) = L_{ik}(0,\ \boldsymbol{H})), \tag{17.27a}$$

whence, for their symmetric and antisymmetric parts

$$L^s_{ik} = \frac{1}{2}(L_{ik} + L_{ki}), \qquad L^a_{ik} = \frac{1}{2}(L_{ik} - L_{ki})$$

we have the symmetry conditions

$$L^s_{ik}(\omega, \boldsymbol{H}) = L^s_{ik}(\omega, -\boldsymbol{H})\,\varepsilon_i\varepsilon_k,$$
$$L^a_{ik}(\omega, \boldsymbol{H}) = -L^a_{ik}(\omega, -\boldsymbol{H})\,\varepsilon_i\varepsilon_k. \tag{17.27b}$$

The symmetry properties (17.27) of the kinetic coefficients are the Onsager reciprocity relations (or Onsager's theorem). Onsager introduced them for the stationary case [85].

In this subsection, we have given a derivation of Onsager's theorem for the case of mechanical perturbations. We shall study it for thermal perturbations in Chapter IV.

The Callen–Welton fluctuation-dissipation theorem for the kinetic coefficient (17.21a) follows from (16.24):

$$L_{ik}(\omega) = -\frac{1}{2\pi\hbar}\int_{-\infty}^{\infty}(e^{\hbar u/\theta}-1)\,J_{\alpha_k\dot{\alpha}_i}(u)\frac{du}{\omega-u+i\varepsilon}, \tag{17.28}$$

where

$$J_{\alpha_k\dot{\alpha}_i}(\omega) = \int_{-\infty}^{\infty}\langle\alpha_k(t)\,\dot{\alpha}_i(t')\rangle\,e^{-i\omega(t-t')}\,dt, \tag{17.28a}$$

or, in another form,

$$L_{ik}(\omega) = -\frac{1}{\pi\hbar}\int_{-\infty}^{\infty}\tanh\frac{\hbar u}{2\theta}\,J_{\{\alpha_k\dot{\alpha}_i\}}(u)\frac{du}{\omega-u+i\varepsilon}, \tag{17.29}$$

where

$$J_{\{\alpha_k\dot{\alpha}_i\}}(\omega) = \frac{1}{2}\left\{J_{\alpha_k\dot{\alpha}_i}(\omega) + J_{\dot{\alpha}_i\alpha_k}(-\omega)\right\} = J_{\alpha_k\dot{\alpha}_i}(\omega)\frac{1}{2}(1+e^{\hbar\omega/\theta}) \tag{17.29a}$$

are the Fourier components of the symmetrized time correlation functions

$$\{\alpha_k(t),\,\dot{\alpha}_i(t')\} = \frac{1}{2}(\langle\alpha_k(t)\,\dot{\alpha}_i(t')\rangle + \langle\dot{\alpha}_i(t')\,\alpha_k(t)\rangle). \tag{17.29b}$$

From (17.29) follow other formulations of the fluctuation-dissipation theorem for kinetic coefficients

$$\overset{*}{L}_{ik}(\omega) - L_{ki}(\omega) = \frac{2}{i\hbar}\tanh\frac{\hbar\omega}{2\theta}\,J_{\{\alpha_i\dot{\alpha}_k\}}(\omega) \tag{17.30}$$

and

$$\{\dot{\alpha}_i(t),\ \alpha_k(t') - \langle\alpha_k\rangle_0\} = \frac{i\hbar}{4\pi}\int\limits_{-\infty}^{\infty}\left(L^*_{ik}(\omega) - L_{ki}(\omega)\right)\coth\frac{\hbar\omega}{2\theta}\, e^{i\omega(t-t')}\, d\omega, \tag{17.30a}$$

analogous to the relations (17.19) and (17.19b) for the generalized susceptibility.

Up to this point, we have considered the general case of the response of a system to the perturbation (17.1). We cite now the Callen–Welton fluctuation-dissipation theorem, Onsager relations and sum rules for the particular case of the electrical conductivity, when the perturbation has the form (15.63). In this case,

$$\sigma_{\alpha\beta}(\omega) = \frac{\tanh(\beta\hbar\omega/2)}{\hbar\omega} J_{\{\alpha,\beta\}}(\omega) - \frac{i}{\pi\hbar}\mathcal{P}\int\limits_{-\infty}^{\infty}\frac{1}{u} J_{\{\alpha,\beta\}}(u)\tanh\frac{\beta\hbar u}{2}\frac{du}{u-\omega}, \tag{17.31}$$

where

$$J_{\{\alpha,\beta\}}(\omega) = \frac{1}{2}\int\limits_{-\infty}^{\infty}\left(\langle j_\alpha(0)\, j_\beta(t)\rangle + \langle j_\beta(t)\, j_\alpha(0)\rangle\right) e^{i\omega t}\, dt$$

$$= \int\limits_{-\infty}^{\infty}\{j_\alpha(0),\ j_\beta(t)\}\, e^{i\omega t}\, dt. \tag{17.31a}$$

(In §15, the quantity j_α was denoted by J_α.)

Thus, the electrical conductivity tensor is related to the equilibrium current fluctuations.

From (17.31), we obtain relations analogous to (17.19a) for the symmetric and antisymmetric parts of the conductivity tensor:

$$\begin{aligned} \operatorname{Re}\sigma^s_{\alpha\beta}(\omega) &= \frac{\tanh(\beta\hbar\omega/2)}{\hbar\omega}\int\limits_{-\infty}^{\infty}\{j_\alpha(0),\ j_\beta(t)\}\cos\omega t\, dt, \\ \operatorname{Im}\sigma^a_{\alpha\beta}(\omega) &= \frac{\tanh(\beta\hbar\omega/2)}{\hbar\omega}\int\limits_{-\infty}^{\infty}\{j_\alpha(0),\ j_\beta(t)\}\sin\omega t\, dt, \end{aligned} \tag{17.32}$$

whence, as $\omega \to 0$, Nyquist's theorem follows:

$$\operatorname{Re}\sigma^s_{\alpha\beta}(0) = \frac{\beta}{2}\int\limits_{-\infty}^{\infty}\{j_\alpha(0)\, j_\beta(t)\}\, dt. \tag{17.32a}$$

The theorem was first obtained by Nyquist as a relation between potential-difference fluctuations and the resistance in linear electrical circuits.

The symmetry properties of the electrical conductivity tensor follow from the symmetry of the Green functions. For example, Onsager reciprocity relations for the electrical conductivity stem from (16.50):

$$\sigma_{\alpha\beta}(\omega, \boldsymbol{H}) = \sigma_{\beta\alpha}(\omega, -\boldsymbol{H}). \tag{17.33}$$

It follows from the sum rule (16.38) that

$$\int_{-\infty}^{\infty} \left\langle\left\langle j_\alpha \middle| \sum_j e_j x_{j\beta} \right\rangle\right\rangle_\omega^r d\omega = \frac{1}{2i\hbar} \left\langle \left[j_\alpha, \sum_j e_j x_{j\beta} \right] \right\rangle = \frac{e^2 N}{2m} \delta_{\alpha\beta}, \tag{17.34}$$

where N is the total number of particles. In (17.34), we have used the commutation relations between the coordinates and momenta:

$$[x_{j\beta}, p_{k\alpha}] = i\hbar\, \delta_{jk}^{\alpha\beta}.$$

The relation (17.34) can be written in the form of a sum rule for $\sigma_{\alpha\beta}(\omega)$:

$$\frac{2}{\pi} \int_{-\infty}^{\infty} \sigma_{\alpha\beta}(\omega)\, d\omega = \frac{e^2 N}{m} \delta_{\alpha\beta}. \tag{17.34a}$$

The dispersion relations for the electrical conductivity tensor have the form of (17.24) and (17.24a).

The theory of thermal radiation (cf. the papers by Rytov [91-93] and Bunkin [94]) is related to the theory of electrical fluctuations.

If the state of the system is far from equilibrium and we may not confine ourselves to the linear response, for the nonlinear susceptibility we must take correlations of higher order into account.†

17.4. Order of the Limits $V \to \infty$, $\varepsilon \to 0$ in the Kinetic Coefficients

In the expressions (17.22a) for the kinetic coefficients

$$L_{ik} = \lim_{\varepsilon \to 0} \lim_{V \to \infty} \int_0^\infty \int_0^\beta e^{-\varepsilon t} \langle \dot{\alpha}_k \dot{\alpha}_i (t + i\hbar\lambda) \rangle\, dt\, d\lambda \tag{17.35}$$

†Concerning nonequilibrium fluctuations, see the paper by Bunkin [95] and the series of articles by Lax [96-99], which are devoted to the theory of noise for the classical and quantum cases and employ the theory of Markovian processes.

we must perform two limiting processes, $V \to \infty$ and $\varepsilon \to 0$. As already observed, the correct order of these limiting processes is that in which $V \to \infty$ first, and then $\varepsilon \to 0$, since this ensures that the retarded solutions of Liouville's equation are selected (cf. Appendices I and III). Starting from the explicit expression (17.35) for the kinetic coefficients, it can be verified that only this order of the limits can lead to a finite result for L_{ik}.

We make use of the spectral expansion (16.15) for the time correlation functions, assuming that the limiting process $V \to \infty$ in the averages $\langle \ldots \rangle$ has already been performed:

$$\langle \dot{\alpha}_k \dot{\alpha}_i (t) \rangle = \frac{1}{2\pi} \int_{-\infty}^{\infty} J_{\dot{\alpha}_k \dot{\alpha}_i}(\omega) \, e^{-i\omega t} \, d\omega. \tag{17.36}$$

Substituting this expression in (17.35), we obtain

$$L_{ik} = \frac{1}{2\pi} \int_0^{\infty} \int_0^{\beta} \int_{-\infty}^{\infty} J_{\dot{\alpha}_k \dot{\alpha}_i}(\omega) \, e^{-i\omega (t + i\hbar\lambda) - \varepsilon t} \, dt \, d\lambda \, d\omega$$

$$= \frac{1}{2\pi} \int_0^{\infty} \int_{-\infty}^{\infty} J_{\dot{\alpha}_k \dot{\alpha}_i}(\omega) \, e^{-i\omega t - \varepsilon t} \, \frac{e^{\beta\hbar\omega} - 1}{\hbar\omega} \, dt \, d\omega,$$

where we have changed the order of integration and omitted the limit sign. Since

$$\int_0^{\infty} e^{-i\omega t - \varepsilon t} \, dt = \frac{1}{i\omega + \varepsilon} = -i \left(\frac{\mathcal{P}}{\omega} + i\pi\delta(\omega) \right) = 2\pi\delta_+(\omega),$$

we obtain for the kinetic coefficients the expression

$$L_{ik} = -\frac{i}{2\pi} \mathcal{P} \int_{-\infty}^{\infty} J_{\dot{\alpha}_k \dot{\alpha}_i}(\omega) \, \frac{e^{\beta\hbar\omega} - 1}{\hbar\omega^2} \, d\omega + \frac{\beta}{2} J_{\dot{\alpha}_k \dot{\alpha}_i}(0).$$

Here, the integral over ω is equal to zero, if there is no magnetic field.

In fact, the correlation functions are then symmetric under time reversal and their spectral intensities are real [cf. (16.46)]. From (16.33) and (17.25), we obtain

$$\begin{aligned} \operatorname{Im} \langle\langle \dot{\alpha}_i \mid \dot{\alpha}_k \rangle\rangle_\omega &= -\frac{1}{2\hbar} (e^{\beta\hbar\omega} - 1) \, J_{\dot{\alpha}_k \dot{\alpha}_i}(\omega), \\ \operatorname{Im} \langle\langle \dot{\alpha}_i \mid \dot{\alpha}_k \rangle\rangle_\omega &= -\operatorname{Im} \langle\langle \dot{\alpha}_i \mid \dot{\alpha}_k \rangle\rangle_{-\omega}. \end{aligned} \tag{17.37}$$

Consequently, the imaginary part of L_{ik} is proportional to the integral of an expression that is odd in ω, and is equal to zero, since

$$\mathscr{P}\frac{1}{2\pi}\int\limits_{-\infty}^{\infty}\operatorname{Im}\langle\langle\dot{\alpha}_i\,|\,\dot{\alpha}_k\rangle\rangle_\omega\frac{d\omega}{\omega^2}=0.$$

In fact, for $\omega \to 0$, it follows from (17.37) that the imaginary part of the Green function tends to zero like ω,

$$\operatorname{Im}\langle\langle\dot{\alpha}_i\,|\,\dot{\alpha}_k\rangle\rangle_\omega \approx -\frac{\beta\omega}{2}J_{\dot{\alpha}_k\dot{\alpha}_i}(0),$$

since it is assumed that the spectral intensity is finite. The integrand has a simple pole and, consequently, the principal-value integral is equal to zero.

Finally, for the kinetic coefficients, we obtain the expression

$$L_{ik}=\frac{\beta}{2}J_{\dot{\alpha}_k\dot{\alpha}_i}(0)=\frac{\beta}{2}\int\limits_{-\infty}^{\infty}\langle\dot{\alpha}_k\dot{\alpha}_i(t)\rangle\,dt=\frac{\beta}{2}\int\limits_{-\infty}^{\infty}\langle\dot{\alpha}_k(t)\,\dot{\alpha}_i\rangle\,dt, \quad (17.38)$$

which is valid in the absence of a magnetic field.

Thus, the kinetic coefficients are finite for systems for which the spectral intensity of the time correlation function (17.36) of the fluxes is finite at $\omega = 0$.†

If, however, we assume that the spectral intensity $J_{\dot{\alpha}_k\dot{\alpha}_i}(\omega)$ has a function δ-function singularity at $\omega = 0$, i.e., if

$$J_{\dot{\alpha}_k\dot{\alpha}_i}(\omega)=C_{ki}\,\delta(\omega)+J'_{ki}(\omega),$$

---

†In taking the limit $\omega \to 0$, we must observe caution, since

$$\int\limits_{-\infty}^{\infty}\langle\dot{\alpha}_k(t)\,\dot{\alpha}_i\rangle\,dt=\lim_{T\to\infty}\left(\langle\alpha_k(T)\,\dot{\alpha}_i\rangle-\langle\alpha_k(-T)\,\dot{\alpha}_i\rangle\right)=0$$

if one can interchange the order of taking the averages and integrating over time, which is not always permissible. In addition, in calculating the kinetic coefficients we must take one further limit, namely, making the wave number **k** corresponding to the Fourier components of the dynamic variables tend to zero. This enables us to define the kinetic coefficients by means of the double limiting cases $\omega \to 0$, $\mathbf{k} \to 0$ and redefine the thermodynamic forces in the corresponding way (for more exactness, cf. §22.8).

Another case when it is clear that the expressions (17.38) for the kinetic coefficients are nonzero is when the averaging is performed not with the equilibrium, but with the quasi-equilibrium distribution

$$\rho_q=\exp\Big\{-\beta\big(\Omega+H-\sum_j\mathscr{F}_j(t)\,\alpha_j\big)\Big\}.$$

For such a generalization of the kinetic coefficients, see in Chapter IV.

where $J'_{ki}(\omega)$ is finite as $\omega \to 0$ and $C_{ki} \neq 0$, the expressions (17.35) for the kinetic coefficients diverge like $1/\varepsilon$. In fact,

$$\int_0^\infty \int_{-\infty}^{\infty} C_{ki}\,\delta(\omega)\,e^{-i\omega t-\varepsilon t}\frac{e^{\beta\hbar\omega}-1}{\beta\hbar\omega}\,dt\,d\omega = \frac{C_{ki}}{\varepsilon} \to \infty.$$

We now show that the order of the limiting processes in which $V \to \infty$ first, and then $\varepsilon \to 0$, is indeed the necessary order.

We shall assume that the limit $V \to \infty$ has not yet been taken and the spectrum is discrete. According to (16.16), the spectral intensity is equal to

$$J_{\dot{\alpha}_k \dot{\alpha}_i}(\omega) = 2\pi Q^{-1} \sum_{\mu,\nu} \dot{\alpha}_k^{\mu\nu}\,\dot{\alpha}_i^{\nu\mu}\,e^{-E_\mu/\theta}\,\delta\left(\frac{E_\mu - E_\nu}{\hbar} - \omega\right), \tag{17.39}$$

where $\dot{\alpha}$ is the matrix element of the flux operator. If the spectrum is discrete, this expression has no meaning at $\omega = 0$, since the δ-function is defined only for continuous arguments. But if the limiting process $V \to \infty$ is performed first, the expression

$$L_{ik} = \beta\pi Q^{-1} \sum_{\mu,\nu} \dot{\alpha}_k^{\mu\nu}\,\dot{\alpha}_i^{\nu\mu}\,e^{-E_\mu/\theta}\,\delta\left(\frac{E_\mu - E_\nu}{\hbar}\right) \tag{17.39a}$$

now has meaning, since the δ-function now depends on a continuous variable.

17.5. Increase of Energy under the Influence of External Mechanical Perturbations

Under the influence of the external driving forces $F_1(t), \ldots, F_n(t)$, the energy and entropy of the system can increase. First we shall consider how the energy of a system with Hamiltonian H changes under the influence of the external perturbation (17.1). The statistical operator satisfies the quantum Liouville equation (8.6):

$$i\hbar\frac{\partial\rho}{\partial t} = [H + H_t^1, \rho]. \tag{17.40}$$

The change of energy of the system under the influence of the perturbation H_t^1 is described by the dynamic variable

$$\frac{dH}{dt} = \frac{1}{i\hbar}[H, H + H_t^1] = \frac{1}{i\hbar}[H, H_t^1], \tag{17.41}$$

since H does not depend explicitly on time. Neither the external field giving rise to the perturbation H_t^1, nor the perturbation itself is included in the system. We obtain the average change of energy by averaging (17.41) with the statistical operator ρ satisfying Eq. (17.40). Consequently,

$$\left\langle \frac{dH}{dt} \right\rangle = \operatorname{Tr}\left(\rho \frac{dH}{dt}\right) = \frac{d}{dt}\langle H \rangle = \frac{1}{i\hbar}\langle [H, H_t^1] \rangle, \qquad (17.41\text{a})$$

where (8.18) has been taken into account. Introducing the notation

$$\dot{H}_t^1 = \frac{1}{i\hbar}[H_t^1, H], \qquad (17.41\text{b})$$

we rewrite (17.41a) in the form

$$\frac{d}{dt}\langle H \rangle = -\langle \dot{H}_t^1 \rangle. \qquad (17.42)$$

The operator $\dot{H}_t^1$ has the meaning of the derivative of the operator with respect only to the time variable appearing in the Heisenberg picture. We use the relation (15.47) from the theory of linear response to write the right-hand side of Eq. (17.42) explicitly. We obtain

$$\frac{d}{dt}\langle H \rangle = -\int_{-\infty}^{\infty} \langle\langle \dot{H}_t^1(t) H_{t'}^1(t') \rangle\rangle \, dt', \qquad (17.43)$$

since, in a state of statistical equilibrium, the fluxes are equal to zero:

$$\langle \dot{H}_t^1 \rangle_0 = -\sum_j F_j(t) \langle \dot{\alpha}_j \rangle_0 = 0.$$

Thus, the rate of change of energy in the system under the influence of the perturbation H_t^1 is determined by the retarded Green function coupling the derivative $\dot{H}_t^1$ of the perturbation with the perturbation itself.

By means of formula (15.51) for the linear response, we write (17.42) in the form

$$\frac{d}{dt}\langle H \rangle = \int_0^{\beta} \int_{-\infty}^{t} \langle \dot{H}_{t'}^1(t' - i\hbar\lambda) \dot{H}_t^1(t) \rangle \, d\lambda \, dt' = \beta \int_{-\infty}^{t} (\dot{H}_{t'}^1(t'), \dot{H}_t^1(t)) \, dt', \qquad (17.44)$$

where we have introduced the notation, already applied earlier,

$$(A(t), B(t')) = \beta^{-1} \int_0^{\beta} \langle A(t) B(t' + i\hbar\lambda) \rangle \, d\lambda \tag{17.44a}$$

for the quantum correlation functions.

Thus, the rate of change (17.44) of the energy of the system is determined by quantum time correlation functions coupling the operators $\dot{H}_t^1$ at different times.

After substituting (17.1) in (17.44), we shall have

$$\frac{d}{dt} \langle H \rangle = \sum_{i,\, k} \int_{-\infty}^{t} L_{ik}(t - t') F_k(t') F_i(t) \, dt', \tag{17.45}$$

where

$$L_{ik}(t - t') = \beta (\dot{\alpha}_k(t'), \dot{\alpha}_i(t)). \tag{17.45a}$$

Taking into account the linear relations (17.20) between the forces and fluxes, we rewrite (17.45) in the form

$$\frac{d}{dt} \langle H \rangle = \sum_i \langle \dot{\alpha}_i \rangle F_i(t) = \langle \dot{\alpha} \rangle \cdot \boldsymbol{F}(t). \tag{17.46}$$

Consequently, the rate of change of the energy of the system under the influence of mechanical perturbations induced by the driving forces F_i is determined by the sum of the products of the forces F_i with their conjugate fluxes $\langle \dot{\alpha}_i \rangle$.

We shall discuss the physical meaning of formula (17.43) and show that it can be written in the form

$$\frac{d \langle H \rangle}{dt} = \sum_{\alpha,\, \beta} \hbar \omega_{\alpha\beta} w_{\alpha\beta}(t) e^{(\Omega - E_\beta)/\theta} \qquad \left(\omega_{\alpha\beta} = \frac{E_\alpha - E_\beta}{\hbar} \right), \tag{17.47}$$

where $w_{\alpha\beta}(t)$ is the probability of the transition $\beta \to \alpha$ in unit time

The meaning of formula (17.47) is obvious:

$$\sum_\alpha \hbar \omega_{\alpha\beta} w_{\alpha\beta}(t)$$

is the probable change of energy in unit time due to quantum transitions from the state α to all possible states β. The sum over β of these expressions with the Gibbs statistical factor $e^{(\Omega - E_\beta)/\theta}$ gives the average statistical rate of change of energy under the influence of the perturbation as a result of all possible transitions.

The probability of a transition $\beta \to \alpha$ in time t under the influence of a perturbation H_t^1 in the first nonvanishing order of perturbation theory (cf., e.g., [100]) is equal to

$$W_{\alpha\beta}(t) = \frac{1}{\hbar^2}\left| \int_{t_0}^{t} \langle \alpha \mid H_{t_1}^1 \mid \beta \rangle e^{i\omega_{\alpha\beta}t_1} dt_1 \right|^2,$$

or, if the perturbation operator is written in the Heisenberg picture,

$$W_{\alpha\beta}(t) = \frac{1}{\hbar^2}\left| \int_{t_0}^{t} \langle \alpha \mid H_{t_1}^1(t_1) \mid \beta \rangle dt_1 \right|^2. \tag{17.48}$$

For the probability of the transition $\beta \to \alpha$ in unit time

$$w_{\alpha\beta}(t) = \frac{dW_{\alpha\beta}(t)}{dt} \tag{17.49}$$

we obtain the expression

$$w_{\alpha\beta}(t) = \frac{1}{\hbar^2} \int_{t_0}^{t} \{\langle \alpha | H_t^1(t) | \beta \rangle \langle \beta | H_{t_1}^1(t_1) | \alpha \rangle + \langle \beta | H_t^1(t) | \alpha \rangle \langle \alpha | H_{t_1}^1(t_1) | \beta \rangle\} dt_1, \tag{17.49a}$$

which is finite as $t_0 \to -\infty$, as we shall assume below.

Substituting (17.49a) in (17.47) and summing over α and β taking into account the completeness of the system of eigenfunctions, we obtain the required formula (17.43).

Using (17.1), we write formula (17.44) in the form

$$\frac{d\langle H \rangle}{dt} = \int_0^{\beta} \int_{-\infty}^{t} \langle \dot{\alpha}(t') \dot{\alpha}(t + i\hbar\lambda) \rangle : \boldsymbol{F}(t) \boldsymbol{F}(t') \, d\lambda \, dt', \tag{17.50}$$

where the symbol : denotes the complete contraction of the tensors.

Expanding the time correlation function as a Fourier integral

$$\langle \dot{\alpha}(t') \dot{\alpha}(t) \rangle = \frac{1}{2\pi} \int_{-\infty}^{\infty} J_{\dot{\alpha}\dot{\alpha}}(\omega) e^{-i\omega(t-t')} d\omega \tag{17.51}$$

and performing the integration over λ, we rewrite (17.50) in the form

$$\frac{d\langle H\rangle}{dt}=\frac{1}{2\pi}\int_{-\infty}^{t}\int_{-\infty}^{\infty}\frac{e^{\beta\hbar\omega}-1}{\hbar\omega}J_{\dot{a}\dot{a}}(\omega):\boldsymbol{F}(t)\boldsymbol{F}(t')e^{-i\omega(t-t')}\,dt'\,d\omega. \tag{17.52}$$

We shall examine how the energy of a system varies under the influence of a periodic perturbation

$$H_t^1=\mathrm{Re}\left\{H^1 e^{\varepsilon t-i\omega t}\right\}, \tag{17.53}$$

where H^1 can be complex. We shall calculate the rate of change of energy by means of (17.43). Substituting (17.53) in (17.43), we obtain

$$\frac{d\langle H\rangle}{dt}=-\frac{1}{2}\,\mathrm{Re}\,\{\langle\langle\dot{H}^{1*}\mid H^1\rangle\rangle_\omega+\langle\langle\dot{H}^1\mid H^1\rangle\rangle_\omega e^{-2i\omega t}\}. \tag{17.54}$$

Thus, the rate of change of energy is made up of two terms – one constant in time, and the other oscillating. The constant term is positive, since

$$-\mathrm{Re}\,\langle\langle\dot{H}^{1*}\mid H^1\rangle\rangle_\omega>0.$$

In fact,

$$-\mathrm{Re}\,\langle\langle\dot{H}^{1*}\mid H^1\rangle\rangle_\omega=\mathrm{Re}\,i\omega\,\langle\langle H^{1*}\mid H^1\rangle\rangle_\omega=$$

$$=-\omega\,\mathrm{Im}\,\langle\langle H^{1*}\mid H^1\rangle\rangle_\omega=\frac{\omega}{2\hbar}\left(e^{\beta\hbar\omega}-1\right)J_{H^1H^{1*}}(\omega)>0,$$

where we have taken into account the relations (16.18), (16.19), and (16.33) and the fact that the function $\omega(e^{\beta\hbar\omega}-1)$ is positive.

Using (17.1a) and (17.21a), we write formula (17.54) in the form

$$\frac{d\langle H\rangle}{dt}=\frac{1}{2}\mathrm{Re}\{\boldsymbol{F}^*\cdot L(\omega)\cdot\boldsymbol{F}+\boldsymbol{F}\cdot L(\omega)\cdot\boldsymbol{F}e^{-2i\omega t}\}. \tag{17.55}$$

The average rate of change of energy is always positive and is equal to

$$\frac{dE}{dt}=\frac{\omega}{2\pi}\int_0^{2\pi/\omega}\frac{d\langle H\rangle}{dt}\,dt=\frac{1}{2}\mathrm{Re}\,\boldsymbol{F}^*\cdot L(\omega)\cdot\boldsymbol{F}>0. \tag{17.56}$$

Those systems whose energy increases under the influence of periodic perturbations are called *dissipative systems*. It is necessary that they possess closely spaced levels, in order that $L(\omega)$ have meaning after the limiting processes $V \to \infty$, $\varepsilon \to 0$.

The average rate of absorption of energy (17.56) can be represented in the form

$$\frac{dE}{dt} = \frac{1}{2}\{L'(\omega)^s : \operatorname{Re} \boldsymbol{FF}^* - L''(\omega)^a : \operatorname{Im} \boldsymbol{FF}^*\}, \tag{17.57}$$

where $L'(\omega)^s$ is the real symmetric part, and $L''(\omega)^a$ is the imaginary antisymmetric part of the tensors of the kinetic coefficients. The other terms give no contribution, since the tensor Re $\mathbf{FF}^*$ is symmetric, and Im $\mathbf{FF}^*$ is antisymmetric.

We now express the energy absorption averaged over a period in terms of the generalized susceptibility. By means of (17.54) and (17.56), we obtain

$$\frac{dE}{dt} = -\frac{\omega}{2} \operatorname{Im} \langle\langle \overset{*}{H^1} \mid H^1 \rangle\rangle_\omega = \frac{\omega}{2} \operatorname{Im} (\boldsymbol{F}^* \cdot \varkappa(\omega) \cdot \boldsymbol{F}) \tag{17.58}$$

or

$$\frac{dE}{dt} = \frac{\omega}{2}\{\varkappa''(\omega)^s : \operatorname{Re} \boldsymbol{FF}^* + \varkappa'(\omega)^a : \operatorname{Im} \boldsymbol{FF}^*\}, \tag{17.59}$$

where $\varkappa''(\omega)^s$ is the imaginary part of the symmetric susceptibility tensor, and $\varkappa'(\omega)^a$ is the real part of the antisymmetric susceptibility tensor.

In the derivation of the fluctuation-dissipation theorems [22, 101], an asymptotic expression is used for the transition probability in unit time under the influence of the perturbation (17.53):

$$w_{\alpha\beta} = \frac{\pi}{2\hbar}\{|\langle\alpha | H^1 | \beta\rangle|^2 \delta(E_\alpha - E_\beta - \hbar\omega) + |\langle\beta | H^1 | \alpha\rangle|^2 \delta(E_\alpha - E_\beta + \hbar\omega)\}; \tag{17.60}$$

this expression is valid for a large time interval t, when the probability (17.48) becomes proportional to the time and the transition probability $w_{\alpha\beta}$ no longer depends on t.

By means of (17.60), we shall calculate the average rate of energy absorption:

$$\frac{d\langle H\rangle}{dt} = \sum_{\alpha,\,\beta} e^{(\Omega - E_\beta)/\theta} \omega_{\alpha\beta} \frac{\pi}{2}\{|\langle\alpha | H^1 | \beta\rangle|^2 \delta(E_\alpha - E_\beta - \hbar\omega) + $$
$$+ |\langle\beta | H^1 | \alpha\rangle|^2 \delta(E_\alpha - E_\beta + \hbar\omega)\} = \frac{\pi}{2\omega} \sum_{\alpha\beta} e^{(\Omega - E_\beta)/\theta} \{|\langle\alpha | \dot{H}^1 | \beta\rangle|^2 \delta(E_\alpha - E_\beta - \hbar\omega) -$$
$$- |\langle\beta | \dot{H}^1 | \alpha\rangle|^2 \delta(E_\alpha - E_\beta + \hbar\omega)\}, \tag{17.61}$$

where $\omega_{\alpha\beta} = (E_\alpha - E_\beta)/\hbar$. Substituting into this the integral representation for the δ-function

$$\delta(E) = \frac{1}{2\pi\hbar} \int_{-\infty}^{\infty} e^{iEt/\hbar}\, dt$$

and performing the summations over α and β, we obtain

$$\frac{d\langle H\rangle}{dt} = \frac{1}{4\hbar\omega} \int_{-\infty}^{\infty} \{\langle \dot{H}^{1*}\dot{H}^{1}(t)\rangle e^{-i\omega t} - \langle \dot{H}^{1}\dot{H}^{1*}(t)\rangle e^{i\omega t}\}\, dt =$$

$$= \frac{1}{4\hbar\omega}\{J_{\dot{H}^{1*}\dot{H}^{1}}(-\omega) - J_{\dot{H}^{1}\dot{H}^{1*}}(\omega)\}. \tag{17.62}$$

Using the properties (16.16a) of the spectral intensities

$$J_{\dot{H}^{1*}\dot{H}^{1}}(-\omega) = e^{\beta\hbar\omega} J_{\dot{H}^{1}\dot{H}^{1*}}(\omega),$$

we transform (17.62) to the form

$$\frac{d\langle H\rangle}{dt} = \frac{e^{\beta\hbar\omega} - 1}{4\hbar\omega} J_{\dot{H}^{1}\dot{H}^{1*}}(\omega) \geqslant 0, \tag{17.62a}$$

i.e., the energy increases.

17.6. Entropy Production

In the preceding subsection, we considered the effect of mechanical perturbations on the change of energy of a system. We now study their influence on the change of entropy.

First of all, it is necessary to define entropy for a nonequilibrium state.

Minus the average of the logarithm of the statistical operator

$$\langle\eta\rangle = -\langle \ln\rho\rangle = -\operatorname{Tr}(\rho\ln\rho), \tag{17.63}$$

which in the equilibrium case is the entropy, cannot describe the entropy of a nonequilibrium state. In fact,

$$\eta = -\ln\rho \tag{17.64}$$

like ρ, satisfies the Liouville equation

$$i\hbar\frac{\partial\eta}{\partial t} = [H + H_t', \eta] \tag{17.65}$$

(cf. §8 of Chapter II). Consequently, η is an integral of motion

$$\frac{d\eta}{dt} = \frac{\partial\eta}{\partial t} + \frac{1}{i\hbar}\left[\eta, H + H_t^1\right] = 0, \tag{17.66}$$

whence it follows that $\langle\eta\rangle$ is constant in time

$$\frac{d}{dt}\langle\eta\rangle = \left\langle\frac{d\eta}{dt}\right\rangle = 0 \tag{17.66a}$$

and cannot possess the properties of the entropy of a nonequilibrium state.

For the moment, we shall not give a general definition of the entropy of a nonequilibrium state (we shall discuss this question in §20), but confine ourselves to defining it for the case of linear response.

We assume that the state of the system remains spatially uniform and stationary in time, i.e., the energy evolved is drawn off. Then it is natural to define the entropy by analogy with the equilibrium state, by the thermodynamic relation (11.24)

$$S = \frac{\langle H\rangle - \mu\langle N\rangle - \Omega}{\theta}, \tag{17.67}$$

but assuming that the averaging is performed over the nonequilibrium state.

The entropy (17.67) is equal to minus the average of the logarithm of the equilibrium distribution (9.42):

$$S = -\langle\ln\rho_0\rangle = -\operatorname{Tr}(\rho\ln\rho_0). \tag{17.67a}$$

The rate of change of the entropy (17.67) with time is equal to

$$\frac{\partial S}{\partial t} = \frac{1}{\theta}\frac{d\langle H\rangle}{dt}, \tag{17.68}$$

since the parameters θ, μ, and Ω characterize a state of statistical equilibrium and do not depend on the time, and the external perturbation does not change the number of particles. Taking (17.45) into account, we write the expression (17.68) in the form

$$\frac{\partial S}{\partial t} = \frac{1}{\theta}\sum_{i,k}\int_{-\infty}^{t} F_i(t)\,L_{ik}(t-t')\,F_k(t')\,dt' = \frac{1}{\theta}\int_{-\infty}^{t} \boldsymbol{F}(t)\cdot L(t-t')\cdot\boldsymbol{F}(t')\,dt', \tag{17.69}$$

For external forces that vary periodically with time [cf. (17.53)] it follows from (17.68) and (17.69) that the average rate of change of entropy over a period is positive

$$\frac{\omega}{2\pi}\int_0^{2\pi/\omega}\frac{\partial S}{\partial t}\,dt=\frac{1}{\theta}\frac{1}{2}\operatorname{Re}\boldsymbol{F}^*\cdot L(\omega)\cdot\boldsymbol{F}>0. \tag{17.70}$$

Thus, entropy is generated in the system, and (17.70) can be called the entropy production.

§ 18. Systems of Charged Particles in an Alternating Electromagnetic Field

In this section, we shall study a system of charged particles in an alternating electromagnetic field as an example of the general theory of linear response. We shall study the connection with retarded Green functions of the dielectric permittivity, magnetic susceptibility and electrical conductivity as functions of frequency and wave number, and their symmetry properties and dispersion relations.

18.1. Dielectric Permittivity and Conductivity

In §§15.4 and 17.3, we considered a system of charged particles in an electric field varying in time but constant in space. We now examine its behavior in an electromagnetic field which varies both in time and in space [40, 41, 52, 56, 64, 102–104].

The Hamiltonian of a system in an external electromagnetic field with vector potential $\mathbf{A}(\mathbf{x}, t)$ and scalar potential $\varphi(\mathbf{x}, t)$ has the form

$$H'=\frac{1}{2m}\int\psi^+(\boldsymbol{x})\left(\frac{\hbar}{i}\nabla-\frac{e}{c}\boldsymbol{A}(\boldsymbol{x},t)\right)^2\psi(\boldsymbol{x})\,d\boldsymbol{x}+ \\ +\int\psi^+(\boldsymbol{x})\psi(\boldsymbol{x})\,e\varphi(\boldsymbol{x},t)\,d\boldsymbol{x}+H_{\mathrm{int}}, \tag{18.1}$$

where e is the particle charge and H_{int} is the operator of the interaction between the particles. For the moment, we do not take into account the interaction of the spin of the particles with the field (cf. §18.3). For simplicity, we assume that the particles have charges of the same sign, for example, if they are electrons, the ions can be regarded as a screening background.

In (18.1) $\psi(\mathbf{x})$ and $\psi^+(\mathbf{x})$ are second-quantized wave functions, i.e., operators acting on the wave functions in occupation-number space and satisfying the commutation relations

$$\begin{gathered}\psi(\boldsymbol{x})\psi^+(\boldsymbol{x}') \pm \psi^+(\boldsymbol{x}')\psi(\boldsymbol{x}) = \delta(\boldsymbol{x}-\boldsymbol{x}'),\\ \psi(\boldsymbol{x})\psi(\boldsymbol{x}') \pm \psi(\boldsymbol{x}')\psi(\boldsymbol{x}) = 0,\\ \psi^+(\boldsymbol{x})\psi^+(\boldsymbol{x}') \pm \psi^+(\boldsymbol{x}')\psi^+(\boldsymbol{x}) = 0,\end{gathered} \tag{18.2}$$

where the plus sign is taken for Fermi statistics and the minus sign for Bose statistics (cf. [105, 25] and the texts on quantum mechanics [100, 106, 107]).

The operators $\psi(\mathbf{x})$ and $\psi^+(\mathbf{x})$ are connected with $a_{\boldsymbol{k}\sigma}$, and $a^+_{\boldsymbol{k}\sigma}$ – the creation and destruction operators for particles in the state $\mathbf{k}, \sigma$ – by the relations

$$\psi(\boldsymbol{x}) = \frac{1}{\sqrt{V}}\sum_{\boldsymbol{k},\sigma} e^{i(\boldsymbol{k}\cdot\boldsymbol{x})}\delta_{\sigma s_z} a_{\boldsymbol{k}\sigma}, \quad \psi^+(\boldsymbol{x}) = \frac{1}{\sqrt{V}}\sum_{\boldsymbol{k},\sigma} e^{-i(\boldsymbol{k}\cdot\boldsymbol{x})}\delta_{\sigma s_z} a^+_{\boldsymbol{k}\sigma}, \tag{18.3}$$

where $a_{\boldsymbol{k}\sigma}$, and $a^+_{\boldsymbol{k}\sigma}$ satisfy the commutation relations

$$\begin{gathered}a_{\boldsymbol{k}\sigma}a^+_{\boldsymbol{k}_1\sigma_1} \pm a^+_{\boldsymbol{k}_1\sigma_1}a_{\boldsymbol{k}\sigma} = \delta_{\boldsymbol{k}\sigma,\,\boldsymbol{k}_1\sigma_1},\\ a_{\boldsymbol{k}\sigma}a_{\boldsymbol{k}_1\sigma_1} \pm a_{\boldsymbol{k}_1\sigma_1}a_{\boldsymbol{k}\sigma} = 0,\\ a^+_{\boldsymbol{k}\sigma}a^+_{\boldsymbol{k}_1\sigma_1} \pm a^+_{\boldsymbol{k}_1\sigma_1}a^+_{\boldsymbol{k}\sigma} = 0;\end{gathered} \tag{18.4}$$

$\delta_{\boldsymbol{k},\boldsymbol{k}_1}$ is the Kronecker symbol, and the argument s_z of $\psi(\mathbf{x})$ will not be written out explicitly. It is easily verified that (18.4) follows from (18.2), and conversely.

The potentials $\mathbf{A}(\mathbf{x}, t)$ and $\varphi(\mathbf{x}, t)$ define the electromagnetic field acting on the particles. As we have already noted in §15.4, we must distinguish two cases – when the Coulomb interaction is taken into account explicitly in the Hamiltonian, and when it is taken into account through a screening field. In the first case, it is necessary to allow for the screening effects, while in the second case this is not necessary, since they are already taken into account through the screening interaction. Below, we shall consider only the first case.

In introducing the dielectric permittivity and magnetic permeability, it is necessary to take into account that the induced charges screen only the longitudinal part of the electric field, since they determine the divergence of the field,† and the induced currents screen only the transverse part of the magnetic field, since they determine the curl of the field. Therefore, the potentials **A**(**x**, t) and φ(**x**, t) determine the electric induction vector

$$\boldsymbol{D} = -\nabla\varphi - \frac{1}{c}\frac{\partial \boldsymbol{A}}{\partial t} \tag{18.5}$$

only for the longitudinal part of **D**, and

$$\mathbf{B} = \operatorname{curl} \mathbf{A} \tag{18.5a}$$

for the transverse part of **B**; incidentally, the field **B** is always transverse, since div **B** = 0.

The current density operator is equal to the functional derivative of the Hamiltonian (18.1) with respect to the vector potential

$$\boldsymbol{j}'(\boldsymbol{x}) = -c\frac{\delta H'}{\delta \boldsymbol{A}(\boldsymbol{x})}, \tag{18.6}$$

and, consequently,

$$\boldsymbol{j}'(\boldsymbol{x}) = \frac{e\hbar}{2mi}\{\psi^+(\boldsymbol{x})\nabla\psi(\boldsymbol{x}) - \nabla\psi^+(\boldsymbol{x})\psi(\boldsymbol{x})\} - \frac{e^2}{mc}\boldsymbol{A}(\boldsymbol{x}, t)\psi^+(\boldsymbol{x})\psi(\boldsymbol{x}). \tag{18.7}$$

The current (18.7) satisfies the charge-conservation equation

$$\frac{\partial\rho(\boldsymbol{x})}{\partial t} = -\operatorname{div}\boldsymbol{j}'(\boldsymbol{x}), \quad \rho(\boldsymbol{x}) = e\psi^+(\boldsymbol{x})\psi(\boldsymbol{x}) = en(\boldsymbol{x}); \tag{18.7a}$$

ρ(**x**) is the charge density.

The total Hamiltonian of the system can be written by separating the linear and quadratic terms in the form

$$H' = H + H_1 + H_2, \tag{18.8}$$

where

$$\begin{aligned} H_1 &= -\frac{1}{c}\int \boldsymbol{j}(\boldsymbol{x})\cdot\boldsymbol{A}(\boldsymbol{x}, t)\,d\boldsymbol{x} + \int \rho(\boldsymbol{x})\varphi(\boldsymbol{x}, t)\,d\boldsymbol{x}, \\ H_2 &= \frac{e^2}{mc^2}\int \rho(\boldsymbol{x})A^2(\boldsymbol{x}, t)\,d\boldsymbol{x}, \end{aligned} \tag{18.8a}$$

†Sometimes, this is not taken into account in textbooks [108], even though a complete clarification of this question was achieved a long time ago in the papers of Ewald [109] (cf. also [110-113]).

H is the Hamiltonian of the system in the absence of the electromagnetic field, and j (x) is the current operator in the absence of the field,

$$j(x) = \frac{e\hbar}{2mi}\{\psi^+(x)\nabla\psi(x) - \nabla\psi^+(x)\psi(x)\}. \tag{18.9}$$

We shall consider the response of the system to the adiabatic switching on of the electromagnetic field.

Let the current be equal to zero in the absence of the field. After the field is switched on, according to (15.47), the current and charge, if we confine ourselves to terms linear in $A(x, t)$ and $\varphi(x, t)$, are equal to

$$J(x, t) = \langle j' \rangle = -\frac{e^2 n}{mc} A(x, t) - \\ -\frac{1}{c}\int\int_{-\infty}^{\infty} \langle\langle j(x, t)\, j(x', t')\rangle\rangle \cdot A(x', t')\, dx'\, dt' + \\ + \int\int_{-\infty}^{\infty} \langle\langle j(x, t)\, \rho(x', t')\rangle\rangle\, \varphi(x', t')\, dx'\, dt', \tag{18.10a}$$

$$\langle\rho(x)\rangle = en - \frac{1}{c}\int\int_{-\infty}^{\infty} \langle\langle \rho(x, t)\, j(x', t')\rangle\rangle \cdot A(x', t')\, dx'\, dt' + \\ + \int\int_{-\infty}^{\infty} \langle\langle \rho(x, t)\, \rho(x', t')\rangle\rangle\, \varphi(x', t')\, dx'\, dt', \tag{18.10b}$$

where, in the integrands, we have retarded Green functions of tensor, vector and scalar types, and $n = \langle\psi^+(x)\psi(x)\rangle$ is the number of particles in unit volume, which, because of the spatial uniformity of the system, does not depend on x.

The relations (18.10a) and (18.10b) take a specially simple form if we expand the functions $A(x, t)$, $\varphi(x, t)$, $J(x, t)$ and $\langle\rho(x)\rangle$ in sums over plane waves

$$A(x, t) = \frac{1}{V}\sum_k \int_{-\infty}^{\infty} A(k, \omega)\, e^{i(k\cdot x) - i\omega t}\, d\omega,$$

$$\varphi(x, t) = \frac{1}{V}\sum_k \int_{-\infty}^{\infty} \varphi(k, \omega)\, e^{i(k\cdot x) - i\omega t}\, d\omega, \tag{18.11}$$

$$J(x, t) = \frac{1}{V}\sum_k \int_{-\infty}^{\infty} J(k, \omega)\, e^{i(k\cdot x) - i\omega t}\, d\omega,$$

write the operators in the momentum representation

$$j(x) = \frac{1}{\sqrt{V}} \sum_k j_k e^{i(k\cdot x)}, \quad \rho(x) = \frac{1}{\sqrt{V}} \sum_k \rho_k e^{i(k\cdot x)}, \tag{18.11a}$$

where

$$j_k = \frac{e\hbar}{m\sqrt{V}} \sum_{q,\sigma} \left(q - \frac{k}{2}\right) a^+_{q-k,\sigma} a_{q\sigma},$$

$$\rho_k = \frac{e}{\sqrt{V}} \sum_{q,\sigma} a^+_{q-k,\sigma} a_{q\sigma}, \tag{18.11b}$$

and expand the Green functions as Fourier integrals. Then the relations (18.10a) and (18.10b) take the form of linear algebraic relations between the Fourier transforms of the current and charge and the Fourier transforms of the vector and scalar potentials:

$$J(k, \omega) = -\frac{e^2 n}{mc} A(k, \omega) - \langle\langle j_k | j_{-k} \rangle\rangle_\omega \cdot \frac{1}{c} A(k, \omega) + \langle\langle j_k | \rho_{-k} \rangle\rangle_\omega \varphi(k, \omega),$$

$$\rho(k, \omega) = -\langle\langle \rho_k | j_{-k} \rangle\rangle_\omega \cdot \frac{1}{c} A(k, \omega) + \langle\langle \rho_k | \rho_{-k} \rangle\rangle_\omega \varphi(k, \omega), \tag{18.12}$$

where we have used the fact that, because of the assumed translational invariance of the problem, the averages $\langle j_k \cdot j_{k'} \rangle$, $\langle j_k \rho_{k'} \rangle$, and $\langle \rho_k \rho_{k'} \rangle$ are nonzero only for $\mathbf{k} + \mathbf{k'} = 0$. In crystals, $\mathbf{k} + \mathbf{k'} = 2\pi\boldsymbol{\tau}$, where $\boldsymbol{\tau}$ is a reciprocal-lattice vector.

The relation (18.12) can be represented in a form such that it does not contain the vector and scalar potential, but contains only the electric and magnetic inductions. We write the tensor and vector Green functions in such a way that their tensor or vector character is expressed explicitly. Since they depend only on the vector k and the system is assumed to be isotropic, the tensor function must depend linearly on the tensor $\mathbf{k}_\alpha \mathbf{k}_\beta$ and on the unit tensor $\delta_{\alpha\beta}$, and the vector function on k_α with coefficients depending only on $|\mathbf{k}|$, i.e.,

$$\chi_{\alpha\beta}(k, \omega) - \frac{e^2 n}{m} \delta_{\alpha\beta} = \langle\langle j^\alpha_k | j^\beta_{-k} \rangle\rangle_\omega = \frac{k_\alpha k_\beta}{k^2} \chi^l(k, \omega) + \left(\delta_{\alpha\beta} - \frac{k_\alpha k_\beta}{k^2}\right) \chi^{tr}(k, \omega),$$

$$\langle\langle j^\alpha_k | \rho_{-k} \rangle\rangle_\omega = \langle\langle (k \cdot j_k) | \rho_{-k} \rangle\rangle_\omega \frac{k_\alpha}{k^2}, \tag{18.13}$$

$$\langle\langle \rho_k | j^\alpha_{-k} \rangle\rangle_\omega = \langle\langle \rho_k | (k \cdot j_{-k}) \rangle\rangle_\omega \frac{k_\alpha}{k^2},$$

where

$$\begin{aligned} \chi^{l}(k, \omega) &= \frac{1}{k^2} \langle\langle (\boldsymbol{k}\cdot\boldsymbol{j}_{\boldsymbol{k}}) \mid (\boldsymbol{k}\cdot\boldsymbol{j}_{-\boldsymbol{k}}) \rangle\rangle_{\omega}, \\ \chi^{tr}(k, \omega) &= \frac{1}{2k^2} \langle\langle [\boldsymbol{k}\times\boldsymbol{j}_{\boldsymbol{k}}] \cdot \mid [\boldsymbol{k}\times\boldsymbol{j}_{-\boldsymbol{k}}] \rangle\rangle_{\omega} \end{aligned} \tag{18.13a}$$

are the longitudinal and transverse parts of the susceptibility tensor

$$\chi_{\alpha\beta}(\boldsymbol{k}, \omega) - \frac{e^2 n}{m}\delta_{\alpha\beta}.$$

The susceptibility $\chi_{\alpha\beta}(\boldsymbol{k}, \omega)$ expresses the linear response of the current to the vector potential.

We obtain a second expression from (18.13a) by calculating the complete contraction of the tensor $\chi_{\alpha\beta}(\boldsymbol{k}, \omega)$ with $\delta_{\alpha\beta} - (\boldsymbol{k}_\alpha \boldsymbol{k}_\beta / k^2)$ noting that

$$\sum_{\alpha\beta} \left(\delta_{\alpha\beta} - \frac{k_\alpha k_\beta}{k^2}\right)\left(\delta_{\beta\alpha} - \frac{k_\beta k_\alpha}{k^2}\right) = 2,$$

and making use of the vector identity

$$[\boldsymbol{k}\times\boldsymbol{j}_{\boldsymbol{k}}] \cdot [\boldsymbol{k}\times\boldsymbol{j}_{-\boldsymbol{k}}] = k^2 (\boldsymbol{j}_{\boldsymbol{k}}\cdot\boldsymbol{j}_{-\boldsymbol{k}}) - (\boldsymbol{k}\cdot\boldsymbol{j}_{\boldsymbol{k}})(\boldsymbol{k}\cdot\boldsymbol{j}_{-\boldsymbol{k}}).$$

We note that the operator $(\mathbf{k}\cdot\mathbf{j}_{\mathbf{k}})$ can be expressed in terms of the time derivatives of $\rho_{\boldsymbol{k}}$ (18.11b). For this, we calculate the commutator of $\rho_{\boldsymbol{k}}$ with the kinetic energy operator T:

$$T = \sum_{\boldsymbol{k},\sigma} \frac{\hbar^2 k^2}{2m} a^{+}_{\boldsymbol{k}\sigma} a_{\boldsymbol{k}\sigma},$$

$$[\rho_{\boldsymbol{k}}, T] = \frac{e\hbar^2}{\sqrt{V}} \sum_{\boldsymbol{q},\sigma} \frac{\boldsymbol{k}}{m}\left(\boldsymbol{q} - \frac{\boldsymbol{k}}{2}\right) a^{+}_{\boldsymbol{q}-\boldsymbol{k},\sigma} a_{\boldsymbol{q}\sigma} = \hbar(\boldsymbol{k}\cdot\boldsymbol{j}_{\boldsymbol{k}}).$$

Consequently, since $\rho_{\boldsymbol{k}}$ commutes with the interaction operator

$$H_{\text{int}} = \frac{1}{2V} \sum_{\substack{\boldsymbol{k}_1\boldsymbol{k}_2\boldsymbol{k} \\ \sigma_1\sigma_2}} \nu(\boldsymbol{k})\, a^{+}_{\boldsymbol{k}_1\sigma_1} a^{+}_{\boldsymbol{k}_2\sigma_2} a_{\boldsymbol{k}_2+\boldsymbol{k},\sigma_2} a_{\boldsymbol{k}_1-\boldsymbol{k},\sigma_1}, \tag{18.14}$$

the equation of motion for ρ_k has the form

$$\dot{\rho}_{\boldsymbol{k}} = \frac{1}{i\hbar}[\rho_{\boldsymbol{k}}, H] = -i(\boldsymbol{k}\cdot\boldsymbol{j}_{\boldsymbol{k}}). \tag{18.15}$$

Repeating the calculation for the commutator between $\dot{\rho}_k$ (18.15) and ρ_{-k}, we obtain the important commutation relation

$$\frac{1}{i\hbar}[\rho_k, \dot{\rho}_{-k}] = \frac{Ne^2}{Vm}k^2, \qquad N = \sum_{q,\sigma} a_{q\sigma}^+ a_{q\sigma}, \tag{18.16}$$

which is called the **longitudinal sum rule.**

Taking (18.13) and (18.15) into account, we obtain for the induced charge (18.12)

$$\rho(k, \omega) = \frac{1}{k^2}\langle\langle\rho_k | \dot{\rho}_{-k}\rangle\rangle_\omega \frac{i}{c}(k\cdot A(k, \omega)) + \langle\langle\rho_k | \rho_{-k}\rangle\rangle_\omega \varphi(k, \omega). \tag{18.17}$$

We express the function $\langle\langle\rho_k | \dot{\rho}_{-k}\rangle\rangle_\omega$ in terms of $\langle\langle\rho_k | \rho_{-k}\rangle\rangle_\omega$ by integrating by parts:

$$\langle\langle\rho_k | \dot{\rho}_{-k}\rangle\rangle_\omega = \int_{-\infty}^{\infty} \langle\langle\rho_k(t)\, \dot{\rho}_{-k}\rangle\rangle e^{i\omega t}\, dt =$$

$$= \frac{1}{i\hbar}\langle[\rho_k, \rho_{-k}]\rangle_\omega + i\omega\langle\langle\rho_k | \rho_{-k}\rangle\rangle_\omega = i\omega\langle\langle\rho_k | \rho_{-k}\rangle\rangle_\omega, \tag{18.18}$$

since the operators ρ_k and ρ_{-k} commute. The term with the commutator in (18.18) appeared because of the differentiation of the discontinuous $\theta(t)$ function. Using (18.18), we rewrite the expression (18.17) for the induced charge in the form

$$\rho(k, \omega) = \frac{1}{k^2}\langle\langle\rho_k | \rho_{-k}\rangle\rangle_\omega\, ik\cdot D(k, \omega), \tag{18.19}$$

where

$$ik\cdot D(k, \omega) = \frac{1}{c}i\omega\,(ik\cdot A(k, \omega) + k^2\varphi(k, \omega)) \tag{18.19a}$$

and, according to (18.5), has the meaning of the Fourier transform of the divergence of the electric induction vector. Thus, we have expressed the induced charge in terms of the longitudinal part of the electric induction vector.

The induced charge is usually expressed in terms of the divergence of the polarization vector $\mathbf{P}(\mathbf{x}, t)$ through the relation

$$\rho_{\text{ind}}(x, t) = -\operatorname{div} P(x, t) = -\frac{1}{4\pi}\operatorname{div}(D(x, t) - E(x, t)), \tag{18.20}$$

or, in Fourier components,

$$\rho(k, \omega) = \frac{1}{4\pi}ik\cdot(E(k, \omega) - D(k, \omega)) = \frac{1}{4\pi}\left(\frac{1}{\varepsilon(k, \omega)} - 1\right)ik\cdot D(k, \omega), \tag{18.20a}$$

where

$$\boldsymbol{k} \cdot \boldsymbol{D}(\boldsymbol{k}, \omega) = \varepsilon(\boldsymbol{k}, \omega)\, \boldsymbol{k} \cdot \boldsymbol{E}(\boldsymbol{k}, \omega), \tag{18.20b}$$

ε (**k**, ω) being the dielectric permittivity determining the connection between the longitudinal components of the electric field and the induction.

We note that it follows from (18.20b) that $\mathbf{D}(\mathbf{k}, \omega) = \varepsilon(\mathbf{k}, \omega)\cdot \mathbf{E}(\mathbf{k}, \omega)$ only for the longitudinal part of the electric field.

Comparing (18.19) with (18.20a), we obtain an expression for the dielectric permittivity as a function of k and ω in terms of Green functions:

$$\varepsilon^{-1}(\boldsymbol{k}, \omega) = 1 + \frac{4\pi}{k^2} \langle\langle \rho_{\boldsymbol{k}} | \rho_{-\boldsymbol{k}} \rangle\rangle_\omega. \tag{18.21}$$

We now examine the expression for the induced current, i.e., the first of the equations of the system (18.12), and, using (18.13) and (18.13a), write it in the form

$$\begin{aligned} \boldsymbol{J}(\boldsymbol{k}, \omega) = -\frac{e^2 n}{mc} \boldsymbol{A}(\boldsymbol{k}, \omega) - \langle\langle \dot\rho_{\boldsymbol{k}} | \dot\rho_{-\boldsymbol{k}} \rangle\rangle_\omega \frac{\boldsymbol{k}}{ck^4}\, \boldsymbol{k}\, \boldsymbol{A}(\boldsymbol{k}, \omega) + \\ + \langle\langle \dot\rho_{\boldsymbol{k}} | \rho_{-\boldsymbol{k}} \rangle\rangle_\omega \frac{i\boldsymbol{k}}{k^2} \varphi(\boldsymbol{k}, \omega) - \\ - \chi^{\mathrm{tr}}(k, \omega) \frac{1}{ck^2} \{k^2 \boldsymbol{A}(\boldsymbol{k}, \omega) - \boldsymbol{k}(\boldsymbol{k} \cdot \boldsymbol{A}(\boldsymbol{k}, \omega))\}, \end{aligned} \tag{18.22}$$

since, according to (18.13a) and (18.15), the longitudinal part of the susceptibility tensor (18.13) is equal to

$$\chi^{\mathrm{l}}(k, \omega) = \frac{1}{k^2} \langle\langle \dot\rho_{\boldsymbol{k}} | \dot\rho_{-\boldsymbol{k}} \rangle\rangle_\omega. \tag{18.23}$$

We express the Green functions $\langle\langle \dot\rho_{\boldsymbol{k}} | \dot\rho_{-\boldsymbol{k}} \rangle\rangle_\omega$ and $\langle\langle \dot\rho_{\boldsymbol{k}} | \rho_{-\boldsymbol{k}} \rangle\rangle_\omega$ in terms of $\langle\langle \rho_{\boldsymbol{k}} | \rho_{-\boldsymbol{k}} \rangle\rangle_\omega$ by integrating them by parts. We obtain

$$\begin{aligned} \langle\langle \dot\rho_{\boldsymbol{k}} | \rho_{-\boldsymbol{k}} \rangle\rangle_\omega &= -i\omega \langle\langle \rho_{\boldsymbol{k}} | \rho_{-\boldsymbol{k}} \rangle\rangle_\omega, \\ \langle\langle \dot\rho_{\boldsymbol{k}} | \dot\rho_{-\boldsymbol{k}} \rangle\rangle_\omega &= -\frac{1}{i\hbar} [\rho_{\boldsymbol{k}}, \dot\rho_{-\boldsymbol{k}}] + \omega^2 \langle\langle \rho_{\boldsymbol{k}} | \rho_{-\boldsymbol{k}} \rangle\rangle_\omega \\ &= -\frac{ne^2}{m} k^2 + \omega^2 \langle\langle \rho_{\boldsymbol{k}} | \rho_{-\boldsymbol{k}} \rangle\rangle_\omega, \end{aligned} \tag{18.23a}$$

where we have used the longitudinal sum rule (18.16). Taking (18.23) and (18.19a) into account, we represent the expression (18.22) for the induced current in the form

$$\begin{aligned} \boldsymbol{J}(\boldsymbol{k}, \omega) = \langle\langle \rho_{\boldsymbol{k}} | \rho_{-\boldsymbol{k}} \rangle\rangle_\omega \frac{i\omega}{k^2} \frac{\boldsymbol{k}}{k^2}\, \boldsymbol{k} \cdot \boldsymbol{D}(\boldsymbol{k}, \omega) - \\ - \left(\chi^{\mathrm{tr}}(k, \omega) + \frac{e^2 n}{m}\right) \frac{1}{ck^2} (k^2 \boldsymbol{A}(\boldsymbol{k}, \omega) - \boldsymbol{k}\boldsymbol{k} \cdot \boldsymbol{A}(\boldsymbol{k}, \omega)), \end{aligned} \tag{18.24}$$

where the longitudinal and transverse parts of the current have been separated.

The transverse part of the current can be identified with the curl of the induced magnetic moment vector **M**(x, t), i.e.,

$$-\frac{1}{k^2c}\left(\chi^{\text{tr}}(k, \omega)+\frac{e^2n}{m}\right)[i\boldsymbol{k}\times\boldsymbol{B}(\boldsymbol{k}, \omega)] = c\,[i\boldsymbol{k}\times\boldsymbol{M}(\boldsymbol{k}, \omega)] =$$

$$=\frac{c}{4\pi}[i\boldsymbol{k}\times(\boldsymbol{B}(\boldsymbol{k}, \omega)-\boldsymbol{H}(\boldsymbol{k}, \omega))] = \frac{c}{4\pi}(1-\mu^{-1}(\boldsymbol{k}, \omega))[i\boldsymbol{k}\times\boldsymbol{B}(\boldsymbol{k}, \omega)],$$

where

$$\boldsymbol{B}(\boldsymbol{k}, \omega) = [i\boldsymbol{k}\times\boldsymbol{A}(\boldsymbol{k}, \omega)] \tag{18.25}$$

is the Fourier transform of the magnetic induction vector (18.5a) and

$$\mu^{-1}(\boldsymbol{k}, \omega) = 1+\frac{4\pi}{c^2k^2}\left(\chi^{\text{tr}}(k, \omega)+\frac{e^2n}{m}\right) \tag{18.26}$$

is the **magnetic permeability**, dependent on k and ω.

Taking (18.21) and (18.26) into account, we write the expression (18.24) for the current in the form

$$\boldsymbol{J}(\boldsymbol{k}, \omega) = -\frac{1}{4\pi}(1-\varepsilon^{-1}(\boldsymbol{k}, \omega))\,i\omega\,\frac{\boldsymbol{k}}{k^2}(\boldsymbol{k}\cdot\boldsymbol{D}(\boldsymbol{k}, \omega))+$$

$$+\frac{c}{4\pi}(1-\mu^{-1}(\boldsymbol{k}, \omega))[i\boldsymbol{k}\times\boldsymbol{B}(\boldsymbol{k}, \omega)]. \tag{18.27}$$

The current (18.27) can also be expressed in terms of the electric field vector **E**(k, ω), by making use of the Maxwell equation

$$\operatorname{curl}\boldsymbol{E}(\boldsymbol{x}, t) = -\frac{1}{c}\frac{\partial\boldsymbol{B}(\boldsymbol{x}, t)}{\partial t}$$

in Fourier components

$$[i\boldsymbol{k}\times\boldsymbol{E}(\boldsymbol{k}, \omega)] = \frac{i\omega}{c}\boldsymbol{B}(\boldsymbol{k}, \omega) \tag{18.28}$$

and eliminating **B**(k, ω) from (18.27). We then obtain

$$\boldsymbol{J}(\boldsymbol{k}, \omega) = \sigma^{\text{l}}(k, \omega)\frac{\boldsymbol{k}}{k^2}(\boldsymbol{k}\cdot\boldsymbol{E}(\boldsymbol{k}, \omega)) - \sigma^{\text{tr}}(k, \omega)\frac{1}{k^2}[\boldsymbol{k}\times[\boldsymbol{k}\times\boldsymbol{E}(\boldsymbol{k}, \omega)]], \tag{18.29}$$

where

$$\sigma^{l}(k, \omega)=\frac{i\omega}{4\pi}(1-\varepsilon(k, \omega)),$$
$$\sigma^{tr}(k, \omega)=\frac{c^2k^2}{4\pi i\omega}(1-\mu^{-1}(k, \omega)) \qquad (18.29a)$$

are the **longitudinal and transverse electric conductivities.**

We examined above the equations (18.12) for the induced current and charge. We shall show that it is sufficient to consider the first of these, the second following from the first if we take the commutation relations between the charge and the current into account.

We take the scalar product of the first equation of the system (18.12) with **k**. Taking (18.15) into account, we obtain

$$\boldsymbol{k}\cdot\boldsymbol{J}(\boldsymbol{k}, \omega)=-\frac{e^2n}{mc}\boldsymbol{k}\cdot\boldsymbol{A}(\boldsymbol{k}, \omega)-\langle\langle\dot{\rho}_{\boldsymbol{k}}|\boldsymbol{j}_{-\boldsymbol{k}}\rangle\rangle_{\omega}\cdot\frac{i}{c}\boldsymbol{A}(\boldsymbol{k}, \omega)+\langle\langle\dot{\rho}_{\boldsymbol{k}}|\rho_{-\boldsymbol{k}}\rangle\rangle_{\omega}i\varphi(\boldsymbol{k}, \omega). \qquad (18.30)$$

By integration by parts, we find

$$\langle\langle\dot{\rho}_{\boldsymbol{k}}|\boldsymbol{j}_{-\boldsymbol{k}}\rangle\rangle_{\omega}=-\frac{1}{i\hbar}\langle[\rho_{\boldsymbol{k}}, \boldsymbol{j}_{-\boldsymbol{k}}]\rangle-i\omega\langle\langle\rho_{\boldsymbol{k}}|\boldsymbol{j}_{-\boldsymbol{k}}\rangle\rangle_{\omega}. \qquad (18.31)$$

Using (18.23) and (18.31), we write (18.30) in the form of a conservation law

$$\boldsymbol{k}\cdot\boldsymbol{J}(\boldsymbol{k}, \omega)=\omega\rho(\boldsymbol{k}, \omega), \qquad (18.30a)$$

where we have taken into account the commutation relation between the Fourier components of the charge and the current,

$$\frac{1}{\hbar}[\rho_{\boldsymbol{k}}, \boldsymbol{j}_{-\boldsymbol{k}}]=\frac{e^2}{m}\frac{N}{V}\boldsymbol{k}, \quad N=\sum_{q\sigma}a^+_{q\sigma}a_{q\sigma}. \qquad (18.32)$$

In fact,

$$\frac{1}{\hbar}[\rho_{\boldsymbol{k}}, \boldsymbol{j}_{-\boldsymbol{k}}]=\frac{e^2}{m}\frac{1}{V}\sum_{\boldsymbol{q}, \sigma}\left(\boldsymbol{q}-\frac{\boldsymbol{k}}{2}\right)(n_{q\sigma}-n_{\boldsymbol{q}+\boldsymbol{k}, \sigma}),$$

which, on replacement of the summation variables in the term with $n_{q+k, \sigma}$, reduces to (18.32).

Thus, we have verified that the average values of the current and charge satisfy exactly the conservation law (18.30a); therefore, by calculating the average current, we also find the induced charge at the same time.

The converse statement is not true. The equation for the current does not follow from the equation for the charge, since it can have a transverse rotational part, and the charge conservation law determines only its longitudinal part.

We obtain an equation for the curl of the current in the Fourier representation. By taking the vector product of **k** with the first equation of the system (18.12) (with **k** on the left), we obtain

$$[\boldsymbol{k} \times \boldsymbol{J}(\boldsymbol{k}, \omega)] = -\frac{1}{c}\left(\frac{e^2 n}{m} + \chi^{\mathrm{tr}}(k, \omega)\right)[\boldsymbol{k} \times \boldsymbol{A}(\boldsymbol{k}, \omega)] =$$

$$= \frac{i}{c}\left(\frac{e^2 n}{m} + \chi^{\mathrm{tr}}(k, \omega)\right)\boldsymbol{B}(\boldsymbol{k}, \omega). \qquad (18.33)$$

According to (18.26), the coefficient in the right-hand side of this equation can be expressed in terms of the magnetic permeability.

In ordinary systems, as $\omega \to 0$, $k \to 0$, the term $\chi^{\mathrm{tr}}(k, \omega)$ almost cancels the term e^2n/m, and their sum gives only a very small diamagnetic effect (L a n d a u d i a m a g n e t i s m [42, 114]). This cancellation is violated only in superconducting systems, because of the existence of a gap in the spectrum of the elementary excitations [115, 116].

We shall examine what the formulas for the dielectric permitttivity (18.21) and the longitudinal susceptibility (18.23) give in the limiting case of a spatially uniform medium, i.e., as $\mathbf{k} \to 0$. In this case, it is necessary to make explicit the undetermined quantities in the formulas. To within terms linear in **k**, we have

$$\rho_{\boldsymbol{k}} = \frac{1}{\sqrt{V}} \int \rho(\boldsymbol{x}) e^{-i(\boldsymbol{k}\cdot\boldsymbol{x})} d\boldsymbol{x} \cong \frac{1}{\sqrt{V}} \int \rho(\boldsymbol{x})\, d\boldsymbol{x} - \frac{i\boldsymbol{k}}{\sqrt{V}} \cdot \int \boldsymbol{x}\rho(\boldsymbol{x})\, d\boldsymbol{x},$$

i.e.,

$$\rho_{\boldsymbol{k}} \cong \frac{eN}{\sqrt{V}} - \frac{i\boldsymbol{k}}{\sqrt{V}} \cdot \boldsymbol{P}_d, \qquad (18.34)$$

where

$$\boldsymbol{P}_d = \int \boldsymbol{x}\rho(\boldsymbol{x})\, d\boldsymbol{x} \qquad (18.34\text{a})$$

is the operator of the total dipole moment. Consequently, as $\mathbf{k} \to 0$, the formula (18.21) takes the form

$$\varepsilon^{-1}(0, \omega) = 1 + \frac{4\pi}{3V} \langle\langle \boldsymbol{P}_d \cdot | \boldsymbol{P}_d \rangle\rangle_\omega, \tag{18.35}$$

where we have used the condition that the medium is isotropic. The dot denotes a scalar product.

From formula (18.23) for $\mathbf{k} \to 0$, we obtain

$$\chi^1(0, \omega) = \frac{1}{3} \langle\langle \boldsymbol{P}_d \cdot | \boldsymbol{P}_d \rangle\rangle_\omega. \tag{18.36}$$

18.2. Symmetry Properties and Dispersion Relations

The susceptibility tensor (18.13), and also the dielectric permittivity (18.21) and magnetic permeability (18.26), possess the symmetry property (17.9), (17.25):

$$\begin{aligned} \chi_{\alpha\beta}(\boldsymbol{k}, \omega) &= \chi^*_{\alpha\beta}(-\boldsymbol{k}, -\omega), \\ \varepsilon(\boldsymbol{k}, \omega) &= \varepsilon^*(-\boldsymbol{k}, -\omega), \\ \mu(\boldsymbol{k}, \omega) &= \mu^*(-\boldsymbol{k}, -\omega), \end{aligned} \tag{18.37}$$

like all generalized susceptibilities and kinetic coefficients. This follows from the fact that $\chi_{\alpha\beta}(\boldsymbol{x} - \boldsymbol{x}', t - t')$ is real.

Taking into account, in addition, the symmetry of the Green functions under inversion of the spatial coordinates $\mathbf{x} \to -\mathbf{x}$, which is equivalent to the replacement $\mathbf{k} \to -\mathbf{k}$, we shall have

$$\langle\langle j^\alpha_{\boldsymbol{k}} | j^\beta_{-\boldsymbol{k}} \rangle\rangle_\omega = \langle\langle j^\alpha_{-\boldsymbol{k}} | j^\beta_{\boldsymbol{k}} \rangle\rangle_\omega, \quad \langle\langle \rho_{\boldsymbol{k}} | \rho_{-\boldsymbol{k}} \rangle\rangle_\omega = \langle\langle \rho_{-\boldsymbol{k}} | \rho_{\boldsymbol{k}} \rangle\rangle_\omega. \tag{18.38}$$

Consequently, the susceptibility tensor (18.13), and also the dielectric permittivity (18.21) and the magnetic permeability (18.26), possess the symmetry properties:

$$\begin{aligned} \chi_{\alpha\beta}(\boldsymbol{k}, \omega) &= \chi^*_{\alpha\beta}(\boldsymbol{k}, -\omega) = \chi_{\alpha\beta}(-\boldsymbol{k}, \omega), \\ \varepsilon(\boldsymbol{k}, \omega) &= \varepsilon^*(\boldsymbol{k}, -\omega) = \varepsilon(-\boldsymbol{k}, \omega), \\ \mu(\boldsymbol{k}, \omega) &= \mu^*(\boldsymbol{k}, -\omega) = \mu(-\boldsymbol{k}, \omega). \end{aligned} \tag{18.37a}$$

Thus, the real parts of $\chi_{\alpha\beta}(\boldsymbol{k}, \omega)$, $\varepsilon(\boldsymbol{k}, \omega)$ and $\mu(\boldsymbol{k}, \omega)$ are symmetric with respect to the replacement $\omega \to -\omega$, while the imaginary parts are antisymmetric:

$$\begin{aligned} \operatorname{Re} \chi_{\alpha\beta}(\boldsymbol{k}, \omega) &= \operatorname{Re} \chi_{\alpha\beta}(\boldsymbol{k}, -\omega), \\ \operatorname{Re} \varepsilon(\boldsymbol{k}, \omega) &= \operatorname{Re} \varepsilon(\boldsymbol{k}, -\omega), \end{aligned}$$

$$\begin{aligned} \operatorname{Re}\mu(\boldsymbol{k},\omega) &= \operatorname{Re}\mu(\boldsymbol{k},-\omega),\\ \operatorname{Im}\chi_{\alpha\beta}(\boldsymbol{k},\omega) &= -\operatorname{Im}\chi_{\alpha\beta}(\boldsymbol{k},-\omega),\\ \operatorname{Im}\varepsilon(\boldsymbol{k},\omega) &= -\operatorname{Im}\varepsilon(\boldsymbol{k},-\omega),\\ \operatorname{Im}\mu(\boldsymbol{k},\omega) &= -\operatorname{Im}\mu(\boldsymbol{k},-\omega). \end{aligned} \tag{18.39}$$

Taking (18.39) into account, we obtain from (16.34) the Kramers–Krönig dispersion relations† for the susceptibility $\chi_{\alpha\beta}(\boldsymbol{k},\omega)$:

$$\begin{aligned} \operatorname{Re}\chi_{\alpha\beta}(\boldsymbol{k},\omega) - \frac{e^2 n}{m}\delta_{\alpha\beta} &= \frac{2}{\pi}\mathscr{P}\int_0^\infty \frac{u}{u^2-\omega^2}\operatorname{Im}\chi_{\alpha\beta}(\boldsymbol{k},u)\,du,\\ \operatorname{Im}\chi_{\alpha\beta}(\boldsymbol{k},\omega) &= -\frac{2\omega}{\pi}\mathscr{P}\int_0^\infty \frac{\operatorname{Re}\chi_{\alpha\beta}(\boldsymbol{k}\ u) - \frac{e^2 n}{m}\delta_{\alpha\beta}}{u^2-\omega^2}\,du. \end{aligned} \tag{18.40}$$

18.3. System of Particles with Spin in an Electromagnetic Field

We shall examine the effect of an electromagnetic field on a system of particles with spin, and calculate the average current and the magnetic moment associated with the presence of the spin.

The interaction of the particle spins with a magnetic field $\mathbf{H}(\mathbf{x}, t)$ is described by the operator

$$H' = -\int \boldsymbol{M}(x)\cdot\boldsymbol{H}(\boldsymbol{x},t)\,d\boldsymbol{x}, \tag{18.41}$$

where

$$\boldsymbol{M}(\boldsymbol{x}) = \sum_s \psi^+(\boldsymbol{x},s)\frac{e\hbar}{2mc}\sigma\psi(\boldsymbol{x},s) \tag{18.41 a}$$

is the operator of the magnetic moment density, and σ_x, σ_y, and σ_x are the Pauli matrices. We assume that the spin of the particles is equal to 1/2.

The perturbation operator (18.41) can be written in the form

$$H' = \int \boldsymbol{M}(\boldsymbol{x})\cdot\operatorname{curl}\boldsymbol{A}(\boldsymbol{x},t)\,d\boldsymbol{x} = -\frac{1}{c}\int \boldsymbol{j}_m(\boldsymbol{x})\cdot\boldsymbol{A}(\boldsymbol{x},t)\,d\boldsymbol{x}, \tag{18.42}$$

†The generalization of the Kramers–Krönig formulas to the relativistic case was carried out by Leontovich [117, 112].

where

$$j_m(x) = c\,\mathrm{curl}\,M(x) \tag{18.42a}$$

is the operator of the current associated with the magnetic moment of the particles. In fact, in accordance with the usual definition (18.6) of the current,

$$j_m(x) = -c\frac{\delta H'}{\delta A(x)}. \tag{18.43}$$

According to (15.47), the perturbation (18.41) changes the magnetic moment

$$\langle M(x)\rangle = \langle M(x)\rangle_0 + \int\int_{-\infty}^{t} \chi_m(xt, x't')\cdot H(x', t')\,dx'\,dt', \tag{18.44}$$

where

$$\chi_m(xt, x't') = -\langle\langle M(x, t)\,M(x', t')\rangle\rangle \tag{18.44a}$$

is the tensor of the magnetic susceptibility associated with the spin.

The relation (18.44) has a specially simple form in the Fourier representation. Putting

$$\begin{aligned}\langle M(x)\rangle &= \langle M(x)\rangle_0 + \frac{1}{V}\sum_k \int_{-\infty}^{\infty} M(k, \omega)\,e^{i(k\cdot x)-i\omega t}\,d\omega,\\ M(x, t) &= \frac{1}{\sqrt{V}}\sum_k M_k(t)\,e^{i(k\cdot x)}\end{aligned} \tag{18.45}$$

and using the spatial uniformity of the system, we obtain

$$M(k, \omega) = \chi_m(k, \omega)\cdot H(k, \omega), \tag{18.46}$$

where

$$\chi_m(k, \omega) = -\langle\langle M_k \mid M_{-k}\rangle\rangle_\omega \tag{18.46a}$$

is the magnetic susceptibility tensor of the spin system in the Fourier representation.

For the case of a magnetic field that is constant in space, formulas (18.46) and (18.46a) go over into (15.82) and (15.83), which were considered earlier. The tensor $\chi_m(k, \omega)$ satisfies symmetry conditions and dispersion relations analogous to (18.39) and (18.40).

18.4. System of Particles with a Dipole Moment

Another case, which is of interest for the theory of dielectrics is that of a system of interacting particles with a dipole moment.

We shall calculate the polarization induced by an electric field in such a system. The interaction of the dipole moments of the particles with the electric field is described by the operator

$$H' = -\int \boldsymbol{P}(\boldsymbol{x}) \cdot \boldsymbol{E}(\boldsymbol{x}, t)\, d\boldsymbol{x}, \tag{18.47}$$

where $\mathbf{P}(\mathbf{x})$ is the operator (or dynamic variable) of the dipole moment density. The perturbation (18.47) induces a dipole moment with density

$$\langle \boldsymbol{P}(\boldsymbol{x})\rangle = \langle \boldsymbol{P}(\boldsymbol{x})\rangle_0 + \int \int_{-\infty}^{t} \alpha(\boldsymbol{x}-\boldsymbol{x}', t-t') \cdot \boldsymbol{E}(\boldsymbol{x}', t')\, d\boldsymbol{x}'\, dt', \tag{18.48}$$

where

$$\alpha(\boldsymbol{x}-\boldsymbol{x}', t-t') = -\langle\langle \boldsymbol{P}(\boldsymbol{x}, t)\ \boldsymbol{P}(\boldsymbol{x}', t')\rangle\rangle \tag{18.48a}$$

is the dielectric polarizability tensor of the system of electric dipoles, and $\langle \boldsymbol{P}(\boldsymbol{x})\rangle_0$ is the dipole moment density in the equilibrium state as $\mathbf{E} \to 0$, which can be nonzero for ferroelectrics. Going over in (18.48) to a Fourier representation of the type (18.45) and using the spatial uniformity of the system, we obtain

$$\boldsymbol{P}(\boldsymbol{k}, \omega) = \alpha(\boldsymbol{k}, \omega) \cdot \boldsymbol{E}(\boldsymbol{k}, \omega), \tag{18.49}$$

where

$$\alpha(\boldsymbol{k}, \omega) = -\langle\langle \boldsymbol{P}_{\boldsymbol{k}} \mid \boldsymbol{P}_{-\boldsymbol{k}}\rangle\rangle_\omega$$

is the dielectric polarizability tensor of the system, as a function of $\mathbf{k}$ and ω.

For a system in a uniform electric field, or for sufficiently long waves, when we can neglect the variation of the field over the correlation length, the connection between the induced moment and the acting field is local,

$$\boldsymbol{P}(\omega) = \alpha(\omega) \cdot \boldsymbol{E}(\omega), \tag{18.50}$$

$$\alpha(\omega) = -\langle\langle \boldsymbol{P} \mid \boldsymbol{P}\rangle\rangle_\omega, \tag{18.50a}$$

where

$$\boldsymbol{P} = \int \boldsymbol{P}(\boldsymbol{x})\, d\boldsymbol{x}$$

is the total electric dipole moment. Writing formula (18.50a) explicitly in terms of the matrix elements of the polarization operator P_i, we obtain

$$\begin{aligned} \operatorname{Re} \alpha_{xx}(\omega) &= Q^{-1} \sum_{k,\,n} e^{-\beta E_k} \frac{2\omega_{nk}}{\omega_{nk}^2 - \omega^2} \left| P_x^{kn} \right|^2, \\ \operatorname{Im} \alpha_{xx}(\omega) &= Q^{-1} \sum_{k,\,n} e^{-\beta E_k} \frac{1}{2} \left| P_x^{kn} \right| \left(\delta(\omega_{nk} - \omega) - \delta(\omega_{nk} + \omega)\right), \end{aligned} \qquad (18.51)$$

where

$$P_x^{kn} = (C_k^* P_x C_n), \qquad \omega_{nk} = (E_n - E_k)/\hbar,$$

C_k and E_k being the wave function and energy of the state k.

Thus, we have obtained a Kramers–Heisenberg formula [100, 106, 107] with statistical averaging.

Chapter IV

The Nonequilibrium Statistical Operator

In Chapter III we studied those nonequilibrium processes which can be represented as the response of the system to external mechanical perturbations. There exist, however, nonequilibrium processes which occur as a result of thermal perturbations, i.e., which are caused by internal inhomogeneities in the system; diffusion, thermal conduction and viscosity are examples. Attempts have occasionally been made to express these in terms of mechanical perturbations. Such an approach is inadequate in that it assumes a prior knowledge of the equations of nonequilibrium thermodynamics and uses an analogy with mechanical perturbations. Besides, the division of perturbations into mechanical and thermal types is, in general, justified only in the linear approximation. In higher approximations, mechanical perturbations create inhomogeneities in the distributions of mass, energy, and momentum and, consequently, lead to the appearance of thermal perturbations.

To develop a statistical thermodynamics of nonequilibrium processes such that thermal perturbations are also included, it is necessary, strictly speaking, to construct statistical ensembles representing the macroscpic conditions in which the systems are found. This turns out to be possible, if we are interested in the behavior of the system for time intervals that are not too small, when the details of the initial state of the system become unimportant and the number of parameters necessary for the description of the state of the system is reduced. This idea for simplifying the description of the system was proposed by Bogolyubov, who used it to construct kinetic equations based on the Liouville equation [1].

In this chapter we shall formulate a statistical theory of irreversible processes using the method of statistical ensembles for nonequilibrium systems [2-5, 170, 184-190], generalizing the usual method of Gibbsian ensembles developed in Chapters I and II. The possibility of carrying over Gibbs' ideas into nonequilibrium statistical mechanics was first noted by Callen and Welton [6] in connection with the fluctuation-dissipation theorem.

In the study of irreversible processes caused by mechanical perturbations, all authors follow one method, namely, a dynamic interpretation of the perturbations under conditions of statistical equilibrium at some initial time (usually at $t = -\infty$). For the study of thermal perturbations, however, many different methods have been proposed. Following Zwanzig [7], these may be divided into the following groups.

1. Indirect methods from the theory of linear response. These are based on representing the effect of thermal perturbations in terms of mechanical perturbations, since the same transport processes can be caused either by external fields or by inhomogeneities in the system [8, 9, 10, 11, 12, 246]. Usually one first calculates the susceptibility to a fictitious external perturbation that could create the given nonequilibrium state, and, from this, the kinetic coefficients by using the fluctuation-dissipation theorem and taking the limit of zero wave number and frequency of the perturbation. In these methods it is usually assumed *a priori* that the macroscopic equations, for example, the Navier– Stokes equation, are valid.

2. Methods using the theory of stochastic processes and the Fokker-Planck equation. These methods, which have their origin in the theory of Brownian motion, were developed chiefly by Kirkwood [13] and Green [14] on the assumption that the processes are Markovian. Uhlenbeck and Ornstein [252], Krylov, and Bogolyubov [169], and Kirkwood and Green succeeded in obtaining the first important results in the general theory of irreversible processes, namely, the relation between the kinetic coefficients and the time correlation functions, in precisely this way. The method of stochastic processes has been developed further by different authors [15-21]. Mori and Kubo have recently improved the method with a

lowance for retardation in the Langevin equation [22–24]. The effect of non-Markovian processes was investigated by Zwanzig [21].

In this group we may also include the work of Helfand [25], in which the Einstein relation $\langle R^2 \rangle = 6Dt$ for the mean square displacement of a Brownian particle in time t (D is the diffusion coefficient) is generalized for other transport processes and used to express the kinetic coefficients in terms of correlation functions.

3. Methods based on a hypothesis about the character of the attenuation, or regression, of the fluctuations. This approach has its origins in the classical papers of Onsager [26], who proposed the hypothesis that the attenuation of the fluctuations follows the same law as the change of the corresponding macroscopic variables.† Taking into account also the reversibility of the microscopic equations of motion, he established his reciprocity relations for the kinetic coefficients (cf. §17.3). This method has been used by Kubo, Yokota, and Nakajima [28] to construct a theory of thermal perturbations in irreversible processes, and later by Felderhof and Oppenheim [12], who studied the space and time dispersion of the kinetic coefficients.

4. Methods based on the use of a local-equilibrium distribution as an initial condition for the Liouville equation. In these it is assumed that, in a weakly nonequilibrium state, a distribution (a local-equilibrium distribution, to be studied in detail in §20) is established in small volumes of the system, and that this distribution is close in form to the equilibrium Gibbsian distribution but with parameters depending on spatial position. Corrections to this distribution are sought. This approach has been developed chiefly by Mori [29–31]. Green [32] obtained expressions for the kinetic coefficients by making use of a local Maxwellian distribution as the initial condition and using the method of Chapman and Enskog to solve the Liouville equation. The connection between the correlator formulas for the kinetic coefficients and the Chapman–Enskog method has been analyzed by Ernst [33]. Klinger [34] has applied Mori's method to the theory of transport processes in semiconductors, and Provotorov [35] has applied it, in combination with Zwanzig's projection-operator method [21], to spin systems. Pelet-

†For Onsager's hypothesis on the attenuation of fluctuations and its region of applicability, see Chapter IV of the monograph by de Groot and Mazur [27], and the references cited therein.

minskii and Yatsenko [36] have developed the method further and have extended its range of applicability to smaller time scales, so that it is now also suitable for the construction of kinetic equations for strongly nonequilibrium states. (See also [248]). The local-equilibrium distribution, in combination with Zwanzig's method, has also been used by Robertson [219] (see also [249]).

5. Methods based on Gibbsian statistical ensembles for nonequilibrium systems. These may be divided into two distinct groups: the method of the nonequilibrium statistical operator, proposed by the author [2-5], which is based on the construction of local integrals of motion, and the method of MacLennan [37-40], which is based on a calculation of the influence of a thermostat in terms of nonconservative forces. The methods [2-5] and [37-40] were developed independently and lead to essentially the same results, although the method of [2-5] evidently possesses greater generality and is simpler to apply. An account of the method of the nonequilibrium statistical operator on the basis of [2-5, 170, 184-190] will be given in this chapter, and its connection with MacLennan's method will be discussed in Appendices II and III.†

The method of [2-5] has been applied by many authors [41-57] to different problems in the theory of irreversible processes. The method is especially simple for the construction of hydrodynamic equations, for example, when internal degrees of freedom are taken into account [4, 41, 228], for the equations of relativistic hydrodynamics [5], and for equations of the type found by Grad (Khazanovich and Savchenko [42]). Buishvili and the author [43], Buishvili [44], and Khutsishvili [45] have applied this method to the theory of nuclear spin diffusion, Buishvili, Bendiashvili, Giorgadze, Zviadadze, and Khutsishvili [46-50] have applied it to the theory of nuclear magnetic resonance and dynamic polarization of nuclei in solids, and Khazanovich has applied it to nuclear magnetic relaxation in liquids [183]. Buishvili and Zviadadze [51] and Grachev [52] have applied it to the theory of spin-lattice relaxation of impurity centers. Kalashnikov has applied the method to the theory of spin-lattice relaxation of conduction electrons [53], and

†In his classification of the different theories of nonequilibrium processes, Zwanzig also mentions methods based on information theory (prediction theory), but refers only to a private communication of Jaynes and Scalapino. The methods of information theory can, in fact, be applied to determine the nonequilibrium statistical operator, as is shown in papers [170, 189] (see §27).

to the theory of hot electrons in semiconductors [54], in which it turned out to be suitable not only for linear dissipative processes but also in the case of strong nonlinearity in the electric field. As was shown by Pokrovskii [55], the method gives the correct result for the rate of exchange of energy between weakly interacting subsystems for a state which is strongly nonlinear in the thermodynamic forces. Pokrovskii [56] has also shown that the method may be applied to obtain generalized kinetic equations of the type found in the work of Peletminskii and Yatsenko [36], by making the appropriate choice of the parameters describing the state of the system. Using the same method, it is possible to obtain not only the usual kinetic equations, but also equations of the Fokker–Planck type, as has been shown by A. G. Bashkirov and the author [57, 190]. (For applications of the method, see also [191-215, 228-234].)

The above classification of methods for studying thermal perturbations is not completely rigid, since, in a number of papers, the indirect method of the theory of linear response is combined with Onsager's hypothesis [12] or with the use of a local-equilibrium distribution as the initial condition [10].

The various methods used in the study of thermal perturbations are described in the review articles of Chester [58], Zwanzig [7], and Ernst et al. [216], and in the monograph by Rice and Gray [59], in which a large bibliography is given. All these methods lead to the same results for the kinetic coefficients, but each has its own region of applicability. Certain authors, however, expressed doubts about the validity of the expressions for the kinetic coefficients in terms of correlation functions [60, 61], but these doubts turned out to be unfounded [62], and the objections were withdrawn by the authors themselves [63, 64]. For a proof of the equivalence of the different methods of constructing nonequilibrium statistical operators, see the paper by the author and Kalashnikov [229].

In most of the papers listed above, apart from [2-5, 37, 38] nonequilibrium corrections to the equilibrium distribution function are calculated. The fundamental question with which we shall be concerned in the following is how to construct, starting from general principles, the complete statistical operator (or, in the classical case, the distribution function) for nonequilibrium processes, i.e., how to generalize the ideas of Gibbsian statistical ensembles to the nonequilibrium case. In the paper [2], a nonequilibrium statistical operator was constructed for stationary processes by generalizing the class of the integrals of motion on which the statistical operator can depend; their meaning will be elucidated in §21. For nonstationary processes, MacLennan [37, 38] has obtained the nonequilibrium statistical operator by another method, by treating the influence of the thermostat in terms of nonconservative forces. The method of local integrals of motion [5] leads to exactly the same distribution as does the method of MacLennan [37, 38].

To construct local integrals of motion, it is necessary to formulate conservation laws for the mechanical quantities in operator form (or in the form of relations for the dynamic variables) and find expressions for the corresponding densities and fluxes of the mechanical quantities. This problem is examined for different systems in §19.

§ 19. Conservation Laws

Conservation laws play a fundamental role in the whole of theoretical physics. The phenomenological thermodynamics of irreversible processes is based on conservation laws for the average values of physical quantities, for example, the number of particles, energy and momentum [27]. The statistical thermodynamics of nonequilibrium processes, which will be described below, also starts from conservation laws, although not for the average values of dynamic quantities but for the dynamic quantities themselves. Thus, the conservation laws will be considered from a microscopic, rather than a macroscopic, point of view.

In this section, we shall study the conservation laws for a system of identical particles with direct interaction, and for a mixture of particles with internal degrees of freedom. These examples make it possible to obtain local conservation laws of sufficiently general form for the energy, momentum, and particle number, and these will serve below as the basis for the construction of nonequilibrium statistical thermodynamics.

19.1. Local Conservation Laws for the Case of Classical Mechanics

We shall consider the conservation laws for energy, momentum, and particle number in local form for the case of classical mechanics. The quantum case will be considered later.

Let a system of identical interacting particles be described by the Hamiltonian (1.2)

$$H = \sum_i \left(\frac{p_i^2}{2m} + \frac{1}{2} \sum_{j \neq i} \phi(|x_i - x_j|) \right), \tag{19.1}$$

where $\phi(|x_i - x_j|)$ is the potential energy of the interaction between the particles, and m is the particle mass. Hamilton's equations (1.1), describing the motion of the particles, have the form

$$x_i = \frac{\partial H}{\partial p_i} = \frac{p_i}{m}, \quad \dot{p}_i = -\frac{\partial H}{\partial x_i} = \sum_{j \neq i} F_{ij}, \tag{19.2}$$

where

$$F_{ij} = -F_{ji} = -\frac{\partial \phi(|x_i - x_j|)}{\partial x_i} \tag{19.2a}$$

is the force of the interaction between particles i and j.

The role of one dynamic variable, the particle-number density, is played by the function

$$n(x) = \sum_i \delta(x_i - x), \tag{19.3}$$

(where the summation is performed over all the particles), since the integral of (19.3) over the volume is equal to the total number of particles, and the average value of the integral over a small volume is equal to the average number of particles in this volume. The particle coordinates x_i in (19.3) change in time in accordance with the equations of motion (19.2); consequently, the time derivative of n(x) is equal to

$$\frac{\partial n(x)}{\partial t} = \sum_i \dot{\delta}(x_i(t) - x) = \sum_i \dot{x}_i \cdot \nabla_i \delta(x_i - x),$$

i.e.,

$$\frac{\partial n(x)}{\partial t} = -\operatorname{div} j(x), \tag{19.4}$$

where

$$j(x) = \sum_i \frac{p_i}{m} \delta(x_i - x) \tag{19.5}$$

is the particle-number flux density. Equation (19.4) is the conservation law for the number of particles in local form.†

As in any field theory, the densities of the mechanical quantities and the densities of their fluxes are not defined uniquely: the densities are defined within the divergence of an arbitrary vector, and the fluxes are defined within the curl of an arbitrary vector. In fact, on integration over the volume, the divergences contribute only a surface integral, while the divergence of the curl part is equal to zero. This indeterminacy is connected with the fact that quantities such as densities are not observable quantities. Only integrals of these over a volume that is small on the macroscopic scale but contains a large number of particles are observable in this case, the surface contribution can be neglected.

We shall now obtain the momentum conservation law in local form. For the momentum density, it is natural to introduce the quantity

$$p(x) = mj(x) = \sum_i p_i \delta(x_i - x). \tag{19.6}$$

The integral of (19.6) over the whole volume is equal to the total momentum **P**,

$$\int p(x)\,dx = \sum_i p_i = P, \tag{19.6a}$$

and the average value of (19.6) gives the momentum flux density.

Differentiating (19.6) with respect to time, we obtain

$$\frac{\partial p(x)}{\partial t} = -\nabla \cdot \sum_i \frac{1}{m} p_i p_i \delta(x_i - x) + \frac{1}{2} \sum_{\substack{i,j \\ (i \neq j)}} F_{ij} (\delta(x_i - x) - \delta(x_j - x)), \tag{19.7}$$

where $\mathbf{p}_i \mathbf{p}_i$ is a tensor; the second summand in (19.7) has been symmetrized using (19.2a).

The equation (19.7) still does not have the form of a local conservation law, since the second term in the right-hand side is not represented in the form of a divergence. In order to trans-

†In accordance with tradition, in the equation of motion for n(**x**) (19.4), we write ∂/∂t, and not d/dt as would follow from (2.20). We shall also use this notation in the remainder of the book.

form it to such a form we remark that, below, we shall be interested not in the dynamic variables themselves, but in integrals of them multiplied by functions which vary little over distances of the order of the effective range of the interaction.

We shall examine the second term in the right-hand side of (19.7)

$$B(x)=\frac{1}{2}\sum_{\substack{i,j\\(i\neq j)}} F_{ij}\left(\delta(x_i-x)-\delta(x_j-x)\right) \tag{19.8}$$

and show how it can be represented in the form of the divergence of a tensor.

We multiply **B(x)** by an arbitrary vector function **A(x)** of the spatial coordinates, varying little over distances of the order of the effective range of the forces, and consider the integral of (19.8) over all space

$$\int A(x)\cdot B(x)\,dx=\frac{1}{2}\sum_{\substack{i,j\\(i\neq j)}} F_{ij}\cdot\left(A(x_i)-A(x_j)\right). \tag{19.9}$$

The force $\mathbf{F}_{ij}$ differs appreciably from zero only for distances of the order of the effective range of the forces, and the function **A(x)**, by assumption, differs little over such distances; therefore, the difference $\mathbf{A}(\mathbf{x}_i)-\mathbf{A}(\mathbf{x}_j)$ can be expanded in a Taylor series in $\mathbf{x}_i-\mathbf{x}_j=\mathbf{x}_{ij}$ with only the first term retained:

$$A(x_i)-A(x_j)\cong\sum_\beta \frac{\partial A}{\partial x_\beta}x_{ij}^\beta.$$

Consequently,

$$\int B(x)\cdot A(x)\,dx=\frac{1}{2}\sum_{i,j,\beta}F_{ij}\cdot\frac{\partial A(x_i)}{\partial x_i^\beta}x_{ij}^\beta$$

$$=\frac{1}{2}\sum_{i,j,\beta}\int F_{ij}\cdot\frac{\partial A(x)}{\partial x^\beta}x_{ij}^\beta\,\delta(x_i-x)\,dx,$$

or, after integration by parts,

$$\int B(x)\cdot A(x)\,dx=-\frac{1}{2}\sum_{i,j}\int A(x)\cdot\sum_\beta\frac{\partial}{\partial x_\beta}F_{ij}\,x_{ij}^\beta\,\delta(x_i-x)\,dx,$$

whence, since **A**(**x**) is arbitrary, we obtain

$$B_\alpha(\boldsymbol{x}) = -\sum_\beta \frac{\partial}{\partial x_\beta} \frac{1}{2} \sum_{i,\,j} F_{ij}^\alpha x_{ij}^\beta \,\delta(\boldsymbol{x}_i - \boldsymbol{x}). \qquad (19.10)$$

Thus, the quantity (19.8) can be represented approximately in the form of the divergence of a tensor, and (19.7) can be written in the form of a local conservation law:

$$\frac{\partial p_\alpha(\boldsymbol{x})}{\partial t} = -\sum_\beta \frac{\partial T_{\beta\alpha}(\boldsymbol{x})}{\partial x_\beta}, \qquad (19.11)$$

where

$$T_{\beta\alpha}(\boldsymbol{x}) = \sum_i \left(\frac{1}{m} p_i^\alpha p_i^\beta + \frac{1}{2} \sum_{j \neq i} F_{ij}^\alpha x_{ij}^\beta \right) \delta(\boldsymbol{x}_i - \boldsymbol{x}) \qquad (19.12)$$

is the symmetric stress tensor. In fact, since the forces are centrally symmetric, the tensor

$$F_{ij}^\alpha x_{ij}^\beta = -\frac{\partial \phi(|\boldsymbol{x}_{ij}|)}{\partial |\boldsymbol{x}_{ij}} \frac{1}{|\boldsymbol{x}_{ij}|} x_{ij}^\alpha x_{ij}^\beta$$

is a symmetric tensor.

We note that the same result for $T_{\beta\alpha}(\boldsymbol{x})$ would have been obtained if we had formally expanded the δ-function in (19.8) in a Taylor series and confined ourselves to two terms of the expansion:

$$\delta(\boldsymbol{x}_j - \boldsymbol{x}) = \delta(\boldsymbol{x}_i - \boldsymbol{x}) + (\boldsymbol{x}_j - \boldsymbol{x}_i) \cdot \nabla_i \delta(\boldsymbol{x}_i - \boldsymbol{x}).$$

The above arguments using the arbitrary slowly varying function give a precise meaning to this expansion of the δ-function. The process of representing Eq. (19.7) in the form of the divergence of a tensor can be continued further, and corrections to $T_{\beta\alpha}(\boldsymbol{x})$ of a higher order of smallness can be found. We shall apply the method described in the quantum case also (cf. §19.2).†

We now examine the energy conservation law in local form. It is natural to define the quantity

$$H(\boldsymbol{x}) = \sum_i \left(\frac{p_i^2}{2m} + \frac{1}{2} \sum_{j \neq i} \phi(|\boldsymbol{x}_i - \boldsymbol{x}_j|) \right) \delta(\boldsymbol{x}_i - \boldsymbol{x}) \qquad (19.13)$$

†In (19.11) and (19.12), an exact representation in the form of a divergence is also possible, but with an extra integration over the parameter in (19.12) [217, 218, 225] (for details, see Appendix A of [218]).

as the dynamic variable of the energy density. It is obvious that

$$H = \int H(\boldsymbol{x})\,d\boldsymbol{x} \tag{19.14}$$

is the Hamiltonian of the system. Differentiating (19.13) and taking into account the equations of motion (19.2), we obtain

$$\frac{\partial H(\boldsymbol{x})}{\partial t} = -\nabla\cdot\sum_i\left(\frac{p_i^2}{2m}+\frac{1}{2}\sum_{j\neq i}\phi(|\boldsymbol{x}_i-\boldsymbol{x}_j|)\right)\frac{\boldsymbol{p}_i}{m}\,\delta(\boldsymbol{x}_i-\boldsymbol{x})+$$
$$+\frac{1}{4}\sum_{\substack{i,\,j\\(i\neq j)}}\frac{1}{m}(\boldsymbol{p}_i+\boldsymbol{p}_j)\cdot\boldsymbol{F}_{ij}(\delta(\boldsymbol{x}_i-\boldsymbol{x})-\delta(\boldsymbol{x}_j-\boldsymbol{x})). \tag{19.15}$$

Following the same procedure for smoothing the operators as was used in the derivation of Eq. (19.11), we obtain

$$\frac{\partial H(\boldsymbol{x})}{\partial t} = -\operatorname{div}\boldsymbol{j}_H(\boldsymbol{x}), \tag{19.16}$$

where

$$\boldsymbol{j}_H(\boldsymbol{x}) = \sum_i\left(\frac{p_i^2}{2m}+\frac{1}{2}\sum_{j\neq i}\phi(|\boldsymbol{x}_i-\boldsymbol{x}_j|)\right)\frac{\boldsymbol{p}_i}{m}\,\delta(\boldsymbol{x}_i-\boldsymbol{x})+$$
$$+\frac{1}{4}\sum_{i,\,j}\frac{1}{m}(\boldsymbol{p}_i+\boldsymbol{p}_j)\cdot\boldsymbol{F}_{ij}\boldsymbol{x}_{ij}\delta(\boldsymbol{x}_i-\boldsymbol{x}) \qquad (\boldsymbol{x}_{ij}=\boldsymbol{x}_i-\boldsymbol{x}_j) \tag{19.17}$$

is the energy flux density vector. In the same approximation, it can also be written in another form:

$$\boldsymbol{j}_H(\boldsymbol{x}) = \sum_i\left(\frac{\boldsymbol{p}_i\boldsymbol{p}_i}{2m}+\frac{1}{2}\sum_{j\neq i}\phi(|\boldsymbol{x}_i-\boldsymbol{x}_j|)\,U+\frac{1}{2}\sum_{j\neq i}\boldsymbol{x}_{ij}\boldsymbol{F}_{ij}\right)\cdot\frac{\boldsymbol{p}_i}{m}\,\delta(\boldsymbol{x}_i-\boldsymbol{x}), \tag{19.17a}$$

where the quantity in the large parentheses is a tensor, and U is the unit tensor.

Thus, the conservation equations for the energy, particle number, and momentum in local form are

$$\begin{aligned}\frac{\partial H(\boldsymbol{x})}{\partial t} &= -\operatorname{div}\boldsymbol{j}_H(\boldsymbol{x}),\\ \frac{\partial n(\boldsymbol{x})}{\partial t} &= -\operatorname{div}\boldsymbol{j}(\boldsymbol{x}),\\ \frac{\partial \boldsymbol{p}(\boldsymbol{x})}{\partial t} &= -\operatorname{Div}T(\boldsymbol{x}),\end{aligned} \tag{19.18}$$

where the densities of the mechanical quantities are defined by the expressions (19.13) for H(**x**), (19.3) for n(**x**), (19.6) for **p**(**x**), and **j**(**x**), (19.17) and (19.17a) for $\mathbf{j}_H(\mathbf{x})$, and (19.12) for T(**x**).

We note that the Fourier components of the functions n(x), H(x), and p (x)

$$\begin{aligned} n_k &= \int n(x) e^{-i k \cdot x} dx, \\ H_k &= \int H(x) e^{-i k \cdot x} dx, \\ p_k &= \int p(x) e^{-i k \cdot x} dx \end{aligned} \tag{19.19}$$

are collective variables. Indeed,

$$\begin{aligned} n_k &= \sum_j e^{-i k \cdot x_j}, \\ H_k &= \sum_j \frac{p_j^2}{2m} e^{-i k \cdot x_j} + \frac{1}{2V} \sum_q \nu(q)(n_q n_{k-q} - n_k), \\ p_k &= \sum p_j e^{-i k \cdot x_j} \qquad \left(\nu(q) = \int \phi(|x|) e^{-i(q\cdot x)} dx\right) \end{aligned} \tag{19.19a}$$

depend symmetrically on the coordinates and momenta of all the particles.

The collective variables are convenient for studying the collective properties of many-particle systems, and especially of systems with long-range (for example, Coulomb) interaction forces They have been applied in the papers [65-68].

19.2. Local Conservation Laws for the Case of Quantum Mechanics

We shall now study the laws of conservation of energy, momentum, and particle number in their local form for a quantum-mechanical system of identical particles with direct interaction between them, following the paper [184].

To obtain a simple, local form of the conservation laws, we carry out a smoothing of the operators over inhomogeneities smaller than the effective range of the forces between the particles, just as we did in the preceding subsection for classical dynamic variables. By a similar method, it is possible to obtain local conservation laws for other systems too.

A quantum-mechanical system of identical particles of mass m with direct interaction with potential ϕ (x) is described by the second quantized Hamiltonian

$$H = \int \psi^+(\boldsymbol{x}) \left\{ -\frac{\hbar^2}{2m} \nabla^2 + \frac{1}{2} \int \phi(\boldsymbol{x} - \boldsymbol{x}') \psi^+(\boldsymbol{x}') \psi(\boldsymbol{x}') \, d\boldsymbol{x}' \right\} \psi(\boldsymbol{x}) \, d\boldsymbol{x}, \qquad (19.20)$$

where the operators $\psi(\mathbf{x})$ and $\psi^+(\mathbf{x}')$, which act on the wave function in the occupation-number representation, satisfy the Fermi or Bose commutation relations

$$\begin{aligned} \psi(\boldsymbol{x}) \psi^+(\boldsymbol{x}') \pm \psi^+(\boldsymbol{x}') \psi(\boldsymbol{x}) &= \delta(\boldsymbol{x} - \boldsymbol{x}'), \\ \psi(\boldsymbol{x}) \psi(\boldsymbol{x}') \pm \psi(\boldsymbol{x}') \psi(\boldsymbol{x}) &= 0, \\ \psi^+(\boldsymbol{x}) \psi^+(\boldsymbol{x}') \pm \psi^+(\boldsymbol{x}') \psi^+(\boldsymbol{x}) &= 0, \end{aligned} \qquad (19.20\text{a})$$

where the plus and minus signs denote Fermi and Bose statistics respectively. If the particles have spin, then, apart from the coordinates $\mathbf{x}$, we must take the spin variables into account. Then in (19.20), in addition to integrating over $\mathbf{x}$ we must perform a sum over the spin variables.

By means of integration by parts, we can write the Hamiltonian (19.20) conveniently in the form

$$H = \int H(\boldsymbol{x}) \, d\boldsymbol{x}, \qquad (19.21)$$

where

$$H(\boldsymbol{x}) = \frac{\hbar^2}{2m} \nabla \psi^+(\boldsymbol{x}) \cdot \nabla \psi(\boldsymbol{x}) + \frac{1}{2} \int \phi(\boldsymbol{x} - \boldsymbol{x}') \psi^+(\boldsymbol{x}) \psi^+(\boldsymbol{x}') \psi(\boldsymbol{x}') \psi(\boldsymbol{x}) \, d\boldsymbol{x}' \qquad (19.21\text{a})$$

is the energy density operator, which is chosen to be Hermitian.

The energy density operator (19.21a) is not completely defined by the condition (19.21) since one can add to $H(\mathbf{x})$ a term representing the divergence of some vector; for example, we can define

$$H(\boldsymbol{x}) = -\frac{\hbar^2}{4m} \{\nabla^2 \psi^+(\boldsymbol{x}) \psi(\boldsymbol{x}) + \psi^+(\boldsymbol{x}) \nabla^2 \psi(\boldsymbol{x})\} + H_{\text{int}}(\boldsymbol{x}) = \frac{\hbar^2}{2m} \left\{ \nabla \psi^+(\boldsymbol{x}) \cdot \nabla \psi(\boldsymbol{x}) - \frac{1}{2} \nabla^2 n(\boldsymbol{x}) \right\} + H_{\text{int}}(\boldsymbol{x}) \qquad (19.21\text{b})$$

or

$$H(\boldsymbol{x}) = -\frac{\hbar^2}{8m} \{\nabla^2 \psi^+(\boldsymbol{x}) \psi(\boldsymbol{x}) - 2 \nabla \psi^+(\boldsymbol{x}) \cdot \nabla \psi(\boldsymbol{x}) + \psi^+(\boldsymbol{x}) \nabla^2 \psi(\boldsymbol{x})\} + H_{\text{int}}(\boldsymbol{x}) = \frac{\hbar^2}{2m} \left\{ \nabla \psi^+(\boldsymbol{x}) \cdot \nabla \psi(\boldsymbol{x}) - \frac{1}{4} \nabla^2 n(\boldsymbol{x}) \right\} + H_{\text{int}}(\boldsymbol{x}), \qquad (19.21\text{c})$$

where

$$n(\boldsymbol{x}) = \psi^+(\boldsymbol{x})\,\psi(\boldsymbol{x}) \tag{19.22}$$

is the particle-number density operator, and

$$H_{\text{int}}(\boldsymbol{x}) = \frac{1}{2}\int \phi(\boldsymbol{x}-\boldsymbol{x}')\,\psi^+(\boldsymbol{x})\,\psi^+(\boldsymbol{x}')\,\psi(\boldsymbol{x}')\,\psi(\boldsymbol{x})\,d\boldsymbol{x}'$$

is the interaction-energy density operator.

This indeterminacy in the definition of the densities of the quantities exists in any field theory. In quantum field theory, the definition (19.21a) is usually applied.

The operator $\psi(\mathbf{x})$ obeys equations of motion stemming from (19.20) and (19.20a):

$$i\hbar\frac{\partial\psi(\boldsymbol{x})}{\partial t} = [\psi(\boldsymbol{x}),\,H] = -\frac{\hbar^2}{2m}\nabla^2\psi(\boldsymbol{x}) + \int \phi(\boldsymbol{x})\,n(\boldsymbol{x}')\,\psi(\boldsymbol{x})\,d\boldsymbol{x}', \tag{19.23}$$

where $n(\mathbf{x})$ is the particle-number density operator (19.22) (see the footnote on p. 244).

The operator $n(\mathbf{x})$ satisfies a conservation law which follows from the equations of motion for $\psi(\mathbf{x})$ and $\psi^+(\mathbf{x})$:

$$\frac{\partial n(\boldsymbol{x})}{\partial t} + \operatorname{div}\boldsymbol{j}(\boldsymbol{x}) = 0, \tag{19.24}$$

where

$$\boldsymbol{j}(\boldsymbol{x}) = \frac{\hbar}{2mi}\{\psi^+(\boldsymbol{x})\nabla\psi(\boldsymbol{x}) - \nabla\psi^+(\boldsymbol{x})\,\psi(\boldsymbol{x})\} \tag{19.24a}$$

is the particle-number flux operator. The expression (19.24) is the quantum analog of the classical particle flux density (19.5).

We shall obtain the equation of conservation of momentum in the local form necessary for the derivation of the hydrodynamic equations. The momentum density operator $\mathbf{p}(\mathbf{x})$ for a one-component system differs from the particle flux operator (19.24a) only by a factor equal to the mass of a particle:

$$\boldsymbol{p}(\boldsymbol{x}) = m\boldsymbol{j}(\boldsymbol{x}). \tag{19.25}$$

The equations of motion for $\mathbf{p}(\mathbf{x})$ have the form

$$\dot{p}_\alpha(\boldsymbol{x}) + \sum_\beta \frac{\partial}{\partial x_\beta}\frac{\hbar^2}{2m}\left\{\frac{\partial\psi^+(\boldsymbol{x})}{\partial x_\beta}\frac{\partial\psi(\boldsymbol{x})}{\partial x_\alpha} + \frac{\partial\psi^+(\boldsymbol{x})}{\partial x_\alpha}\frac{\partial\psi(\boldsymbol{x})}{\partial x_\beta} - \frac{1}{2}\frac{\partial^2 n(\boldsymbol{x})}{\partial x_\beta\,\partial x_\alpha}\right\} =$$

$$= -\int \frac{\partial\phi(\boldsymbol{x}-\boldsymbol{x}')}{\partial x_\alpha}\,\psi^+(\boldsymbol{x})\,\psi^+(\boldsymbol{x}')\,\psi(\boldsymbol{x}')\,\psi(\boldsymbol{x})\,d\boldsymbol{x}' = -B_\alpha(\boldsymbol{x}) \tag{19.26}$$

$$(\alpha = 1,\ 2,\ 3).$$

The equation (19.26) is not yet in the usual form of a conservation law, since the operator $B_\alpha(\mathbf{x})$ is not represented in the form of the divergence of a tensor, but, as we shall see below, it is possible to do this with good accuracy for real systems with a short effective interaction range between the molecules.† In fact, we are interested not in the operator $B_\alpha(\mathbf{x})$ itself, but in integrals of it, of the type $\int \boldsymbol{A}(\boldsymbol{x}) \cdot \boldsymbol{B}(\boldsymbol{x})\, d\boldsymbol{x}$, where $\mathbf{A}(\mathbf{x})$ is some arbitrary vector function of the space coordinates that varies little over distances of the order of the range of the forces of interaction between the particles; we shall consider such integrals in §20 and below.

We have

$$\int A_\alpha(\boldsymbol{x})\, B_\alpha(\boldsymbol{x})\, d\boldsymbol{x} = \int A_\alpha(\boldsymbol{x}) \frac{\partial\phi(\boldsymbol{x}-\boldsymbol{x}')}{\partial x_\alpha} F(\boldsymbol{x}, \boldsymbol{x}')\, d\boldsymbol{x}\, d\boldsymbol{x}' =$$

$$= \frac{1}{2}\int (A_\alpha(\boldsymbol{x}) - A_\alpha(\boldsymbol{x}')) \frac{\partial\phi(\boldsymbol{x}-\boldsymbol{x}')}{\partial x_\alpha} F(\boldsymbol{x}, \boldsymbol{x}')\, d\boldsymbol{x}\, d\boldsymbol{x}',$$

where

$$F(\boldsymbol{x}, \boldsymbol{x}') = \psi^+(\boldsymbol{x})\,\psi^+(\boldsymbol{x}')\,\psi(\boldsymbol{x}')\,\psi(\boldsymbol{x})$$

is an operator symmetric with respect to $\mathbf{x}$ and $\mathbf{x}'$; we assume that the function $\phi(\mathbf{x}-\mathbf{x}')$ is radially symmetric.

Using the slow variation of the function $A_\alpha(\mathbf{x}')$, we expand it in a series in $\mathbf{x}-\mathbf{x}' = \mathbf{x}_1$, up to terms of first order, and then integrate by parts. Then

$$\int A_\alpha(\boldsymbol{x})\, B_\alpha(\boldsymbol{x})\, d\boldsymbol{x} \cong \frac{1}{2}\sum_\beta \int d\boldsymbol{x}\, \frac{\partial A_\alpha(\boldsymbol{x})}{\partial x_\beta} \int d\boldsymbol{x}_1\, x_{1\beta} \frac{\partial\phi(\boldsymbol{x}_1)}{\partial x_{1\alpha}} F(\boldsymbol{x}, \boldsymbol{x}-\boldsymbol{x}_1) =$$

$$= -\frac{1}{2}\int d\boldsymbol{x}\, A_\alpha(\boldsymbol{x}) \sum_\beta \frac{\partial}{\partial x_\beta} \int d\boldsymbol{x}'\, (x_\beta - x'_\beta) \frac{\partial\phi(\boldsymbol{x}-\boldsymbol{x}')}{\partial x_\alpha} F(\boldsymbol{x}, \boldsymbol{x}').$$

Since $A_\alpha(\mathbf{x})$ is arbitrary, we obtain

$$B_\alpha(\boldsymbol{x}) = -\sum_\beta \frac{\partial}{\partial x_\beta} \frac{1}{2} \int (x_\beta - x'_\beta)(x_\alpha - x'_\alpha) \frac{1}{r} \frac{d\phi(r)}{dr} \times$$

$$\times\, \psi^+(\boldsymbol{x})\,\psi^+(\boldsymbol{x}')\,\psi(\boldsymbol{x}')\,\psi(\boldsymbol{x})\, d\boldsymbol{x}', \qquad r = |\boldsymbol{x}-\boldsymbol{x}'|;$$

$B_\alpha(\mathbf{x})$ is, therefore, the divergence of a tensor. If this expression is taken into account, the equation of motion (19.26) for the momentum density takes the form of a local conservation law:

$$\frac{\partial p_\alpha(\boldsymbol{x})}{\partial t} + \sum_\beta \frac{\partial}{\partial x_\beta} T_{\beta\alpha}(\boldsymbol{x}) = 0, \qquad (19.27)$$

†See the footnote on p. 246.

where

$$T_{\beta\alpha}(x) = \frac{\hbar^2}{2m}\left\{\frac{\partial\psi^+(x)}{\partial x_\beta}\frac{\partial\psi(x)}{\partial x_\alpha} + \frac{\partial\psi^+(x)}{\partial x_\alpha}\frac{\partial\psi(x)}{\partial x_\beta} - \frac{1}{2}\frac{\partial^2 n(x)}{\partial x_\beta\,\partial x_\alpha}\right\} -$$
$$-\frac{1}{2}\int (x_\beta - x'_\beta)(x_\alpha - x'_\alpha)\frac{1}{r}\frac{d\phi(r)}{dr}\psi^+(x)\psi^+(x')\psi(x')\psi(x)\,dx' \qquad (19.27a)$$

is the stress tensor operator, analogous to the classical expression (19.12).

It follows from(19.27a) that the stress tensor must be symmetric:

$$T_{\alpha\beta}(x) = T_{\beta\alpha}(x);$$

this is connected with the radial symmetry of the interaction forces.

Using the same method it is also possible to take account of the higher terms of the expansion in $\mathbf{x} - \mathbf{x}'$; in the expression for $T_{\beta\alpha}(x)$ this leads to terms with higher derivatives of $\psi(\mathbf{x})$ which, for short-range forces, are very small (cf. the footnote on p. 246).

We now obtain a conservation equation for the energy density, since we shall need this later to study energy transport. By means of the commutation relations or the equations of motion (19.23), we obtain for the energy density operator (19.21a)

$$\frac{\partial H(x)}{\partial t} + \operatorname{div} j'_H(x) = B(x), \qquad (19.28)$$

where

$$j'_H(x) = \frac{\hbar^3}{4m^2 i}\{\nabla\psi^+(x)\nabla^2\psi(x) - \nabla^2\psi^+(x)\nabla\psi(x)\} +$$
$$+\frac{1}{2}\int \phi(x - x')\psi^+(x')j(x)\psi(x')\,dx',$$

$$B(x) = -\frac{1}{2}\int \nabla\phi(x - x')\cdot\{\psi^+(x')j(x)\psi(x') +$$
$$+\psi^+(x)j(x')\psi(x)\}\,dx'. \qquad (19.28a)$$

Following the method of smoothing over small inhomogeneities used above in the derivation of the equation of conservation of momentum in local form, we shall represent B(**x**) approximately in the form of the divergence of a vector. To do this, we write the expression

$$\int A(\boldsymbol{x}) B(\boldsymbol{x})\, d\boldsymbol{x} = -\frac{1}{4} \int A(\boldsymbol{x}) (\nabla\phi(\boldsymbol{x}-\boldsymbol{x}') - \nabla'\phi(\boldsymbol{x}-\boldsymbol{x}')) \cdot \boldsymbol{F}(\boldsymbol{x}, \boldsymbol{x}')\, d\boldsymbol{x}\, d\boldsymbol{x}'$$

$$= -\frac{1}{4} \int (A(\boldsymbol{x}) - A(\boldsymbol{x}')) \nabla\phi(\boldsymbol{x}-\boldsymbol{x}') \cdot \boldsymbol{F}(\boldsymbol{x}, \boldsymbol{x}')\, d\boldsymbol{x}\, d\boldsymbol{x}',$$

where

$$\boldsymbol{F}(\boldsymbol{x}, \boldsymbol{x}') = \psi^{+}(\boldsymbol{x}') \boldsymbol{j}(\boldsymbol{x}) \psi(\boldsymbol{x}') + \psi^{+}(\boldsymbol{x}) \boldsymbol{j}(\boldsymbol{x}') \psi(\boldsymbol{x})$$

is an operator symmetric with respect to **x** and **x'**. A(**x**) is an arbitrary function varying slowly over distances of the order of the range of the interaction between the particles.

Using the slowness of the variation of the function A(**x'**), we expand it in a series in $\mathbf{x}_1 = \mathbf{x} - \mathbf{x}'$ up to terms of first order. After integrating by parts, we have

$$\int A(\boldsymbol{x}) B(\boldsymbol{x})\, d\boldsymbol{x} \cong -\frac{1}{4} \int \sum_{\alpha} \frac{\partial A(\boldsymbol{x})}{\partial x_\alpha} x_{1\alpha} \nabla\phi(\boldsymbol{x}_1) \cdot \boldsymbol{F}(\boldsymbol{x}, \boldsymbol{x}-\boldsymbol{x}_1)\, d\boldsymbol{x}\, d\boldsymbol{x}_1$$

$$= \int d\boldsymbol{x}\, A(\boldsymbol{x}) \sum_{\alpha} \frac{\partial}{\partial x_\alpha} \frac{1}{4} \int d\boldsymbol{x}_1\, x_{1\alpha} \nabla\phi(\boldsymbol{x}_1) \cdot \boldsymbol{F}(\boldsymbol{x}, \boldsymbol{x}-\boldsymbol{x}_1).$$

Since A(**x**) is arbitrary, we obtain

$$B(\boldsymbol{x}) = \sum_{\alpha} \frac{\partial}{\partial x_\alpha} \frac{1}{4} \int x_{1\alpha} \nabla\phi(\boldsymbol{x}_1) \cdot \boldsymbol{F}(\boldsymbol{x}, \boldsymbol{x}-\boldsymbol{x}_1)\, d\boldsymbol{x}_1$$

$$= \operatorname{div} \frac{1}{4} \int (\boldsymbol{x}-\boldsymbol{x}') \nabla\phi(\boldsymbol{x}-\boldsymbol{x}') \cdot (\psi^{+}(\boldsymbol{x}') \boldsymbol{j}(\boldsymbol{x}) \psi(\boldsymbol{x}') + \psi^{+}(\boldsymbol{x}) \boldsymbol{j}(\boldsymbol{x}') \psi(\boldsymbol{x}))\, d\boldsymbol{x}'.$$

Thus, we have represented B(**x**) approximately in the form of the divergence of a vector. The equation (19.28) takes the form of a local energy conservation law

$$\frac{\partial H(\boldsymbol{x})}{\partial t} + \operatorname{div} \boldsymbol{j}_H(\boldsymbol{x}) = 0, \tag{19.29}$$

where

$$\boldsymbol{j}_H(\boldsymbol{x}) = -\frac{\hbar^3}{4m^2 i} \{\nabla^2\psi^{+}(\boldsymbol{x}) \nabla\psi(\boldsymbol{x}) - \nabla\psi^{+}(\boldsymbol{x}) \nabla^2\psi(\boldsymbol{x})\} + \frac{1}{2} \int \phi(\boldsymbol{x}-\boldsymbol{x}') \psi^{+}(\boldsymbol{x}') \boldsymbol{j}(\boldsymbol{x}) \psi(\boldsymbol{x}')\, d\boldsymbol{x}' - \frac{1}{4} \int (\boldsymbol{x}-\boldsymbol{x}') \nabla\phi(\boldsymbol{x}-\boldsymbol{x}') \cdot (\psi^{+}(\boldsymbol{x}') \boldsymbol{j}(\boldsymbol{x}) \psi(\boldsymbol{x}') + \psi^{+}(\boldsymbol{x}) \boldsymbol{j}(\boldsymbol{x}') \psi(\boldsymbol{x}))\, d\boldsymbol{x}' \tag{19.29a}$$

is the energy flux operator, corresponding to the classical expression (19.17a).

Thus, we have obtained a complete system of conservation laws in local form for the energy, particle number and momentum for the case of a quantum-mechanical system of identical particles:

$$\frac{\partial H(\boldsymbol{x})}{\partial t} = -\operatorname{div} \boldsymbol{j}_H(\boldsymbol{x}),$$
$$\frac{\partial n(\boldsymbol{x})}{\partial t} = -\operatorname{div} \boldsymbol{j}(\boldsymbol{x}), \tag{19.30}$$
$$\frac{\partial \boldsymbol{p}(\boldsymbol{x})}{\partial t} = -\operatorname{Div} T(\boldsymbol{x}).$$

This system is analogous to the classical equations (19.18), but the densities of the mechanical quantities in it are quantum-mechanical operators. For example, $H(\mathbf{x})$ is defined by Eq. (19.21a), $n(\mathbf{x})$ by Eq. (19.22), $p(\mathbf{x})$ by Eq. (19.25), $\mathbf{j}_H(\mathbf{x})$ by Eq. (19.29a), $\mathbf{j}(\mathbf{x})$ by Eq.(19.24a), and $T(\mathbf{x})$ by Eq. (19.27a).

The conservation laws (19.30) can be written conveniently in the form of one equation:

$$\frac{\partial P_m(\boldsymbol{x})}{\partial t} + \nabla \cdot j_m(\boldsymbol{x}) = 0 \qquad (m = 0,\ 1,\ 2), \tag{19.31}$$

where

$$P_0(\boldsymbol{x}) = H(\boldsymbol{x}), \qquad P_1(\boldsymbol{x}) = \boldsymbol{p}(\boldsymbol{x}), \qquad P_2(\boldsymbol{x}) = n(\boldsymbol{x}) \tag{19.31a}$$

are the densities of the mechanical quantities, and

$$j_0(\boldsymbol{x}) = \boldsymbol{j}_H(\boldsymbol{x}), \qquad j_1(\boldsymbol{x}) = T(\boldsymbol{x}), \qquad j_2(\boldsymbol{x}) = \boldsymbol{j}(\boldsymbol{x}) \tag{19.31b}$$

are the flux densities.

In the more general case when we cannot neglect the inhomogeneities of the system over distances of the order of the effective interaction range between the particles, there is no need to introduce $\mathbf{j}_m(\mathbf{x})$ explicitly, and the balance equation for the mechanical quantities can be written in the form

$$\frac{\partial P_m(\boldsymbol{x}, t)}{\partial t} = \frac{1}{i\hbar}[P_m(\boldsymbol{x}, t),\ H]. \tag{19.31c}$$

Among the conservation equations (19.30) for the mechanical quantities, we have not written out the law of conservation of angular momentum. This is not accidental, since, for our case of centrally symmetric forces, the law of conservation of angular momentum follows from the law of conservation of momentum. In fact, we shall introduce the angular momentum density tensor $m_{\alpha\beta}(x)$, defined by

$$m_{\alpha\beta}(x) = x_\alpha p_\beta(x) - x_\beta p_\alpha(x), \tag{19.32}$$

where **p(x)** is the momentum density operator. We assume that the particles do not have intrinsic angular momentum; otherwise, it would be necessary to include it in (19.32). The tensor $m_{\alpha\beta}(x)$ satisfies the equation of motion

$$\frac{\partial m_{\alpha\beta}(x)}{\partial t} = x_\alpha \frac{\partial p_\beta(x)}{\partial t} - x_\beta \frac{\partial p_\alpha(x)}{\partial t} = -\sum_\gamma \left\{ x_\alpha \frac{\partial T_{\gamma\beta}(x)}{\partial x_\gamma} - x_\beta \frac{\partial T_{\gamma\alpha}(x)}{\partial x_\gamma} \right\} =$$
$$= -\sum_\gamma \frac{\partial}{\partial x_\gamma} \{ x_\alpha T_{\gamma\beta}(x) - x_\beta T_{\gamma\alpha}(x) \} + T_{\alpha\beta}(x) - T_{\beta\alpha}(x).$$

In the present case of central forces, the tensor $T_{\alpha\beta}(x)$ is symmetric, and the equation for $m_{\alpha\beta}(x)$ already has the form of a conservation law:

$$\frac{\partial m_{\alpha\beta}(x)}{\partial t} + \sum_\gamma \frac{\partial}{\partial x_\gamma} \{ x_\alpha T_{\gamma\beta}(x) - x_\beta T_{\gamma\alpha}(x) \} = 0. \tag{19.33}$$

Thus, the law of conservation of angular momentum in the case of central forces follows from the law of conservation of momentum. For noncentral forces, the tensor $T_{\alpha\beta}(x)$ is nonsymmetric and the angular momentum $\int m_{\alpha\beta}(x)\,dx$ is not conserved; however, all this means is that it is necessary to take into account the contribution of internal degrees of freedom to the total angular momentum, for example, to include the angular momentum associated with the rotation of the molecules or with the spin of the particles (cf. [41]). To take the noncentral forces into account without introducing the internal degrees of freedom would be inconsistent.†

19.3. The Virial Theorem for the Nonuniform Case

The virial theorem, both classical and quantum, for the case of statistical equilibrium was considered in Chapters I and II in

†In this case, it is possible to symmetrize the tensor $T_{\alpha\beta}(x)$, by making use of the indeterminacy in its definition [69].

§§5.3 and 11.3. We now discuss how it can be generalized for the spatially nonuniform case.

We shall start, in the quantum case, from the expression (19.27a) for the stress tensor [in the classical case, one must use the expression (19.12)]. Below, we consider the quantum-mechanical case.

If there are no hydrodynamical fluxes in the gas (or liquid), then the stress tensor $\langle T_{\alpha\beta}(x)\rangle$ coincides with the pressure tensor $\langle P_{\alpha\beta}(x)\rangle$. But if there are fluxes with average velocity

$$v(x) = \langle j(x)\rangle / \langle n(x)\rangle, \tag{19.34}$$

then, to define the pressure tensor, it is necessary to go over to the moving coordinate system in which the hydrodynamic velocity is equal to zero, and define the pressure tensor in this system.

The transition to a system moving with velocity $\mathbf{v}(\mathbf{x})$ can be realized by means of a canonical transformation of the operators†

$$\psi(x) = \psi'(x)\, e^{i\varphi(x)}, \qquad v(x) = \frac{\hbar}{m}\nabla\varphi(x). \tag{19.35}$$

In this case, $T_{\alpha\beta}(x)$ goes over into $T'_{\alpha\beta}(x)$, which we shall call the pressure tensor operator $P_{\alpha\beta}(x)$:

$$P_{\alpha\beta}(x) = T_{\alpha\beta}(x) - m\left(v_\alpha j_\beta(x) + v_\beta j_\alpha(x)\right) + mn(x)\, v_\alpha v_\beta.$$

By analogy with ordinary hydrodynamics, we can define the pressure operator as one-third of the trace of the tensor $P_{\alpha\beta}(x)$:

$$p(x) = \frac{1}{3}\sum_{\alpha} P_{\alpha\alpha}(x). \tag{19.37}$$

† The canonical transformation (19.35) effects a transition to an inertial accompanying coordinate frame and is valid in the case of an irrotational velocity field. To go over to a noninertial accompanying frame in the general case of an arbitrary velocity field, it is convenient to make use of a Lagrangian coordinate frame. In the case of classical mechanics, such a transformation can be effected by means of the generating function [235]

$$\mathcal{F}(\mathbf{r}_i^0, \mathbf{p}_i; t) = -\sum_i (\mathbf{r}_i^0 + \mathbf{S}(\mathbf{r}_i^0, t))\cdot(\mathbf{p}_i - m\mathbf{v}(\mathbf{r}_i^0, t)).$$

Here, $\mathbf{p}_i$ is the momentum of the i-th particle, $\mathbf{r}_i^0$ is its Lagrangian coordinate, connected with the Eulerian coordinate $\mathbf{r}_i$ by the relation $\mathbf{r}_i = \mathbf{r}_i^0 + \mathbf{S}(\mathbf{r}_i^0, t)$, $\mathbf{S}(\mathbf{r}_i^0, t)$ is the field of the macroscopic deformations, which is given as a function of the initial points $\mathbf{r}_i^0$ and the time t, and $\mathbf{v}(\mathbf{r}_i^0, t) = \partial\mathbf{S}(\mathbf{r}_i^0, t)/\partial t$ is the Lagrangian velocity field. In the quantum case, this transformation is effected by a certain unitary operator [235].

Consequently, p(x) is a scalar operator, equal to

$$p(\boldsymbol{x}) = \frac{\hbar^2}{3m}\left\{\nabla\psi^+(\boldsymbol{x})\cdot\nabla\psi(\boldsymbol{x}) - \frac{1}{4}\nabla^2 n(\boldsymbol{x})\right\} - \\ -\frac{1}{6}\int(\boldsymbol{x}-\boldsymbol{x}')\cdot\nabla\phi(\boldsymbol{x}-\boldsymbol{x}')\,\psi^+(\boldsymbol{x})\,\psi^+(\boldsymbol{x}')\,\psi(\boldsymbol{x}')\,\psi(\boldsymbol{x})\,d\boldsymbol{x}' - \\ -\frac{2}{3}mv(\boldsymbol{x})\cdot\boldsymbol{j}(\boldsymbol{x}) + \frac{1}{3}mn(\boldsymbol{x})\,v^2(\boldsymbol{x}). \qquad (19.37a)$$

We obtain the virial theorem for the nonuniform case by averaging (19.37a):

$$\langle p(\boldsymbol{x})\rangle = \frac{\hbar^2}{3m}\left\{\langle\nabla\psi^+(\boldsymbol{x})\cdot\nabla\psi(\boldsymbol{x})\rangle - \frac{1}{4}\nabla^2\langle n(\boldsymbol{x})\rangle\right\} - \frac{2}{3}m\frac{v^2(\boldsymbol{x})}{2} - \\ -\frac{1}{6}\int(\boldsymbol{x}-\boldsymbol{x}')\cdot\nabla\phi(\boldsymbol{x}-\boldsymbol{x}')\,\langle\psi^+(\boldsymbol{x})\,\psi^+(\boldsymbol{x}')\,\psi(\boldsymbol{x}')\,\psi(\boldsymbol{x})\rangle\,d\boldsymbol{x}', \qquad (19.37b)$$

where $\langle\ldots\rangle$ represents averaging in which $\langle j(x)\rangle = \langle n(x)\rangle v(x)$.

It is easy to see that the average pressure for a spatially uniform state with $\mathbf{v}(\mathbf{x}) = 0$ satisfies the virial theorem in its usual form (11.15). Indeed, in this case,

$$\langle n(\boldsymbol{x})\rangle_0 = \text{const} \quad \text{and} \quad \nabla^2\langle n(\boldsymbol{x})\rangle_0 = 0,$$

where $\langle\ldots\rangle_0$ represents averaging over the equilibrium state; therefore, the average equilibrium pressure is equal to

$$p = \langle p(\boldsymbol{x})\rangle_0 = \frac{2}{3}\frac{\hbar^2}{2m}\langle\nabla\psi^+(\boldsymbol{x})\cdot\nabla\psi(\boldsymbol{x})\rangle - \\ -\frac{1}{6}\int(\boldsymbol{x}-\boldsymbol{x}')\cdot\nabla\phi(\boldsymbol{x}-\boldsymbol{x}')\,F_2(\boldsymbol{x}-\boldsymbol{x}')\,d\boldsymbol{x}', \qquad (19.38)$$

where

$$F_2(\boldsymbol{x}-\boldsymbol{x}') = \langle\psi^+(\boldsymbol{x})\,\psi^+(\boldsymbol{x}')\,\psi(\boldsymbol{x}')\,\psi(\boldsymbol{x})\rangle_0 \qquad (19.38a)$$

is the equilibrium pair correlation function.

In formula (19.38), the first term is equal to two-thirds of the average kinetic energy density, and the second term is equal to one-third of the virial of the interaction, i.e., (19.38) coincides with (11.15).

In §§5.3 and 11.3, the virial theorem was proved by a method using an infinitesimal increase (5.9b), (11.10b) in scale of the coordinates and momenta; thus, it is an exact theorem. It may appear strange that we have obtained it from an approximate expression for $T_{\alpha\beta}(x)$. We shall convince ourselves that there is no contradiction here, and that for the uniform state the higher terms of the expansion give no contribution to $\langle T_{\alpha\beta}(x)\rangle_0$. By the same method as that of §19.2, taking into account all the terms in the expansion, we obtain

$$T_{\alpha\beta}(x) = \frac{\hbar^2}{2m}\left\{\frac{\partial\psi^+(x)}{\partial x_\alpha}\frac{\partial\psi(x)}{\partial x_\beta} + \frac{\partial\psi^+(x)}{\partial x_\beta}\frac{\partial\psi(x)}{\partial x_\alpha} - \frac{1}{2}\frac{\partial^2 n(x)}{\partial x_\alpha \partial x_\beta}\right\} -$$
$$\sum_{n\geqslant 1}\frac{1}{n!}\frac{\partial^{n-1}}{\partial x_\beta^{n-1}}\int \frac{\partial\phi(x-x')}{\partial x_\alpha}(x_\beta - x'_\beta)^n \psi^+(x)\psi^+(x')\psi(x')\psi(x)\,dx'.$$

For a uniform state, the pair correlation function (19.38a) depends only on the difference $\mathbf{x}-\mathbf{x}'$, and the average value of the integrals in the right-hand side of this expression, therefore, does not depend on $\mathbf{x}$. Consequently, in the sum over n in $\langle T_{\alpha\beta}(x)\rangle_0$, only the one term n = 1 remains, and this gives the virial.

19.4. Conservation Laws for a Mixture of Gases or Liquids [185]

In the preceding subsection, we studied the conservation laws for particle number, energy, and momentum for the case of a system of identical particles with direct interaction. We now examine the conservation laws for a mixture of different gases or liquids, confining ourselves to the case when there are no chemical reactions between the components and no excitation of the internal degrees of freedom of the molecules. This example is of interest in that it enables us to investigate the mutual exchange of energy and momenta between the components, since these quantities are no longer integrals of motion for each component.

The Hamiltonian of a system of interacting molecules of l types has the form

$$H = \sum_i^l \int \psi_i^+(x)\left\{-\frac{\hbar^2}{2m_i}\nabla^2 + \sum_k \frac{1}{2}\int \phi_{ik}(x-x')\psi_k^+(x')\psi_k(x')\,dx'\right\}\psi_i(x)\,dx, \tag{19.39}$$

where $\phi_{ik}(x-x')$ is the potential energy of the interaction of particles of types i and k, assumed to be radially symmetric, and the second-quantized operators $\psi_i(\mathbf{x})$ and $\psi_k(\mathbf{x})$ for each component satisfies the commutation relations of Fermi or Bose statistics

$$\psi_i(\boldsymbol{x})\,\psi_i^+(\boldsymbol{x}') \pm \psi_i^+(\boldsymbol{x}')\,\psi_i(\boldsymbol{x}) = \delta(\boldsymbol{x}-\boldsymbol{x}'),$$
$$\psi_i(\boldsymbol{x})\,\psi_i(\boldsymbol{x}') \pm \psi_i(\boldsymbol{x}')\,\psi_i(\boldsymbol{x}) = 0,$$

depending on the parity of the spin of the molecules, and commute for $i \neq k$, i.e., for different components.

We write the Hamiltonian (19.39) in the form

$$H = \sum_i \int H_i(\boldsymbol{x})\,d\boldsymbol{x}, \tag{19.39a}$$

where

$$H_i(\boldsymbol{x}) = \frac{\hbar^2}{2m_i}\nabla\psi_i^+(\boldsymbol{x})\cdot\nabla\psi_i(\boldsymbol{x}) + \\ + \sum_k \frac{1}{2}\int \phi_{ik}(\boldsymbol{x}-\boldsymbol{x}')\,\psi_i^+(\boldsymbol{x})\,\psi_k^+(\boldsymbol{x}')\,\psi_k(\boldsymbol{x}')\,\psi_i(\boldsymbol{x})\,d\boldsymbol{x}' \tag{19.39b}$$

is the energy density of the i-th component with allowance for the interaction with the other components.

The number of particles of the i-th component is represented by the operator

$$N_i = \int n_i(\boldsymbol{x})\,d\boldsymbol{x}, \quad n_i(\boldsymbol{x}) = \psi_i^+(\boldsymbol{x})\,\psi_i(\boldsymbol{x}) \tag{19.40}$$

and is conserved, since

$$\frac{\partial n_i(\boldsymbol{x})}{\partial t} + \operatorname{div}\boldsymbol{j}_i(\boldsymbol{x}) = 0, \tag{19.41}$$

where

$$\boldsymbol{j}_i(\boldsymbol{x}) = \frac{\hbar}{2m_i i}\left\{\psi_i^+(\boldsymbol{x})\,\nabla\psi_i(\boldsymbol{x}) - \nabla\psi_i^+(\boldsymbol{x})\,\psi_i(\boldsymbol{x})\right\} \tag{19.41a}$$

is the flux density operator for particles of the i-th component. This is a consequence of the fact that there are no chemical reactions between the components.

The equation of conservation of energy of the i-th component in local form is

$$\frac{\partial H_i(\boldsymbol{x})}{\partial t} + \operatorname{div}\boldsymbol{j}_{H_i}(\boldsymbol{x}) = J_{H_i}(\boldsymbol{x}), \tag{19.42}$$

where

$$j_{H_i}(x) = \frac{\hbar^3}{4m_i^2 i}\{\nabla\psi_i^+(x)\nabla^2\psi_i(x) - \nabla^2\psi_i^+(x)\nabla\psi_i(x)\} +$$
$$+ \sum_k \frac{1}{2}\int \phi_{ki}(x-x')\psi_k^+(x')j_i(x)\psi_k(x')\,dx' -$$
$$- \sum_k \frac{1}{8m_k}\int (x-x')\nabla\phi_{ki}(x-x')\cdot\{\psi_k^+(x')j_i(x)\psi_k(x') +$$
$$+ \psi_i^+(x')j_k(x)\psi_i(x') + \psi_k^+(x)j_i(x')\psi_k(x) + \psi_i^+(x)j_k(x')\psi_i(x)\}\,dx' \quad (19.42a)$$

is the energy flux density operator of the i-th component, and

$$J_{H_i}(x) = -\sum_m \frac{1}{4}\int \nabla\phi_{mi}(x-x')\cdot\{n_m(x')j_i(x) + n_i(x)j_m(x') -$$
$$- n_i(x')j_m(x) - n_m(x)j_i(x')\}\,dx' \quad (19.42b)$$

is an operator representing the rate of change of energy of the i-th component as a result of its interaction with the other components. In obtaining Eqs. (19.42)-(19.42b), we have made use of the method of smoothing operators over small inhomogeneities described in §19.2.

The operator $J_{H_i}(x)$ in Eq. (19.42) satisfies the relation

$$\sum_i J_{H_i}(x) = 0, \quad (19.42c)$$

which is a consequence of the conservation of the total energy. This operator cannot be represented in the form of the divergence of a vector, since $H_i = \int H_i(x)\,dx$ is not an integral of motion.

The total energy density

$$H(x) = \sum_i H_i(x) \quad (19.43)$$

satisfies a conservation law in the usual form

$$\frac{\partial H(x)}{\partial t} + \operatorname{div} j_H(x) = 0, \quad (19.44)$$

where

$$j_H(x) = \sum_i j_{H_i}(x) \quad (19.44a)$$

is the total energy flux density. The rate of change of energy of the i-th component is equal to the integral of (19.42) over the volume:

$$\frac{\partial H_i}{\partial t} = J_{H_i}, \tag{19.45}$$

where

$$J_{H_i} = -\sum_m \frac{1}{2} \int \nabla \phi_{mi}(\boldsymbol{x} - \boldsymbol{x}') \cdot (n_m(\boldsymbol{x}')\boldsymbol{j}_i(\boldsymbol{x}) + n_i(\boldsymbol{x})\boldsymbol{j}_m(\boldsymbol{x}'))\, d\boldsymbol{x}\, d\boldsymbol{x}' \tag{19.45a}$$

is the operator of the total energy flux for the i-th component, with

$$\sum_i J_{H_i} = 0. \tag{19.45b}$$

The equations (19.45) and (19.45b) describe the exchange of energy between the components of the mixture. Using them, we can study the relaxation of this process, if the energy transfer occurs slowly (for example, if the masses of the components are very different).

The equations of conservation of momentum for the i-th component and of the total momentum in local form are

$$\begin{aligned} \frac{\partial \boldsymbol{p}_i(\boldsymbol{x})}{\partial t} + \operatorname{Div} T_i(\boldsymbol{x}) &= \boldsymbol{f}_i(\boldsymbol{x}), \\ \frac{\partial \boldsymbol{p}(\boldsymbol{x})}{\partial t} + \operatorname{Div} T(\boldsymbol{x}) &= 0, \end{aligned} \tag{19.46}$$

where

$$p_i(\boldsymbol{x}) = m_i j_i(\boldsymbol{x}), \qquad \boldsymbol{p}(\boldsymbol{x}) = \sum_i m_i \boldsymbol{j}_i(\boldsymbol{x}) \tag{19.46a}$$

are the densities of the momentum of the i-th component and of the total momentum,

$$\begin{aligned} T_i^{\alpha\beta}(\boldsymbol{x}) = \frac{\hbar^2}{2m_i} \left\{ \frac{\partial \psi_i^+(\boldsymbol{x})}{\partial x_\alpha} \frac{\partial \psi_i(\boldsymbol{x})}{\partial x_\beta} + \frac{\partial \psi_i^+(\boldsymbol{x})}{\partial x_\beta} \frac{\partial \psi_i(\boldsymbol{x})}{\partial x_\alpha} - \frac{1}{2} \frac{\partial^2 n_i(\boldsymbol{x})}{\partial x_\alpha\, \partial x_\beta} \right\} - \\ - \sum_j \frac{1}{4} \int (x_\beta - x'_\beta)(x_\alpha - x'_\alpha) \frac{1}{r} \frac{d\phi_{ji}(r)}{dr} \times \\ \times \{\psi_j^+(\boldsymbol{x})\, \psi_i^+(\boldsymbol{x}')\, \psi_i(\boldsymbol{x}')\, \psi_j(\boldsymbol{x}) + \psi_j^+(\boldsymbol{x}')\, \psi_i^+(\boldsymbol{x})\, \psi_i(\boldsymbol{x})\, \psi_j(\boldsymbol{x}')\}\, d\boldsymbol{x}', \\ T^{\alpha\beta}(\boldsymbol{x}) = \sum_i T_i^{\alpha\beta}(\boldsymbol{x}) \end{aligned} \tag{19.46b}$$

are the operator of the stress tensor of the i-th component and the total stress tensor, and

$$\boldsymbol{f}_i(\boldsymbol{x}) = -\sum_{j \neq i} \frac{1}{2} \int \nabla\phi_{ji}(\boldsymbol{x}-\boldsymbol{x}')\,(n_j(\boldsymbol{x}')\,n_i(\boldsymbol{x}) - n_j(\boldsymbol{x})\,n_i(\boldsymbol{x}'))\,d\boldsymbol{x}' \qquad (19.46c)$$

is the operator of the density of the frictional force between the i-th component and the other components. The sum of all the frictional forces $\boldsymbol{f}_i(\mathbf{x})$ is equal to zero:

$$\sum_i \boldsymbol{f}_i(\boldsymbol{x}) = 0. \qquad (19.46d)$$

The total momentum of the i-th component

$$\boldsymbol{P}_i = \int \boldsymbol{p}_i(\boldsymbol{x})\,d\boldsymbol{x} \qquad (19.47)$$

is not conserved, since

$$\frac{\partial \boldsymbol{P}_i}{\partial t} = \boldsymbol{F}_i, \qquad (19.47a)$$

where

$$\boldsymbol{F}_i = \int \boldsymbol{f}_i(\boldsymbol{x})\,d\boldsymbol{x} =$$

$$= -\sum_{j \neq i} \frac{1}{2} \int \nabla\phi_{ji}(\boldsymbol{x}-\boldsymbol{x}')\,\{n_j(\boldsymbol{x})\,n_i(\boldsymbol{x}') - n_j(\boldsymbol{x}')\,n_i(\boldsymbol{x})\}\,d\boldsymbol{x}\,d\boldsymbol{x}' \qquad (19.47b)$$

is the friction force between the i-th component and the other components. The total momentum $\boldsymbol{P} = \sum \boldsymbol{P}_i$ is, of course, conserved, since

$$\sum_i \boldsymbol{F}_i = 0. \qquad (19.47c)$$

The equations (19.42), (19.41), and (19.46) can be written conveniently in the form of one matrix equation:

$$\frac{\partial P_{mi}(\boldsymbol{x})}{\partial t} + \nabla \cdot j_{mi}(\boldsymbol{x}) = J_{mi}(\boldsymbol{x}), \qquad (19.48)$$

where

$$\begin{aligned} &P_{0i}(\boldsymbol{x}) = H_i(\boldsymbol{x}), && P_{1i}(\boldsymbol{x}) = \boldsymbol{p}_i(\boldsymbol{x}), && P_{2i}(\boldsymbol{x}) = n_i(\boldsymbol{x}), \\ &j_{0i}(\boldsymbol{x}) = \boldsymbol{j}_{H_i}(\boldsymbol{x}), && j_{1i}(\boldsymbol{x}) = T_i(\boldsymbol{x}), && j_{2i}(\boldsymbol{x}) = \boldsymbol{j}_i(\boldsymbol{x}), \\ &J_{0i}(\boldsymbol{x}) = J_{H_i}(\boldsymbol{x}), && J_{1i}(\boldsymbol{x}) = \boldsymbol{f}_i(\boldsymbol{x}), && J_{2i}(\boldsymbol{x}) = 0, \end{aligned} \qquad (19.48a)$$

i.e., $P_{mi}(\mathbf{x})$ is the matrix of the densities of the mechanical quantities (energy, momentum, and particle number), $j_{mi}(\mathbf{x})$ is the matrix of the fluxes, and $J_{mi}(\mathbf{x})$ is the matrix of the sources.

19.5. Conservation Laws for a System of Particles with Internal Degrees of Freedom

If the gas or liquid consists of complex molecules, then excitation of the internal degrees of freedom, for example, vibrational, rotational, or other degrees of freedom, is possible. Exchange of energy between the internal and translational degrees of freedom may be difficult, and then relaxation phenomena, i.e., the slow establishment of equilibrium between the external and internal degrees of freedom, are possible. In order to study these phenomena, we must formulate the laws of conservation of particle number, energy, and momentum for subsystems with specified internal states of the molecules; following paper [4], we shall do this in this subsection.

We denote by y the aggregate of variables $y_1, y_2, \ldots,$ describing the internal degrees of freedom of a molecule and use $\mathbf{x}$ to denote the coordinates of its center of gravity. The Hamiltonian of the system has the form

$$H = \int \psi^+(\boldsymbol{x}, y)\Big\{-\frac{\hbar^2\nabla^2}{2m} + H_{\text{in}}(y) + \\ + \frac{1}{2}\int \psi^+(\boldsymbol{x}', y')\,\phi(\boldsymbol{x}y, \boldsymbol{x}'y')\,\psi(\boldsymbol{x}', y')\,d\boldsymbol{x}'\,dy'\Big\}\,\psi(\boldsymbol{x}, y)\,d\boldsymbol{x}\,dy, \tag{19.49}$$

where $\phi(\boldsymbol{x}y, \boldsymbol{x}'y') = \phi(\boldsymbol{x}'y', \boldsymbol{x}y)$ is the operator of the energy of interaction between the molecules, and $H_{\text{in}}(y)$ is the Hamiltonian of the internal degrees of freedom, with

$$H_{\text{in}}(y)\,\varphi_i(y) = E_i\varphi_i(y), \quad \psi(\boldsymbol{x}, y) = \frac{1}{\sqrt{V}}\sum_{ki} a_{ki}\,e^{i\,\boldsymbol{k}\cdot\boldsymbol{x}}\,\varphi_i(y), \tag{19.49a}$$

where $\varphi_i(y)$ and E_i are the eigenfunction and energy of the internal state i, and a_{ki} are second-quantized operators in the occupation-number space of k and i.

We introduce the second-quantized operators $\psi_i(\mathbf{x})$ describing the subsystem with given quantum number i:

$$\psi_i(\boldsymbol{x}) = \frac{1}{\sqrt{V}}\sum_k a_{ki}\,e^{i\,\boldsymbol{k}\cdot\boldsymbol{x}}, \quad \psi(\boldsymbol{x}, y) = \sum_i \varphi_i(y)\,\psi_i(\boldsymbol{x}). \tag{19.49b}$$

The operators $\psi_i(x)$ satisfy the commutation relations

$$\begin{aligned}\psi_i(x)\psi_j^+(x') \pm \psi_j^+(x')\psi_i(x) = \delta_{ij}\delta(x-x'),\\ \psi_i(x)\psi_j(x') \pm \psi_j(x')\psi_i(x) = 0,\end{aligned} \tag{19.49c}$$

where the plus sign is taken for odd spin of the molecules, and the minus sign for even spin. The relations (19.49c) follow from the commutation relations for a_{ki}.

Taking (19.49a) and (19.49b) into account, we write the Hamiltonian (19.49) in the form

$$H = \sum_i \int \psi_i^+(x)\left(-\frac{\hbar^2}{2m}\nabla^2 + E_i\right)\psi_i(x)\,dx + \\ + \frac{1}{2}\sum_{ijkl}\int \psi_i^+(x)\psi_j^+(x')\,\phi_{ij}^{kl}(x, x')\,\psi_k(x')\,\psi_l(x)\,dx\,dx', \tag{19.50}$$

where the function $\phi_{ij}^{kl}(x, x')$ has the form

$$\phi_{ij}^{kl}(x, x') = \int \varphi_i^*(y)\,\varphi_j^*(y')\,\phi(xy, x'y')\,\varphi_k(y')\,\varphi_l(y)\,dy\,dy' \tag{19.50a}$$

and possesses the symmetry properties

$$\phi_{ij}^{kl}(x, x') = \phi_{ji}^{lk}(x', x), \quad \phi_{ij}^{kl}(x, x')^* = \phi_{lk}^{ji}(x, x'), \tag{19.50 b}$$

which follow from the symmetry of the function $\phi(xy, x'y')$ with respect to the replacement $x \to x'$, $y \to y'$ and the Hermiticity of the interaction operator.

The function $\phi_{ij}^{kl}(x, x')$ plays the role of the potential of the interaction between molecules in states k and l that transforms them into molecules in states i and j. We may imagine that a chemical reaction occurs between the molecules, in accordance with the scheme

$$(k) + (l) \rightleftarrows (i) + (j).$$

The function $\phi_{ij}^{kl}(x, x')$ can be estimated from the effective cross-section of the inelastic collision with the transition $k, l \to i, j$. The Hamiltonian (19.50) is similar to the Hamiltonian (19.39) of the gas mixture, with the difference that in (19.50) the internal energy E_i of the molecules and the possibility of transitions $k, l \rightleftarrows i, j$ in the collisions are taken into account.

The Hamiltonian in the form (19.50) was used in [4]. It can be regarded as a starting model for a system of particles with internal degrees of freedom; it could be written down immediately (the above arguments are no more than indicative). A more detailed consideration of the internal degrees of freedom is given in the papers of Pokrovskii [41, 209], and in the thorough review of Dahler and Hoffmann [228], where references to other papers are given.

The operator $\psi_i(\mathbf{x})$ satisfies the equation of motion

$$i\hbar\frac{\partial\psi_i(x)}{\partial t}=\left\{-\frac{\hbar^2}{2m}\nabla^2+E_i\right\}\psi_i(\boldsymbol{x})+$$

$$+\frac{1}{2}\sum_{jkl}\int\psi_j^+(\boldsymbol{x}')\left(\phi_{ij}^{kl}(\boldsymbol{x},\boldsymbol{x}')+\phi_{ji}^{kl}(\boldsymbol{x}',\boldsymbol{x})\right)\psi_k(\boldsymbol{x}')\psi_l(\boldsymbol{x})\,d\boldsymbol{x}'. \qquad (19.51)$$

The number of particles in state i

$$n_i(\boldsymbol{x})=\psi_i^+(\boldsymbol{x})\psi_i(\boldsymbol{x}) \qquad (19.52)$$

is not conserved, since in the collisions transitions from one internal state to another are possible. The operator $n_i(\mathbf{x})$ satisfies the balance equation

$$\frac{\partial n_i(\boldsymbol{x})}{\partial t}+\operatorname{div}\boldsymbol{j}_i(\boldsymbol{x})=J_i(\boldsymbol{x}), \qquad (19.53)$$

where

$$J_i(\boldsymbol{x})=\frac{1}{2i\hbar}\sum_{jkl}\int\left\{\psi_i^+(\boldsymbol{x})\psi_j^+(\boldsymbol{x}')\left(\phi_{ij}^{kl}(\boldsymbol{x},\boldsymbol{x}')+\phi_{ji}^{kl}(\boldsymbol{x}',\boldsymbol{x})\right)\psi_k^+(\boldsymbol{x}')\psi_l(\boldsymbol{x})-\right.$$

$$\left.-\psi_l^+(\boldsymbol{x})\psi_k^+(\boldsymbol{x}')\left(\phi_{ij}^{kl}(\boldsymbol{x},\boldsymbol{x}')^*+\phi_{ji}^{kl}(\boldsymbol{x}',\boldsymbol{x})^*\right)\psi_j(\boldsymbol{x}')\psi_i(\boldsymbol{x})\right\}d\boldsymbol{x}' \qquad (19.53\text{a})$$

is the operator of the "reaction" rate of formation of particles in state i. The particle flux operator $\mathbf{j}_i(\mathbf{x})$ has the usual form (19.41a).

The total density of particles in all the internal states

$$n(\boldsymbol{x})=\sum_i n_i(\boldsymbol{x})$$

is conserved, since, using (19.50a), we have

$$\sum_i J_i(\boldsymbol{x})=0. \qquad (19.53\text{b})$$

On the other hand, the number of particles in state i

$$N_i = \int n_i(\boldsymbol{x})\, d\boldsymbol{x} \tag{19.54}$$

is not conserved, since

$$\frac{\partial N_i}{\partial t} = \int J_i(\boldsymbol{x})\, d\boldsymbol{x} = J_i, \tag{19.54a}$$

where J_i is the operator of the rate of formation of particles in state i, which is nonzero.

The internal energy density in state i

$$H_i(\boldsymbol{x}) = E_i n_i(\boldsymbol{x}) \tag{19.55}$$

satisfies the balance equation

$$\frac{\partial H_i(\boldsymbol{x})}{\partial t} + \operatorname{div} E_i \boldsymbol{j}_i(\boldsymbol{x}) = E_i J_i(\boldsymbol{x}), \tag{19.55a}$$

and the total energy in state i

$$H_i = \int H_i(\boldsymbol{x})\, d\boldsymbol{x} \tag{19.56}$$

satisfies the equation

$$\frac{\partial H_i}{\partial t} = E_i J_i. \tag{19.57}$$

The conservation equations (19.53), (19.55a), (19.54a), and (19.57) enable us to investigate the relaxation of the internal degrees of freedom; we shall study this in §23.

The complete set of conservation laws in the case of a system with internal degrees of freedom can be written in the matrix form (19.48), where i is the index indicating the internal degrees of freedom and

$$J_{2i}(\boldsymbol{x}) = J_i(\boldsymbol{x}), \tag{19.58}$$

which, as before, is nonzero, i.e., there are sources not only of energy and momentum, but also of particles. This is the most general form of the conservation laws.

§20. The Local-Equilibrium Distribution

To determine the thermodynamic functions of nonequilibrium states, it is necessary to construct the corresponding statistical ensemble representing systems in a state differing from the equilibrium state.

Sometimes, this is done by means of switching on the auxiliary field that would make the thermodynamic state an equilibrium state while leaving it nonuniform, as was done in Leontovich's text on statistical physics [70]. However, no auxiliary field can produce nonuniformity of temperature, unless we introduce the rather artificial procedure of switching on a gravitational field satisfying the general theory of relativity [9]. Therefore, we shall make use of another method, based on the introduction of Gibbsian local-equilibrium distributions.

20.1. The Statistical Operator and Distribution Functions for Local-Equilibrium Systems

The concept of a Gibbsian statistical ensemble can be carried over to nonequilibrium stationary systems in the following way.

By a Gibbsian statistical ensemble in this case, we shall mean an aggregate of systems under identical external stationary conditions, i.e., having the same types of contact with thermostats and semipermeable partitions and possessing all the possible values of the microscopic parameters consistent with the given values of the macroscopic parameters. The latter may be given within small well-defined limits of the order of the possible fluctuations, rather than exactly.

In a system situated in stationary external conditions, a certain stationary distribution will be established; we shall call this a stationary local-equilibrium distribution. If the external conditions depend on time, the local-equilibrium distribution will be nonstationary. For a precise definition of a local-equilibrium ensemble, it is necessary to define the distribution function for the statistical operator corresponding to it.

Let the nonequilibrium state be specified by a nonuniform distribution of energy and particle number; corresponding to the densities of these are the operators H(x) and n(x) [cf. (19.21a) and (19.22)] or their Fourier components

$$H_k = \int e^{-i\,k\cdot x} H(x)\,dx,$$
$$n_k = \int e^{-i\,k\cdot x} n(x)\,dx, \qquad (20.1)$$

which, for a one-component system, have the form

$$H_k = \sum_q \frac{\hbar^2}{2m} q \cdot (q+k)\, a_q^+ a_{q+k} + \frac{1}{2V} \sum_{k_1'+k_2'=k_1+k_2+k} \nu(k_2'-k_2)\, a_{k_1}^+ a_{k_2}^+ a_{k_2'} a_{k_1'}, \tag{20.1a}$$

$$n_k = \sum_q a_q^+ a_{q+k} \qquad \left(H_k^+ = H_{-k},\ n_k^+ = n_{-k}\right).$$

We shall assume that there are enough of these variables to describe the macroscopic state of the system.

In the classical case, H_k and n_k are the collective variables (19.19a).

We note that the zero Fourier components of the energy density and particle number density are integrals of motion:

$$H_k|_{k=0} = H_0 = H, \qquad n_k|_{k=0} = n_0 = N. \tag{20.1b}$$

Consequently, for sufficiently small k they are almost integrals of motion.

The simplest method of constructing the local-equilibrium statistical operator (or distribution function) is based on information theory, the connection of which with statistical mechanics was discussed in §§4 and 10 (cf. [71, 72]).

The statistical operator or distribution function is determined from the maximum of the information entropy, which is equal to (10.1) in the quantum case

$$S_i = -\langle \ln \rho \rangle = -\operatorname{Tr}(\rho \ln \rho) \quad (\operatorname{Tr}\rho = 1), \tag{20.2}$$

or equal to (4.5) in the classical case

$$S_i = -\langle \ln f \rangle = -\int f \ln f \, d\Gamma \tag{20.2a}$$

under the supplementary conditions that the average Fourier components of the energy density and particle-number density remain constant on variation of ρ or f

$$\langle H_k \rangle = \text{const}, \quad \langle n_k \rangle = \text{const} \tag{20.2b}$$

and that the normalization be constant

$$\langle 1 \rangle = \text{const.} \tag{20.2c}$$

Here the brackets denote either quantum or classical averaging.

As usual, we seek the absolute extremum of the function

$$S' = -\operatorname{Tr}(\rho \ln \rho) - \sum_k \beta_{-k} \operatorname{Tr}(\rho H_k) + \sum_k \nu_{-k} \operatorname{Tr}(\rho n_k) - \lambda \operatorname{Tr} \rho,$$

where β_{-k}, ν_{-k}, and λ are Lagrangian multipliers determined from Eqs. (20.2b) and (20.2c). The extremum condition for S', i.e., that its variation with respect to ρ vanish, gives the statistical operator of the local-equilibrium distribution

$$\rho_l = Q_l^{-1} \exp\left\{-\sum_k (\beta_{-k} H_k - \nu_{-k} n_k)\right\}, \tag{20.3}$$

where

$$Q_l = \operatorname{Tr} \exp\left\{-\sum_k (\beta_{-k} H_k - \nu_{-k} n_k)\right\} \tag{20.3a}$$

is the corresponding partition function.

For the classical case, by precisely the same method, we obtain the local-equilibrium distribution function

$$f_l = Q_l^{-1} \exp\left\{-\sum_k (\beta_{-k} H_k - \nu_{-k} n_k)\right\}, \tag{20.4}$$

where

$$Q_l = \int \exp\left\{-\sum_k (\beta_{-k} H_k - \nu_{-k} n_k)\right\} d\Gamma. \tag{20.4a}$$

In their external form, (20.3) and (20.4) are equivalent; the difference lies only in the fact that in (20.3) H_k and n_k are operators, while in (20.4) they are functions of the coordinates and momenta of the particles.

Going over from the Fourier components H_k and n_k to the operators $H(x)$ and $n(x)$ of the energy density and particle-number density, we write (20.3) and (20.3a) in the form

$$\begin{aligned} \rho_l &= Q_l^{-1} \exp\left\{-\int \beta(\boldsymbol{x}) \left[H(\boldsymbol{x}) - \mu(\boldsymbol{x})\, n(\boldsymbol{x})\right] d\boldsymbol{x}\right\}, \\ Q_l &= \operatorname{Tr} \exp\left\{-\int \beta(\boldsymbol{x}) \left[H(\boldsymbol{x}) - \mu(\boldsymbol{x})\, n(\boldsymbol{x})\right] d\boldsymbol{x}\right\}, \end{aligned} \tag{20.5}$$

where

$$\beta(\boldsymbol{x}) = \sum_k \beta_k e^{i\,\boldsymbol{k}\cdot\boldsymbol{x}}, \qquad \beta(\boldsymbol{x})\,\mu(\boldsymbol{x}) = \sum_k \nu_k e^{i\,\boldsymbol{k}\cdot\boldsymbol{x}}, \tag{20.6}$$

$\beta(\mathbf{x})$ plays the role of the local inverse temperature, and $\mu(\mathbf{x})$ that of the local chemical potential.

For the classical case, (20.4) and (20.4a) can be written in the same form as (20.5):

$$\begin{aligned} f_l &= Q_l^{-1} \exp\left\{-\int \beta(\boldsymbol{x})\,[H(\boldsymbol{x}) - \mu(\boldsymbol{x})\,n(\boldsymbol{x})]\,d\boldsymbol{x}\right\}, \\ Q_l &= \int \exp\left\{-\int \beta(\boldsymbol{x})\,[H(\boldsymbol{x}) - \mu(\boldsymbol{x})\,n(\boldsymbol{x})]\,d\boldsymbol{x}\right\} d\Gamma. \end{aligned} \tag{20.7}$$

In the special case when the temperature and chemical potential are spatially uniform, (20.7) and (20.5) go over into the grand canonical distribution (3.30) and (9.42).

We have shown that (20.3) corresponds to an extremum of the information entropy. We shall prove now that this extremum corresponds to a maximum, by making use of the inequality (10.2)

$$\operatorname{Tr}(\rho \ln \rho) \geqslant \operatorname{Tr}(\rho \ln \rho_l), \tag{20.8}$$

which holds for any two statistical operators. The equality sign holds only for $\rho = \rho_l$.

Substituting (20.3) into (20.8) we obtain

$$S_{\mathrm{i}} = -\operatorname{Tr}(\rho \ln \rho) \leqslant \ln Q_l + \sum_k (\beta_{-k}\langle H_k\rangle - \nu_{-k}\langle n_k\rangle), \tag{20.9}$$

where

$$\langle H_k\rangle = \operatorname{Tr}(\rho H_k) = \operatorname{Tr}(\rho_l H_k) = \langle H_k\rangle_l, \qquad \langle n_k\rangle = \langle n_k\rangle_l. \tag{20.9a}$$

If we take (20.9a) into account, the inequality (20.9) can be rewritten in the form

$$S_{\mathrm{i}} = -\operatorname{Tr}(\rho \ln \rho) \leqslant -\operatorname{Tr}(\rho_l \ln \rho_l), \tag{20.8a}$$

where the equality sign holds only for $\rho = \rho_l$.

Consequently, the local-equilibrium distribution (20.3) [like (20.4)] corresponds to the maximum of the information entropy under the supplementary conditions that the average Fourier components of the energy and particle number be constant and that the normalization be conserved.

In the general case, for systems with conservation laws in the matrix form (19.48), the local-equilibrium distribution has the form

$$\rho_l = Q_l^{-1} \exp\left\{ -\sum_{i,m} \int F_{im}(\boldsymbol{x}, t)\, P_{mi}(\boldsymbol{x})\, d\boldsymbol{x} \right\}. \qquad (20.10)$$

The possibility of introducing a local-equilibrium distribution is connected with the fact that there exist two relaxation times, of a different order of magnitude [1, 30]: the relaxation time τ for the establishment of statistical equilibrium in the whole system (this time depends on the volume of the system), and another, much shorter relaxation time $\tau_r \ll \tau$, which determines the time for establishment of equilibrium in a volume that is macroscopically small but contains a large number of particles; this time does not depend on the volume of the whole system. A local-equilibrium state is first established in a time τ_r in these small volumes, and then tends slowly to a Gibbsian distribution, with characteristic time τ, if there are no external forces impeding this.

The kinetic theory of gases is also based on the existence of relaxation times of different orders of magnitude – the collision time, the time τ_r between collisions, and the time τ taken to establish equilibrium in the whole volume. This idea was first expressed, and developed systematically as the basis of approximations, in the papers of Bogolyubov on dynamic problems in statistical physics [1].

The two scales for the relaxation times do not always exist. For extremely dilute gases, τ_r may become of the order of τ, and the local-equilibrium distribution loses its meaning.

The local-equilibrium distribution is sometimes introduced by means of nonrigorous intuitive arguments [30]. We shall give a brief description of these arguments. We shall assume that, in a time τ_r in a macroscopically small volume ΔV about the point $\mathbf{x}$, a "quasi-Gibbsian" distribution with local temperature $T(\mathbf{x}) = \beta^{-1}(\mathbf{x})$ and chemical potential $\mu(\mathbf{x})$ is established. It is proportional to the Gibbs factor

$$\exp\left\{ -\beta(\boldsymbol{x}) \left(\int_{\Delta V} H(\boldsymbol{x})\, d\boldsymbol{x} - \mu(\boldsymbol{x}) \int_{\Delta V} n(\boldsymbol{x})\, d\boldsymbol{x} \right) \right\}.$$

Assuming that the distributions of this type in different volumes ΔV are statistically independent, we multiply these operators and arrive at the local-equilibrium distribution (20.5). The weak point in this argument is that the operators H(**x**) at different points do not commute, and the product of the exponentials is not equal to the exponential of the sum.

We shall elucidate now the physical meaning of the parameters β_k and ν_k. They can be expressed in terms of $\langle H_k\rangle_l$, and $\langle n_k\rangle_l$ from the equations

$$\langle H_k\rangle_l = \mathrm{Tr}\,(\rho_l H_k), \qquad \langle n_k\rangle_l = \mathrm{Tr}\,(\rho_l n_k). \tag{20.11}$$

Differentiating $\ln Q_l$ with respect to β_{-k} and ν_{-k}, we obtain the relations

$$\langle H_k\rangle_l = -\left(\frac{\partial \ln Q_l}{\partial \beta_{-k}}\right)_\nu, \qquad \langle n_k\rangle_l = \left(\frac{\partial \ln Q_l}{\partial \nu_{-k}}\right)_\beta, \tag{20.12}$$

which are analogous to the thermodynamic equalities (3.33a).

We introduce the entropy of the local-equilibrium distribution by the relation

$$S = -\,\mathrm{Tr}\,(\rho_l \ln \rho_l) = \ln Q_l + \sum_k (\beta_{-k}\langle H_k\rangle_l - \nu_{-k}\langle n_k\rangle_l) =$$
$$= \ln Q_l + \int \beta(\boldsymbol{x})\left(\langle H(\boldsymbol{x})\rangle_l - \mu(\boldsymbol{x})\langle n(\boldsymbol{x})\rangle_l\right) d\boldsymbol{x}. \tag{20.13}$$

This can be regarded as a function of $\langle H_k\rangle_l$ and $\langle n_k\rangle_l$, if we assume that β_k and ν_k are expressed in terms of $\langle H_k\rangle_l$ and $\langle n_k\rangle_l$ from the solution of the system (20.12). Then we shall have

$$\beta_{-k} = \frac{\partial S}{\partial \langle H_k\rangle_l}, \qquad \nu_{-k} = -\frac{\partial S}{\partial \langle n_k\rangle_l}, \tag{20.14}$$

since on variation of (20.13), the remaining terms cancel each other by virtue of (20.12).

The equalities (20.12) and (20.13) can be also written in the form of relations in functional derivatives:

$$\langle H(\boldsymbol{x})\rangle_l - \mu(\boldsymbol{x})\langle n(\boldsymbol{x})\rangle_l = -\left(\frac{\delta \ln Q_l}{\delta \beta(\boldsymbol{x})}\right)_{\mu(\boldsymbol{x})},$$
$$\langle n(\boldsymbol{x})\rangle_l = \beta^{-1}(\boldsymbol{x})\left(\frac{\delta \ln Q_l}{\delta \mu(\boldsymbol{x})}\right)_{\beta(\boldsymbol{x})} \tag{20.14a}$$

and

$$\beta(x) = \frac{\delta S}{\delta \langle H(x)\rangle_l}, \qquad \mu(x)\beta(x) = -\frac{\delta S}{\delta \langle n(x)\rangle_l}. \tag{20.14b}$$

The thermodynamic relations (20.12)-(20.14b) can be regarded as the generalization to the nonequilibrium case of the usual thermodynamic relations (11.7), (11.24), and (11.25). This indicates that $\beta(\mathbf{x}) = [T(\mathbf{x})]^{-1}$ can be interpreted as the inverse temperature at the point **x** and $\mu(\mathbf{x})$ as the chemical potential at the same point.

The role of the partition function is played by the functional $Q_l[\beta(x), \mu(x)]$ (20.5), (20.7), and that of the entropy by the functional $S(\langle H(x)\rangle_l, \langle n(x)\rangle_l)$; thus, for a local-equilibrium state, the thermodynamic functionals play the role of the thermodynamic functions.

The local-equilibrium distribution is easily generalized for an l-component system with a nonuniform distribution of momentum density and particle-number density. In this case, in addition to the Fourier coefficients H_k of the energy density operator and n_k^α of the particle-number density operator of the α-th component, it is also necessary to use the Fourier coefficient $\mathbf{p}_k$ of the momentum density operator:

$$\boldsymbol{p}_k = \int e^{-i\boldsymbol{k}\cdot\boldsymbol{x}} \boldsymbol{p}(\boldsymbol{x})\,d\boldsymbol{x}, \qquad \boldsymbol{p}(\boldsymbol{x}) = \sum_\alpha m_\alpha \boldsymbol{j}_\alpha(\boldsymbol{x}). \tag{20.15}$$

In the second-quantization representation,

$$\boldsymbol{p}_k = \sum_{\alpha,\, q_\alpha} \hbar\left(\boldsymbol{q}_\alpha + \frac{\boldsymbol{k}}{2}\right) a^+_{q_\alpha} a_{q_\alpha + k}. \tag{20.15a}$$

For $\mathbf{k} = 0$, we have $H_0 = H$, $n_0^\alpha = N_\alpha$, and $\mathbf{p}_0 = \mathbf{p}$, where N_α is the number of particles of sort α and **P** is the total momentum. All these quantities are integrals of motion, and, consequently, for small **k** they change slowly. This is clear if only from the fact that the conservation laws (19.30) in the momentum representation have the form

$$\frac{\partial H_k}{\partial t} = -\boldsymbol{k}\cdot\boldsymbol{j}_k^H, \qquad \frac{\partial \boldsymbol{p}_k}{\partial t} = -\boldsymbol{k}\cdot T_k, \qquad \frac{\partial n_k^\alpha}{\partial t} = -\boldsymbol{k}\cdot\boldsymbol{j}_k,$$

where the right-hand sides contain the small vector **k**.

Choosing as our starting point the operators H_k, n_k^α, and p_k, we determine the statistical operator for a local-equilibrium state, by analogy with (20.3) from the extremum of the information entropy (20.2) under the supplementary conditions

$$\langle H_k \rangle = \text{const}, \quad \langle n_k^\alpha \rangle = \text{const}, \quad \langle p_k \rangle = \text{const}. \tag{20.16}$$

Then

$$\rho_l = Q_l^{-1} \exp\left\{ -\sum_k \left(\beta_{-k} H_k - \sum_\alpha \nu_{-k}^\alpha n_k^\alpha - \gamma_{-k} \cdot p_k \right) \right\}, \tag{20.17}$$

or

$$\rho_l = Q_l^{-1} \exp\left\{ -\int \beta(x) \left[H(x) - \sum_\alpha \left(\mu_\alpha(x) - \frac{m_\alpha}{2} v^2(x) \right) n_\alpha(x) - v(x) \cdot p(x) \right] dx \right\}, \tag{20.17a}$$

where we have introduced the notation

$$\begin{gathered} \left(\mu_\alpha(x) - \frac{m_\alpha}{2} v^2(x) \right) \beta(x) = \nu_\alpha(x) = \sum_k \nu_k^\alpha e^{i\, k \cdot x}, \\ v(x) \beta(x) = \gamma(x) = \sum_k \gamma_k e^{i(kx)} \end{gathered} \tag{20.17b}$$

and where

$$\begin{aligned} Q_l = \operatorname{Tr} \exp\left\{ -\int \beta(x) \left[H(x) - \sum_\alpha \left(\mu_\alpha(x) - \frac{m_\alpha}{2} v^2(x) \right) n_\alpha(x) - v(x) \cdot p(x) \right] dx \right\} = \\ = \operatorname{Tr} \exp\left\{ -\sum_k \left(\beta_{-k} H_k - \sum_\alpha \nu_{-k}^\alpha n_k^\alpha - \gamma_{-k} \cdot p_k \right) \right\} \end{aligned} \tag{20.17c}$$

is the partition function, which is a functional of the parameters (20.17b).

In the expression (20.17a), $\beta(\mathbf{x})$ is the inverse temperature, $\mu_\alpha(\mathbf{x})$ is the chemical potential of component α, and $\mathbf{v}(\mathbf{x})$ is mass velocity.

The physical meaning of this choice of parameters is easily understood. The average of the logarithm of ρ_l determines the entropy. In order to eliminate systematic motion of the liquid, we must choose the system of coordinates moving together with the element of liquid, with velocity $\mathbf{v}(\mathbf{x})$. In this coordinate system, we define the statistical operator

$$\rho_l = Q_l^{-1} \exp\left\{ - \int \beta(x) \left[H'(x) - \sum_\alpha \mu_\alpha(x) n'_\alpha(x) \right] dx \right\}, \qquad (20.18)$$

where $H'(\mathbf{x})$ and $n'_\alpha(\mathbf{x}) = n_\alpha(\mathbf{x})$ are the densities of the energy and of the number of particles of sort α in the moving system.

By means of the canonical transformation (19.35), we transform to the laboratory system, where the energy density is $H(\mathbf{x})$ and the momentum density is $\mathbf{p}(\mathbf{x})$:

$$\begin{aligned} H'(x) &= H(x) - v(x) \cdot p(x) + \frac{v^2(x)}{2} \rho(x), \\ p'(x) &= p(x) - \rho(x) v(x), \end{aligned} \qquad (20.18a)$$

where

$$\rho(x) = \sum_\alpha m_\alpha n_\alpha(x) \qquad (20.18b)$$

is the mass density operator.

Substituting (20.18a) into (20.18), we arrive at (20.17a).

We define the mass velocity $\mathbf{v}(\mathbf{x})$ by the relation

$$v(x) = \frac{\langle p(x) \rangle_l}{\langle \rho(x) \rangle_l}, \qquad (20.19)$$

and so

$$\langle p'(x) \rangle_l = \mathrm{Tr}(\rho_l p(x)) = 0, \qquad (20.19a)$$

i.e., the average momentum in the accompanying system is equal to zero.

It follows from (20.19) that the variational derivative of $\ln Q_l$ with respect to $\mathbf{v}(\mathbf{x})$ is equal to zero:

$$\frac{\delta \ln Q_l}{\delta v(x)} = \beta(x) \{ \langle p(x) \rangle_l - \langle \rho(x) \rangle_l v(x) \} = 0. \qquad (20.19b)$$

If the average hydrodynamic velocities $\mathbf{v}_\alpha(\mathbf{x})$ and the temperatures $T_\alpha(\mathbf{x}) = \beta_\alpha^{-1}(\mathbf{x})$ are assumed to be different for different components, then the statistical operator for a local-equilibrium state with such a "separation" of the temperatures and velocities of the components can be defined in the form

$$\rho_l = Q_l^{-1} \exp\left\{-\sum_\alpha \int \beta_\alpha(\boldsymbol{x})\left[H'_\alpha(\boldsymbol{x}) - \mu_\alpha(\boldsymbol{x})\, n'_\alpha(\boldsymbol{x})\right] d\boldsymbol{x}\right\}, \qquad (20.20)$$

where for each component we choose its accompanying coordinate system, moving with velocity $\mathbf{v}_\alpha(\mathbf{x})$. In this case, transforming to the laboratory system, we write the statistical operator (20.20) in the form

$$\rho_l = Q_l^{-1} \exp\left\{-\sum_\alpha \int \beta_\alpha(\boldsymbol{x})\left[H_\alpha(\boldsymbol{x}) - \right.\right.$$

$$\left.\left. - \left(\mu_\alpha(\boldsymbol{x}) - \frac{m_\alpha}{2} v_\alpha^2(\boldsymbol{x})\right) n_\alpha(\boldsymbol{x}) - \boldsymbol{v}_\alpha(\boldsymbol{x}) \cdot \boldsymbol{p}_\alpha(\boldsymbol{x})\right] d\boldsymbol{x}\right\}, \qquad (20.20a)$$

where we have

$$\frac{\delta \ln Q_l}{\delta \boldsymbol{v}_\alpha(\boldsymbol{x})} = \beta_\alpha(\boldsymbol{x})\{\langle \boldsymbol{p}_\alpha(\boldsymbol{x})\rangle_l - \langle \rho_\alpha(\boldsymbol{x})\rangle_l\, \boldsymbol{v}_\alpha(\boldsymbol{x})\} = 0$$
$$(\rho_\alpha(\boldsymbol{x}) = m_\alpha n_\alpha(\boldsymbol{x})). \qquad (20.20b)$$

If the molecules are not spherical, it is necessary to take into account the transfer of angular momentum in the collisions. Just as we introduced earlier the average velocity $\mathbf{v}(\mathbf{x})$ of the hydrodynamic motion, we can also introduce the average angular velocity $\boldsymbol{\omega}(\mathbf{x})$ of the rotational motion. For this, it is necessary to define first the statistical operator in the system rotating locally with angular velocity $\boldsymbol{\omega}(\mathbf{x})$ equal to the average angular velocity of the rotational motion of the particles in the vicinity of the given point. In this system, any systematic rotation of the particles is compensated by the motion of the coordinate system (cf. [41]). The distribution function and the hydrodynamic equations for systems of molecules with intrinsic angular momentum was treated earlier by Grad [73] and Curtiss [74]; the latter generalized the Boltzmann kinetic equation and the Chapman–Enskog theory to the case of molecules which do not possess radial symmetry.

The calculation of average values by means of ρ_l is a fairly complicated problem, although in the local approximation, when the parameters $\beta(\mathbf{x})$ and $\mu_\alpha(\mathbf{x})$ vary little over distances of the order of the correlation length of the quantities H(x) and n(x), the principal term in the calculation of the averages is very simple. It is equal to

$$\langle A\rangle_l = (\langle A\rangle_0)_{\beta=\beta(x),\ \ldots,\ \mu_\alpha=\mu_\alpha(x)},$$

i.e., in the equilibrium averages, we must replace the equilibrium parameters by their values depending on spatial position. This relation can be proved by selecting and summing the appropriate terms of the perturbation-theory series. For a method of calculating local-equilibrium averages, see the paper by Kuni and Storonkin [236].

20.2. Thermodynamic Equalities

We obtain the thermodynamic equalities for a nonuniform system by taking the variation of the partition function (20.17c):

$$Q_l = \operatorname{Tr} \exp\left\{-\sum_m \int F_m(\boldsymbol{x})\, P_m(\boldsymbol{x})\, d\boldsymbol{x}\right\}, \tag{20.17d}$$

where

$$\begin{aligned} F_0(\boldsymbol{x}) &= \beta(\boldsymbol{x}), & P_0(\boldsymbol{x}) &= H(\boldsymbol{x}),\\ F_1(\boldsymbol{x}) &= -\beta(\boldsymbol{x})\, v(\boldsymbol{x}), & P_1(\boldsymbol{x}) &= p(\boldsymbol{x}),\\ F_{\alpha+1}(\boldsymbol{x}) &= -\beta(\boldsymbol{x})\left(\mu_\alpha(\boldsymbol{x}) - \frac{m_\alpha}{2} v^2(\boldsymbol{x})\right), & P_{\alpha+1}(\boldsymbol{x}) &= n_\alpha(\boldsymbol{x})\\ &(\alpha = 1,\ 2,\ \ldots), \end{aligned} \tag{20.17e}$$

with respect to the local parameters $F_m(x)$, whence we shall have

$$\frac{\delta \ln Q_l}{\delta F_m(\boldsymbol{x})} = -\langle P_m(\boldsymbol{x})\rangle_l, \tag{20.21}$$

or, in more explicit form,

$$\begin{aligned} -\frac{\delta \ln Q_l}{\delta\beta(\boldsymbol{x})} &= \langle H'(\boldsymbol{x})\rangle_l - \sum_\alpha \mu_\alpha(\boldsymbol{x})\langle n_\alpha(\boldsymbol{x})\rangle_l,\\ \frac{\delta \ln Q_l}{\delta\mu_\alpha(\boldsymbol{x})} &= \beta(\boldsymbol{x})\langle n_\alpha(\boldsymbol{x})\rangle_l,\\ \frac{\delta \ln Q_l}{\delta v(\boldsymbol{x})} &= 0, \end{aligned} \tag{20.21a}$$

where H'(x) is the energy density in the accompanying system.

The relations (20.21a) are a natural generalization of the thermodynamic equalities (11.25) which hold in the case of statistical equilibrium.

From the relations (20.21a) for the variational derivatives, we obtain for the total variation of Q_l

$$\delta \ln Q_l = \int \left\{ -(\langle H'(\boldsymbol{x})\rangle_l - \sum_\alpha \mu_\alpha(\boldsymbol{x}) \langle n_\alpha(\boldsymbol{x})\rangle_l)\, \delta\beta(\boldsymbol{x}) + \right.$$
$$\left. + \beta(\boldsymbol{x}) \sum_\alpha \langle n_\alpha(\boldsymbol{x})\rangle_l\, \delta\mu_\alpha(\boldsymbol{x}) \right\} d\boldsymbol{x}. \tag{20.21b}$$

Writing

$$\ln Q_l = -\int \beta(\boldsymbol{x})\, \Omega(\boldsymbol{x})\, d\boldsymbol{x}, \tag{20.21c}$$

we obtain

$$\int \left\{ \delta(\beta(\boldsymbol{x})\, \Omega(\boldsymbol{x})) - \left(\langle H'(\boldsymbol{x})\rangle_l - \sum_\alpha \mu_\alpha(\boldsymbol{x}) \langle n_\alpha(\boldsymbol{x})\rangle_l\right) \delta\beta(\boldsymbol{x}) + \right.$$
$$\left. + \beta(\boldsymbol{x}) \sum_\alpha \langle n_\alpha(\boldsymbol{x})\rangle_l\, \delta\mu_\alpha(\boldsymbol{x}) \right\} d\boldsymbol{x} = 0,$$

whence, since the volume is arbitrary, it follows that

$$\delta(\beta(\boldsymbol{x})\, \Omega(\boldsymbol{x})) = \langle H'(\boldsymbol{x})\rangle_l\, \delta\beta(\boldsymbol{x}) - \sum_\alpha \langle n_\alpha(\boldsymbol{x})\rangle_l\, \delta(\beta(\boldsymbol{x})\, \mu_\alpha(\boldsymbol{x})). \tag{20.22}$$

Equation (20.22) is called the Gibbs–Duhem relation.

We introduce the entropy density $S(\boldsymbol{x})$ by the relation

$$\beta^{-1}(\boldsymbol{x})\, S(\boldsymbol{x}) = \langle H'(\boldsymbol{x})\rangle_l - \sum_\alpha \mu_\alpha(\boldsymbol{x}) \langle n_\alpha(\boldsymbol{x})\rangle_l - \Omega(\boldsymbol{x}), \tag{20.23}$$

which is analogous to (11.24), and rewrite the thermodynamic equality (20.22) in the form

$$-\delta\Omega(\boldsymbol{x}) = S(\boldsymbol{x})\, \delta\beta^{-1}(\boldsymbol{x}) + \sum_\alpha \langle n_\alpha(\boldsymbol{x})\rangle_l\, \delta\mu_\alpha(\boldsymbol{x}) \tag{20.24}$$

or

$$\beta^{-1}(\boldsymbol{x})\, \delta S(\boldsymbol{x}) = \delta\langle H'(\boldsymbol{x})\rangle_l - \sum_\alpha \mu_\alpha(\boldsymbol{x})\, \delta\langle n_\alpha(\boldsymbol{x})\rangle_l. \tag{20.24a}$$

We introduce quantities calculated for unit mass, namely,

$$s(\boldsymbol{x}) = \frac{S(\boldsymbol{x})}{\langle \rho(\boldsymbol{x})\rangle_l} \tag{20.23a}$$

which is the entropy per unit mass,

$$u(x) = \frac{\langle H'(\boldsymbol{x})\rangle_l}{\langle \rho(\boldsymbol{x})\rangle_l} \tag{20.23b}$$

which is the energy per unit mass in the moving system, and

$$c_\alpha(\boldsymbol{x}) = \frac{m_\alpha \langle n_\alpha(\boldsymbol{x})\rangle_l}{\langle \rho(\boldsymbol{x})\rangle_l} \tag{20.23c}$$

which is the relative mass concentration of particles of component α. Then the thermodynamic equality (20.24a) takes the more usual form

$$T(\boldsymbol{x})\,\delta s(\boldsymbol{x}) = \delta u(\boldsymbol{x}) + p(\boldsymbol{x})\,\delta v(\boldsymbol{x}) - \sum_\alpha \frac{\mu_\alpha(\boldsymbol{x})}{m_\alpha}\,\delta c_\alpha(\boldsymbol{x}), \tag{20.24b}$$

where

$$v(\boldsymbol{x}) = \frac{1}{\langle \rho(\boldsymbol{x})\rangle_l} \tag{20.24c}$$

is the specific volume per unit mass, and

$$p(\boldsymbol{x}) = -\,\Omega(\boldsymbol{x}) = T(\boldsymbol{x})\,S(\boldsymbol{x}) - \langle H'(\boldsymbol{x})\rangle + \sum_\alpha \mu_\alpha(\boldsymbol{x})\langle n_\alpha(\boldsymbol{x})\rangle_l \tag{20.24d}$$

is the pressure.

The thermodynamic equality (20.24b) is analogous to the equality (5.8) and (20.24d) is analogous to the relations (3.34), and (5.27) of equilibrium thermodynamics.

We have considered the variation of quantities in the accompanying system moving with the center of mass of the element with velocity $\mathbf{v}(\mathbf{x})$; consequently, (20.24b) can be written in the form of a relation in total derivatives:

$$T(\boldsymbol{x})\frac{ds(\boldsymbol{x})}{dt} = \frac{du(\boldsymbol{x})}{dt} + p(\boldsymbol{x})\frac{dv(\boldsymbol{x})}{dt} - \sum_\alpha \mu_\alpha(\boldsymbol{x})\frac{dc_\alpha(\boldsymbol{x})}{dt}, \tag{20.24e}$$

where

$$\frac{d}{dt} = \frac{\partial}{\partial t} + \boldsymbol{v}(\boldsymbol{x})\cdot\nabla.$$

The relation (20.24e) is usually postulated in nonequilibrium thermodynamics as an expression of the local-equilibrium hypothesis [27].

20.3. Fluctuations in a Local-Equilibrium Ensemble [3]

The local-equilibrium distribution (20.17a) makes it possible to express fluctuations of the densities of energy, particle number, and momentum in terms of the variations of the average values of the densities of these physical quantities with respect to the local parameters $\beta(\mathbf{x})$, $\mu_\alpha(\mathbf{x})$, and $\mathbf{v}(\mathbf{x})$, just as in §§6 and 12, the fluctuations of the mechanical quantities for the grand ensemble were expressed in terms of derivatives of their average values with respect to the corresponding parameters.

We shall calculate the variation of the average value of the arbitrary operator $A(\mathbf{x})$ with respect to $\beta(\mathbf{x})$, $\mu_\alpha(\mathbf{x})$, and $\mathbf{v}(\mathbf{x})$. We have

$$\langle A(\boldsymbol{x})\rangle_l = Q_l^{-1}\,\mathrm{Tr}\left\{e^{-\int \beta(\boldsymbol{x})\tilde{H}(\boldsymbol{x})\,dx} A(\boldsymbol{x})\right\}, \tag{20.25}$$

where

$$\tilde{H}(\boldsymbol{x}) = H(\boldsymbol{x}) - \sum_\alpha \left[\mu_\alpha(\boldsymbol{x}) - \frac{m_\alpha}{2} v^2(\boldsymbol{x})\, n_\alpha(\boldsymbol{x})\right] - \boldsymbol{v}(\boldsymbol{x})\cdot\boldsymbol{p}(\boldsymbol{x}) =$$

$$= H'(\boldsymbol{x}) - \sum_\alpha \mu_\alpha(\boldsymbol{x})\, n_\alpha(\boldsymbol{x}). \tag{20.25a}$$

Noting that, for any operator, the variation of the exponential is equal to (12.13), and taking into account, that the variation of Q_l gives the thermodynamic relations (20.21), we obtain

$$\begin{aligned}
\frac{\delta\langle A(\boldsymbol{x})\rangle_l}{\delta\beta(\boldsymbol{x}')} &= -\,(A(\boldsymbol{x}),\ \tilde{H}(\boldsymbol{x}')),\\
\frac{\delta\langle A(\boldsymbol{x})\rangle_l}{\delta\mu_\alpha(\boldsymbol{x}')} &= \beta(\boldsymbol{x}')\,(A(\boldsymbol{x}),\ n_\alpha(\boldsymbol{x}')),\\
\frac{\delta\langle A(\boldsymbol{x})\rangle_l}{\delta v(\boldsymbol{x}')} &= \beta(\boldsymbol{x}')\,(A(\boldsymbol{x}),\ \boldsymbol{p}(\boldsymbol{x}') - \rho(\boldsymbol{x}')\,\boldsymbol{v}(\boldsymbol{x}')),
\end{aligned} \tag{20.26}$$

where for the quantum correlation functions we have introduced the notation

$$(A(\boldsymbol{x}),\, B(\boldsymbol{x}')) = \int_0^1 \langle (A(\boldsymbol{x}) - \langle A(\boldsymbol{x})\rangle_l)\,(B(\boldsymbol{x}',\, i\tau) - \langle B(\boldsymbol{x}')\rangle_l)\rangle_l\, d\tau, \tag{20.26a}$$

$$B(i\tau) = e^{-\tau\int \beta(\boldsymbol{x})\tilde{H}(\boldsymbol{x})\,dx}\, B e^{\tau\int \beta(\boldsymbol{x})\tilde{H}(\boldsymbol{x})\,dx} \tag{20.26b}$$

The notation of (20.26a) and (20.26b) is analogous to that of (12.18) and (12.16) used earlier in §12. In the limiting case of classical mechanics, the quantum correlation functions (20.26a) go over into classical correlation functions:

$$(A(\boldsymbol{x}),\, B(\boldsymbol{x}')) = \langle (A(\boldsymbol{x}) - \langle A(\boldsymbol{x})\rangle_l)\,(B(\boldsymbol{x}') - \langle B(\boldsymbol{x}')\rangle_l)\rangle_l. \tag{20.26c}$$

From (20.26) we find the fluctuations of the energy, particle-number, and momentum densities:

$$\begin{aligned} \frac{-\delta \langle H(\boldsymbol{x})\rangle_l}{\delta\beta(\boldsymbol{x}')} &= (H(\boldsymbol{x}),\ \tilde{H}(\boldsymbol{x}')), \\ \frac{\delta \langle n_\alpha(\boldsymbol{x})\rangle_l}{\delta\mu_\beta(\boldsymbol{x}')} &= \beta(\boldsymbol{x}')(n_\alpha(\boldsymbol{x}),\ n_\beta(\boldsymbol{x}')), \\ \frac{\delta \langle \boldsymbol{p}(\boldsymbol{x})\rangle_l}{\delta \boldsymbol{v}(\boldsymbol{x}')} &= \beta(\boldsymbol{x}')(\boldsymbol{p}(\boldsymbol{x}),\ \boldsymbol{p}(\boldsymbol{x}') - \rho(\boldsymbol{x}')\boldsymbol{v}(\boldsymbol{x}')). \end{aligned} \tag{20.27}$$

In the same way, one can also express other variational derivatives in terms of correlation functions.

We shall consider the more general case of a nonequilibrium ensemble in which the state is determined not only by $H(\mathbf{x})$, $n_\alpha(\mathbf{x})$, and $\mathbf{p}(\mathbf{x})$, but also by quantities $\xi_k(\mathbf{x})$ which, generally speaking, are not densities of integrals of motion. In this case, we may again construct a statistical operator, which is constrained to give average values of $\xi_k(\mathbf{x})$ equal to the averages $\langle \xi_k(\boldsymbol{x})\rangle_l$ over the local-equilibrium state.

The statistical operator corresponding to the maximum of the information entropy (20.2) and with given $\langle \xi_k(\boldsymbol{x})\rangle$, has the form

$$\rho_l = Q_l^{-1} \exp\left\{ - \int \beta(\boldsymbol{x}) \left[H(\boldsymbol{x}) - \sum_\alpha \left(\mu_\alpha(\boldsymbol{x}) - \frac{m_\alpha}{2} v^2(\boldsymbol{x})\, n_\alpha(\boldsymbol{x}) \right) - \boldsymbol{v}(\boldsymbol{x}) \cdot \boldsymbol{p}(\boldsymbol{x}) + \sum_k a_k(\boldsymbol{x})\, \xi_k(\boldsymbol{x}) \right] d\boldsymbol{x} \right\}, \tag{20.28}$$

where $a_k(\mathbf{x})$ is the auxiliary field which makes $\langle \xi_k(\boldsymbol{x})\rangle_l \neq 0$. nonzero. The average value $\langle \xi_k(\boldsymbol{x})\rangle$ corresponding to the state (20.28) is equal to

$$\langle \xi_k(\boldsymbol{x})\rangle_l = -\frac{1}{\beta(\boldsymbol{x})} \frac{\delta \ln Q_l}{\delta a_k(\boldsymbol{x})}, \tag{20.29}$$

where the variation is taken with $\beta(\mathbf{x})$, $\mu_\alpha(\mathbf{x})$, and the other $a_{k'}(\mathbf{x})$ $(k' \neq k)$ held constant.

The statistical operator (20.28) can be written conveniently in the more symmetric form

$$\rho = \exp\left\{ -\Phi[F_0(\boldsymbol{x}), \ldots, F_n(\boldsymbol{x})] - \sum_{i=0}^{n} \int F_i(\boldsymbol{x}) P_i(\boldsymbol{x})\, d\boldsymbol{x} \right\}, \tag{20.30}$$

where

$$\begin{aligned} F_0(\boldsymbol{x}) &= \beta(\boldsymbol{x}), & P_0(\boldsymbol{x}) &= H(\boldsymbol{x}), \\ F_1(\boldsymbol{x}) &= -\beta(\boldsymbol{x})\,\boldsymbol{v}(\boldsymbol{x}), & P_1(\boldsymbol{x}) &= \boldsymbol{p}(\boldsymbol{x}), \\ F_{\alpha+1}(\boldsymbol{x}) &= -\beta(\boldsymbol{x})\Big[\mu_\alpha(\boldsymbol{x}) - \frac{m_\alpha}{2} v^2(\boldsymbol{x})\Big], & P_{\alpha+1}(\boldsymbol{x}) &= n_\alpha(\boldsymbol{x}) \\ F_i(\boldsymbol{x}) &= \beta(\boldsymbol{x})\, a_i(\boldsymbol{x}), & P_i(\boldsymbol{x}) &= \xi_i(\boldsymbol{x}) \end{aligned} \tag{20.30a}$$

$$(\alpha = 1, 2, \ldots, l; \quad i = l+1, \ldots, n)$$

and

$$\Phi[F_0(\boldsymbol{x}), \ldots, F_n(\boldsymbol{x})] = \ln Q \tag{20.30b}$$

is the Massieu–Planck functional.

It was shown in §§6 and 12 that in the theory of fluctuations it is very convenient to apply the Massieu–Planck function (6.4), (12.7). The Massieu–Planck functional (20.30b) also turns out to be very convenient for the study of fluctuations in spatially nonuniform systems [3].

We write the statistical operator (20.30) in the form

$$\rho = \exp\left\{-S - \sum_{i=0}^{n} \int F_i(\boldsymbol{x})\,(P_i(\boldsymbol{x}) - \langle P_i(\boldsymbol{x})\rangle_l)\,d\boldsymbol{x}\right\}, \tag{20.31}$$

where

$$S = \Phi + \sum_{i=0}^{n} \int F_i(\boldsymbol{x})\,\langle P_i(\boldsymbol{x})\rangle_l\,d\boldsymbol{x} \tag{20.31a}$$

is the entropy regarded as a functional of $\langle P_i(\boldsymbol{x})\rangle_l$. The relation (20.31a) is the analog for thermodynamic functionals of the Legendre transformation (3.10b) of thermodynamic functions.

From the normalization condition for the statistical operator in the form (20.30)

$$\Phi = \ln \operatorname{Tr} \exp\left\{-\sum_{i=0}^{n} \int F_i(\boldsymbol{x})\,P_i(\boldsymbol{x})\,d\boldsymbol{x}\right\}, \tag{20.32}$$

we obtain, by varying it with respect to $F_i(\mathbf{x})$, the thermodynamic equalities

$$\frac{\delta\Phi}{\delta F_i(\boldsymbol{x})} = -\langle P_i(\boldsymbol{x})\rangle_l \qquad (i = 0, 1, \ldots, n). \tag{20.33}$$

Analogously, from the normalization condition for the statistical operator in the form (20.31)

$$S = \ln \operatorname{Tr} \exp\left\{ -\sum_{i=0}^{n} \int F_i(\boldsymbol{x}) \left(P_i(\boldsymbol{x}) - \langle P_i(\boldsymbol{x})\rangle_l\right) d\boldsymbol{x} \right\}, \tag{20.34}$$

we obtain, by varying it with respect to $\langle P_i(\boldsymbol{x})\rangle_l$, the thermodynamic equalities

$$\frac{\delta S}{\delta \langle P_i(\boldsymbol{x})\rangle_l} = F_i(\boldsymbol{x}). \tag{20.35}$$

We shall express the fluctuations in terms of the variation of the average value $\langle P_i(\boldsymbol{x})\rangle_l$ with respect to $F_m(\mathbf{x})$. We have

$$\delta \langle P_i(\boldsymbol{x})\rangle_l = \operatorname{Tr}\left(P_i(\boldsymbol{x})\,\delta\rho\right).$$

Using (20.30), (12.13), and (20.33), we obtain for the variation $\delta\rho$

$$\delta\rho = -\sum_{i=0}^{n} \int \int_0^1 e^{-A\tau} \left(P_i(\boldsymbol{x}') - \langle P_i(\boldsymbol{x}')\rangle_l\right) e^{A\tau} e^{-A}\, \delta F_i(\boldsymbol{x}')\, d\tau\, d\boldsymbol{x}',$$

where

$$A = \Phi + \sum_i \int F_i(\boldsymbol{x}) P_i(\boldsymbol{x})\, d\boldsymbol{x}. \tag{20.36}$$

Consequently,

$$\delta \langle P_i(\boldsymbol{x})\rangle_l = -\sum_m \int \left(P_i(\boldsymbol{x}),\ P_m(\boldsymbol{x}')\right) \delta F_m(\boldsymbol{x}')\, d\boldsymbol{x}', \tag{20.37}$$

where

$$\left(P_i(\boldsymbol{x}),\ P_m(\boldsymbol{x}')\right) = \int_0^1 \langle P_i(\boldsymbol{x})\, e^{-A\tau} \left(P_m(\boldsymbol{x}') - \langle P_m(\boldsymbol{x}')\rangle_l\right) e^{A\tau}\rangle_l\, d\tau. \tag{20.37a}$$

It follows from (20.37) that the quantum correlation function (20.37a) can be expressed in terms of the variational derivative of $\langle P_i(\boldsymbol{x})\rangle_l$ with respect to $F_m(\mathbf{x}')$:

$$\frac{\delta \langle P_i(\boldsymbol{x})\rangle_l}{\delta F_m(\boldsymbol{x}')} = -\left(P_i(\boldsymbol{x}),\ P_m(\boldsymbol{x}')\right). \tag{20.37b}$$

We shall generalize the Einstein thermodynamic theory of fluctuations [75, 76], described in §6 for the equilibrium case, to the case of a local-equilibrium state, following paper [3]. For the local-equilibrium distribution (20.30), we can introduce a macroscopic functional or a macroscepic distribution function similar to the function (6.11).

We introduce the macroscopic distribution function W for the Fourier components P_k^i of the variables $P_i(x)$

$$P_k^i = \int P_i(x) e^{-i k\cdot x} dx, \quad F_i^k = \frac{1}{V}\int F_i(x) e^{-i k\cdot x} dx \quad (i = 0, 1, \ldots, n), \tag{20.38}$$

which gives the probability that the parameters $\ldots P_k^0 \ldots P_k^n \ldots$ lie in the regions $\ldots \Delta P_k^0 \ldots \Delta P_k^n \ldots$ about the points $\ldots P_k^0 \ldots P_k^n \ldots$:

$$W \Delta P_k^0 \ldots \Delta P_k^n = \Omega \Delta P_k^0 \ldots \Delta P_k^n \exp\left\{-\Phi - \sum_{k,\, i} F_i^{-k} P_k^i\right\} = \\ = \Omega \Delta P_k^0 \ldots \Delta P_k^n \exp\left\{-\Phi - \sum_i \int F_i(x) P_i(x)\, dx\right\}, \tag{20.39}$$

where the quantities P_k^i are now to be regarded not as operators but as ordinary functions, although we are using for them the same notation as for the corresponding operators. For brevity, we do not attach subscripts to the k vectors.

The quantity $\Omega \Delta P_k^0 \ldots \Delta P_k^n \ldots$ has the meaning of the number of micro-states in the region $\ldots \Delta P_k^0 \ldots \Delta P_k^n \ldots$. It can be estimated in terms of the entropy s of the microcanonical ensemble in which the parameters $\ldots P_k^0 \ldots P_k^n \ldots$ are specified to be in the regions $\ldots \Delta P_k^0 \ldots \Delta P_k^n \ldots$:

$$s = \ln \frac{\Omega}{\Omega_0}, \tag{20.40}$$

where Ω_0 is a constant which, for the present, is unimportant; we shall determine it later from the normalization condition for W.

Taking (20.40) and (20.34) into account, we write the macroscopic distribution function (20.39) in the form

$$W = \Omega_0 \exp\left\{-\Phi + s - \sum_i \int F_i(x) P_i(x)\, dx\right\} \\ = \Omega_0 \exp\left\{s - S - \sum_i \int F_i(x)(P_i(x) - \langle P_i(x)\rangle_l)\, dx\right\}, \tag{20.39a}$$

where S is the entropy in the local-equilibrium grand canonical ensemble (20.31a).

Because of the equivalence of the statistical ensembles, which was proved in §13, the entropy in the local-equilibrium grand canonical ensemble is the same function of $\langle P_k^0\rangle, \ldots, \langle P_k^n\rangle$, as the entropy in the local-equilibrium microcanonical ensemble is of $P_k^0, \ldots, P_k^n$, i.e., S and s are the same functions, but of different variables. Therefore, it is convenient to expand s − S in a functional series in $\Delta P_i(x) = P_i(x) - \langle P_i(x)\rangle_l$ and confine ourselves, because the fluctuations are small, to the second-order terms:

$$s - S = \sum_i \int F_i(x)\,\Delta P_i(x)\,dx + \\ + \frac{1}{2}\sum_{im}\int\int \frac{\delta^2 S}{\delta\langle P_i(x_1)\rangle_l\,\delta\langle P_m(x_2)\rangle_l}\,\Delta P_i(x_1)\,\Delta P_m(x_2)\,dx_1\,dx_2. \tag{20.41}$$

Substituting (20.41) into (20.39a) and taking into account that the linear terms cancel, we obtain

$$W = A\exp\left\{\frac{1}{2}\sum_{im}\int\int \frac{\delta^2 S}{\delta\langle P_i(x_1)\rangle_l\,\delta\langle P_m(x_2)\rangle_l}\,\Delta P_i(x_1)\,\Delta P_m(x_2)\,dx_1\,dx_2\right\}, \tag{20.42}$$

or

$$W = A\exp\left\{-\frac{1}{2}\sum_{im}\int\int f_{im}(x_1, x_2)\,\Delta P_i(x_1)\,\Delta P_m(x_2)\,dx_1\,dx_2\right\}, \tag{20.42a}$$

where

$$f_{im}(x_1, x_2) = -\frac{\delta^2 S}{\delta\langle P_i(x_1)\rangle_l\,\delta\langle P_m(x_2)\rangle_l} \tag{20.42b}$$

is a function describing the correlation of the fluctuations in space; we shall study this function below.

Formulas (20.41)-(20.42b) are the direct generalization of formulas (6.14)-(6.17a).

Particularly simple relations for the fluctuations hold when the parameters $F_i(x)$ and the quantities $P_i(x)$ are expressed in the Fourier representation (20.38). Then

$$\rho = \exp\left\{-\Phi - \sum_{i,k} F_i^{-k} P_k^i\right\}. \tag{20.43}$$

The entropy (20.31a) is now no longer a functional, but a function of the Fourier components $\langle P_k^i \rangle$.

In lieu of the thermodynamic equalities (20.33) in variational derivatives, we obtain relations in ordinary derivatives

$$\frac{\partial \Phi}{\partial F_i^{-k}} = -\langle P_k^i \rangle_l \tag{20.44}$$

and, for the quantum correlation functions, we obtain expressions

$$(P_{k_1}^i, P_{k_2}^m) = -\frac{\partial \langle P_{k_1}^i \rangle_l}{\partial F_m^{-k_2}} = -\frac{\partial \langle P_{k_2}^m \rangle_l}{\partial F_i^{-k_1}}, \tag{20.44a}$$

analogous to formulas (20.33) and (20.37b).

The macroscepic distribution function (20.42) in the Fourier representation has the form

$$W = A \exp\left\{ \frac{1}{2} \sum_{\substack{im \\ k_1 k_2}} \frac{\partial^2 S}{\partial \langle P_{k_1}^i \rangle_l \partial \langle P_{k_2}^m \rangle_l} \Delta P_{k_1}^i \Delta P_{k_2}^m \right\}, \tag{20.45}$$

where $\Delta P_k^i = P_k^i - \langle P_k^i \rangle_l$. As before, we regard P_k^i as an ordinary variable rather than as a dynamic variable. Thus, for the Fourier components P_k^i, the Gaussian distribution (20.45) is approximately valid.

The exponent of (20.45) can be expressed in terms of a correlation function. In fact,

$$\frac{\partial S}{\partial \langle P_k^i \rangle_l} = F_i^{-k}, \tag{20.45a}$$

where S is the entropy (20.31a). Consequently,

$$-\sum_{m' k_2'} \frac{\partial^2 \Phi}{\partial F_i^{-k_1} \partial F_{m'}^{-k_2'}} \frac{\partial^2 S}{\partial \langle P_{k_2'}^{m'} \rangle_l \partial \langle P_{k_2}^{m} \rangle_l} = \sum_{m' k_2'} \frac{\partial \langle P_{k_1}^i \rangle_l}{\partial F_{m'}^{-k_2'}} \frac{\partial F_{m'}^{-k_2'}}{\partial \langle P_{k_2}^m \rangle_l} = \frac{\partial \langle P_{k_1}^i \rangle_l}{\partial \langle P_{k_2}^m \rangle_l} = \delta_{k_1 k_2}^{im} \tag{20.45b}$$

which is analogous to formula (6.24). Thus, the matrices of the second derivatives of Φ and S are mutually reciprocal.

The formula (20.45) describes the correlation of the Fourier components (20.38) and thus expresses a relation between the fluctuations at different points. If we take into account only the correlation between the Fourier coefficients P_{k}^{i} and $P_{-k}^{m} = P_{k}^{m*}$, i.e., use the fact that the state is almost spatially uniform, then the formula (20.45) can be represented approximately in the form

$$W = A \exp\left\{ -\frac{1}{2V} \sum_{i,\,m,\,k} f_{k}^{im} \Delta P_{k}^{i} \Delta P_{-k}^{m} \right\}, \tag{20.46}$$

where

$$f_{k}^{im} = \int f_{im}(\boldsymbol{x})\, e^{-i\boldsymbol{x}\cdot\boldsymbol{k}}\, d\boldsymbol{x} = -V \frac{\partial^2 S}{\partial \langle P_{k}^{i}\rangle_l\, \partial \langle P_{-k}^{m}\rangle_l}, \quad f_{k}^{im} = f_{-k}^{mi}. \tag{20.46a}$$

This spatial correlation effect lies at the basis of the theory of fluctuations near the critical point (the theory of Onstein and Zernike [77]). This theory was developed further by Klein and Tisza [78] (see also the reviews [237, 238]); we shall study it in the following subsection, following the paper [3].

20.4. Critical Fluctuations [3]

By means of the probability distribution function (20.45), (20.46), we can investigate the fluctuations close to the critical point, where they can increase sharply.

First we shall examine some exact relations for the fluctuations.

The distribution function (20.45) can be written conveniently in the form of the exponential of the complete contraction of a product of matrices with elements i, m:

$$W = A \exp\left\{ -\frac{1}{2V} \sum_{k_1 k_2} f_{k_1 k_2} : \Delta P_{k_1} \Delta P_{-k_2} \right\}, \tag{20.47}$$

where $f_{k_1 k_2}$ is a matrix with elements

$$f_{k_1 k_2}^{im} = \frac{1}{V} \int f_{im}(\boldsymbol{x}_1, \boldsymbol{x}_2)\, e^{-i k_1 \cdot x_1 + i k_2 \cdot x_2}\, d\boldsymbol{x}_1\, d\boldsymbol{x}_2 = -V \frac{\partial^2 S}{\partial \langle P_{k_1}^{i}\rangle_l\, \partial \langle P_{-k_2}^{m}\rangle_l}, \tag{20.47a}$$

ΔP_{k} is a vector with components ΔP_{k}^{i}, and the symbol : denotes the complete contraction of the product of matrices. In the coordinate representation, the function (20.47) has the form

$$W = A \exp\left\{ -\frac{1}{2} \int f(\boldsymbol{x}_1, \boldsymbol{x}_2) : \Delta P(\boldsymbol{x}_1)\, \Delta P(\boldsymbol{x}_2)\, d\boldsymbol{x}_1\, d\boldsymbol{x}_2 \right\}. \qquad (20.47b)$$

The distribution function (20.47) can be written in an even more compact form:

$$W = A \exp\left\{ -\frac{1}{2V} f : \Delta P \Delta P^* \right\}, \qquad (20.47c)$$

where f is the matrix with elements $f^{im}_{\boldsymbol{k}_1\boldsymbol{k}_2}$, ΔP is the matrix with elements $\Delta P^i_{\boldsymbol{k}}$, and the symbol : denotes the contraction over all the indices im, k_1k_2.

It is well-known [cf. formula (6.22)] that the fluctuations calculated by means of a Gaussian distribution function are expressed in terms of the matrix inverse to f, i.e.,

$$g = \frac{1}{V} \langle \Delta P\, \Delta P^* \rangle = f^{-1}, \qquad (20.48)$$

or, in explicit form,

$$\sum_{\boldsymbol{k}_1} f_{\boldsymbol{k}\boldsymbol{k}_1}\, g_{\boldsymbol{k}_1\boldsymbol{k}'} = \delta_{\boldsymbol{k}\boldsymbol{k}'}, \qquad (20.48a)$$

where $\delta_{\boldsymbol{k}\boldsymbol{k}'}$ is the Kronecker symbol. In the left-hand side of (20.48a), a contraction over all indices except i and m is also implied, while the right-hand side is to be understood as a unit matrix with indices i and m that are not written out explicitly. Consequently, (20.48a) coincides with (20.45b).

Equation (20.48a) in the x-representation is the integral equation

$$\int f(\boldsymbol{x}, \boldsymbol{x}_1)\, g(\boldsymbol{x}_1, \boldsymbol{x}')\, d\boldsymbol{x}_1 = \delta(\boldsymbol{x} - \boldsymbol{x}'), \qquad (20.49)$$

where

$$g(\boldsymbol{x}_1, \boldsymbol{x}') = \frac{1}{V} \sum_{\boldsymbol{k}_1\boldsymbol{k}_2} e^{i\boldsymbol{k}_1\cdot\boldsymbol{x}_1 - i\boldsymbol{k}_2\cdot\boldsymbol{x}'}\, g_{\boldsymbol{k}_1\boldsymbol{k}_2} = (P(\boldsymbol{x}_1), P(\boldsymbol{x}')). \qquad (20.49a)$$

Equation (20.49) can be satisfied, for example, if f and g have δ-function singularities when their arguments coincide (similar singularities in correlation functions are well known in the theory of fluctuations). Consequently, this equation can be conveniently regularized by separating out the δ-function singularities. In place of f and g, we introduce the functions f_1 and g_1, which no longer have such singularities:

$$f(x, x_1) = A(x)(\delta(x - x_1) - f_1(x, x_1)),$$
$$g(x_1, x') = A_1(x')(\delta(x_1 - x') + g_1(x_1, x')). \qquad (20.50)$$

We shall find the function $A_1(x)$ by integrating the second of the relations (20.50) twice over a small volume ΔV about the point $x = x_1$; we choose the dimensions of this volume in such a way that the contribution from g_1 can be neglected. Then

$$\left\langle \int_{\Delta V} \Delta P(x')\,dx' \int_{\Delta V} \Delta P(x')\,dx' \right\rangle = \int_{\Delta V} A_1(x')\,dx',$$

whence, applying a theorem on averages, we find

$$A_1(x) = (\Delta V^{-1}) \left\langle \int_{\Delta V} \Delta P(x)\,dx \int_{\Delta V} \Delta P(x)\,dx \right\rangle.$$

For the volume ΔV we can choose the volume per particle, $\Delta V = \langle n(x) \rangle^{-1}$; in this case, it is clear that the condition that ΔV be small will be satisfied.

Substituting (20.50) into (20.49), we obtain an integral equation for g_1:

$$g_1(x, x') = f_1(x, x') + \int f_1(x, x_1)\, g_1(x_1, x')\,dx_1, \qquad (20.51)$$
$$A(x)\,A_1(x) = 1.$$

This is the Ornstein–Zernike integral equation for the nonuniform case [3]. The function f_1 is usually called the direct correlation function. The function g_1 satisfies a normalization condition which is obtained from the second equation (20.50) by means of a double integration over x_1 and x':

$$\int\int A_1(x')\, g_1(x_1, x')\,dx_1\,dx' = \langle \Delta P\,\Delta P \rangle - \int A_1(x)\,dx, \qquad (20.51a)$$

where

$$\Delta P = \int \Delta P(x)\,dx, \qquad \langle \Delta P\,\Delta P \rangle = -\frac{\partial \langle P \rangle}{\partial \overline{F}}, \qquad \overline{F} = \frac{1}{V} \int F(x)\,dx.$$

For the uniform case, f_1 and g_1 depend only on the difference of the arguments, and Eq. (20.51a) goes over into the usual Ornstein–Zernike equation [77-79]:

$$g_1(\boldsymbol{x}-\boldsymbol{x}') = f_1(\boldsymbol{x}-\boldsymbol{x}') + \int f_1(\boldsymbol{x}-\boldsymbol{x}_1)\, g_1(\boldsymbol{x}_1-\boldsymbol{x}')\, d\boldsymbol{x}_1 \tag{20.52}$$

with the normalization condition

$$\int g_1(\boldsymbol{x})\, d\boldsymbol{x} = -\frac{1}{V}\frac{\partial\langle P\rangle}{\partial \bar{F}}\frac{1}{A} - 1. \tag{20.52a}$$

The spatially uniform case can be treated by starting from (20.46) directly:

$$W = A\exp\left\{-\frac{1}{2V}\sum_{\boldsymbol{k}} f_{\boldsymbol{k}} : \Delta P_{\boldsymbol{k}}\, \Delta P^*_{\boldsymbol{k}}\right\}, \tag{20.53}$$

where $f_{\boldsymbol{k}}$ is a matrix with elements $f^{im}_{\boldsymbol{k}}$, with

$$f_{\boldsymbol{k}} = \int f(\boldsymbol{x})\, e^{-i\,\boldsymbol{k}\cdot\boldsymbol{x}}\, d\boldsymbol{x} = -V\frac{\partial^2 S}{\partial\langle P_{\boldsymbol{k}}\rangle_l\, \langle\partial P^*_{\boldsymbol{k}}\rangle_l} = f_{\boldsymbol{k}\boldsymbol{k}}. \tag{20.53a}$$

In place of (20.48a), we shall have the very simple relation

$$f_{\boldsymbol{k}} g_{-\boldsymbol{k}} = 1, \tag{20.54}$$

which is equivalent to the integral equation

$$\int f(\boldsymbol{x}-\boldsymbol{x}_1)\, g(\boldsymbol{x}_1-\boldsymbol{x}')\, d\boldsymbol{x}_1 = \delta(\boldsymbol{x}-\boldsymbol{x}'). \tag{20.54a}$$

After regularization of this equation, we again arrive at the usual Ornstein–Zernike equation.

We shall now examine the solution of the Ornstein–Zernike equation. In the solution of Eq. (20.52), the assumption is made that the function $f(\mathbf{x})$ falls off sufficiently rapidly with distance, in such a way that its even moments $\int \boldsymbol{x}^{2n} f(\boldsymbol{x})\, d\boldsymbol{x}$ are finite and its odd moments equal to zero because of the spatial isotropy. The function $g(\mathbf{x})$ is expanded in a Taylor series up to the second-order terms, and the integral equation is reduced to a differential equation [79, 237, 238]. The same results can be obtained even more simply by expanding $f_{\boldsymbol{k}}$ in (20.53) in a series in k^2 and keeping only a few terms, as we shall do below.

We shall study the behavior of the correlation function of the fluctuations

$$\langle \Delta P_i(\boldsymbol{x})\, \Delta P_m(\boldsymbol{x}')\rangle$$

at large distances $|\mathbf{x}-\mathbf{x}'|$ in the case of spatial uniformity. For this, we expand f_k^{im} (20.53a) in a series in powers of k^2, confining ourselves to terms up to and including k^4. The expansion in small k means that we are considering the behavior of the correlation function at large $|\mathbf{x}-\mathbf{x}'|$. The coefficients of the odd powers of k are equal to zero because of the spatial isotropy of $f(\mathbf{x})$. We have

$$f_k^{im} = a_{im} + b_{im}k^2 + c_{im}k^4, \tag{20.55}$$

where

$$a_{im} = \int f_{im}(\boldsymbol{x})\, d\boldsymbol{x}, \quad b_{im} = -\frac{1}{3!}\int f_{im}(\boldsymbol{x})\, \boldsymbol{x}^2\, d\boldsymbol{x},$$
$$c_{im} = \frac{1}{5!}\int f_{im}(\boldsymbol{x})\, \boldsymbol{x}^4\, d\boldsymbol{x}. \tag{20.55a}$$

Here we are assuming that the integrals (20.55a) converge, i.e., that $f_{\mathrm{im}}(\mathbf{x})$ has the character of a short-range function.

At the critical point, the zeroth terms of this expansion, i.e., the matrix

$$f_0 = a = \int f(\boldsymbol{x})\, d\boldsymbol{x} = -V\frac{\partial^2 S}{\partial\langle P_0\rangle\, \partial\langle P_0\rangle} = -V\frac{\partial^2 S}{\partial\langle P\rangle\, \partial\langle P\rangle} \tag{20.55b}$$

becomes singular, corresponding to the stability limit. In this case, the determinant of the matrix f_0^{im} goes to zero,

$$|f_0^{im}| = 0, \tag{20.55c}$$

and the matrix inverse to f_0, which, according to (20.48), determines the fluctuations, tends to infinity. Therefore, if we did not take into account the correlations between the fluctuations at different points, they would tend to infinity. In reality, the fluctuations only increases strongly at the critical point, and this growth is suppressed by the correlation between the fluctuations. We note that, at the critical point, the expansion (22.55) in powers of k^2 becomes difficult to justify, and a nonanalytic dependence on $|\mathbf{k}|$ is possible. The Ornstein–Zernike theory, strictly speaking, refers not to the critical point itself, but to its neighborhood.

The average value of the fluctuations is expressed in terms of the matrix inverse to $f_{\mathbf{k}}$, i.e.,

$$\langle \Delta P_k^i \, \Delta P_{-k}^m \rangle = V \left(a + bk^2 + ck^4\right)_{im}^{-1}. \tag{20.56}$$

The Ornstein–Zernike theory follows from (20.56) with c = 0. The case c ≠ 0 also presents no difficulties.

For the correlation function

$$g_{im}(\boldsymbol{x} - \boldsymbol{x}') = \frac{1}{V} \sum_k \langle \Delta P_k^i \, \Delta P_{-k}^m \rangle \, e^{i\,\boldsymbol{k}.(\boldsymbol{x}-\boldsymbol{x}')} \tag{20.57}$$

we obtain the expression

$$g_{im}(\boldsymbol{x} - \boldsymbol{x}') = \sum_k (a + bk^2 + ck^4)_{im}^{-1} \, e^{i\,\boldsymbol{k}.(\boldsymbol{x}-\boldsymbol{x}')}, \tag{20.57a}$$

which satisfies the differential equation

$$(a - b\nabla^2 + c\nabla^4)\, g(\boldsymbol{x}) = V\,\delta(\boldsymbol{x}), \tag{20.58}$$

where g(x) is a matrix with elements $g_{im}(x)$. The differential equation (20.58) with c = 0 is treated in the Ornstein–Zernike theory. We can find g(x) by solving this equation, but it is simpler to calculate the integral (20.57a) directly.

In the case (c = 0) treated by Ornstein and Zernike, we obtain for the correlation function of the fluctuations in the diagonal representation

$$g_{ii}(r) = \frac{V}{2\pi^2}\frac{1}{r}\int_0^\infty (a_{ii} + b_{ii}k^2)^{-1} \sin(kr)\, k\, dk = \frac{V b_{ii}^{-1}}{4\pi r} e^{-\left(\sqrt{a_{ii}/b_{ii}}\, r\right)},$$
$$\text{if } a_{ii}/b_{ii} > 0, \tag{20.59}$$

i.e., an exponential fall-off with distance. Here, a_{ii}, b_{ii}, and c_{ii} are coefficients of the expansion of f_k in the diagonal representation.

At the critical point, i.e., for $a_{ii} = 0$, we obtain

$$g_{ii}(r) \cong \frac{V}{4\pi b_{ii}}\frac{1}{r}, \tag{20.59a}$$

i.e., the fluctuations fall off very slowly.

In the more general case when $c_{ii} \neq 0$, we obtain

$$g_{ii}(r) = \frac{V}{2\pi^2}\frac{1}{r}\int_0^\infty (a_{ii} + b_{ii}k^2 + c_{ii}k^4)^{-1}\sin(kr)\,k\,dk =$$

$$= \frac{V}{4\pi}\frac{1}{r}\,\mathrm{Re}\left\{(b_{ii}^2 - 4a_{ii}c_{ii})^{-1/2}(e^{-a_1 r} - e^{-a_2 r})\right\}, \qquad (20.59b)$$

where

$$a_1^2 = -\frac{1}{2}c_{ii}^{-1}\left(b_{ii} - \sqrt{b_{ii}^2 - 4a_{ii}c_{ii}}\right),$$

$$a_2^2 = -\frac{1}{2}c_{ii}^{-1}\left(b_{ii} + \sqrt{b_{ii}^2 - 4a_{ii}c_{ii}}\right),$$

if $\mathrm{Re}\,a_1 > 0$ and $\mathrm{Re}\,a_2 > 0$.

Thus, the fluctuations fall off exponentially and oscillate. At the critical point, $a_{ii} = 0$, $a_1 = 0$, and $a_2^2 = -b_{ii}/c_{ii}$; Consequently,

$$g(r) = \frac{V}{4\pi b_{ii}}\frac{1}{r}\left(1 - \cos\sqrt{\frac{b_{ii}}{c_{ii}}}\,r\right).$$

An experimental investigation [79] of the fluctuations in solid binary solutions, close to the critical point of separation of the mixture, points to the existence of attenuating and oscillating fluctuations in the correlation functions coupling the densities of the different components.

Critical fluctuations were considered in the work of Klein and Tisza [78], who divided the system into small cells and took the continuum limit; this is a somewhat arbitrary device. The meaning of this method lies in the use of a well-defined, but perhaps not very successful, representation for the functionals, which, in the equilibrium case, are the thermodynamic functions. As we saw above, the Fourier representation for the thermodynamic functionals is considerably simpler and more natural and reduces them to functions.

The theory of critical fluctuations described has a semi-phenomenological character, since it uses the macroscopic distribution function. We have given an account of it as a simple example of the application of the local-equilibrium distribution function. The ideas of the Ornstein–Zernike theory turn out to be useful in the theory of transport phenomena [238, 233].

20.5. Absence of Dissipative Processes in a Local-Equilibrium State

We shall show that in the local-equilibrium state (20.17a) there are no dissipative processes, i.e., no thermal conduction, diffusion or viscosity. We shall calculate the average values of the fluxes of energy $\mathbf{j}_H(\mathbf{x})$, particle number $\mathbf{j}_\alpha(\mathbf{x})$, and momentum $T(\mathbf{x})$ over the local-equilibrium distribution.

We first put $\mathbf{v}(\mathbf{x}) = 0$. Then the statistical operator (or distribution function) is equal to

$$\rho_{l'} = Q_{l'}^{-1} \exp\left\{ - \int \beta(\boldsymbol{x}) \left[H(\boldsymbol{x}) - \sum_i \mu_i(\boldsymbol{x})\, n_i(\boldsymbol{x}) \right] d\boldsymbol{x} \right\}, \qquad (20.60)$$

where $H(\mathbf{x})$ is expressed by formula (19.13) in the classical case and by (19.21a) in the quantum case. The expression (20.60) contains only the energy density (19.13), (19.21a) and the particle-number density (19.3), (19.22), which are invariant under reversal of the momenta, i.e., they are even quantities. Therefore, $\rho_{l'}$ is also invariant under this operation, i.e., is an even quantity. On the other hand, under this transformation the α components of the fluxes of particle number $j^\alpha(\mathbf{x})$ and energy $j_H^\alpha(\mathbf{x})$ and the off-diagonal elements of the tensor $T_{\alpha\beta}(x)$, which depend on p^α and p^β, change their sign; consequently, the average of these quantities over (20.60) is equal to zero. The local-equilibrium average of the integral part of the off-diagonal elements of $T_{\alpha\beta}(x)$ goes to zero because of the fact that it contains a product of $(\boldsymbol{x}_\beta - \boldsymbol{x}'_\beta)(\boldsymbol{x}_\alpha - \boldsymbol{x}'_\alpha)$ with the two-particle distribution function, which depends only on $|\mathbf{x} - \mathbf{x}'|$ in the local approximation. This can be seen from (19.5), (19.17a), and (19.12) in the classical case, and from (19.24a), (19.29a), and (19.27a) in the quantum case.

Consequently, on averaging over $\rho_{l'}$, the quantities $\mathbf{j}(\mathbf{x})$, $\mathbf{j}_H(\mathbf{x})$, and the off-diagonal elements of $T_{\alpha\beta}(x)$ go to zero, since inside the trace (or integral) there is a product of quantities of different parity:

$$\begin{aligned} \langle \boldsymbol{j}(\boldsymbol{x}) \rangle_{l'} &= \operatorname{Tr}(\rho_{l'} \boldsymbol{j}(\boldsymbol{x})) = 0, \\ \langle \boldsymbol{j}_H(\boldsymbol{x}) \rangle_{l'} &= \operatorname{Tr}(\rho_{l'} \boldsymbol{j}_H(\boldsymbol{x})) = 0, \\ \langle T_{\alpha\beta}(\boldsymbol{x}) \rangle_{l'} &= \operatorname{Tr}(\rho_{l'} T_{\alpha\beta}(\boldsymbol{x})) = \delta_{\alpha\beta} \langle T_{\alpha\alpha}(\boldsymbol{x}) \rangle_{l'} = \delta_{\alpha\beta} \langle p(\boldsymbol{x}) \rangle_{l'}, \end{aligned} \qquad (20.61)$$

where we have introduced

$$\langle p(\boldsymbol{x})\rangle_{l'} = \frac{1}{3}\sum_{\alpha}\langle T_{\alpha\alpha}(\boldsymbol{x})\rangle_{l'},$$

which has the meaning of the pressure (19.37).

Now let $\mathbf{v}(\mathbf{x}) \neq 0$. We go over to the coordinate system moving with velocity $\mathbf{v}(\mathbf{x})$ by means of the transformation

$$\boldsymbol{p}_i = \boldsymbol{p}_i' + m\boldsymbol{v}(\boldsymbol{x}),$$

in the classical case, and by means of the canonical transformation (19.35) in the quantum case.† For $\mathbf{v}(\mathbf{x})$ we choose the average mass velocity, i.e.,

$$\boldsymbol{v}(\boldsymbol{x}) = \frac{\langle \boldsymbol{p}(\boldsymbol{x})\rangle_l}{\langle \rho(\boldsymbol{x})\rangle_l}. \tag{20.62}$$

In this case, the densities of energy, momentum, and particle number are transformed as follows:

$$\begin{aligned} H(\boldsymbol{x}) &= H'(\boldsymbol{x}) + \boldsymbol{v}(\boldsymbol{x})\cdot\boldsymbol{p}'(\boldsymbol{x}) + \frac{1}{2}\rho(\boldsymbol{x})\,v^2(\boldsymbol{x}),\\ \boldsymbol{p}(x) &= \boldsymbol{p}'(\boldsymbol{x}) + \rho(\boldsymbol{x})\,\boldsymbol{v}(\boldsymbol{x}), \quad n_i(\boldsymbol{x}) = n_i'(\boldsymbol{x}), \end{aligned} \tag{20.63}$$

where

$$\rho(\boldsymbol{x}) = \sum_i m_i n_i(\boldsymbol{x}). \tag{20.63a}$$

In the new variables, the statistical operator takes the form

$$\rho_l = Q_l^{-1}\exp\left\{-\int \beta(\boldsymbol{x})\left[H'(\boldsymbol{x}) - \sum_i \mu_i(\boldsymbol{x})\,n_i'(\boldsymbol{x})\right]d\boldsymbol{x}\right\}, \tag{20.64}$$

where $\mathrm{H}'(\mathbf{x})$ and $\mathrm{n}_i'(\mathbf{x})$ have the same form as before, but with $\psi(\mathbf{x})$ replaced by $\psi'(\mathbf{x})$, or, in the classical case, with $\mathbf{p}_i$ replaced by $\mathbf{p}_i'$. In the moving system, we have

$$\begin{aligned} &\langle \boldsymbol{j}_H'(\boldsymbol{x})\rangle_l = 0, \quad \langle \boldsymbol{j}_k'(\boldsymbol{x})\rangle_l = 0,\\ &\langle T_{\alpha\beta}'(\boldsymbol{x})\rangle_l = \langle p(\boldsymbol{x})\rangle_l\,\delta_{\alpha\beta}, \end{aligned} \tag{20.65}$$

†See the footnote to p. 255.

where p(x) is the hydrostatic pressure operator (19.37), which we have left unprimed since it is precisely in the system moving with the mass velocity that the hydrostatic pressure is defined.

When we go over to the coordinate system moving with velocity $\mathbf{v}(\mathbf{x})$, the fluxes of the particle number (19.24a), energy (19.29a), and momentum (19.27a) are transformed according to the formulas

$$j_i(x) = j'_i(x) + n_i(x)\,v(x),$$
$$j_H(x) = j'_H(x) + \left\{ H'(x) + \rho(x)\frac{v^2(x)}{2} + (v(x)\cdot p'(x)) \right\} v(x) + \frac{v^2(x)}{2}\,p'(x) + T'(x)\cdot v(x) + \frac{\hbar^2}{4m}\nabla n \cdot \nabla v(x),$$
$$T(x) = T'(x) + v(x)\,p'(x) + p'(x)\,v(x) + \rho(x)\,v(x)\,v(x). \qquad (20.66)$$

In the formula (20.66) for $\mathbf{j}_H(\mathbf{x})$, the term $\frac{\hbar^2}{4m}\nabla n(x)\cdot\nabla v(x)$ has a purely quantum origin and is very small in real cases, when $\mathbf{v}(\mathbf{x})$ and $\langle n(\mathbf{x})\rangle$ vary little over distances of the order of the de Broglie wavelength corresponding to the average energy of the particles; therefore, we shall omit the term with the velocity gradient. The transformation formulas for the densities and fluxes then take the form

$$H(x) = H'(x) + v(x)\,p'(x) + \frac{1}{2}\rho(x)\,v^2(x),$$
$$n_i(x) = n'_i(x), \quad j_i(x) = j'_i(x) + n_i(x)\,v(x),$$
$$p(x) = p'(x) + \rho(x)\,v(x),$$
$$j_H(x) = j'_H(x) + \left\{ H'(x) + \rho(x)\frac{v^2(x)}{2} + (p'(x)\cdot v(x)) \right\} v(x) + \frac{v^2(x)}{2}\,p'(x) + T'(x)\cdot v(x),$$
$$T(x) = T'(x) + v(x)\,p'(x) + p'(x)\,v(x) + \rho(x)\,v(x)\,v(x), \qquad (20.67)$$

which are valid in both the quantum and classical cases.

Since the relations (20.65) are valid in the system moving with the local velocity $\mathbf{v}(\mathbf{x})$, by averaging (20.67) over the local-equilibrium distribution we obtain

$$\langle H(x)\rangle_l = \langle H'(x)\rangle_l + \frac{1}{2}\langle \rho(x)\rangle_l\, v^2(x),$$
$$\langle n_i(x)\rangle_l = \langle n'_i(x)\rangle_l, \quad \langle j_i(x)\rangle_l = \langle n_i(x)\rangle_l\, v(x),$$
$$\langle p(x)\rangle_l = \langle \rho(x)\rangle_l\, v(x), \qquad (20.68)$$
$$\langle j_H(x)\rangle_l = \left\{ h(x) + \langle \rho(x)\rangle_l \frac{v^2(x)}{2} \right\} v(x),$$
$$\langle T(x)\rangle_l = \langle p(x)\rangle_l\, U + \langle \rho(x)\rangle_l\, v(x)\,v(x),$$

where

$$h(\boldsymbol{x}) = \langle H'(\boldsymbol{x})\rangle_l + \langle p(\boldsymbol{x})\rangle_l \tag{20.68a}$$

is the enthalpy density, $\langle p(\boldsymbol{x})\rangle_l$ is the average pressure, and U is the unit tensor.

It can be seen from formulas (20.68) that in a local-equilibrium state we have only those fluxes which are characteristic for an ideal liquid [82]. The fluxes of energy and particle number are proportional to the mass velocity, and the momentum flux to a bilinear function of the mass velocity. They do not depend on gradients of the thermodynamic parameters, and have a convective character.

The local-equilibrium distribution (20.17a) is determined from the condition that the information entropy be a maximum at given values of $\langle H(\boldsymbol{x})\rangle_l$, $\langle \boldsymbol{p}(\boldsymbol{x})\rangle_l$, and $\langle n_i(\boldsymbol{x})\rangle_l$. This means that there is arbitrariness in the specification of these quantities or of the corresponding thermodynamic (and hydrodynamic) parameters $\beta(\boldsymbol{x}, t)$, $\boldsymbol{v}(\boldsymbol{x}, t)$, and $\mu_i(\boldsymbol{x}, t)$. We shall specify them in such a way that they satisfy the hydrodynamic equations of an ideal liquid. The fluxes (20.68) correspond to precisely this case. We put

$$\begin{aligned} \frac{\partial\langle H(\boldsymbol{x})\rangle_l}{\partial t} &= -\nabla\cdot\langle \boldsymbol{j}_H(\boldsymbol{x})\rangle_l, \\ \frac{\partial\langle \boldsymbol{p}(\boldsymbol{x})\rangle_l}{\partial t} &= -\nabla\cdot\langle T(\boldsymbol{x})\rangle_l, \\ \frac{\partial\langle n_i(\boldsymbol{x})\rangle_l}{\partial t} &= -\nabla\cdot\langle \boldsymbol{j}_i(\boldsymbol{x})\rangle_l \end{aligned} \tag{20.69}$$

or, in more compact notation,

$$\frac{\partial\langle P_m(\boldsymbol{x})\rangle_l}{\partial t} = -\nabla\cdot\langle j_m(\boldsymbol{x})\rangle_l. \tag{20.69a}$$

These equations cannot be obtained simply by averaging the conservation laws (19.18) over ρ_l, since ρ_l does not satisfy Liouville's equation, and, therefore, for ρ_l the derivatives of averages of operators are not equal, generally speaking, to the averages of their derivatives.

Using the relations (20.68) and the balance equations for the total mass and the kinetic energy

$$\frac{\partial}{\partial t}\langle\rho(\boldsymbol{x})\rangle_l = -\operatorname{div}\langle\rho(\boldsymbol{x})\rangle_l\,\boldsymbol{v}(\boldsymbol{x}, t),$$

$$\frac{\partial}{\partial t}\langle\rho(\boldsymbol{x})\rangle_l\,\frac{v^2(\boldsymbol{x}, t)}{2} = -\operatorname{div}\left\{\langle\rho(\boldsymbol{x})\rangle_l\,\frac{v^2(\boldsymbol{x}, t)}{2}\,\boldsymbol{v}(\boldsymbol{x}, t)\right\} - \boldsymbol{v}(\boldsymbol{x}, t)\cdot\nabla p(\boldsymbol{x}),$$

we write the hydrodynamic equations (20.69) of an ideal liquid in the form

$$\frac{du}{dt} = \frac{\partial u}{\partial t} + \boldsymbol{v} \cdot \nabla u = -(u + p) \operatorname{div} \boldsymbol{v},$$
$$\frac{dn_i}{dt} = \frac{\partial n_i}{\partial t} + \boldsymbol{v} \cdot \nabla n_i = -n_i \operatorname{div} \boldsymbol{v}, \tag{20.69b}$$
$$\frac{d\boldsymbol{v}}{dt} = \frac{\partial \boldsymbol{v}}{\partial t} + \boldsymbol{v} \cdot \nabla \boldsymbol{v} = -\rho^{-1} \nabla p,$$

where

$$u = \langle H'(\boldsymbol{x})\rangle_l, \quad n_i = \langle n_i'(\boldsymbol{x})\rangle_l = \langle n_i(\boldsymbol{x})\rangle_l,$$
$$\boldsymbol{v} = \boldsymbol{v}(\boldsymbol{x}, t), \quad \rho = \langle \rho(\boldsymbol{x})\rangle_l, \quad p = p(\boldsymbol{x}), \tag{20.69c}$$

i.e., u is the energy density in the accompanying system.

We shall show that, for the local-equilibrium distribution, it follows from (20.69) that the entropy

$$S = -\langle \ln \rho_l \rangle_l = \Phi + \sum_m \int F_m(\boldsymbol{x}, t) \langle P_m(\boldsymbol{x})\rangle_l \, d\boldsymbol{x}, \tag{20.70}$$

where

$$\Phi = \ln \operatorname{Tr} \exp\left\{ -\sum_m \int F_m(\boldsymbol{x}, t) P_m(\boldsymbol{x}) \, d\boldsymbol{x} \right\}, \tag{20.70a}$$

is conserved, i.e., cannot be produced in the system, but only flows into or out of the system through its surface.

The time derivative of the entropy (20.70) is equal to

$$\frac{\partial S}{\partial t} = \sum_m \int F_m(\boldsymbol{x}, t) \frac{\partial}{\partial t} \langle P_m(\boldsymbol{x})\rangle_l \, d\boldsymbol{x}, \tag{20.71}$$

since it follows from (20.70a) that

$$\frac{\partial \Phi}{\partial t} = -\sum_m \int \frac{\partial F_m(\boldsymbol{x}, t)}{\partial t} \langle P_m(\boldsymbol{x})\rangle_l \, d\boldsymbol{x}. \tag{20.71a}$$

Using the hydrodynamic equations (20.69a), we write (20.71) in the form

$$\frac{\partial S}{\partial t} = -\sum_m \int F_m(\boldsymbol{x}, t) \nabla \cdot \langle j_m(\boldsymbol{x})\rangle_l \, d\boldsymbol{x} =$$
$$= -\sum_m \int F_m(\boldsymbol{x}, t) \langle j_m(\boldsymbol{x})\rangle_l \cdot d\sigma + \sum_m \int \nabla F_m(\boldsymbol{x}, t) \cdot \langle j_m(\boldsymbol{x})\rangle_l \, d\boldsymbol{x}, \tag{20.71b}$$

where the integral in the first sum in the right-hand side of the equality is taken over the surface of the system. The second sum in (20.71b) is equal to

$$\sum_m \nabla F_m(\boldsymbol{x}, t) \cdot \langle j_m(\boldsymbol{x})\rangle_l =$$

$$= \left\{\langle \boldsymbol{j}_H(\boldsymbol{x})\rangle_l - \langle T(\boldsymbol{x})\rangle_l \cdot \boldsymbol{v}(\boldsymbol{x}, t) + \langle \rho(\boldsymbol{x})\rangle_l \frac{v^2(\boldsymbol{x}, t)}{2} \boldsymbol{v}(\boldsymbol{x}, t)\right\} \cdot \nabla\beta(\boldsymbol{x}, t) -$$
$$- \beta(\boldsymbol{x}, t)\{\langle T(\boldsymbol{x})\rangle_l - \langle\rho(\boldsymbol{x})\rangle_l \boldsymbol{v}(\boldsymbol{x}, t)\boldsymbol{v}(\boldsymbol{x}, t)\} : \nabla\boldsymbol{v}(\boldsymbol{x}, t) -$$
$$- \sum_i \boldsymbol{v}(\boldsymbol{x}, t)\langle n_i(\boldsymbol{x})\rangle_l \cdot \nabla\beta(\boldsymbol{x}, t)\mu_i(\boldsymbol{x}, t). \qquad (20.71\text{c})$$

The latter sum in (20.71c) can be expressed in terms of the gradient of the function

$$\Phi(\boldsymbol{x}) = \beta(\boldsymbol{x}, t)\, p(\boldsymbol{x}, t), \qquad (20.71\text{d})$$

where $p(x, t)$ is the pressure, by making use of the Gibbs–Duhem relation (20.22) and (20.24d):

$$\delta\Phi(\boldsymbol{x}) = \delta(\beta(\boldsymbol{x}, t)\, p(\boldsymbol{x}, t)) = -\langle H'(\boldsymbol{x})\rangle_l\, \delta\beta(\boldsymbol{x}, t) + \sum_i \langle n_i(\boldsymbol{x})\rangle_l\, \delta\nu_i(\boldsymbol{x}, t),$$

$$\nu_i(\boldsymbol{x}\ t,) = \beta(\boldsymbol{x}, t)\mu_i(\boldsymbol{x}, t), \quad \Phi = \ln Q_l = \int \Phi(\boldsymbol{x})\, d\boldsymbol{x}, \qquad (20.72)$$

whence it follows that

$$\frac{\partial\Phi(\boldsymbol{x})}{\partial\beta(\boldsymbol{x}, t)} = -\langle H'(\boldsymbol{x})\rangle_l, \qquad \frac{\partial\Phi(\boldsymbol{x})}{\partial\nu_i(\boldsymbol{x}, t)} = \langle n_i(\boldsymbol{x})\rangle_l. \qquad (20.72\text{a})$$

The gradient of $\Phi(x)$ as a function of $\beta(x, t)$ and $\nu_i(x, t)$ is equal to

$$\nabla\Phi(\boldsymbol{x}) = \frac{\partial\Phi(\boldsymbol{x})}{\partial\beta(\boldsymbol{x}, t)}\nabla\beta(\boldsymbol{x}, t) + \sum_i \frac{\partial\Phi(\boldsymbol{x})}{\partial\nu_i(\boldsymbol{x}, t)}\nabla\nu_i(\boldsymbol{x}, t)$$
$$= -\langle H'(\boldsymbol{x})\rangle_l \nabla\beta(\boldsymbol{x}, t) + \sum_i \langle n_i(\boldsymbol{x})\rangle_l \nabla\nu_i(\boldsymbol{x}, t). \qquad (20.72\text{b})$$

By making use of this relation, we write (20.71c) in the form

$$\sum_m \nabla F_m(\boldsymbol{x}, t) \cdot \langle j_m(\boldsymbol{x})\rangle_l = \left\{\langle \boldsymbol{j}_H(\boldsymbol{x})\rangle_l -\right.$$
$$\left. - \left(\langle H'(\boldsymbol{x})\rangle_l - \langle\rho(\boldsymbol{x})\rangle_l \frac{v^2(\boldsymbol{x}, t)}{2}\right)\boldsymbol{v}(\boldsymbol{x}, t) - \langle T(\boldsymbol{x})\rangle_l \cdot \boldsymbol{v}(\boldsymbol{x}, t)\right\} \cdot \nabla\beta(\boldsymbol{x}, t) -$$
$$- \{\langle T(\boldsymbol{x})\rangle_l - U p(\boldsymbol{x}) - \langle\rho(\boldsymbol{x})\rangle_l \boldsymbol{v}(\boldsymbol{x}, t)\boldsymbol{v}(\boldsymbol{x}, t)\} : \beta(\boldsymbol{x}, t)\nabla\boldsymbol{v}(\boldsymbol{x}, t) -$$
$$- \operatorname{div}(\boldsymbol{v}(\boldsymbol{x}, t)\beta(\boldsymbol{x}, t)\, p(\boldsymbol{x})) = -\operatorname{div}(\boldsymbol{v}(\boldsymbol{x}, t)\beta(\boldsymbol{x}, t)\, p(\boldsymbol{x})), \qquad (20.73)$$

since the factors multiplying $\nabla\beta(\boldsymbol{x}, t)$ and $\nabla v(\boldsymbol{x}, t)$ go to zero by virtue of (20.68).

In addition,

$$\sum_m F_m(\boldsymbol{x}, t)\langle j_m(\boldsymbol{x})\rangle_l + \beta(\boldsymbol{x}, t)\, v(\boldsymbol{x}, t)\, p(\boldsymbol{x}) = \langle j_s(\boldsymbol{x})\rangle_l = v(\boldsymbol{x}, t)\, S(\boldsymbol{x}), \quad (20.74)$$

since it follows from (20.23) that

$$\beta^{-1}(\boldsymbol{x}, t)\, S(\boldsymbol{x}) = \langle H'(\boldsymbol{x})\rangle_l + p(\boldsymbol{x}) - \sum_i \langle n_i(\boldsymbol{x})\rangle_l \mu_i(\boldsymbol{x}, t). \quad (20.75)$$

Taking (20.73) and (20.74) into account, we write the balance equation (20.71) for the entropy in the form of a surface integral:

$$\frac{\partial S}{\partial t} = -\int v(\boldsymbol{x}, t)\, S(\boldsymbol{x}) \cdot d\sigma, \quad (20.76)$$

or, in local form,

$$\frac{\partial S(\boldsymbol{x})}{\partial t} = -\operatorname{div}(v(\boldsymbol{x}, t)\, S(\boldsymbol{x})). \quad (20.76a)$$

Thus, the entropy in a local-equilibrium state can change only as a result of its flowing into or out of the volume of the system. Consequently, in a local-equilibrium state, dissipative processes are absent and although the state describes nonequilibrium processes, these processes are reversible.

It is easy to convince oneself that in any quasi-equilibrium state which is described by the operator

$$\rho_q = \exp\left\{-\Phi - \sum_m P_m F_m\right\} = \exp\left\{-S(t, 0)\right\},$$

the entropy production is equal to zero. In fact,

$$\langle \dot{S}(t, 0)\rangle_q = \operatorname{Tr}\left(\dot{S}(t,0)\exp\left\{-S(t,0)\right\}\right) = \operatorname{Tr}\left\{\frac{1}{i\hbar}\,[S(t,0), H]\exp\left\{-S(t,0)\right\}\right\} =$$

since

$$\operatorname{Tr}\left\{\frac{\partial S(t,0)}{\partial t}\exp\left\{-S(t,0)\right\}\right\} = \frac{\partial}{\partial t}\operatorname{Tr}\exp\left\{-S(t,0)\right\} = 0$$

and the operators $S(t, 0)$ and $\exp\{-S(t, 0)\}$ commute.

Thus, the local-equilibrium distribution enables us to define the thermodynamic functions of nonequilibrium states and obtain thermodynamic equalities for nonuniform systems; however, it does not enable us to describe transport processes. This is connected with the fact that ρ_l is not a solution of Liouville's equation, since $H(\mathbf{x})$, $n(\mathbf{x})$, and $\mathbf{p}(\mathbf{x})$ are not integrals of motion. We note, however, that these quantities are defined only within the divergence of a vector or tensor, and that we can make use of this fact to construct a statistical operator satisfying Liouville's equation [2-5]. This question will be treated in the next section, in which the local-equilibrium distribution is modified to enable it to describe irreversible transport processes.

§ 21. Statistical Operator for Nonequilibrium Systems [2-5]

The existence of exact (i.e., valid for any interaction) expressions for the kinetic coefficients in terms of the equilibrium time correlation functions, which we studied in Chapter III, prompts the thought that these expressions might be obtained by generalizing the statistical operator to nonequilibrium states and expanding it in small gradients of the thermodynamic parameters. We shall show, following the papers of the author [2-5, 184-186], that it is possible to generalize the Gibbsian method described in Chapters I and II to the nonequilibrium case and to construct a nonequilibrium statistical operator or distribution function which will enable us to obtain transport equations and calculate the kinetic coefficients in terms of the correlation functions and which, for the equilibrium case, goes over to one of the Gibbsian distributions. In the linear approximation, these distributions lead to the results of the linear response theory described in Chapter III.

Bogolyubov's idea of a hierarchy of relaxation times in nonequilibrium statistical mechanics [1] has very great significance for the following development and will now be described.†

If the initial distribution is arbitrary, in the initial stage the state of the system may be very far from equilibrium and, to describe it, it may be necessary to specify a large number of distribution functions; we then need not only one- and two-particle

†For a very clear account of these ideas of Bogolyubov, see the article by Uhlenbeck in the book [83].

functions, but also functions of higher order, which vary rapidly with time in accordance with Liouville's equation.

However, for many systems with a large number of particles, "synchronization" of the distribution functions sets in very rapidly (for example, in a time of the order of the collision time for low-density or weakly interacting gases); this is the kinetic stage, when all the distribution functions are completely defined by specifying the one-particle distribution function. For this stage, it is possible, starting from Liouville's equation, to construct a kinetic equation for the one-particle distribution function.

For larger time scales (for gases, appreciably larger than the time between collisions), the number of parameters necessary for the description of the state of the system is still further reduced, and the hydrodynamic stage arises. This stage can be described by the hydrodynamic equations (together with the thermal-conduction equation), i.e., by only a few moments of the distribution function (by the average number of particles, the average energy and the average velocity). The distribution function then depends on time only through these parameters.

Below, we shall show that it is possible to describe the hydrodynamic stage by means of a certain nonequilibrium distribution function or statistical operator, depending on time through its parameters, and that this is possible not only for dilute gases and systems with weak interactions, but also for the more general case of condensed matter.

Gibbs' construction of the equilibrium statistical ensembles, which we studied in Chapters I and II, is based on the Liouville theorem, according to which the time derivative of the statistical operator (or distribution function) is equal to zero if the statistica operator is a function only of the integrals of motion. The Gibbsia distribution (3.30), (9.42) is obtained by assuming that these integrals are the additive integrals of the motion, namely, the energ and particle number. Below we shall start, as in the equilibrium case, from the Liouville theorem, but we shall generalize the clas of integrals of motion on which the statistical operator or distribution function can depend.

21.1. The Nonequilibrium Statistical Operator

We attempt to construct a statistical operator for a nonequilibrium system (for example, for a liquid) consisting of l components, taking into account transport of energy, particles, and momentum.

We shall consider a system whose state is macroscopically defined by giving the fields of the temperature, chemical potential, and velocity (i.e., the densities of energy, momentum, and particle number) as functions of the space and time coordinates. We can regard this system as being in thermal, material, and mechanical contact with a combination of thermostats and reservoirs maintaining the given distribution of parameters.

We assume that the chosen parameters are sufficient to characterize the state of the system macroscopically. If this is not so (for example, if it is impossible to describe the system by a single temperature, or if in general the concept of the temperature of the system loses its meaning), it is necessary to choose another, more complete set of quantities characterizing the state of the system.

The construction of the Gibbsian grand ensemble is based (cf. §§3.4 and 9.4) on the laws of conservation of energy and particle number. We too shall start from the laws of conservation of energy, particle number, and momentum, but in local form; these were discussed in detail in §19:

$$\frac{\partial H(\boldsymbol{x}, t)}{\partial t} + \operatorname{div} \boldsymbol{j}_H(\boldsymbol{x}, t) = 0,$$
$$\frac{\partial n_i(\boldsymbol{x}, t)}{\partial t} + \operatorname{div} \boldsymbol{j}_i(\boldsymbol{x}, t) = 0 \qquad (i = 1, 2, \ldots, l), \tag{21.1}$$
$$\frac{\partial \boldsymbol{p}(\boldsymbol{x}, t)}{\partial t} + \operatorname{Div} T(\boldsymbol{x}, t) = 0,$$

$$\boldsymbol{p}(\boldsymbol{x}, t) = \sum_i m_i \boldsymbol{j}_i(\boldsymbol{x}, t), \tag{21.1a}$$

where $H(\mathbf{x})$, $n_i(\mathbf{x})$, and $\mathbf{p}(\mathbf{x})$ are the operators for the density of the energy, particle number, and momentum, $\mathbf{j}_H(\mathbf{x})$ is the energy flux density, $\mathbf{j}_i(\mathbf{x})$ is the particle-number flux density, and $T(\mathbf{x})$ is the momentum flux tensor or stress tensor. The form of these operators is assumed to be known. For example, for a many-component mixture of different particles with direct interaction in the quantum case, they are expressed by formulas (19.39b), (19.43), (19.40), (19.41a), (19.42a), (19.44a), and (19.46b), and for a one-component system in the classical case, by formulas (19.13), (19.5), (19.6), (19.17), and (19.12).

In the quantum case, all operators are studied in the Heisenberg picture, for example,

$$H(\boldsymbol{x}, t) = e^{iHt/\hbar} H(\boldsymbol{x}) e^{-iHt/\hbar} \qquad (H(\boldsymbol{x}, 0) = H(\boldsymbol{x})), \tag{21.2}$$

where H does not depend on the time. In the case of classical mechanics, the Heisenberg picture (21.2) is replaced by the action of the evolution operator (cf. §2.3)

$$H(\boldsymbol{x}, t) = e^{-iLt} H(\boldsymbol{x}), \tag{21.2a}$$

where **L** is the Liouville operator (2.16).

If the system under consideration consists of subsystems with a small interaction which leads only to slow exchange of energy or momentum, then we can study the balance equations for the energy and momentum for each subsystem. Then the right-hand sides of Eqs. (21.1), written out for each component separately, will contain source terms expressing the exchange of energy and momentum between the subsystems [cf. Eqs. (19.42) and (19.46)].

If excitation of the internal degrees of freedom (or chemical reactions) is possible during the molecular collisions, then the equation for the density of particles with a given quantum number (or of molecules of a given type) will contain in the right-hand side a term representing a source of the particles with the given quantum number, this source being due to the excitation of the internal degrees of freedom or to the fluxes of the molecules formed in the chemical reaction [cf. Eq. (19.53)].

It is convenient to write the conservation laws (21.1) in a more compact form, analogous to (19.48):

$$\frac{\partial P_m(\boldsymbol{x}, t)}{\partial t} + \nabla \cdot j_m(\boldsymbol{x}, t) = 0 \qquad (m = 0, 1, \ldots, l+1), \tag{21.3}$$

where

$$\begin{aligned} P_0(\boldsymbol{x}) &= H(\boldsymbol{x}), & j_0(\boldsymbol{x}) &= \boldsymbol{j}_H(\boldsymbol{x}), \\ P_1(\boldsymbol{x}) &= \boldsymbol{p}(\boldsymbol{x}), & j_1(\boldsymbol{x}) &= T(\boldsymbol{x}), \\ P_{i+1}(\boldsymbol{x}) &= n_i(\boldsymbol{x}), & j_{i+1}(\boldsymbol{x}) &= \boldsymbol{j}_i(\boldsymbol{x}) \quad (i = 1, 2, \ldots, l). \end{aligned} \tag{21.3a}$$

It follows from the definitions (21.3a) that the densities $P_m(x)$ may be scalars or vectors, and the fluxes $j_m(x)$, vectors or tensors. The dot after the nabla operator denotes the scalar product, i.e., the divergence of the vector or tensor.

Below we shall consider only the quantum case, as the classical case is completely analogous.

To construct a statistical operator describing nonequilibrium processes, we shall make use of the facts that in irreversible processes there are different time scales of relaxation, as mentioned at the beginning of this section, and that we shall be interested in the state of the system for time scales that are not too short.

We shall assume that in a small interval of time τ a nonequilibrium distribution is established which depends on time only through its parameters and comparatively slowly, with characteristic time $\tau_1 \gg \tau$. Then we shall seek the statistical operator ρ for $t \gg \tau$ as an integral of the quantum Liouville equation (8.6)

$$\frac{\partial \rho}{\partial t} + \frac{1}{i\hbar}[\rho,\ H] = 0,$$

in which the partial derivative of ρ denotes differentiation with respect to time on which the parameters F_m appearing in ρ depend. Depending on the choice of the parameters F_m and the operators P_m, such an approach is possible in both the hydrodynamic and kinetic stages. The exposition of the remainder of this chapter will be concerned with the hydrodynamic stage.

To construct the statistical operator ρ describing a nonequilibrium state of the system, we use a set of operators $B_m(\mathbf{x}, t)$ depending on the position $\mathbf{x}$ and time t through the values of the parameters $F_m(\mathbf{x}, t)$.

Let the $B_m(\mathbf{x}, t)$ satisfy the equation

$$\frac{\partial B_m(\boldsymbol{x},\ t)}{\partial t} + \frac{1}{i\hbar}[B_m(\boldsymbol{x},\ t),\ H] = 0. \tag{21.4}$$

Then, if $\rho(t)$ is a functional of the $B_m(\mathbf{x}, t)$ as functions of $\mathbf{x}$, i.e., if

$$\rho(t) = \rho\{\ldots B_m(\boldsymbol{x},\ t)\ldots\}, \tag{21.5}$$

then $\rho(t)$ satisfies Liouville's equation. In fact, the functional (21.5) can be regarded as a function of the Fourier components of the $B_m(\mathbf{x}, t)$ with respect to $\mathbf{x}$; the latter also satisfy Eq. (21.4) and consequently, so too does ρ itself.

To construct the quantities $B_m(x, t)$, we shall start from the balance equations for the energy, particle number, and momentum in the differential form (21.1), or, in the more compact form (21.3).

We shall construct operators which depend on time through $F_m(x, t)$ by taking the invariant part of the operators $F_m(x, t + t_1)$ $P_m(x, t_1)$ for the case of evolution with Hamiltonian H, i.e.,

$$B_m(x, t) = \widetilde{F_m(x, t) P_m(x)} = \varepsilon \int_{-\infty}^{0} e^{\varepsilon t_1} F_m(x, t + t_1) P_m(x, t_1)\, dt_1 \qquad (\varepsilon \to 0), \tag{21.6}$$

where the parameters $F_m(x, t)$ have the meaning of the thermodynamic parameters (20.17), which depend on time:

$$\begin{aligned} F_0(x, t) &= \beta(x, t), \\ F_1(x, t) &= -\beta(x, t)\, v(x, t), \\ F_{i+1}(x, t) &= -\beta(x, t)\left(\mu_i(x, t) - \frac{m_i}{2} v^2(x, t)\right); \end{aligned} \tag{21.6a}$$

$\beta(x, t)$ is the inverse temperature, $\mu_i(x, t)$ is the chemical potential, and $v(x, t)$ is the mass velocity [their meaning will be elucidated below; cf. (21.15) and (21.16)]. The time argument of the operator $P_m(x, t_1)$ refers to the Heinsenberg picture (21.2) with H independent of time; for the present, we confine ourselves to this case. The parameter ε will tend to zero, although after the thermodynamic limiting process.

The operation of taking the invariant part, which smooths the oscillating terms, is used in the formal theory of scattering to impose the boundary conditions which exclude the advanced solutions of the Schrödinger equation [84, 85] (see Appendix I); we too select the retarded solutions (in this case, of the Liouville equation) in this way.

If the parameters $F_m(x, t)$ do not depend on time, then,

$$\begin{aligned} B_m(x) &= F_m(x)\, \widetilde{P_m(x)}, \\ \widetilde{P_m(x)} &= \varepsilon \int_{-\infty}^{0} e^{\varepsilon t} P_m(x, t)\, dt = P_m(x) + \int_{-\infty}^{0} e^{\varepsilon t}\, \nabla \cdot j_m(x, t)\, dt, \end{aligned} \tag{21.7}$$

where $\widetilde{P_m(x)}$ are local integrals of motion which differ only by divergences from the corresponding densities $P_m(x)$ and therefore also have the meaning of densities of the energy, momentum, and particle number. We shall show that they are, in fact, local integrals of motion for $\varepsilon \to 0$.

In the Heisenberg picture, Eqs. (21.7) have the form

$$\widetilde{P_m(x, t)} = P_m(x, t) + \int_{-\infty}^{0} e^{\varepsilon t_1} \nabla \cdot j_m(x, t + t_1)\, dt_1. \qquad (21.7a)$$

Differentiating (21.7a) with respect to time, and using the conservation laws (21.3), we obtain

$$\frac{\partial \widetilde{P_m(x, t)}}{\partial t} = \frac{\partial P_m(x, t)}{\partial t} + \int_{-\infty}^{0} e^{\varepsilon t_1} \nabla \cdot \frac{\partial j_m(x, t + t_1)}{\partial t}\, dt_1 = \varepsilon \int_{-\infty}^{0} e^{\varepsilon t_1} \dot{P}_m(x, t + t_1)\, dt_1, \qquad (21.8)$$

i.e., the derivative of (21.7) is expressed in terms of the same integral as that which occurs in the right-hand side of (21.7), but with one more factor ε. Consequently, $\dfrac{\partial \widetilde{P_m(x, t)}}{\partial t}$ goes to zero as $\varepsilon \to 0$, if the integrals in the right-hand side of (21.8) are finite; consequently, $\widetilde{P_m(x, t)}$ are local integrals of motion for $\varepsilon \to 0$.

We shall call $\widetilde{P_m(x, t)}$ integrals of motion although they are conserved only in the limit $\varepsilon \to 0$ (which is taken after the thermodynamic limit) and it would be more correct to call them "quasi-integrals of motion." Below, we shall not make this qualification but shall call them local integrals of motion. We remark that the operations of taking the invariant part and of taking the thermodynamic limit do not commute.†

Integrating by parts, we can write the operators (21.6) in the form

$$B_m(x, t) = F_m(x, t)\, P_m(x) - \int_{-\infty}^{0} e^{\varepsilon t_1} \left\{ F_m(x, t + t_1)\, \dot{P}_m(t_1) + \frac{\partial F_m(x, t + t_1)}{\partial t} P_m(x, t_1) \right\} dt_1. \qquad (21.9)$$

† We may note the close analogy between this property and the Van Hove condition [86] of the "diagonal singularity" of the matrix elements, which also arises when the volume of the system (or the number of particles) tends to infinity.

They satisfy the equation

$$\frac{\partial B_m(\boldsymbol{x}, t)}{\partial t} + \frac{1}{i\hbar}[B_m(\boldsymbol{x}, t), H] =$$
$$= \varepsilon \int_{-\infty}^{0} e^{\varepsilon t_1} \left\{ F_m(\boldsymbol{x}, t+t_1)\dot{P}_m(\boldsymbol{x}, t_1) + \frac{\partial F_m(\boldsymbol{x}, t+t_1)}{\partial t} P_m(\boldsymbol{x}, t_1) \right\} dt_1,$$

where in the right-hand side we have operators of the same type as those in (21.9), but multiplied by the parameter ε. Consequently, for $\varepsilon \to 0$, the $B_m(\mathbf{x}, t)$ satisfy Eq. (21.4).

For the functional (2.15), we choose the same form as for the local-equilibrium state (20.17a):

$$\rho(t) = Q^{-1} \exp\left\{ - \sum_m \int B_m(\boldsymbol{x}, t)\, d\boldsymbol{x} \right\},$$
$$Q = \operatorname{Tr} \exp\left\{ - \sum_m \int B_m(\boldsymbol{x}, t)\, d\boldsymbol{x} \right\} \tag{21.10}$$

or, in explicit form,

$$\rho(t) = Q^{-1} \exp\left\{ - \sum_m \varepsilon \int \int_{-\infty}^{0} e^{\varepsilon t} F_m(\boldsymbol{x}, t+t_1) P_m(\boldsymbol{x}, t_1)\, d\boldsymbol{x}\, dt_1 \right\}$$
$$= Q^{-1} \exp\left\{ - \sum_m \int \left\{ F_m(\boldsymbol{x}, t) P_m(\boldsymbol{x}) - \int_{-\infty}^{0} e^{\varepsilon t_1} (F_m(\boldsymbol{x}, t+t_1) P_m(\boldsymbol{x}, t_1) + \right.\right.$$
$$\left.\left. + \frac{\partial F_m(\boldsymbol{x}, t+t_1)}{\partial t_1} P_m(\boldsymbol{x}, t_1))\, dt_1 \right\} d\boldsymbol{x} \right\}, \tag{21.10a}$$

where in the calculation of the averages the limit $\varepsilon \to 0$ must be taken after the thermodynamic limit. For the distribution function in the classical case, we take the analogous expression

$$f(t) = Q^{-1} \exp\left\{ - \sum_m \int B_m(\boldsymbol{x}, t)\, d\boldsymbol{x} \right\},$$
$$Q = \int \exp\left\{ - \sum_m \int B_m(\boldsymbol{x}, t)\, d\boldsymbol{x} \right\} d\Gamma, \tag{21.10b}$$

which, unlike (21.10), is not an operator, but a function of the coordinates and momenta of all the particles.

We shall verify that (21.10) does indeed satisfy Liouville's equation (21.3) for $\varepsilon \to 0$.

We recall that, if ρ satisfies Liouville's equation, then

$$\eta = -\ln\rho \tag{21.11}$$

also satisfies Liouville's equation [cf. (8.24) and (2.23)].

In our case,

$$\frac{\partial\eta}{\partial t} + \frac{1}{i\hbar}[\eta, H] = \varepsilon \sum_m \int \int_{-\infty}^{0} e^{\varepsilon t_1} \Big\{ F_m(\boldsymbol{x}, t+t_1) \dot{P}_m(\boldsymbol{x}, t_1) + \frac{\partial F_m(\boldsymbol{x}, t+t_1)}{\partial t_1} \left(P_m(\boldsymbol{x}, t_1) - \langle P_m(\boldsymbol{x}, t_1) \rangle \right) \Big\} dt_1\, d\boldsymbol{x}, \tag{21.12}$$

i.e., the right-hand side of (21.12) contains operators of the same type as those in the second term of the exponent of (21.10a). Consequently, $\ln\rho$ satisfies Liouville's equation in the limit $\varepsilon \to 0$.

Using the conservation laws (21.3), integrating by parts and neglecting the surface integrals, we represent the statistical operator (21.10a) in the form

$$\rho(t) = Q^{-1} \exp\Big\{ -\sum_m \int \Big[F_m(\boldsymbol{x}, t) P_m(\boldsymbol{x}) - \int_{-\infty}^{0} e^{\varepsilon t_1} \Big(\nabla F_m(\boldsymbol{x}, t+t_1) \cdot j_m(\boldsymbol{x}, t_1) + \frac{\partial F_m(\boldsymbol{x}, t+t_1)}{\partial t_1} P_m(\boldsymbol{x}, t_1) \Big) dt_1 \Big] d\boldsymbol{x} \Big\}. \tag{21.10c}$$

If $j_m(x)$ is a tensor, it is assumed to be symmetric.

The statistical operator (21.10c) is not yet an exact solution of the usual Liouville equation for $\varepsilon \to 0$, since we have neglected the fluxes of energy, particles, and momentum through the surface of the system. These terms correspond to the "nonconservative forces" which MacLennan introduces to describe the influence of the thermostat [37–40] (cf. Appendices II and III).

In the stationary case, (21.10a) goes over to the distribution

$$\rho = Q^{-1} \exp\Big\{ -\sum_m \int F_m(\boldsymbol{x}) \widetilde{P_m(\boldsymbol{x})}\, d\boldsymbol{x} \Big\}$$
$$= Q^{-1} \exp\Big\{ -\sum_m \int \Big[F_m(\boldsymbol{x}) P_m(\boldsymbol{x}) - \int_{-\infty}^{0} e^{\varepsilon t} F_m(\boldsymbol{x}) \dot{P}_m(\boldsymbol{x}, t)\, dt \Big] d\boldsymbol{x} \Big\}, \tag{21.10d}$$

where the $\widetilde{P_m(\boldsymbol{x})}$ are also densities of integrals of motion, differing from the $P_m(\boldsymbol{x})$ only by divergences; (21.10c) goes over into the expression

$$\rho = Q^{-1} \exp \left\{ -\sum_m \int \left[F_m(\boldsymbol{x}) P_m(\boldsymbol{x}) - \int_{-\infty}^{0} e^{\varepsilon t} j_m(\boldsymbol{x}, t) \cdot \nabla F_m(\boldsymbol{x}) \, dt \right] d\boldsymbol{x} \right\}, \tag{21.10e}$$

which was obtained in paper [2].

It is more convenient, however, to treat the stationary case as the limit of the nonstationary case when $\partial F_m / \partial t \to 0$. Cancellation of the terms representing the free evolution of the operators then occurs more naturally.

Substituting into (21.10c) the expressions (21.6a) for the parameters $F_m(\mathbf{x}, t)$ and (21.3a) for the densities $P_m(\mathbf{x})$, we write the statistical operator (21.10c) in the explicit form

$$\begin{aligned} \rho(t) = Q^{-1} \exp \Bigg\{ &- \int \beta(\boldsymbol{x}, t) \Big[H(\boldsymbol{x}) - \sum_i \Big(\mu_i(\boldsymbol{x}, t) - \\ &\qquad - \frac{m_i}{2} v^2(\boldsymbol{x}, t)\, n_i(\boldsymbol{x}) \Big) - \boldsymbol{v}(\boldsymbol{x}, t) \cdot \boldsymbol{p}(\boldsymbol{x}) \Big] d\boldsymbol{x} + \\ &+ \int \int_{-\infty}^{0} e^{\varepsilon t_1} \Big(\boldsymbol{j}_H(\boldsymbol{x}, t_1) \cdot \nabla \beta(\boldsymbol{x}, t + t_1) + H(\boldsymbol{x}, t_1) \frac{\partial \beta(\boldsymbol{x}, t + t_1)}{\partial t_1} - \\ &- \sum_i \boldsymbol{j}_i(\boldsymbol{x}, t_1) \cdot \nabla \beta(\boldsymbol{x}, t + t_1) \Big(\mu_i(\boldsymbol{x}, t + t_1) - \frac{m_i}{2} v^2(\boldsymbol{x}, t + t_1) \Big) + \\ &+ \sum_i n_i(\boldsymbol{x}, t_1) \frac{\partial}{\partial t_1} \beta(\boldsymbol{x}, t + t_1) \Big(\mu_i(\boldsymbol{x}, t + t_1) - \frac{m_i}{2} v^2(\boldsymbol{x}, t + t_1) \Big) - \\ &- T(\boldsymbol{x}, t_1) : \nabla \beta(\boldsymbol{x}, t + t_1)\, \boldsymbol{v}(\boldsymbol{x}, t + t_1) - \\ &- \boldsymbol{p}(\boldsymbol{x}, t_1) \cdot \frac{\partial}{\partial t_1} \beta(\boldsymbol{x}, t + t_1)\, \boldsymbol{v}(\boldsymbol{x}, t + t_1) \Big) dt_1 \, d\boldsymbol{x} \Bigg\}. \end{aligned} \tag{21.10f}$$

Here, ∇ and $\partial / \partial t_1$ act on all the thermodynamic parameters, β, μ_i, and $\mathbf{v}$, standing to the right of them. The statistical operator (21.10f) was obtained by MacLennan [37, 38] by another method based on the introduction of nonconservative forces describing the influence of the thermostat (see Appendix II). For other derivations of (21.10a), see §27 and in [170, 186–189]. A very simple derivation of the nonequilibrium statistical operator (21.10a) by means of the imposition of boundary conditions of the retarded type on the solutions of Liouville's equation is given in Appendix III.

We shall examine particular cases of the distribution (21.10f).

If all the parameters $F_m(\mathbf{x})$ are constant in space, then (21.10f) goes over into the Gibbsian distribution for the grand ensemble of a system moving as a whole with velocity $\mathbf{v}$:

$$\rho_0 = Q_0^{-1} \exp\left\{-\sum_m F_m P_m\right\}$$
$$= Q_0^{-1} \exp\left\{-\beta\left(H - \sum_i \mu_i N_i - \boldsymbol{v} \cdot \boldsymbol{P} + \frac{1}{2} v^2 \sum_i m_i N_i\right)\right\}. \qquad (21.13)$$

If in (21.10f) we neglect the term with the fluxes, we obtain the statistical operator of the local-equilibrium distribution

$$\rho_l = Q_l^{-1} \exp\left\{-\sum_m F_m(\boldsymbol{x}, t)\, P_m(\boldsymbol{x})\, d\boldsymbol{x}\right\}$$
$$= Q_l^{-1} \exp\left\{-\int \beta(\boldsymbol{x}, t)\left[H(\boldsymbol{x}) - \sum_i \left(\mu_i(\boldsymbol{x}, t) - \frac{m_i}{2} v^2(\boldsymbol{x}, t)\right) n_i(\boldsymbol{x}) - \right.\right.$$
$$\left.\left. - \boldsymbol{v}(\boldsymbol{x}, t) \cdot \boldsymbol{p}(\boldsymbol{x})\right] d\boldsymbol{x}\right\}, \qquad (21.14)$$

which was used by Mori [29-31] as the initial condition for the solution of Liouville's equation. The local-equilibrium statistical operator is not a solution of the Liouville equation and does not describe irreversible processes, although for small gradients of the thermodynamic parameters it differs little from (21.10f). The latter circumstance is the reason for the success of Mori's theory.

21.2. Physical Meaning of the Parameters

We shall discuss now in more detail the meaning of the parameters $F_m(\mathbf{x}, t)$ which appear in the statistical operator (21.10)-(21.10c).

We choose the parameters $F_m(\mathbf{x}, t)$ in such a way that they have the meaning of the thermodynamic parameters (21.6a) conjugate to the $\langle P_m(\mathbf{x})\rangle$ and satisfy the thermodynamic equalities (20.21). For this, it is sufficient to require that the average values of $P_m(\mathbf{x})$ over the distribution (21.10a), (21.10c) be equal to the averages over the local-equilibrium distribution, i.e., that

$$\langle P_m(\boldsymbol{x})\rangle = \langle P_m(\boldsymbol{x})\rangle_l; \qquad (21.15)$$

this gives the conditions for the determination of the parameters

$F_m(\mathbf{x}, t)$. Indeed, we then have

$$\frac{\delta \ln Q_l}{\delta F_m(\boldsymbol{x}, t)} = -\langle P_m(\boldsymbol{x})\rangle_l = -\langle P_m(\boldsymbol{x})\rangle, \tag{21.16}$$

where

$$Q_l = \operatorname{Tr} \exp \left\{ -\sum_m \int F_m(\boldsymbol{x}, t)\, P_m(\boldsymbol{x})\, d\boldsymbol{x} \right\}. \tag{21.16a}$$

A similar definition of the thermodynamic parameters is well known in the kinetic theory of gases. In the theory of irreversible processes, Green [14], Mori [30], MacLennan [37-40], and many other authors follow this procedure.

Such a definition of the parameters $F_m(\mathbf{x}, t)$ corresponds to the introduction of the concept of thermodynamic functions of nonequilibrium states [70, 14], by which we mean the thermodynamic functions of the local-equilibrium state that is characterized by the same quantities $\langle P_m(\mathbf{x})\rangle$ that characterize the given nonequilibrium state. The local-equilibrium state can be regarded as an equilibrium state in fictitious external fields [9].

The introduction of the statistical operator (21.10a)-(21.10f) depending on the local parameters $F_m(\mathbf{x}, t)$ is based on the assumption that there are enough of these parameters to describe the macroscopic state of the system and that the fluctuations of the local mechanical quantities are not too great. We may formulate the condition for the applicability of the method in the following way.

If the macroscopic state of the system can be described by the parameters $F_m(\mathbf{x}, t)$, then the corresponding statistical operator has the form (21.10a).

We can apply the statistical operator to describe the hydrodynamic stage if the parameters $F_m(\mathbf{x}, t)$ have the meaning of (21.6a), although this is not a necessary requirement. The method is also applicable with another, more general choice of parameters. For example, we can choose them in such a way that the statistical operator is also suitable for the description of the kinetic stage, and derive the generalized kinetic equations [56] (cf. § 25).

If the fluctuations are large, the state is characterized not only by the average values of the mechanical quantities but also by their variances; the latter too could be regarded as thermodynamic parameters characterizing the macroscopic state of the system. The extent to which the statistical ensemble is representative is characterized by the completeness of the set of parameters. For example, a statistical ensemble which is characterized only by average values of the velocities is not representative for the description of turbulent motions. For this, it is necessary to introduce a random field of velocities and their correlations.

21.3. The Meaning of Local Integrals of Motion

We now make a few more remarks on the meaning of the local integrals of motion in the form (21.6) and (21.7).

If we assume that for the operator $P_m(\boldsymbol{x}, t)$ the limit $\lim\limits_{t \to -\infty} P_m(\boldsymbol{x}, t)$ exists, then the operation of taking the invariant part would coincide with the operation of time averaging, i.e.,

$$\lim_{\varepsilon \to 0} \varepsilon \int_{-\infty}^{0} e^{\varepsilon t} P_m(\boldsymbol{x}, t)\, dt = \lim_{T \to \infty} \frac{1}{T} \int_{-T}^{0} P_m(\boldsymbol{x}, t)\, dt = P_m(\boldsymbol{x}, -\infty).$$

According to the Tauber theorems [88, 239], if one of these limits exists and $P_m(\boldsymbol{x}, t)$ has a lower bound, then the second limit also exists and is equal to the first.

In fact, $\lim\limits_{t \to -\infty} P_m(\boldsymbol{x}, t)$ is not fully defined and there exists only the limit of the expressions

$$\langle A \rangle = Q^{-1}\, \mathrm{Tr}\left\{A \exp\left(-\sum_m \int \underset{\sim}{F_m(\boldsymbol{x}, t)\, P_m(\boldsymbol{x})\, d\boldsymbol{x}}\right)\right\}$$

(where first $V \to \infty$, and then $\varepsilon \to 0$) for any operator A representing an observable quantity; therefore, the Tauber theorems are inapplicable, and it is impossible, generally speaking, to replace the taking of the invariant part by a time averaging. However, if we do make this replacement, we again obtain local integrals of motion, but with another definition of the meaning of the improper integrals, namely,

$$\overline{P_m(\boldsymbol{x})} = \frac{1}{T} \int_{-T}^{0} P_m(\boldsymbol{x}, t)\, dt = P_m(\boldsymbol{x}) + \int_{-T}^{0} \left(1 + \frac{t}{T}\right) \nabla \cdot j_m(\boldsymbol{x}, t)\, dt, \qquad (21.17)$$

i.e., integrals in the sense of Cesaro, rather than in the sense of Abel as in (21.7).

The quantities (21.17) are also conserved for $T \to \infty$. In fact,

$$\lim_{T\to\infty} \overline{P_m(\boldsymbol{x})} = \lim_{T\to\infty} \frac{1}{T} \int_{-T}^{0} \dot{P}_m(\boldsymbol{x}, t)\, dt = \lim_{T\to\infty} \frac{P_m(\boldsymbol{x}, 0) - P_m(\boldsymbol{x}, -T)}{T} = 0.$$

Application of the local integrals of motion in the form (21.17) is not convenient, as we shall see in §22.3, because a further supplementary definition of the meaning of the integrals obtained is required.

The formula (21.17) corresponds to the usual time smoothing of the dynamic variables, and (21.7) to a causal time smoothing of them.

The choice of the local integrals of motion in the form (21.6), (21.7) is not the only one possible. It is possible, for example, to choose local integrals of motion not of the retarded type, like (21.6), (21.7), but of the advanced type, i.e.,

$$\widetilde{P'_m(\boldsymbol{x})} = P_m(\boldsymbol{x}) - \int_0^{\infty} e^{-\varepsilon t}\, \nabla \cdot j(\boldsymbol{x}, t)\, dt, \tag{21.18}$$

or to choose a superposition of the retarded and advanced solutions.

The quantity (21.18) is conserved for $\varepsilon \to 0$, in the same way as is (21.7a), since

$$\frac{\partial \widetilde{P'_m(\boldsymbol{x}, t)}}{\partial t} = -\varepsilon \int_0^{\infty} e^{-\varepsilon t'}\, \nabla j_m(\boldsymbol{x}, t + t')\, dt'$$

which is analogous to the relation (21.8).

As we shall see in §22.4, the local integrals of motion of the advanced type (21.18) give not an increase but a decrease of the local entropy production, and it is therefore necessary to reject the choice (21.18) in constructing the statistical operator.†

---

†The choice of the advanced solution leads to the conjugate thermal conduction equation

$$-c\,\frac{\partial T^+}{\partial t} = \varkappa \nabla^2 T^+$$

with a minus sign before the time derivative, unlike in the usual thermal conduction equation. The conjugate temperature T^+ is convenient as an auxiliary concept in the formulation of the variational principle [87]. Conjugate equations can be introduced analogously for other quantities too.

The choice of local integrals of motion of the retarded type corresponds, as has already been mentioned above, to imposing boundary conditions that exclude the advanced solutions of Liouville's equation, i.e., to imposing causality conditions. This is intimately related to the choice of boundary conditions in formal scattering theory. This connection becomes especially obvious if we compare these boundary conditions with those in formal scattering theory as described by Gell-Mann and Goldberger [84, 85], where retarded solutions of the Schrödinger equation are also selected by using the passage to the limit $\varepsilon \to 0$ in integrals of the type (21.6), (21.7) after taking the limit $V \to \infty$ (see Appendices I and III).

We shall show that the nonequilibrium statistical operator (21.10a) corresponds to the invariant part of the logarithm of the local-equilibrium operator (21.14), i.e., that

$$\ln \rho = \widetilde{\ln \rho_l} = \varepsilon \int\limits_{-\infty}^{0} e^{\varepsilon t_1} \ln \rho_l (t + t_1, \ t_1)\, dt_1, \tag{21.19}$$

where $\varepsilon \to 0$ after $V \to \infty$. The first argument of ρ_l denotes the time dependence through the parameters, and the second denotes the Heisenberg-picture time dependence. We have

$$\ln \rho_l (t + t_1, t_1) = -\Phi_l (t + t_1) - \sum_m \int F_m (\boldsymbol{x}, \ t + t_1)\, P_m (\boldsymbol{x}, \ t_1)\, d\boldsymbol{x}, \tag{21.20}$$

where

$$\Phi_l (t + t_1) = \ln \operatorname{Tr} \exp \left\{ - \sum_m \int F_m (\boldsymbol{x}, t + t_1)\, P_m (\boldsymbol{x})\, d\boldsymbol{x} \right\}. \tag{21.21}$$

Using (21.20), we write (21.19) in the form

$$\rho = \exp \left\{ -\Phi - \sum_m \varepsilon \int\limits_{-\infty}^{0} \int e^{\varepsilon t_1} F_m (\boldsymbol{x}, t + t_1)\, P_m (\boldsymbol{x}, t_1)\, dt_1\, d\boldsymbol{x} \right\}, \tag{21.22}$$

where

$$\Phi = \varepsilon \int\limits_{-\infty}^{0} e^{\varepsilon t_1}\, \Phi_l (t + t_1)\, dt_1 =$$

$$= \varepsilon \int\limits_{-\infty}^{0} dt_1\, e^{\varepsilon t_1} \ln \operatorname{Tr} \exp \left\{ - \sum_m \int F_m (\boldsymbol{x}, t + t_1)\, P_m (\boldsymbol{x})\, d\boldsymbol{x} \right\}. \tag{21.23}$$

On the other hand, from the requirement that the normalization is conserved after the invariant part is taken, we obtain

$$\Phi = \ln \operatorname{Tr} \exp\left\{-\sum_m \varepsilon \int_{-\infty}^{0} \int e^{\varepsilon t_1} F_m(\boldsymbol{x}, t+t_1)\, P_m(\boldsymbol{x}, t_1)\, dt_1\, d\boldsymbol{x}\right\}. \tag{21.24}$$

It follows from (21.23) that

$$\delta\Phi = -\varepsilon \int_{-\infty}^{0} \int e^{\varepsilon t_1} \langle P_m(\boldsymbol{x})\rangle_l^{t+t_1}\, \delta F_m(\boldsymbol{x}, t+t_1)\, dt_1\, d\boldsymbol{x}, \tag{21.25}$$

and from (21.24) that

$$\delta\Phi = -\varepsilon \int_{-\infty}^{0} \int e^{\varepsilon t_1} \langle P_m(\boldsymbol{x}, t_1)\rangle^t\, \delta F_m(\boldsymbol{x}, t+t_1)\, dt\ \ d\boldsymbol{x}. \tag{21.25a}$$

Consequently, the fulfillment of the condition

$$\langle P_m(\boldsymbol{x}, t_1)\rangle^t = \langle P_m(\boldsymbol{x})\rangle_l^{t+t_1} \tag{21.26}$$

ensures the conservation of the normalization, since in this case the variations (21.25) and (21.25a) coincide. For the case of statistical equilibrium, (21.23) and (21.24) also coincide.

Thus, the conditions (21.15), which were applied earlier in order to satisfy the thermodynamic equalities, can be obtained from the condition that the normalization be conserved after the invariant part of the local-equilibrium operator is taken (cf. [184, 185, 188]).

We shall give a further simple interpretation of the nonequilibrium statistical operator (21.19). We write it in the form

$$\ln\rho = \int_{-\infty}^{t} \frac{1}{\tau} e^{-(t-t_0)/\tau} e^{-iH(t-t_0)/\hbar} \ln\rho_l(t_0, 0)\, e^{iH(t-t_0)/\hbar}\, dt_0,$$

where $\tau = 1/\varepsilon$ and $t_0 = t + t_1$. This means that, starting from time t_0, the logarithm of the statistical operator evolves freely with the Hamiltonian H, up to time t, after which the system undergoes a random transition under the influence of the interaction with the surroundings (the thermostat), with probability $W(t, t_0) = (1/\tau)\times \exp\{-(t-t_0)/\tau\}$. The quantity τ defines the average time interval between the random impetuses, which obey the Poisson distribution $W(t, t_0)$. The true distribution is obtained after averaging over all initial times with probability $W(t, t_0)$.

§ 22. Tensor, Vector, and Scalar Processes. The Equations of Hydrodynamics, Thermal Conduction, and Diffusion in a Multicomponent Fluid

In this section, we shall obtain the linear relations between the fluxes and thermodynamic forces, the entropy production, and the transport equations for the example of a many-component fluid. We shall study tensor, vector, and scalar transport processes, when the fluxes and thermodynamic forces are tensors (shear viscosity), vectors (thermal conduction, diffusion, thermal diffusion, and the Dufour effect), or scalars (bulk viscosity).

22.1. Transport Processes in a Multicomponent Fluid. The Statistical Operator

We shall consider processes involving the transport of energy, momentum, and particle number in an isotropic multicomponent system (liquid or gas), when the statistical operator has the form (21.10c):

$$\rho = Q^{-1}\exp\left\{-\sum_m \int \left[F_m(\boldsymbol{x}, t)\,P_m(\boldsymbol{x}) - \right.\right.$$

$$\left.\left. - \int_{-\infty}^{0} e^{\varepsilon t_1}\left(j_m(\boldsymbol{x}, t_1)\cdot\nabla F_m(\boldsymbol{x}, t+t_1) + P_m(\boldsymbol{x}, t_1)\,\frac{\partial F_m(\boldsymbol{x}, t+t_1)}{\partial t}\right)dt_1\right]d\boldsymbol{x}\right\}, \tag{22.1}$$

where

$$F_0(\boldsymbol{x}, t) = \beta(\boldsymbol{x}, t), \qquad P_0(\boldsymbol{x}) = H(\boldsymbol{x}),$$

$$F_1(\boldsymbol{x}, t) = -\beta(\boldsymbol{x}, t)\,\boldsymbol{v}(\boldsymbol{x}, t), \qquad P_1(\boldsymbol{x}) = \boldsymbol{p}(\boldsymbol{x}),$$

$$F_{i+1}(\boldsymbol{x}, t) = -\beta(\boldsymbol{x}, t)\left(\mu_i(\boldsymbol{x}, t) - \frac{m_i}{2}v^2(\boldsymbol{x}, t)\right), \quad P_{i+1}(\boldsymbol{x}) = n_i(\boldsymbol{x}), \tag{22.1a}$$

$$j_0(\boldsymbol{x}) = \boldsymbol{j}_H(\boldsymbol{x}), \qquad j_1(\boldsymbol{x}) = T(\boldsymbol{x}), \qquad j_{i+1}(\boldsymbol{x}) = \boldsymbol{j}_i(\boldsymbol{x}) \qquad (i \geqslant 1).$$

In addition to gradients of the thermodynamic forces, the exponential in (22.1) also contains time derivatives of the thermodynamic forces. The latter can also be expressed in terms of gradients by means of the hydrodynamic equations [38].

Confining ourselves to the local approximation, we shall assume that the pressure p(**x**) at the point **x** is a function of the values β (**x**) and ν_i(**x**) at this same point, i.e., that p(**x**) = p[β (**x**), ..., ν_i(**x**),...]; consequently,

$$\nabla p = \frac{\partial p}{\partial \beta} \nabla \beta + \sum_i \frac{\partial p}{\partial \nu_i} \nabla \nu_i. \tag{22.2}$$

This assumption may be violated near critical points (cf. §20.4). If we take (22.2) into account, the last equation of the system (20.69b), i.e., Euler's equation, takes the form

$$\begin{gathered} \frac{dv}{dt} = -\frac{1}{\langle \rho \rangle} \frac{\partial p}{\partial \beta} \nabla \beta - \frac{1}{\langle \rho \rangle} \sum_i \frac{\partial p}{\partial \nu_i} \nabla \nu_i, \\ \langle \rho \rangle = \langle \rho (x) \rangle, \end{gathered} \tag{22.3}$$

where

$$\frac{d}{dt} = \frac{\partial}{\partial t} + v \cdot \nabla$$

is a total time derivative. We shall make use of the thermodynamic equalities (20.72a), which, when (20.71d) and (20.69c) are taken into account, have the form

$$\frac{\partial \beta p}{\partial \beta} = -u, \qquad \frac{\partial \beta p}{\partial \nu_i} = n_i, \tag{22.4}$$

or

$$\frac{\partial p}{\partial \beta} = -\frac{u+p}{\beta}, \qquad \frac{\partial p}{\partial \nu_i} = \frac{n_i}{\beta}, \tag{22.4a}$$

where

$$u = \langle H'(x) \rangle_l$$

is the energy density, and

$$n_i = \langle n_i (x) \rangle_l$$

is the particle-number density, and write (22.3) in the form

$$\frac{dv}{dt} = \frac{u+p}{\langle \rho \rangle \beta} \nabla \beta - \sum_i \frac{n_i}{\langle \rho \rangle \beta} \nabla \nu_i. \tag{22.5}$$

Thus, the time derivative of the velocity is expressed in terms of gradients of the thermodynamic parameters β and ν_i and of the velocity $\mathbf{v}$.

Analogously, $\partial\beta/\partial t$ can also be expressed in terms of these gradients. As before, confining ourselves to the local approximation, i.e., assuming that $\beta(\mathbf{x}) = \beta[u(\mathbf{x}), \ldots, n_i(\mathbf{x}), \ldots]$, we obtain

$$\frac{d\beta}{dt} = \frac{\partial\beta}{\partial u}\frac{du}{dt} + \sum_i \frac{\partial\beta}{\partial n_i}\frac{dn_i}{dt}, \tag{22.6}$$

or, using the first and second equations of the system (20.69b),

$$\frac{d\beta}{dt} = -(u+p)\frac{\partial\beta}{\partial u}\operatorname{div}\boldsymbol{v} - \sum_i \frac{\partial\beta}{\partial n_i} n_i \operatorname{div}\boldsymbol{v}$$

$$= \left\{\frac{\partial p}{\partial\beta}\frac{\partial\beta}{\partial u} - \sum_i \frac{\partial p}{\partial\nu_i}\frac{\partial\beta}{\partial n_i}\right\}\beta\operatorname{div}\boldsymbol{v}, \tag{22.6a}$$

where we have taken the thermodynamic equalities (22.4a) into account.

According to (20.35),

$$\beta = \frac{\partial S}{\partial u}, \qquad -\nu_i = \frac{\partial S}{\partial n_i}, \tag{22.7}$$

where S is the entropy as a function of u and n_i. Consequently, β and ν_i are connected by the thermodynamic equality

$$\frac{\partial\beta}{\partial n_i} = -\frac{\partial\nu_i}{\partial u} = \frac{\partial^2 S}{\partial u\,\partial n_i}; \tag{22.7a}$$

we can therefore rewrite (22.6a) in the form

$$\frac{d\beta}{dt} = \left\{\frac{\partial p}{\partial\beta}\frac{\partial\beta}{\partial u} + \sum_i \frac{\partial p}{\partial\nu_i}\frac{\partial\nu_i}{\partial u}\right\}\beta\operatorname{div}\boldsymbol{v} = \beta\left(\frac{\partial p}{\partial u}\right)_n \operatorname{div}\boldsymbol{v}, \tag{22.8}$$

i.e., the partial time derivative of β is expressed in terms of $\mathbf{v}\cdot\nabla\beta$ and div $\mathbf{v}$.

Postulating, further, that $\nu_i(\mathbf{x}) = \nu_i(u(\mathbf{x}), \ldots, n_k(\mathbf{x}), \ldots)$, we obtain

$$\frac{d\nu_i}{dt} = \frac{\partial\nu_i}{\partial u}\frac{du}{dt} + \sum_k \frac{\partial\nu_i}{\partial n_k}\frac{dn_k}{dt} = -\frac{\partial\nu_i}{\partial u}(u+p)\operatorname{div}\boldsymbol{v} - \sum_k \frac{\partial\nu_i}{\partial n_k} n_k \operatorname{div}\boldsymbol{v}, \tag{22.9}$$

or, taking into account (22.4a), (22.7), and the thermodynamic equality

$$\frac{\partial \nu_i}{\partial n_k} = \frac{\partial \nu_k}{\partial n_i}, \tag{22.9a}$$

which is derived analogously to (22.7a), we obtain

$$\frac{d\nu_i}{dt} = \left\{\frac{\partial p}{\partial \beta}\frac{\partial \nu_i}{\partial u} - \sum_k \frac{\partial p}{\partial \nu_k}\frac{\partial \nu_i}{\partial n_k}\right\}\beta \operatorname{div} \boldsymbol{v} = \beta\left\{\frac{\partial p}{\partial \beta}\frac{\partial \nu_i}{\partial u} - \sum_k \frac{\partial p}{\partial \nu_k}\frac{\partial \nu_k}{\partial n_i}\right\}\operatorname{div} \boldsymbol{v} =$$
$$= -\beta\left\{\frac{\partial p}{\partial \beta}\frac{\partial \beta}{\partial n_i} + \sum_k \frac{\partial p}{\partial \nu_k}\frac{\partial \nu_k}{\partial n_i}\right\}\operatorname{div} \boldsymbol{v} = -\beta\left(\frac{\partial p}{\partial n_i}\right)_u \operatorname{div} \boldsymbol{v}. \tag{22.9b}$$

Finally, for the time derivatives of the thermodynamic parameters in the ideal-liquid hydrodynamic approximation, we shall have

$$\begin{aligned} \frac{d\beta}{dt} &= \beta\left(\frac{\partial p}{\partial u}\right)_n \operatorname{div} \boldsymbol{v}, \\ \frac{d\nu_i}{dt} &= -\beta\left(\frac{\partial p}{\partial n_i}\right)_u \operatorname{div} \boldsymbol{v}, \\ \frac{d\boldsymbol{v}}{dt} &= \frac{u+p}{\langle\rho\rangle\beta}\nabla\beta - \sum_i \frac{n_i}{\langle\rho\rangle\beta}\nabla\nu_i. \end{aligned} \tag{22.10}$$

We now express the sum

$$\sum_m j_m(\boldsymbol{x}) \cdot \nabla F_m(\boldsymbol{x}, t),$$

appearing in the exponent in formula (22.1), in terms of $\nabla\beta$, $\nabla\nu_i = \nabla\beta\mu_i$, and $\nabla\boldsymbol{v}$. We obtain

$$\sum_m j_m(\boldsymbol{x}) \cdot \nabla F_m = \boldsymbol{j}_H(\boldsymbol{x}) \cdot \nabla\beta - \sum_i \boldsymbol{j}_i(\boldsymbol{x}) \cdot \nabla\beta\left(\mu_i - \frac{m_i v^2}{2}\right) - T(\boldsymbol{x}) : \nabla\beta\boldsymbol{v}$$
$$= \left(\boldsymbol{j}_H(\boldsymbol{x}) - T(\boldsymbol{x}) \cdot \boldsymbol{v} + \boldsymbol{p}(\boldsymbol{x})\frac{v^2}{2}\right) \cdot \nabla\beta - \sum_i \boldsymbol{j}_i(\boldsymbol{x}) \cdot \nabla\nu_i -$$
$$-\beta\left(T(\boldsymbol{x}) - \boldsymbol{v}\boldsymbol{p}(\boldsymbol{x})\right) : \nabla\boldsymbol{v}, \tag{22.11}$$

or

$$\sum_m j_m(\boldsymbol{x}) \cdot \nabla F_m =$$
$$= -\frac{\boldsymbol{p}(\boldsymbol{x})}{\langle\rho\rangle} \cdot \beta\nabla p + \left(\boldsymbol{j}_H(\boldsymbol{x}) - T(\boldsymbol{x}) \cdot \boldsymbol{v} + \boldsymbol{p}(\boldsymbol{x})\frac{v^2}{2} - \frac{u+p}{\langle\rho\rangle}\boldsymbol{p}(\boldsymbol{x})\right) \cdot \nabla\beta -$$
$$-\sum_i \left(\boldsymbol{j}_i(\boldsymbol{x}) - \frac{\langle n_i\rangle}{\langle\rho\rangle}\boldsymbol{p}(\boldsymbol{x})\right) \cdot \nabla\nu_i - \beta\left(T(\boldsymbol{x}) - \boldsymbol{v}\boldsymbol{p}(\boldsymbol{x})\right) : \nabla\boldsymbol{v}, \tag{22.11a}$$

since, taking into account (22.2), (22.4a) for the gradient of the pressure, we have

$$\nabla p = -\frac{u+p}{\beta}\nabla\beta + \sum_i \frac{\langle n_i\rangle}{\beta}\nabla\nu_i. \qquad (22.11b)$$

This form is convenient, since, on averaging over a local-equilibrium state, the coefficient of $\nabla\nu_i$ goes to zero,

$$\langle \boldsymbol{j}_i(\boldsymbol{x})\rangle_l - \frac{\langle n_i\rangle}{\langle\rho\rangle}\langle \boldsymbol{p}(\boldsymbol{x})\rangle_l = \langle \boldsymbol{j}_i(\boldsymbol{x})\rangle_l - \langle n_i\rangle \boldsymbol{v}(\boldsymbol{x}) = 0.$$

We shall express

$$\sum_m P_m(\boldsymbol{x})\frac{\partial F_m(\boldsymbol{x},t)}{\partial t},$$

i.e., the second term in the exponent in formula (22.1), in terms of $\nabla\beta$, $\nabla\nu_i$, and $\nabla\boldsymbol{v}$. We obtain

$$\sum_m P_m(\boldsymbol{x})\frac{\partial F_m}{\partial t} = H(\boldsymbol{x})\frac{\partial\beta}{\partial t} - \boldsymbol{p}(\boldsymbol{x})\cdot\frac{\partial\beta\boldsymbol{v}}{\partial t} - \sum_i n_i(\boldsymbol{x})\frac{\partial}{\partial t}\beta\left(\mu_i - \frac{m_i v^2}{2}\right)$$

$$= H'(\boldsymbol{x})\frac{\partial\beta}{\partial t} - \sum_i n_i(\boldsymbol{x})\frac{\partial\nu_i}{\partial t} - \beta\boldsymbol{p}'(\boldsymbol{x})\frac{\partial\boldsymbol{v}}{\partial t}, \qquad (22.12)$$

where

$$H'(\boldsymbol{x}) = H(\boldsymbol{x}) - \boldsymbol{p}(\boldsymbol{x})\cdot\boldsymbol{v} + \rho(\boldsymbol{x})\frac{v^2}{2}, \qquad \boldsymbol{p}'(\boldsymbol{x}) = \boldsymbol{p}(\boldsymbol{x}) - \rho(\boldsymbol{x})\boldsymbol{v} \qquad (22.12a)$$

are the densities of energy and momentum in the system moving with the mass velocity **v** [cf. (20.66), (20.67)]. We substitute the expressions for $\partial\beta/\partial t$, $\partial\nu_i/\partial t$, and $\partial\mathbf{v}/\partial t$ from (22.10) into (22.12). Taking (22.11b) into account, we obtain

$$\sum_m P_m(\boldsymbol{x})\frac{\partial F_m}{\partial t} = \frac{\boldsymbol{p}(\boldsymbol{x})}{\langle p\rangle}\cdot\beta\nabla p - \left(H'(\boldsymbol{x}) - \frac{\rho(\boldsymbol{x})}{\langle\rho\rangle}(u+p)\right)\boldsymbol{v}\cdot\nabla\beta +$$

$$+\sum_i\left(n_i(\boldsymbol{x}) - \langle n_i\rangle\frac{\rho(\boldsymbol{x})}{\langle\rho\rangle}\right)\boldsymbol{v}\cdot\nabla\nu_i +$$

$$+\beta\left(H'(\boldsymbol{x})\left(\frac{\partial p}{\partial u}\right)_n U + \sum_i n_i(\boldsymbol{x})\left(\frac{\partial p}{\partial n_i}\right)_u U + \boldsymbol{p}'(\boldsymbol{x})\boldsymbol{v}\right):\nabla\boldsymbol{v}, \qquad (22.12b)$$

where U is the unit tensor. Combining (22.11a) with (22.12b), we note that the terms with the pressure gradient cancel each other,

$$\sum_m\left(j_m(\boldsymbol{x})\cdot\nabla F_m + P_m(\boldsymbol{x})\frac{\partial F_m}{\partial t}\right) =$$

$$= \boldsymbol{j}_Q(\boldsymbol{x})\cdot\nabla\beta - \beta j^1(\boldsymbol{x}):\nabla\boldsymbol{v} - \sum_i j^i_d(\boldsymbol{x})\cdot\nabla\nu_i = \sum_m j^m(\boldsymbol{x})\cdot X_m(\boldsymbol{x}), \qquad (22.13)$$

where

$$
\begin{aligned}
j^0(\boldsymbol{x}) &= \boldsymbol{j}_Q(\boldsymbol{x}) = \boldsymbol{j}'_H(\boldsymbol{x}) - \frac{u+p}{\langle\rho\rangle}\boldsymbol{p}'(\boldsymbol{x}),\\
j^1(\boldsymbol{x}) &= T'(\boldsymbol{x}) - \left(\frac{\partial p}{\partial u}\right)_n H'(\boldsymbol{x})\,U - \sum_i \left(\frac{\partial p}{\partial n_i}\right)_u n_i(\boldsymbol{x})\,U, \qquad (22.13a)\\
j^{i+1}(\boldsymbol{x}) &= \boldsymbol{j}^i_d(\boldsymbol{x}) = \boldsymbol{j}'_i(\boldsymbol{x}) - \frac{\langle n_i\rangle}{\langle\rho\rangle}\boldsymbol{p}'(\boldsymbol{x}) \quad (i \geqslant 1)
\end{aligned}
$$

are the operators of the thermal, viscous, and diffusional fluxes, and

$$
\begin{aligned}
X_0(\boldsymbol{x}, t) &= \nabla\beta(\boldsymbol{x}, t),\\
X_1(\boldsymbol{x}, t) &= -\beta(\boldsymbol{x}, t)\,\nabla\boldsymbol{v}(\boldsymbol{x}, t), \qquad (22.13b)\\
X_{i+1}(\boldsymbol{x}, t) &= -\nabla\nu_i(\boldsymbol{x}, t) = -\nabla\beta(\boldsymbol{x}, t)\,\mu_i(\boldsymbol{x}, t) \quad (i \geqslant 1)
\end{aligned}
$$

are the corresponding thermodynamic forces.

Taking (22.13)-(22.13b) into account, we write the statistical operator (22.1) in the form

$$
\rho = Q^{-1}\exp\left\{-\sum_m \int \left(F_m(\boldsymbol{x}, t)\,P_m(\boldsymbol{x}) - \int_{-\infty}^{0} e^{\varepsilon t_1}\, j^m(\boldsymbol{x}, t_1)\cdot X_m(\boldsymbol{x}, t+t_1)\,dt_1\right)d\boldsymbol{x}\right\}. \qquad (22.14)
$$

This expression, unlike (22.1), is approximate, being correct to within gradients of the thermodynamic parameters in its exponent, since the time derivatives of the thermodynamic parameters were eliminated by means of the hydrodynamic equations (22.10) of an ideal liquid. In higher approximations, if we eliminate the time derivatives by means of hydrodynamic equations with viscosity, thermal conduction and diffusion, the exponent of (22.14) will contain terms with higher spatial derivatives of the F_m. Another approximation made in deriving (22.14) was the application of thermodynamic relations in local form.

22.2. Linear Relations between the Fluxes and Thermodynamic Forces

If the thermodynamic forces are small, then by means of (22.14) we can obtain linear relations, which are nonlocal and with time retardation, for the average fluxes, relating the latter to the thermodynamic forces.

We write the statistical operator (22.14) in the form

$$\rho = Q^{-1} e^{-A-B}, \tag{22.15}$$

where

$$A = \sum_m \int F_m(\boldsymbol{x}, t)\, P_m(\boldsymbol{x})\, d\boldsymbol{x},$$

$$B = -\sum_m \int \int_{-\infty}^{0} e^{\varepsilon t_1} j^m(\boldsymbol{x}, t_1) \cdot X_m(\boldsymbol{x}, t + t_1)\, d\boldsymbol{x}\, dt_1,$$

and expand e^{-A-B} in a series in powers of B following the same method as in §12. For this, it is convenient to introduce an operator K(τ) by means of the relation

$$e^{-(A+B)\tau} = K(\tau)\, e^{-A\tau}, \tag{22.16}$$

which is equivalent to the operator equation

$$K(\tau) = 1 - \int_0^{\tau} K(\tau_1)\, e^{-A\tau_1} B e^{A\tau_1}\, d\tau_1 \tag{22.17}$$

with the initial condition

$$K(0) = 1.$$

Iterating this equation, we obtain in the approximation linear in B

$$e^{-A-B} = e^{-A} - \int_0^1 e^{-A\tau} B e^{A\tau} e^{-A}\, d\tau,$$

$$\rho = \frac{e^{-A} - \int_0^1 e^{-A\tau} B e^{A\tau} e^{-A}\, d\tau}{\mathrm{Tr}\, e^{-A} - \int_0^1 \mathrm{Tr}\,(e^{-A\tau} B e^{A\tau} e^{-A})\, d\tau} \cong$$

$$\cong \left\{ 1 - \int_0^1 (e^{-A\tau} B e^{A\tau} - \langle e^{-A\tau} B e^{A\tau} \rangle_l)\, d\tau \right\} \rho_l, \tag{22.18}$$

where

$$\rho_l = e^{-A} / \mathrm{Tr}\, e^{-A} \tag{22.18a}$$

and

$$\langle \ldots \rangle_l = \operatorname{Tr}(\rho_l \ldots) \tag{22.18b}$$

denotes averaging with the local-equilibrium distribution (21.14), (22.18a).

By means of (22.18) for the average values of the fluxes, we obtain

$$\langle j^m(\boldsymbol{x})\rangle = \langle j^m(\boldsymbol{x})\rangle_l + \sum_n \int \int_{-\infty}^{t} e^{\varepsilon(t'-t)} (j^m(\boldsymbol{x}),\ j^n(\boldsymbol{x}',\ t'-t)) \cdot X_n(\boldsymbol{x}',\ t')\, dt'\, d\boldsymbol{x}', \tag{22.19}$$

where

$$(j^m(\boldsymbol{x}),\ j^n(\boldsymbol{x}',\ t)) = \beta^{-1} \int_0^{\beta} \langle j^m(\boldsymbol{x})(j^n(\boldsymbol{x}',\ t,\ i\tau) - \langle j^n(\boldsymbol{x}',\ t)\rangle_l)\rangle_l\, d\tau \tag{22.19a}$$

are the quantum time correlation functions,

$$j^n(\boldsymbol{x}',\ t,\ i\tau) = e^{-\beta^{-1}A\tau}\, j^n(\boldsymbol{x}',\ t)\, e^{\beta^{-1}A\tau}. \tag{22.19b}$$

The linear relations (22.19) between the fluxes and the thermodynamic forces are retarded and nonlocal.

Suppose the thermodynamic forces depend periodically on the time, with frequency ω,

$$X_n(\boldsymbol{x}',\ t') = X_n(\boldsymbol{x}') \cos \omega t'.$$

Then

$$\langle j^m(\boldsymbol{x})\rangle = \langle j^m(\boldsymbol{x})\rangle_l + \sum_n \operatorname{Re} e^{i\omega t} \int \int_{-\infty}^{t} e^{\varepsilon(t'-t)} \times$$

$$\times (j^m(\boldsymbol{x}),\ j^n(\boldsymbol{x}',\ t'-t)) \cdot X_n(\boldsymbol{x}')\, e^{i\omega(t'-t)}\, dt'\, d\boldsymbol{x}'. \tag{22.19c}$$

The Fourier components of the quantum correlation functions are the kinetic coefficients in (22.19c), i.e., taking retardation into account leads to dispersion of the kinetic coefficients [89, 90, 233].

If we neglect the retardation in (22.19), i.e., assume that $X_n(\mathbf{x}',\ t')$ changes little in the attenuation time of the correlation between the fluxes, then we can take the thermodynamic forces at time $t' = t$ outside the integral over the time. We then obtain linear relations between the thermodynamic forces and fluxes that are without retardation, but nonlocal in character:

$$\langle j^m(\boldsymbol{x})\rangle = \langle j^m(\boldsymbol{x})\rangle_l + \sum_n \int L_{mn}(\boldsymbol{x}, \boldsymbol{x}') \cdot X_n(\boldsymbol{x}', t)\, d\boldsymbol{x}', \tag{22.20}$$

where

$$L_{mn}(\boldsymbol{x}, \boldsymbol{x}') = \int_{-\infty}^{0} e^{\varepsilon t} (j^m(\boldsymbol{x}), j^n(\boldsymbol{x}', t))\, dt \tag{22.20a}$$

are the kinetic coefficients.

In the expressions (22.20b) for the kinetic coefficients in the linear approximation, we can replace the averaging over the local-equilibrium distribution by averaging over the equilibrium statistical operator ρ_0 (21.13) with the inclusion of variables $F_m(\mathbf{x})$ depending on position, or even with spatial average values of these parameters. For the kinetic coefficients, we then obtain the expressions

$$L_{mn}(\boldsymbol{x}, \boldsymbol{x}') = \beta^{-1} \int_0^{\beta} \int_{-\infty}^{0} e^{\varepsilon t} \langle j^m(\boldsymbol{x}) (j^n(\boldsymbol{x}', t + i\hbar\tau) - \langle j^n(\boldsymbol{x}')\rangle_0)\rangle_0\, d\tau\, dt, \tag{22.20b}$$

where

$$\langle \ldots \rangle_0 = \mathrm{Tr}\,(\rho_0 \ldots).$$

We assume that the fluxes commute with N_n, the total number of particles of type n:

$$[j^m, N_n] = 0,$$

which is usually the case.

The expressions (22.20b) for the kinetic coefficients differ from those obtained by Mori [29-31] using ordinary rather than causal averaging,

$$L_{mn}(\boldsymbol{x}, \boldsymbol{x}') = \beta^{-1} \int_0^{\beta} \int_0^{T} \left(1 - \frac{t}{T}\right) \langle j^n(\boldsymbol{x}) (j^m(\boldsymbol{x}', t + i\hbar\tau) - \langle j^m(\boldsymbol{x}')\rangle_0)\rangle_0\, d\tau\, dt$$

only in the fact that our improper integrals are defined in the sense of Abel, and those of Mori in the sense of Cesaro. This is easily seen by replacing $t \to -t$ and $\tau \to \beta - \tau$ and taking (16.17) into account.

If we neglect the nonlocal character, i.e., assume that the thermodynamic forces vary little over the correlation length over which $L_{mn}(x, x')$ differs appreciably from zero, then in (22.20) we can take $X_n(x', t)$ at the point $x' = x$ outside the integral over space. Then

$$\langle j^m(\boldsymbol{x})\rangle = \langle j^m(\boldsymbol{x})\rangle_l + \sum_n L_{mn}(\boldsymbol{x}) \cdot X_n(\boldsymbol{x}), \tag{22.20c}$$

where

$$L_{mn}(\boldsymbol{x}) = \int L_{mn}(\boldsymbol{x}, \boldsymbol{x}')\, d\boldsymbol{x}'. \tag{22.20d}$$

From (22.20) for the average thermal, viscous, and diffusional fluxes, we obtain

$$\begin{aligned} \langle \boldsymbol{j}_Q(\boldsymbol{x})\rangle &= \langle \boldsymbol{j}'_H(\boldsymbol{x})\rangle = \sum_n \int L_{0n}(\boldsymbol{x}, \boldsymbol{x}') \cdot X_n(\boldsymbol{x}', t)\, d\boldsymbol{x}', \\ \langle T'(\boldsymbol{x})\rangle &= \langle T'(\boldsymbol{x})\rangle_l + \sum_n \int L_{1n}(\boldsymbol{x}, \boldsymbol{x}') \cdot X_n(\boldsymbol{x}', t)\, d\boldsymbol{x}', \\ \langle \boldsymbol{j}^i_d(\boldsymbol{x})\rangle &= \langle \boldsymbol{j}'_i(\boldsymbol{x})\rangle = \sum_n \int L_{in}(\boldsymbol{x}, \boldsymbol{x}') \cdot X_n(\boldsymbol{x}', t)\, d\boldsymbol{x}' \qquad (i \geqslant 1), \end{aligned} \tag{22.20e}$$

since, according to (20.65),

$$\langle \boldsymbol{j}'_H(\boldsymbol{x})\rangle_l = \langle \boldsymbol{j}'_i(\boldsymbol{x})\rangle_l = 0.$$

In the local approximation, the linear relations have an algebraic rather than an integral character:

$$\begin{aligned} \langle \boldsymbol{j}_Q(\boldsymbol{x})\rangle &= \langle \boldsymbol{j}'_H(\boldsymbol{x})\rangle = \sum_n L_{0n}(\boldsymbol{x}) \cdot X_n(\boldsymbol{x}, t), \\ \langle T'(\boldsymbol{x})\rangle &= \langle T'(\boldsymbol{x})\rangle_l + \sum_n L_{1n}(\boldsymbol{x}) \cdot X_n(\boldsymbol{x}, t), \\ \langle \boldsymbol{j}^i_d(\boldsymbol{x})\rangle &= \langle \boldsymbol{j}'_i(\boldsymbol{x})\rangle = \sum_n L_{in}(\boldsymbol{x}) \cdot X_n(\boldsymbol{x}, t) \qquad (i \geqslant 1). \end{aligned} \tag{22.20f}$$

Noting that in (22.20a) the averaging is performed over the local-equilibrium state,

$$\rho_l = Q_l^{-1} \exp\left\{ -\int \beta(\boldsymbol{x}, t)\left(H'(\boldsymbol{x}) - \sum_i \mu_i(\boldsymbol{x}, t)\, n'_i(\boldsymbol{x})\right) d\boldsymbol{x} \right\}, \tag{22.21}$$

we may omit the primes in formulas (22.13a) and (22.21), putting

$$j^0(\boldsymbol{x}) = \boldsymbol{j}_Q(\boldsymbol{x}) = \boldsymbol{j}_H(\boldsymbol{x}) - \frac{u+p}{\langle \rho \rangle} \boldsymbol{p}(\boldsymbol{x}),$$

$$j^1(\boldsymbol{x}) = T(\boldsymbol{x}) - \left(\frac{\partial p}{\partial u}\right)_n H(\boldsymbol{x}) U - \sum_i \left(\frac{\partial p}{\partial n_i}\right)_u n_i(\boldsymbol{x}) U, \tag{22.22}$$

$$j^{i+1}(\boldsymbol{x}) = \boldsymbol{j}_d^i(\boldsymbol{x}) = \boldsymbol{j}_i(\boldsymbol{x}) - \frac{\langle n_i \rangle}{\langle \rho \rangle} \boldsymbol{p}(\boldsymbol{x}),$$

and replace the averaging with (22.21) in (22.20a) by averaging with

$$\rho_l = Q_l^{-1} \exp\left\{ - \int \beta(\boldsymbol{x}, t)\left(H(\boldsymbol{x}) - \sum_i \mu_i(\boldsymbol{x}, t) n_i(\boldsymbol{x})\right) d\boldsymbol{x} \right\}. \tag{22.21a}$$

In the case when the thermodynamic forces are constant in space, the linear relations (22.20) can be written in the form of relations between the total fluxes and the thermodynamic forces:

$$\langle J^m \rangle = \langle J^m \rangle_l + \sum_n V L_{mn} \cdot X_n, \tag{22.23}$$

where

$$J^m = \int j^m(\boldsymbol{x})\, d\boldsymbol{x} \tag{22.23a}$$

are the total fluxes, and

$$L_{mn} = \frac{1}{V} \int_{-\infty}^{0} e^{\varepsilon t} (J^m, J^n(t))\, dt \tag{22.23b}$$

are the kinetic coefficients.

22.3. Onsager's Reciprocity Relations

The kinetic coefficients (22.20b) can be expressed in terms of the two-time retarded Green functions (15.48). In fact, introducing a further integration over t, we write (22.20b) in the form

$$L_{mn}(\boldsymbol{x}, \boldsymbol{x}') = \beta^{-1} \int_{-\infty}^{0} \int_{-\infty}^{t} \int_{0}^{\beta} e^{\varepsilon t} \left\langle j^m(\boldsymbol{x}) \frac{d}{dt'} j^n(\boldsymbol{x}', t' + i\hbar\tau)\right\rangle_0 dt\, dt'\, d\tau,$$

since we assume that the correlation between the fluxes vanishes as $t \to -\infty$, i.e.,

$$\lim_{t \to -\infty} \langle j^m(\boldsymbol{x}) j^n(\boldsymbol{x}', t)\rangle_0 = \langle j^m(\boldsymbol{x})\rangle_0 \langle j^n(\boldsymbol{x}')\rangle_0.$$

Carrying out the integration over τ, we obtain

$$L_{mn}(x, x') =$$

$$= \beta^{-1} \int_{-\infty}^{0} \int_{-\infty}^{t} e^{\varepsilon t} \frac{1}{i\hbar} \{\langle j^m(x) j^n(x', t' + i\hbar\beta)\rangle - \langle j^m(x) j^n(x', t')\rangle\} dt\, dt',$$

where we have dropped the subscript 0 from the brackets $\langle\ldots\rangle$.

In the first term in the integrand, we can change the order of the operators by a time displacement of $i\hbar\beta$, using the identity (16.17), which in our case gives

$$\langle j^m(x) j^n(x', t + i\hbar\beta)\rangle = \langle j^n(x', t) j^m(x)\rangle,$$

whence we obtain

$$L_{mn}(x, x') = -\beta^{-1} \int_{-\infty}^{0} \int_{-\infty}^{t} e^{\varepsilon t} \frac{1}{i\hbar} \langle [j^m(x), j^n(x', t')]\rangle dt\, dt' =$$

$$= -\beta^{-1} \int_{-\infty}^{0} \int_{-\infty}^{t} \varepsilon^{\varepsilon t} \langle\langle j^m(x) j^n(x', t')\rangle\rangle^r dt\, dt', \qquad (22.24)$$

where in the integrand we have a retarded two-time Green function of the type (16.1). (22.23b) can also be represented in an analogous form:

$$L_{mn} = -\frac{1}{\beta V} \int_{-\infty}^{0} \int_{-\infty}^{t} e^{\varepsilon t} \langle\langle J^m J^n(t')\rangle\rangle^r dt\, dt'. \qquad (22.24a)$$

The Onsager relations for the kinetic coefficients follow directly from the expressions (22.24) and (22.24a) and from the invariance of the Hamiltonian under time reversal, $t \to -t$, with simultaneous reversal of the magnetic field, $H \to -H$. It follows from the symmetry property (16.50) of the Green functions that

$$\langle\langle j^n(x) j^m(x', t')\rangle\rangle_H = \langle\langle j^m(x') j^n(x, t')\rangle\rangle_{-H},$$

since the flux operators are Hermitian; therefore, the Onsager reciprocity relations

$$\begin{aligned} L_{mn}(x, x', H) &= L_{nm}(x', x, -H) \\ L_{mn}(H) &= L_{nm}(-H) \end{aligned} \qquad (22.25)$$

are valid for the kinetic coefficients.

If the system is rotating with constant angular velocity ω, the rotation induces centrifugal and Coriolis forces, and since the latter change sign on reversal of the velocities, on time reversal it is necessary to reverse the direction of the angular velocity. Consequently, the Onsager reciprocity relations in this case have the form

$$L_{mn}(\boldsymbol{x}, \boldsymbol{x}', \omega) = L_{nm}(\boldsymbol{x}', \boldsymbol{x}, -\omega), \qquad L_{mn}(\omega) = L_{nm}(-\omega). \tag{22.25a}$$

We shall study the condition under which the kinetic coefficients (22.20b) have finite values. Since these expressions and (17.35) have the same form, apart from the factor β^{-1}, as is clear after the replacements $t \to -t$ and $\tau \to \beta - \tau$, the formula (22.20b) can be transformed in the same way as was (17.35) in §17.4. Assuming that there is no magnetic field, we obtain

$$L_{mn} = \frac{1}{2V} \int_{-\infty}^{\infty} \langle J^m(t)(J^n - \langle J^n \rangle)\rangle\, dt = \frac{1}{2V} J_{mn}(0), \tag{22.26}$$

i.e., the kinetic coefficients are proportional to the spectral intensity of the correlation function of the fluxes at $\omega = 0$ and are finite when this quantity is finite, i.e., for dissipative processes.†

Earlier [see (21.17)], we pointed out the formal possibility of defining the improper integrals over time differently, in the sense of the passage to the limit $T \to \infty$,

$$\int_{-T}^{0} \left(1 + \frac{t}{T}\right) \nabla \cdot j_m(\boldsymbol{x}, t)\, dt.$$

We shall show that this definition of the integrals is considerably less convenient than (22.20b). For the kinetic coefficients, we would obtain the expressions of Mori [30]

$$L_{mn} = (\beta V)^{-1} \int_0^{\beta} \int_{-T}^{0} \left(1 + \frac{t}{T}\right) \langle J^m (J^n(t + i\hbar\tau) - \langle J^n \rangle)\rangle\, d\tau\, dt. \tag{22.27}$$

Assuming that the limit $V \to \infty$ has already been taken, we calculate the limit of (22.27) as $T \to \infty$. For this it is sufficient to consider the integral, in the sense of (22.27), of one harmonic of

†This occurs for dissipative systems in which the fluctuations of the fluxes are an ergodic process, in the terminology customary in the theory of random processes.

the quantum correlation function, i.e.,

$$\int_{-T}^{0}\left(1+\frac{t}{T}\right)e^{-i\omega t}\,dt=-\frac{1}{i\omega}+\frac{1}{T\omega^2}\left(1-e^{i\omega T}\right). \tag{22.27a}$$

Using this equality and the spectral representation (16.15), we obtain for the kinetic coefficient (22.23b)

$$L_{mn}=\frac{1}{2\pi V}\int_{-\infty}^{\infty}J_{mn}(\omega)\,\frac{e^{\hbar\beta\omega}-1}{\hbar\beta\omega}\,\frac{d\omega}{-i\omega}, \tag{22.28}$$

since the contribution due to the second term of (22.27a) tends to zero as $T\to\infty$. The integrand in (22.28) has a pole at $\omega=0$, since

$$J_{mn}(\omega)\frac{e^{\hbar\beta\omega}-1}{\hbar\beta\omega}$$

is finite as $\omega\to 0$ ($J_{mn}(0)$ is assumed finite). Therefore, the integral in (22.28) is, strictly speaking, not defined so long as the integration contour is not chosen. In order to give this integral a definite meaning, we displace the integration contour over ω from the real axis by $i\varepsilon$ into the upper half-plane, making the replacement $\omega\to\omega+i\varepsilon$ ($\varepsilon>0$, ω real). Then

$$L_{mn}=\frac{1}{2\pi V}\int_{-\infty}^{\infty}J_{mn}(\omega)\,\frac{e^{\hbar\beta\omega}-1}{\hbar\beta\omega}\,\frac{d\omega}{-i(\omega+i\varepsilon)}=\frac{1}{2V}J_{mn}(0), \tag{22.29}$$

i.e., we obtain the same result as when the integral is defined in the sense of (22.20b).

We note that the displacement of the integration contour over ω by $i\varepsilon$ into the upper half-plane is equivalent to the introduction of a damping factor $\exp(\varepsilon t)$ into the time integrals. Thus, in doing this, we are again returning to the definition of the integrals in the sense of (22.20b), and the introduction of the factor $(1+(t/T))$ is found to be superfluous.

The definition of the integrals in the sense of (22.27) is equivalent to the time smoothing procedure (21.17), which has often been applied in the statistical mechanics of nonequilibrium processes, for example, in the work of Kirkwood [13]. The discussion given above demonstrates the inadequacy of this procedure and the advantage of the causal time smoothing (21.6).

22.4. Entropy Production in Nonequilibrium Processes

The meaning of the thermodynamic functions of nonequilibrium states has already been discussed in §21.2 in choosing the parameters $F_m(x, t)$. The quantity $-\langle \ln \rho \rangle$, where ρ is given by formula (21.10), cannot be chosen as the entropy, since $\ln \rho$ satisfies Liouville's equation and the entropy would be conserved rather than increase. We shall define the entropy of the nonequilibrium state (21.10), (21.10c) as the entropy of the corresponding local-equilibrium state

$$\rho_l = Q_l^{-1} \exp\left\{-\sum_m \int F_m(\boldsymbol{x}, t)\, P_m(\boldsymbol{x})\, d\boldsymbol{x}\right\}, \tag{22.30}$$

which is characterized by the same values of the average densities [cf. (21.15)], i.e., we put

$$S = -\langle \ln \rho_l \rangle_l = -\langle \ln \rho_l \rangle \tag{22.31}$$

or

$$S = \Phi + \sum_m \int F_m(\boldsymbol{x}, t) \langle P_m(\boldsymbol{x}) \rangle_l\, d\boldsymbol{x} =$$

$$= \Phi + \sum_m \int F_m(\boldsymbol{x}, t) \langle P_m(\boldsymbol{x}) \rangle\, d\boldsymbol{x}, \tag{22.32}$$

where

$$\Phi = \ln Q_l \tag{22.33}$$

is the Massieu–Planck function of the local-equilibrium state.

We have selected the local-equilibrium state from the condition that the information entropy be a maximum for arbitrary given values of $\langle P_m(\boldsymbol{x}) \rangle_l$ [cf. (20.16)]; this arbitrariness is removed by the conditions (21.15).

According to (21.31), the entropy of a nonequilibrium state is the entropy of that equilibrium state, in auxiliary fields $F_m(x, t)$, with the same distribution of densities $\langle P_m(\boldsymbol{x}) \rangle$ of the mechanical quantities, i.e., $\langle n(\boldsymbol{x}) \rangle$, $\langle H(\boldsymbol{x}) \rangle$, and $\langle \boldsymbol{p}(\boldsymbol{x}) \rangle$, as in the given nonequilibrium state. This interpretation of the thermodynamic functions of nonequilibrium states was given a long time ago by Leontovich [70], in the theory of fluctuations, for a state with nonuniform density $\langle n(\boldsymbol{x})\rangle$. A nonuniform distribution of energy $\langle H(\boldsymbol{x})\rangle$ may be regarded, following Luttinger [9], as the result of the influence of a gravitational field (taking the general theory of relativity into account), and nonuniformity of $\langle \boldsymbol{p}(\boldsymbol{x})\rangle$ as the consequence of the influence of a magnetic field through a vector potential.

The definition (22.32) of the entropy ensures that the thermodynamic equalities are satisfied. In fact, we have

$$\frac{\delta\Phi}{\delta F_m(\boldsymbol{x},\, t)} = -\langle P_m(\boldsymbol{x})\rangle_l = -\langle P_m(\boldsymbol{x})\rangle; \tag{22.34}$$

consequently,

$$\frac{\delta S}{\delta\langle P_m(\boldsymbol{x})\rangle} = F_m(\boldsymbol{x},\, t), \tag{22.35}$$

where $S = S(\ldots\langle P_m(\boldsymbol{x})\rangle\ldots)$, which confirms the correctness of the definition (22.31).

We shall calculate the change with time of the entropy (22.32):

$$\frac{\partial S}{\partial t} = \frac{\partial\Phi}{\partial t} + \sum_m \int \frac{\partial F_m(\boldsymbol{x},\, t)}{\partial t}\langle P_m(\boldsymbol{x})\rangle\, d\boldsymbol{x} + \sum_m \int F_m(\boldsymbol{x},\, t)\langle \dot{P}_m(\boldsymbol{x})\rangle\, d\boldsymbol{x}.$$

Differentiating (22.33), we find

$$\frac{\partial\Phi}{\partial t} = -\sum_m \int \frac{\partial F_m(\boldsymbol{x},\, t)}{\partial t}\langle P_m(\boldsymbol{x})\rangle\, d\boldsymbol{x}.$$

Consequently,

$$\frac{\partial S}{\partial t} = \sum_m \int F_m(\boldsymbol{x},\, t)\langle \dot{P}_m(\boldsymbol{x})\rangle\, d\boldsymbol{x}. \tag{22.36}$$

Using the conservation laws and integrating (22.36) by parts, we obtain

$$\frac{\partial S}{\partial t} = -\sum_m \int F_m(\boldsymbol{x},\, t)\langle j_m(\boldsymbol{x})\rangle \cdot d\sigma + \sum_m \int \langle j_m(\boldsymbol{x})\rangle \cdot \nabla F_m(\boldsymbol{x},\, t)\, d\boldsymbol{x}, \tag{22.36a}$$

where $d\sigma$ is an element of surface. Thus, the entropy can change, even if the surface integral in (22.36a) is equal to zero, i.e., unlike the energy, momentum, and particle number the entropy in the system is not conserved.

We introduce the entropy density $S(\mathbf{x})$ using (20.21c) and (20.24d):

$$S = \int S(\boldsymbol{x})\, d\boldsymbol{x}, \quad \Phi = \int \beta(\boldsymbol{x}, t)\, p(\boldsymbol{x}, t)\, d\boldsymbol{x}, \tag{22.37}$$

$$S(\boldsymbol{x}) = \sum_m F_m(\boldsymbol{x}, t) \langle P_m(\boldsymbol{x})\rangle + \beta(\boldsymbol{x}, t)\, p(\boldsymbol{x}, t). \tag{22.37a}$$

Then, using (20.73), from (22.36a) we obtain the entropy balance equation

$$\frac{\partial S(\boldsymbol{x})}{\partial t} = -\operatorname{div} \boldsymbol{j}_S(\boldsymbol{x}) + \sigma(\boldsymbol{x}), \tag{22.38}$$

where

$$\boldsymbol{j}_S(\boldsymbol{x}) = \sum_m F_m(\boldsymbol{x}, t) \langle \boldsymbol{j}_m(\boldsymbol{x})\rangle + \beta(\boldsymbol{x}, t)\, \boldsymbol{v}(\boldsymbol{x}, t)\, p(\boldsymbol{x}) \tag{22.38a}$$

is the entropy flux density, and

$$\sigma(\boldsymbol{x}) = \sum_m (\langle \boldsymbol{j}_m(\boldsymbol{x})\rangle - \langle \boldsymbol{j}_m(\boldsymbol{x})\rangle_l) \cdot \nabla F_m(\boldsymbol{x}, t) \tag{22.38b}$$

is the local entropy production, i.e., the density of its sources. The quantities $S(\mathbf{x})$, $\mathbf{j}_S(\mathbf{x})$, and $\sigma(\mathbf{x})$ depend also on t, but we shall omit this argument for brevity. According to (20.38b), the entropy production is equal to a sum of products of the thermodynamic forces with their conjugate fluxes.

We write the local entropy production (22.38b) in terms of the thermodynamic forces X_m (22.13b):

$$\sigma(\boldsymbol{x}) = \sum_m (\langle \boldsymbol{j}_m(\boldsymbol{x})\rangle - \langle \boldsymbol{j}_m(\boldsymbol{x})\rangle_l) \cdot \nabla F_m =$$

$$= \langle \boldsymbol{j}'_H(\boldsymbol{x})\rangle \cdot \nabla\beta - \beta(\langle T'(\boldsymbol{x})\rangle - \langle T'(\boldsymbol{x})\rangle_l) : \nabla \boldsymbol{v} - \sum_{i=1}^{l} \langle \boldsymbol{j}'_i(\boldsymbol{x})\rangle \cdot \nabla \nu_i =$$

$$= \sum_m (\langle j^m(\boldsymbol{x})\rangle - \langle j^m(\boldsymbol{x})\rangle_l) \cdot X_m(\boldsymbol{x}), \tag{22.39}$$

where we have used the relation (22.11) and the fact that the averages of $n(\mathbf{x})$, $H(\mathbf{x})$, and $\mathbf{p}(\mathbf{x})$ are equal in the state (22.1) and in the local-equilibrium state.

Using (20.74), we see that the average entropy flux, according to (22.38a), is equal to

$$j_S(x, t) = \sum_m F_m(x, t)\langle j_m(x)\rangle + \beta(x, t)\, v(x, t)\, p(x) =$$
$$= S(x)\, v(x, t) + \beta(x, t)\langle j_Q(x)\rangle -$$
$$- \sum_i \langle j_d^i(x)\rangle \beta(x, t)\left(\mu_i(x, t) - \frac{m_i v^2(x, t)}{2}\right), \qquad (22.40)$$

where S(x) is the entropy density (22.37a), and the average thermal diffusional and viscous fluxes are given by formula (22.20e) or (22.20f). The first term in (22.40) represents the convectional flux, while the remaining terms represent the contribution of irreversible transport processes.

Substituting the linear relations (22.20) into (22.39), we obtain for the local entropy production the expression

$$\sigma(x) = \sum_{m, n} \int L_{mn}(x, x') : X_n(x', t)\, X_m(x, t)\, dx'. \qquad (22.41)$$

We shall show that the total entropy production is positive, i.e., that

$$\int \sigma(x)\, dx \geqslant 0. \qquad (22.42)$$

In fact, (22.42) can be written in the form

$$\int \sigma(x)\, dx = \int_{-\infty}^{0} e^{\varepsilon t}(C, C(t))\, dt \geqslant 0 \qquad (C^+ = C), \qquad (22.42a)$$

where we have introduced the notation

$$C = \sum_m \int j^m(x) \cdot X_m(x, t)\, dx. \qquad (22.42b)$$

That (22.42a) is positive follows from the fact that the spectral intensity of self-adjoint operators is positive [cf. (16.18a)]. Transforming (22.42), analogously to (17.38), we obtain

$$\int \sigma(x)\, dx = \frac{1}{2} \int_{-\infty}^{\infty} \langle C(C(t) - \langle C\rangle_0)\rangle_0\, dt > 0, \qquad (22.43)$$

since $C = C^+$.

In the local approximation, when the thermodynamic forces vary little over distances of the order of the correlation length, as is usually assumed in hydrodynamics, not only the total, but also the local entropy production is positive. In this case, we have

$$\sigma = \sum_{m,\, n} L_{mn} : X_n X_m. \tag{22.44}$$

Thus, the entropy production is a positive-definite form of the thermodynamic forces.

Consequently, choosing the retarded local integrals of motion (21.6), (21.7a) leads to the law of increase of entropy. However, as already noted above, this choice is not the only possible choice. If in place of the integrals of motion of the retarded type (21.7a), we choose integrals of motion of the advanced type (21.18), then in place of formula (22.20) we obtain

$$\langle j_m(\boldsymbol{x})\rangle = \langle j_m(\boldsymbol{x})\rangle_l - \sum_n \int L_{mn}(\boldsymbol{x},\ \boldsymbol{x}') \cdot X_n(\boldsymbol{x}',\ t)\, d\boldsymbol{x}'. \tag{22.45}$$

In this case, for the entropy production, in place of (22.41) we shall have

$$\sigma(\boldsymbol{x}) = -\sum_{mn} \int L_{mn}(\boldsymbol{x},\ \boldsymbol{x}') : X_n(\boldsymbol{x}',\ t)\, X_m(\boldsymbol{x},\ t)\, d\boldsymbol{x}'$$

or, in the local case,

$$\sigma = -\sum_{mn} L_{mn} : X_n X_m,$$

i.e., the same expressions as earlier, but with the opposite sign; consequently, the entropy decreases rather than increases. Therefore, under ordinary conditions, it makes sense to choose the local integrals of motion to be of the retarded, rather than of the advanced type. However, cases of nonequilibrium systems are possible for which fluxes reflected from the boundaries are important. For the description of such systems on the microscopic level, a superposition of integrals of motion of the type (21.7a) and (21.18) are found, generally speaking, to be appropriate. An analogous situation is also found in radiation problems, when standing waves are obtained by means of a superposition of retarded and advanced potentials.

Nonequilibrium thermodynamics can give a new approach to the very old question of the heat death of the universe.

The entropy of an isolated closed system increases, as follows from (22.42). If we regard the universe as an isolated closed system and assume that the results of thermodynamics are applicable to it, then the universe must tend to a state of statistical equilibrium, i.e., to heat death, which in reality does not occur. This paradox has occupied scientists since the time of Boltzmann. It is, of course, not justified to treat the universe as an isolated closed thermodynamic system. The derivations of thermodynamics refer to large, but finite systems, situated in given external conditions; it is, therefore, better to treat the observable part of the universe as a large but finite nonisolated system. Here, too, a paradox remains, since it follows from irreversible thermodynamics that the local entropy production (22.44) is positive, and thermodynamics indicates only one possible process – entropy increase. The reverse process – local decrease of entropy not involving entropy transport – is not permitted by thermodynamics, and it seems incomprehensible why the universe does not tend to statistical equilibrium.

This paradox is still not resolved at the present time, and different hypotheses have been proposed to explain it, for example, the fluctuational hypothesis of Boltzmann [91, 92], according to which the universe is a gigantic fluctuation from the state of statistical equilibrium. The weakness of this hypothesis lies in the fact that the probability of fluctuations from the state of statistical equilibrium is extremely small and decreases exponentially, since it is described by the Gaussian distribution (6.16). Therefore, the appearance, as a result of fluctuations, of states differing greatly from equilibrium states is very improbable.

Considerably more convincing are the hypotheses based on the general theory of relativity [92, 93], according to which the universe must be regarded not as a closed system but as a system in a varying gravitational field with a time-dependent metric tensor. Indeed, if gravitation is taken into account, a uniform distribution of mass is unstable and does not correspond to the maximum entropy. Therefore, the formation of stars and galaxies from uniformly distributed matter may occur with increase of entropy [94] and, thus, the increase of entropy does not contradict the evolution of the universe.

The derivation given in this section of the increase of entropy of a system is based on the linear relations between the thermodynamic forces and fluxes, which are valid only for small deviations from equilibrium. For certain very simple cases, one can prove the increase of entropy for strongly nonequilibrium states too (cf. §23). However, as was noted in § 15.3, there may be cases in which there is absolutely no one-to-one connection between the thermodynamic forces and the fluxes, namely, when there exists a feedback mechanism of the type (15.61) in nonlinear nonequilibrium thermodynamics. We have already given examples of this in §15.3.

If a state of statistical equilibrium of the universe is unstable, like an unstable center in the theory of nonlinear vibrations, and there exists a feedback mechanism, then fluctuations in it will grow and it will go over into a strongly nonequilibrium but stable self-oscillating state, similar to the limit cycle of the theory of nonlinear vibrations. The heat-death paradox could not now be extended to this model of the universe with feedback. An oscillating model of the universe, in which oscillations without increase and with increase of entropy† are possible, is known in relativistic astrophysics (cf. [94], Chapter 20). The first case corresponds to a self-oscillating regime. These problems are still in the development stage and are far from completely solved.

22.5. Tensor, Vector, and Scalar Processes. Thermal Conduction, Diffusion, Thermal Diffusion, the Dufour Effect, and Shear and Bulk Viscosity

The entropy production (22.39) can be written conveniently in a somewhat different form by decomposing the viscous stress tensor $\langle\pi\rangle = \langle T(x)\rangle - \langle T(x)\rangle_l$ and the tensor ∇v into a part with zero trace and a divergence multiplied by the unit tensor U:

$$\langle\pi\rangle = \langle\Pi\rangle U + \langle\overset{\circ}{\pi}\rangle, \tag{22.46}$$

$$\nabla v = \frac{1}{3} U \operatorname{div} v + \overset{\circ}{\nabla v} \tag{22.46a}$$

where

$$\langle\Pi\rangle = \frac{1}{3}\langle\pi\rangle U = \frac{1}{3}\sum_{\alpha}\langle\pi_{\alpha\alpha}\rangle. \tag{22.46b}$$

†Here, by the entropy of the universe we mean the entropy of the corresponding model of the universe.

The tensors $\langle\overset{\circ}{\pi}\rangle$ and $\nabla^{\circ} v$ have zero trace. The complete contraction of the tensors (22.46) and (22.46a) is equal to

$$\langle\pi\rangle : \nabla \boldsymbol{v} = \langle\overset{\circ}{\pi}\rangle : \nabla^{\circ} v + \langle\Pi\rangle \operatorname{div} \boldsymbol{v}. \tag{22.46c}$$

Noting that $\langle\pi\rangle$ is a symmetric tensor, we obtain

$$\langle\pi\rangle : \nabla \boldsymbol{v} = \langle\overset{\circ}{\pi}\rangle : (\nabla^{\circ} \boldsymbol{v})^{s} + \langle\Pi\rangle \operatorname{div} \boldsymbol{v} \tag{22.46d}$$

(the superscript s denotes the symmetric part of the tensor), since the complete contraction of a symmetric and antisymmetric tensor is equal to zero.

It follows from (22.22) that the average diffusional fluxes are connected by the relation

$$\sum_i m_i \langle \boldsymbol{j}_d^i(\boldsymbol{x})\rangle = \sum_i \left(m_i \langle \boldsymbol{j}_i(\boldsymbol{x})\rangle - \frac{m_i \langle n_i\rangle}{\langle\rho\rangle} \langle \boldsymbol{p}(\boldsymbol{x})\rangle \right) = 0, \tag{22.47}$$

since $\langle \boldsymbol{p}(\boldsymbol{x})\rangle$ is the total momentum. Taking (22.46d) into account and eliminating the l-th diffusional flux by means of (22.47), we obtain for the entropy production the expression

$$\sigma = \langle \boldsymbol{j}_Q(\boldsymbol{x})\rangle \cdot \nabla\beta - \sum_{i=1}^{l-1} \langle \boldsymbol{j}_d^i(\boldsymbol{x})\rangle \cdot m_i \nabla \left(\frac{\nu_i}{m_i} - \frac{\nu_l}{m_l} \right) - \beta \langle\overset{\circ}{\pi}\rangle : (\nabla\overset{\circ}{\boldsymbol{v}})^{s} - \beta \langle\Pi\rangle \operatorname{div} \boldsymbol{v}. \tag{22.48}$$

According to the nature of the fluxes and the thermodynamic forces, irreversible processes can be divided into three groups: *vector processes* [the first two terms in (22.48)], associated with the transport of energy and matter, a *tensor process* [the third term in (22.48)], representing the shear viscosity, and a *scalar process* [the fourth term in (22.48)], describing the bulk viscosity.

For an isotropic medium, the linear relations between the forces and fluxes can be simplified if we take into account that, according to Curie's theorem, fluxes and thermodynamic forces of different tensor dimensions cannot be connected with each other (cf. the proof of this theorem in [27]).

Consequently, in this case, the linear relations can be written separately for vector processes

$$\langle \boldsymbol{j}_Q\rangle = -L_{00} \cdot \frac{\operatorname{grad} T}{T^2} - \sum_{i=1}^{l} L_{0i} \cdot \operatorname{grad}\left(\frac{\mu_i}{T}\right),$$

$$\langle j_d^i \rangle = -L_{i0} \cdot \frac{\operatorname{grad} T}{T^2} - \sum_{j=1}^{l} L_{ij} \cdot \operatorname{grad}\left(\frac{\mu_j}{T}\right), \tag{22.49}$$

tensor processes

$$\langle \overset{\circ}{\pi} \rangle = -\frac{L_{11}^{(1)}}{T} \cdot (\nabla^\circ v)^s \tag{22.49a}$$

and scalar processes

$$\langle \Pi \rangle = \frac{1}{3} \sum_{\alpha=1}^{3} \langle \pi_{\alpha\alpha} \rangle = -\frac{L_{11}^{(2)}}{T} \operatorname{div} \boldsymbol{v}, \tag{22.49b}$$

where the kinetic coefficients are equal to

$$\begin{aligned}
L_{00} &= \int \int_{-\infty}^{0} e^{\varepsilon t} \left(j_Q(\boldsymbol{x}),\ j_Q(\boldsymbol{x}', t)\right) d\boldsymbol{x}'\, dt, \\
L_{0i} &= \int \int_{-\infty}^{0} e^{\varepsilon t} \left(j_Q(\boldsymbol{x}),\ j_d^i(\boldsymbol{x}', t)\right) d\boldsymbol{x}'\, dt, \\
L_{i0} &= \int \int_{-\infty}^{0} e^{\varepsilon t} \left(j_d^i(\boldsymbol{x}),\ j_Q(\boldsymbol{x}', t)\right) d\boldsymbol{x}'\, dt, \\
L_{ij} &= \int \int_{-\infty}^{0} e^{\varepsilon t} \left(j_d^i(\boldsymbol{x}),\ j_d^j(\boldsymbol{x}', t)\right) d\boldsymbol{x}'\, dt, \\
L_{11}^{(1)} &= \int \int_{-\infty}^{0} e^{\varepsilon t} \left(\overset{\circ}{T}(\boldsymbol{x}),\ \overset{\circ}{T}(\boldsymbol{x}', t)\right) d\boldsymbol{x}'\, dt,
\end{aligned} \tag{22.50}$$

$$L_{11}^{(2)} = \int \int_{-\infty}^{0} e^{\varepsilon t} \Big(p(\boldsymbol{x}),\ p(\boldsymbol{x}', t) - \left(\frac{\partial p}{\partial u}\right)_n H(\boldsymbol{x}', t) - \sum_i \left(\frac{\partial p}{\partial n_i}\right)_u n_i(\boldsymbol{x}', t)\Big) d\boldsymbol{x}'\, dt. \tag{22.50a}$$

Here, $j_Q(\boldsymbol{x})$ is the heat flux density, determined by formula (22.22), $j_d^i(\boldsymbol{x})$ is the diffusion flux density (22.22), and $\overset{\circ}{T}(\boldsymbol{x})$ is the divergenceless part of the stress tensor (19.46b),

$$T(\boldsymbol{x}) = \overset{\circ}{T}(\boldsymbol{x}) + U p(\boldsymbol{x}), \tag{22.51}$$

where p(**x**) is the pressure operator

$$p(\boldsymbol{x}) = \frac{1}{3} \sum_{\alpha} T_{\alpha\alpha}(\boldsymbol{x}), \tag{22.51a}$$

and U is the unit tensor.

In the formulas (22.50), the averaging is performed with the local-equilibrium distribution (22.21). The brackets denote the quantum correlation functions (22.19a). In the case of classical mechanics, the flux density operators must be replaced by the corresponding dynamic variables (cf. § 19.1), and the quantum correlation functions by classical correlation functions.

As a consequence of the Onsager reciprocity relations (22.25), we have for the kinetic coefficients (22.23b) in the absence of a magnetic field the symmetry relations

$$L_{01} = L_{10}, \qquad L_{ij} = L_{ji}, \tag{22.52}$$

or, if there is a magnetic field H,

$$L_{01}(\boldsymbol{H}) = L_{10}(-\boldsymbol{H}), \qquad L_{ij}(\boldsymbol{H}) = L_{ji}(-\boldsymbol{H}). \tag{22.52a}$$

For the case under consideration of an isotropic medium, the relations (22.49)-(22.49b) and the expressions (22.50) can be still further simplified. In this case, correlation functions constructed from vectors or tensors have the form of scalars multiplied by unit tensors:

$$\begin{gathered} L_{00}^{\mu\nu} = L_0\delta_{\mu\nu}, \qquad L_{0i}^{\mu\nu} = L_{i0}^{\mu\nu} = L_i\delta_{\mu\nu}, \\ L_{11}^{(1)\,\mu\nu\mu_1\nu_1} = L_1^{(1)}\frac{1}{2}\left\{\delta_{\mu\mu_1}\delta_{\nu\nu_1} + \delta_{\mu\nu_1}\delta_{\nu\mu_1} - \frac{2}{3}\delta_{\mu\nu}\delta_{\mu_1\nu_1}\right\}. \end{gathered} \tag{22.52b}$$

We have added the last term in the right-hand side of the last expression in order to satisfy the properties

$$\sum_{\mu} L^{\mu\mu\mu_1\nu_1} = \sum_{\mu_1} L^{\mu\nu\mu_1\mu_1} = 0,$$

since the correlator is constructed from tensors with zero trace.

We shall find the scalar functions L_0, L_i, and L_1 by calculating the trace (i.e., the contraction) of the tensors in the left- and right-hand sides of the relations (22.52b):

$$\begin{aligned} L_0 &= \frac{1}{3}\operatorname{Tr} L_{00} = L_{00}^{xx}, \\ L_i &= \frac{1}{3}\operatorname{Tr} L_{0i} = \frac{1}{3}\operatorname{Tr} L_{i0} = L_{i0}^{xx}, \\ L_1^{(1)} &= \frac{1}{5}\operatorname{Tr} L_{11}^{(1)} = 2L_{11}^{xyxy} \qquad (x \neq y). \end{aligned} \tag{22.52c}$$

In fact,

$$\frac{1}{2}\sum_{\mu\nu}\left\{\delta_{\mu\nu}\delta_{\nu\mu} + \delta_{\mu\mu}\delta_{\nu\nu} - \frac{2}{3}\delta_{\mu\nu}\delta_{\mu\nu}\right\} = 5.$$

Substituting (22.52b) into (22.49)-(22.49b), we obtain

$$\langle j_Q \rangle = -L_0\frac{\operatorname{grad} T}{T^2} - \sum_{i=1}^{l} L_i \operatorname{grad}\left(\frac{\mu_i}{T}\right),$$

$$\langle j_d^i \rangle = -L_i \frac{\operatorname{grad} T}{T^2} - \sum_{j=1}^{l} L_{ij} \operatorname{grad}\left(\frac{\mu_j}{T}\right),$$

$$\langle \mathring{\pi}_{\alpha\beta} \rangle = -\frac{L_1^{(1)}}{2T}\left\{\frac{\partial v_\alpha}{\partial x_\beta} + \frac{\partial v_\beta}{\partial x_\alpha} - \frac{2}{3}\delta_{\alpha\beta} \operatorname{div} \boldsymbol{v}\right\}, \qquad (22.53)$$

$$\langle \Pi \rangle = -\frac{L_{11}^{(2)}}{T} \operatorname{div} \boldsymbol{v},$$

where the kinetic coefficients have a scalar character and are equal to

$$L_0 = \frac{1}{3} \int \int_{-\infty}^{0} e^{\varepsilon t} (\boldsymbol{j}_Q(\boldsymbol{x}) \cdot \boldsymbol{j}_Q(\boldsymbol{x}', t))\, d\boldsymbol{x}'\, dt,$$

$$L_{0i} = \frac{1}{3} \int \int_{-\infty}^{0} e^{\varepsilon t} (\boldsymbol{j}_Q(\boldsymbol{x}) \cdot \boldsymbol{j}_d^i(\boldsymbol{x}', t))\, d\boldsymbol{x}'\, dt,$$

$$L_{ij} = \frac{1}{3} \int \int_{-\infty}^{0} e^{\varepsilon t} (\boldsymbol{j}_d^i(\boldsymbol{x}) \cdot \boldsymbol{j}_d^j(\boldsymbol{x}', t))\, d\boldsymbol{x}'\, dt, \qquad (22.53a)$$

$$L_1^{(1)} = \frac{1}{5} \int \int_{-\infty}^{0} e^{\varepsilon t} (\mathring{T}(\boldsymbol{x}) : \mathring{T}(\boldsymbol{x}', t))\, d\boldsymbol{x}'\, dt;$$

the dot denotes the scalar product of vectors, and two dots denote the complete contraction of tensors.

Rewriting (22.53) in another, more usual notation, we obtain

$$\langle j_Q \rangle = -\lambda \nabla T - \sum_i L_i \operatorname{grad}\left(\frac{\mu_i}{T}\right),$$

$$\langle j_d^i \rangle = -L_i \frac{\operatorname{grad} T}{T^2} - \sum_{j=1}^{l} L_{ij} \operatorname{grad}\left(\frac{\mu_j}{T}\right),$$

$$\langle \mathring{\pi}_{\alpha\beta} \rangle = -\eta\left\{\frac{\partial v_\alpha}{\partial x_\beta} + \frac{\partial v_\beta}{\partial x_\alpha} - \frac{2}{3} \operatorname{div} \boldsymbol{v}\, \delta_{\alpha\beta}\right\}, \qquad (22.54)$$

$$\langle \Pi \rangle = \langle p \rangle - \langle p \rangle_l = -\zeta \operatorname{div} \boldsymbol{v},$$

where

$$\lambda = \frac{L_0}{T^2} = \frac{1}{3T^2} \int\int_{-\infty}^{0} e^{\varepsilon t} (\boldsymbol{j}_Q(\boldsymbol{x}) \cdot \boldsymbol{j}_Q(\boldsymbol{x}', t))\, d\boldsymbol{x}' dt = \frac{1}{T^2} \int\int_{-\infty}^{0} e^{\varepsilon t} (j_Q^x(\boldsymbol{x}),\ j_Q^x(\boldsymbol{x}', t))\, d\boldsymbol{x}'\, d \qquad (22.54a)$$

is the thermal conductivity coefficient,

$$\eta = \frac{L_1^{(1)}}{2T} = \frac{1}{10T} \int \int_{-\infty}^{0} e^{\varepsilon t} (\mathring{T}(\boldsymbol{x}) : \mathring{T}(\boldsymbol{x}', t))\, d\boldsymbol{x}'\, dt$$

$$= \frac{1}{T} \int \int_{-\infty}^{0} e^{\varepsilon t} (T_{xy}(\boldsymbol{x}),\ T_{xy}(\boldsymbol{x}', t))\, d\boldsymbol{x}'\, dt \qquad (22.54b)$$

is the shear viscosity coefficient, and

$$\zeta = \frac{L_{11}^{(2)}}{T} = \frac{1}{T} \int \int_{-\infty}^{0} e^{\varepsilon t} \left(p(\boldsymbol{x}),\ p(\boldsymbol{x}', t) - \left(\frac{\partial p}{\partial u}\right)_n H(\boldsymbol{x}', t) - \right.$$
$$\left. - \sum_i \left(\frac{\partial p}{\partial n_i}\right)_u n_i(\boldsymbol{x}', t) \right) d\boldsymbol{x}'\, dt \qquad (22.54c)$$

is the bulk viscosity coefficient.

The kinetic coefficients λ, η, and ζ are positive.

Noting that, on the basis of (22.52b), (22.52c), (22.54), (22.54b), and (22.54c), the tensor $L_{11} = L_{11}^{(1)} + L_{11}^{(2)} U$ is equal to

$$L_{11}^{\mu\nu\mu_1\nu_1} = T\eta \left\{ \delta_{\mu\mu_1}\delta_{\nu\nu_1} + \delta_{\mu\nu_1}\delta_{\nu\mu_1} - \frac{2}{3}\delta_{\mu\nu}\delta_{\mu_1\nu_1} \right\} + T\zeta\delta_{\mu\nu}\delta_{\mu_1\nu_1}, \qquad (22.54d)$$

and putting $\mu = \nu = \mu_1 = \nu_1$, we obtain another expression for the bulk viscosity:

$$\zeta + \frac{4}{3}\eta = \frac{1}{T} L_{11}^{xxxx} = \frac{1}{T} \int \int_{-\infty}^{0} e^{\varepsilon t} (T_{xx}(\boldsymbol{x}),\ T_{xx}(\boldsymbol{x}', t) -$$
$$- \left(\frac{\partial p}{\partial u}\right)_n H(\boldsymbol{x}', t) - \sum_i \left(\frac{\partial p}{\partial n_i}\right)_u n_i(\boldsymbol{x}', t))\, d\boldsymbol{x}'\, dt. \qquad (22.54e)$$

The coefficients L_i describe the transport of matter due to the temperature gradient, i.e., thermal diffusion (or the Soret effect), and the transport of heat due to a concentration gradient, i.e., the Dufour effect. Such processes are called *cross-processes*. They will be considered in more detail in §22.7 with the example of a binary mixture. The coefficients L_{ij} describe the transport of matter due to a concentration gradient, i.e., ordinary diffusion (cf. §22.7).

The expressions (22.50), (22.53a), and (22.54a–d) for the kinetic coefficients were first obtained by Green [14] by a method from the theory of stochastic processes for the classical case, on the basis of the microcanonical ensemble. In this case, we can omit the terms with $(\partial p/\partial u)_n$ and $(\partial p/\partial n_i)_u$ in the expression (22.54c) for the bulk viscosity. In fact, all the kinetic coefficients can be expressed in terms of the total fluxes J^m:

$$L_{mn} = \frac{1}{V} \int \int \int_{-\infty}^{0} e^{\varepsilon t} (j^m(\boldsymbol{x}),\ j^n(\boldsymbol{x}', t))\, d\boldsymbol{x}\, d\boldsymbol{x}'\, dt = \frac{1}{V} \int_{-\infty}^{0} e^{\varepsilon t} (J^m,\ J^n(t))\, dt, \qquad (22.5$$

where

$$J^m = \int j^m(\boldsymbol{x})\,d\boldsymbol{x}.$$

In (22.54e), we have taken into account that the average correlator depends only on $\mathbf{x} - \mathbf{x}'$ and one can introduce a further integration over $\mathbf{x}$, with the factor $1/V$. The formula (22.54c) for the bulk viscosity then takes the form

$$\zeta = \frac{V}{T}\int e^{\varepsilon t}\left(p,\ p(t) - \left(\frac{\partial p}{\partial u}\right)_n \frac{H}{V} - \sum_i \left(\frac{\partial p}{\partial n_i}\right)_u \frac{N_i}{V}\right) dt, \qquad (22.54g)$$

where

$$p = \frac{1}{V}\int p(\boldsymbol{x})\,d\boldsymbol{x}$$

is the average pressure operator, and

$$H = \int H(\boldsymbol{x})\,d\boldsymbol{x}, \qquad N_i = \int n_i(\boldsymbol{x})\,d\boldsymbol{x}$$

are the total Hamiltonian and total number of particles of type i. If the averaging is performed over the microcanonical ensemble, as in Green's paper [14], then H and N_i do not experience fluctuations, and, consequently, in the correlator (22.54g), the terms with $(\partial p/\partial u)_n$ and $(\partial p/\partial n_i)_u$ can be omitted. Then

$$\zeta = \frac{V}{T}\int_{-\infty}^{0} e^{\varepsilon t}(p,\ p(t))\,dt. \qquad (22.54h)$$

Such an expression was obtained by Green. However, the microcanonical ensemble is inconvenient for calculations, since in any case one has to take into account the supplementary conditions that H and N_i be constant; therefore, formula (22.54g) is more effective.

The expressions (22.50), (22.53a), and (22.54a)-(22.54c) were obtained for the quantum case by Mori [29-31], who integrated Liouville's equation with the initial condition in the form of the local-equilibrium grand canonical distribution (21.14). His expressions for the kinetic coefficients contained integrals in the sense of Cesaro (22.27) rather than in the sense of Abel (22.23b) (this difference was discussed in §22.3). In addition, in the expression

(22.54g) for the bulk viscosity, he did not include the terms with $(\partial p/\partial u)_n$ and $(\partial p/\partial n_i)_u$. This inaccuracy was corrected later by Mori himself [95], using the projection-operator method and collective variables. This result has also been confirmed by other authors [38, 9, 10, 62].

Since the work of Green and Mori, formulas for the kinetic coefficients in the form of correlation functions of the fluxes have been obtained by many other authors by means of the different methods, discussed at the beginning of this chapter, for taking thermal perturbations into account, or by means of combinations of these methods.

The formula (22.54b) for the shear viscosity was obtained by an indirect linear-response method by Montroll [8], by treating the viscous flow created by a change of the vessel dimensions. This idea was expressed earlier by Feynman in an unpublished lectures.

All the formulas for the kinetic coefficients have been obtained by Kadanoff and Martin [10], and also by Luttinger [9], by means of different variants of the indirect linear-response method.

Kadanoff and Martin considered a state of a liquid, weakly nonequilibrium in the temperature, chemical potential, and velocity, and introduced fictitious mechanical perturbations (the difference between the exponents of the local-equilibrium and equilibrium distributions) bringing the system to the same state. This perturbation is switched off instantaneously at $t = 0$, and the system then develops in accordance with the equations of hydrodynamics.

The average values of the mechanical quantities (the densities of the energy, momentum, and particle number) at $t \leq 0$ can be expressed in terms of the perturbation and the susceptibility, the latter being expressed in terms of correlation functions or Green functions. From a comparison of these expressions with the solutions of the hydrodynamic equations, we can express the susceptibility in terms of the kinetic coefficients. Inversion of these relations gives formulas for the kinetic coefficients in terms of correlation functions.

The method of Luttinger [9] is very close to that of Kadanoff and Martin, although, compared with the latter authors, Luttinger aims at a closer analogy between the auxiliary mechanical perturbations and the real fields. The perturbation of the local temperature is associated with a gravitational field giving rise to the same nonuniformity in the energy density, the perturbation of the chemical potential is associated with an electric potential field, and the perturbation of the velocity is associated with a magnetic field described by a vector potential. These fields can create the same distributions of energy, mass, and momentum in an equilibrium state as those in the given nonequilibrium state.

Other papers in which formulas are derived for kinetic coefficients in terms of correlation functions in the case of thermal perturbations have been mentioned already at the beginning of this chapter, and we refer the reader to the literature cited there.

The formulas for the kinetic coefficients in terms of correlation functions in the case of gases of sufficiently low density, when the Boltzmann kinetic equation is applicable, lead to the same expressions for the kinetic coefficients as those given by the Chapman–Enskog theory [96] in the linear approximation in the gradients of the thermodynamic parameters [32, 97-99, 240]. This result is understandable, since the normal solution of the Boltzmann equation (studied in the Chapman–Enskog theory) is based on assumptions similar to those which were used in constructing the nonequilibrium distribution, namely, that for a time sufficiently long compared with the time between collisions, the distribution function begins to depend on time only through its thermodynamic parameters.†

In obtaining the Chapman–Enskog formulas from (22.53a), we can omit the terms with the interaction potential in the expressions (22.22), (19.12), (19.17a), (19.27a), and (19.29a) for the fluxes. For liquids, these terms are essential. For monatomic ideal gases, according to the Chapman–Enskog theory, the bulk viscosity coefficient is equal to zero, $\zeta = 0$.

In higher approximations in powers of the density, formulas (22.50) leads to the same results as the theory of Bogolyubov [1, 100], which is based on solving a chain of equations for the distribution functions by expanding in powers of the density. For the higher approximations, three-body collisions and collisions of

†Such solutions are called normal solutions. As shown in a paper [241] by the author and Khon'kin, they can be obtained if an appropriate boundary condition for the Boltzmann equation is formulated.

higher order must be taken into account, and this is a very complicated problem. For the connection between the calculation of the kinetic coefficients and the solution of the generalized kinetic equation of Bogolyubov, see in [33]. For the theory of transport processes, see also [216-226, 251].

22.6. Transport Processes in a One-Component Fluid. The Thermal-Conduction Equation and the Navier–Stokes Equation

In a one-component liquid (or gas), the diffusion flux density operator $\mathbf{j}_d(\mathbf{x})$ (22.13a) vanishes identically, since

$$\boldsymbol{j}(\boldsymbol{x}) - \frac{\langle n \rangle}{m \langle n \rangle} \boldsymbol{p}(\boldsymbol{x}) = 0; \tag{22.55}$$

consequently, diffusion and thermal diffusion processes and the Dufour effect are absent, since

$$L_i = L_{ij} = 0. \tag{22.56}$$

In this case, there remain only the processes of thermal conduction and viscosity, which we shall consider below. (If the one-component fluid is a mixture of isotopes, then equalization of the isotopic composition, i.e., a self-diffusion process, can occur in it.)

According to (20.68), in a local-equilibrium state the densities of the fluxes of energy, momentum, and particle number are equal to

$$\begin{gathered}\langle \boldsymbol{j}_H \rangle_l = \left(u + p + \rho \frac{v^2}{2}\right) \boldsymbol{v},\\ \langle T_{\alpha\beta} \rangle_l = p\delta_{\alpha\beta} + \rho v_\alpha v_\beta,\\ \langle \boldsymbol{j} \rangle_l = \rho \boldsymbol{v}, \quad \rho = \langle \rho(\boldsymbol{x}) \rangle_l, \quad u = \langle H'(\boldsymbol{x}) \rangle_l,\end{gathered} \tag{22.57}$$

while the corresponding entropy, entropy flux, to (20.74), has the form

$$\boldsymbol{j}_S = S(\boldsymbol{x}) \boldsymbol{v}, \tag{22.57a}$$

where $S(\mathbf{x})$ is the entropy density (22.37a).

To study the thermal conduction process, we can consider the transport equation for either the energy or the entropy, the entropy transport equation in a local-equilibrium state having the very simple form

$$\frac{\partial S(\boldsymbol{x})}{\partial t} = -\operatorname{div} \boldsymbol{j}_S = -\operatorname{div}(S(\boldsymbol{x})\,\boldsymbol{v}) \tag{22.58}$$

[cf. (22.38) and (22.38a)].

The linear relations (22.54) between the thermodynamic forces and fluxes in a one-component fluid take the form

$$\begin{aligned} \langle \boldsymbol{j}_Q\rangle &= -\lambda \nabla T, \\ \langle \mathring{\pi}_{\alpha\beta}\rangle &= -\eta\left\{\frac{\partial v_\alpha}{\partial x_\beta} + \frac{\partial v_\beta}{\partial x_\alpha} - \frac{2}{3}\delta_{\alpha\beta}\operatorname{div}\boldsymbol{v}\right\}, \\ \langle \Pi\rangle &= \langle p\rangle - \langle p\rangle_l = -\zeta \operatorname{div}\boldsymbol{v}, \end{aligned} \tag{22.59}$$

where $\langle \mathring{\pi}\rangle$ and $\langle \Pi\rangle$ are the divergenceless and divergence parts of the viscous stress tensor,

$$\langle T_{\alpha\beta}\rangle - \langle T_{\alpha\beta}\rangle_l = \langle \mathring{\pi}_{\alpha\beta}\rangle + \langle \Pi\rangle\,\delta_{\alpha\beta}. \tag{22.59a}$$

The first equation of the system (22.59) is sometimes called Fick's first law.

The conservation laws for the mass, energy, and momentum and the entropy balance equation have the form

$$\begin{aligned} &\frac{\partial \rho}{\partial t} + \operatorname{div}\rho\boldsymbol{v} = 0, \\ &\frac{\partial \langle H(\boldsymbol{x})\rangle}{\partial t} + \operatorname{div}\langle \boldsymbol{j}_H\rangle = 0, \\ &\frac{\partial \langle p_\alpha\rangle}{\partial t} + \sum_\beta \frac{\partial \langle T_{\beta\alpha}\rangle}{\partial x_\beta} = 0, \\ &\frac{\partial S(\boldsymbol{x})}{\partial t} + \operatorname{div}\boldsymbol{j}_S = \sigma, \end{aligned} \tag{22.60}$$

where

$$\sigma = \langle \boldsymbol{j}_Q\rangle \cdot \nabla\beta - \beta\langle \mathring{\pi}\rangle : (\mathring{\nabla}\boldsymbol{v})^s - \beta\langle \Pi\rangle \operatorname{div}\boldsymbol{v} \tag{22.61}$$

is the entropy production, and

$$\boldsymbol{j}_S = S(\boldsymbol{x})\,\boldsymbol{v} + \beta\langle \boldsymbol{j}_Q\rangle \tag{22.62}$$

is the entropy flux, since the other terms in the formulas (22.48) and (22.40) go to zero.

Taking the linear relations (22.59) into account, we write the entropy production (22.61) and the entropy flux (22.62) in the form

$$\sigma = \frac{\lambda}{T^2}(\nabla T)^2 + \frac{2\eta}{T}(\nabla \overset{\circ}{v})^s : (\nabla \overset{\circ}{v})^s + \frac{\zeta}{T}(\operatorname{div} v)^2, \tag{22.62a}$$

$$j_S = S(x)\, v - \frac{\lambda}{T}\nabla T, \tag{22.62b}$$

and the balance equations for the energy, momentum, and entropy in the form

$$\frac{\partial}{\partial t}\left(u + \rho\frac{v^2}{2}\right) + \operatorname{div}\left(u + p + \frac{\rho v^2}{2}\right)v = \nabla(\lambda \nabla T)$$
$$+ 2\nabla \cdot \left(\eta\, (\nabla \overset{\circ}{v})^s \cdot v\right) + \nabla \cdot \left(\zeta v \cdot \operatorname{div} v\right),$$

$$\frac{\partial \rho v_\alpha}{\partial t} + \sum_\beta \frac{\partial}{\partial x_\beta}\rho v_\alpha v_\beta + \frac{\partial p}{\partial x_\alpha} = \tag{22.63}$$
$$= \sum_\beta \frac{\partial}{\partial x_\beta}\eta\left(\frac{\partial v_\alpha}{\partial x_\beta} + \frac{\partial v_\beta}{\partial x_\alpha} - \frac{2}{3}\delta_{\alpha\beta}\operatorname{div} v\right) + \frac{\partial}{\partial x_\alpha}\zeta \operatorname{div} v,$$

$$\frac{\partial S(x)}{\partial t} + \operatorname{div}(S(x)\, v) = \nabla \cdot \left(\frac{\lambda}{T}\nabla T\right) + \frac{\lambda}{T^2}(\nabla T)^2 + \frac{2\eta}{T}(\nabla \overset{\circ}{v})^s : (\nabla \overset{\circ}{v})^s + \frac{\zeta}{T}(\operatorname{div} v)^2.$$

In Eqs. (22.63), the kinetic coefficients depend on the thermodynamic parameters, for example, on the pressure and temperature, and, consequently, for a spatially nonuniform state, may depend on position. However, this dependence is usually small and these kinetic coefficients can be regarded as constant and placed to the left of the gradient symbol. Then,

$$\frac{\partial}{\partial t}\left(u + \rho\frac{v^2}{2}\right) + \operatorname{div}\left(u + p + \rho\frac{v^2}{2}\right)v = \lambda\nabla^2 T + 2\eta\nabla \cdot \left((\nabla \overset{\circ}{v})^s \cdot v\right) + \zeta\nabla \cdot \left(v \operatorname{div} v\right),$$

$$\rho\left(\frac{\partial v}{\partial t} + v \cdot \nabla v\right) + \nabla p = \eta\nabla^2 v + \left(\zeta + \frac{1}{3}\eta\right)\nabla \operatorname{div} v, \tag{22.63a}$$

$$\rho\left(\frac{\partial s}{\partial t} + v \cdot \nabla s\right) = \frac{\lambda}{T}\nabla^2 T + \frac{2\eta}{T}(\nabla \overset{\circ}{v})^s : (\nabla \overset{\circ}{v})^s + \frac{\zeta}{T}(\operatorname{div} v)^2,$$

where

$$s(x) = S(x)/\rho$$

is the entropy per unit mass.

The second equation of the system (22.63a) is the Navier–Stokes equation, and the first (or third) is the heat transport equation. Because of (22.37a), the third equation follows from the first two and from the law of conservation of mass.

If the flow velocity of the fluid is much less than the velocity of sound, then the change in the pressure as a result of the motion is very small and we can neglect the change of the density and of other thermodynamic quantities to which it gives rise. Consequently, in calculating derivatives of the thermodynamic quantities we can regard the pressure as constant and

$$\frac{\partial s}{\partial t} = \left(\frac{\partial s}{\partial T}\right)_p \frac{\partial T}{\partial t}, \qquad \nabla s = \left(\frac{\partial s}{\partial T}\right)_p \nabla T. \tag{22.64}$$

Taking into account that

$$T\left(\frac{\partial s}{\partial T}\right)_p = C_p \tag{22.64a}$$

is the specific heat at constant pressure, we obtain

$$\frac{\partial s}{\partial t} = \frac{C_p}{T}\frac{\partial T}{\partial t}, \qquad \nabla s = \frac{C_p}{T}\nabla T. \tag{22.64b}$$

Consequently, the heat transport equation for incompressible motion of the fluid, when div $\mathbf{v} = 0$, takes the form

$$\frac{\partial T}{\partial t} + v \cdot \nabla T = \chi \nabla^2 T + \frac{2\nu}{C_p} (\nabla \overset{\circ}{v})^s : (\nabla \overset{\circ}{v})^s, \tag{22.65}$$

where

$$\chi = \lambda/\rho C_p \tag{22.65a}$$

is the thermal diffusivity coefficient, and

$$\nu = \eta/\rho \tag{22.65b}$$

is the kinematic viscosity. In a stationary fluid, when the heat transport is due exclusively to the thermal conduction mechanism, Eq. (22.65) takes the form

$$\frac{\partial T}{\partial t} = \chi \nabla^2 T \tag{22.65c}$$

and is called the *thermal conduction equation* or *Fourier's equation.*

22.7. Transport Processes in a Binary Mixture. Thermal Conduction, Diffusion, and Cross Effects

We shall consider transport processes in a two-component liquid (or gas) in the absence of velocity gradients. In this case, only vector processes occur (thermal conduction, diffusion, and the cross effects, i.e., thermal diffusion and Dufour effect), and the linear relations (22.54) between the thermodynamic forces and fluxes take the form

$$\langle j_Q \rangle = -\frac{L_0}{T^2}\nabla T - \sum_{i=1}^{2} L_i \nabla\left(\frac{\mu_i}{T}\right),$$
$$\langle j_d^i \rangle = -L_i \frac{1}{T^2}\nabla T - \sum_{j=1}^{2} L_{ij} \nabla\left(\frac{\mu_j}{T}\right) \qquad (i=1,\ 2). \tag{22.66}$$

The diffusion fluxes $\langle j_d^1 \rangle$, and $\langle j_d^2 \rangle$ are connected by the relation

$$\sum_i m_i L_i = 0, \qquad \sum_i m_i L_{ij} = 0. \tag{22.67}$$

[for any ∇T and $\nabla(\mu_i/T)$] which follows from the last relation in (22.22); consequently, the kinetic coefficients L_i and L_{ij} satisfy the relations

$$\sum_{i=1}^{2} m_i \langle j_d^i \rangle = 0 \tag{22.68}$$

In the linear relations (22.66), it is sufficient to consider only one diffusional flux, for example, $\langle j_d^1 \rangle = \langle j_d \rangle$, since the second flux can be found from Eq. (22.67). Taking (22.68) into account, we write (22.66) in the form

$$\langle j_Q \rangle = -\frac{L_0}{T^2}\nabla T - L_1 m_1 \nabla\left(\frac{\mu}{T}\right),$$
$$\langle j_d \rangle = -\frac{L_1}{T^2}\nabla T - L_{11} m_1 \nabla\left(\frac{\mu}{T}\right), \tag{22.66a}$$

where we have introduced the chemical potential [82]

$$\mu = \frac{\mu_1}{m_1} - \frac{\mu_2}{m_2}. \tag{22.66b}$$

The chemical potential (22.66b) occurs in the thermodynamic equality (20.42b). In fact,

$$ds = \beta\, du(x) + \beta p\, dv(x) - \beta \sum_i \frac{\mu_i}{m_i}\, dC_i$$
$$= \beta\, du(x) + \beta p\, dv(x) - \beta\left(\frac{\mu_1}{m_1} - \frac{\mu_2}{m_2}\right) dC, \tag{22.69}$$

where $C_1 + C_2 = 1$, $C = C_1 = m_1 n_1/\rho$ is the mass concentration, $u(x) = \langle H'(x)\rangle/\rho$ is the energy density per unit mass, and $v(x) = 1/\rho$ is the specific volume per unit mass.

It follows from (22.69) that

$$\mu = -T\left(\frac{\partial s}{\partial C}\right)_{u,\,v}. \tag{22.69a}$$

We go over in the thermodynamic functions to the variables p, T, and C. Then,

$$\nabla\mu = \left(\frac{\partial\mu}{\partial C}\right)_{p,\,T}\nabla C + \left(\frac{\partial\mu}{\partial T}\right)_{C,\,p}\nabla T + \left(\frac{\partial\mu}{\partial p}\right)_{C,\,T}\nabla p. \tag{22.70}$$

Substituting (22.70) into (22.66a), we obtain

$$\begin{aligned}\langle j_Q\rangle &= -\lambda\nabla T + \frac{L_1}{L_{11}}\langle j_d\rangle,\\ \langle j_d\rangle &= -\frac{1}{m_1}\rho D\left(\nabla C + \frac{K_T}{T}\nabla T + \frac{K_p}{p}\nabla p\right),\end{aligned} \tag{22.71}$$

where

$$\frac{L_1}{L_{11}} = m_1\left(K_T\left(\frac{\partial\mu}{\partial C}\right)_{p,\,T} - T^2\frac{\partial}{\partial T}\left(\frac{\mu}{T}\right)_{C,\,p}\right), \tag{22.71a}$$

$$\lambda = \frac{L_0}{T^2} - \frac{L_1^2}{L_{11}T^2} \tag{22.72}$$

is the thermal conductivity coefficient in a binary mixture,

$$D = \frac{L_{11}m_1^2}{\rho T}\left(\frac{\partial\mu}{\partial C}\right)_{p,T} \tag{22.73}$$

is the diffusion coefficient,

$$\frac{\rho K_T D}{T m_1} = \frac{L_1}{T^2} + L_{11}m_1\frac{\partial}{\partial T}\left(\frac{\mu}{T}\right)_{C,\,p}, \tag{22.74}$$

$K_T D$ is the thermal diffusion coefficient, K_T is the thermal diffusion ratio,

$$K_p = p\left(\frac{\partial\mu}{\partial p}\right)_{C,T}\Big/\left(\frac{\partial\mu}{\partial C}\right)_{p,T}, \tag{22.75}$$

and $K_p D$ is the pressure-diffusion coefficient.

It follows from the second equation of the system (22.71) that the diffusional flux is induced by gradients of the concentration, temperature, and pressure (ordinary diffusion, thermal diffusion, and pressure diffusion). The latter process is important if strong pressure gradients are created, for example, in a centrifuge.

The thermal diffusion coefficient is proportional to the product of $C_1 = C$ and $C_2 = 1 - C$, unlike the diffusion coefficient, which does not depend on the concentrations in the first approximation. We therefore introduce the thermal diffusion constant

$$\alpha = \frac{K_T}{C(1-C)}. \tag{22.76}$$

The diffusion and thermal conduction in our case, when $\mathbf{v} = 0$, are determined by the equations

$$\begin{aligned} \frac{\partial \langle n_1 \rangle}{\partial t} + \operatorname{div} \langle \boldsymbol{j}_d \rangle = 0, \\ \frac{\partial S(\boldsymbol{x})}{\partial t} + \operatorname{div} \boldsymbol{j}_S = \sigma, \end{aligned} \tag{22.77}$$

where, according to (22.40) and (22.39),

$$\boldsymbol{j}_S = \langle \boldsymbol{j}_Q \rangle \beta - \sum_i \langle \boldsymbol{j}_d^i \rangle \nu_i = \beta (\boldsymbol{j}_Q - \boldsymbol{j}_d m_1 \mu), \tag{22.77a}$$

$$\sigma = \langle \boldsymbol{j}_Q \rangle \cdot \nabla \beta - \sum_i \langle \boldsymbol{j}_d^i \rangle \cdot \nabla \nu_i. \tag{22.77b}$$

Substituting (22.71) and (22.77a) into (22.77) and neglecting terms with a pressure gradient and terms of higher order in the gradients, we obtain the system of equations

$$\begin{aligned} \frac{\partial C}{\partial t} &= D \left(\nabla^2 C + \frac{K_T}{T} \nabla^2 T \right), \\ \frac{\partial T}{\partial t} &= \chi \nabla^2 T + \frac{K_T}{C_p} \left(\frac{\partial \mu}{\partial C} \right)_{p,T} \frac{\partial C}{\partial t}, \end{aligned} \tag{22.78}$$

which determines the distribution of concentration and temperature in a binary mixture. In the particular case when the temperature is constant, we obtain

$$\frac{\partial C}{\partial t} = D \nabla^2 C, \tag{22.79}$$

i.e., the usual diffusion equation.

22.8. Another Choice of Thermodynamic Forces

In §22.1 we started from the nonequilibrium statistical operator in the form (22.1) and chose the thermodynamic forces in the form (22.13b) as gradients of the thermodynamic parameters. There are also other possible choices for the thermodynamic forces.

We shall start from a nonequilibrium statistical operator in the form (21.10a)

$$\rho = Q^{-1}\exp\Big\{-\sum_m \int F_m(\boldsymbol{x}, t)\, P_m(\boldsymbol{x})\, d\boldsymbol{x} + \sum_m \int \int_{-\infty}^{0} e^{\varepsilon t_1}\Big(F_m(\boldsymbol{x}, t+t_1)\, \dot{P}_m(\boldsymbol{x}, t_1) + \frac{\partial F_m(\boldsymbol{x}, t+t_1)}{\partial t} P_m(\boldsymbol{x}, t_1)\Big)\, d\boldsymbol{x}\, dt_1\Big\} \qquad (22.80)$$

and choose as the thermodynamic forces the Fourier components of the parameters $F_m(\mathbf{x}, t)$ with respect to the space variables.

The statistical operator in the form (22.80) is sometimes more convenient than (22.1), since it does not require a knowledge of explicit expressions for the fluxes $j_m(\mathbf{x}, t)$, for which one cannot make a completely unique choice. This is especially important for a system with long-range forces, when it is impossible to perform a smoothing of the operators over a small range of action of the forces, as in §19.

The thermodynamic parameters $F_m(\mathbf{x}, t)$ in the linear approximation in the velocities have the form

$$\begin{aligned} F_0(\boldsymbol{x}, t) &= \beta(\boldsymbol{x}, t), \\ F_1(\boldsymbol{x}, t) &= -\beta(\boldsymbol{x}, t)\, \boldsymbol{v}(\boldsymbol{x}, t) \equiv -\boldsymbol{u}(\boldsymbol{x}, t), \\ F_{i+1}(\boldsymbol{x}, t) &= -\beta(\boldsymbol{x}, t)\Big(\mu_i(\boldsymbol{x}, t) - \frac{v^2(\boldsymbol{x}, t)}{2}\Big) \cong \\ &\cong -\beta(\boldsymbol{x}, t)\, \mu_i(\boldsymbol{x}, t) \equiv -\nu_i(\boldsymbol{x}, t). \end{aligned} \qquad (22.81)$$

We shall expand the operators $P_m(\mathbf{x})$ and the parameters $F_m(\mathbf{x})$ in Fourier integrals. Then the statistical operator (22.80) takes the form

$$\rho = Q^{-1}\exp\Big\{-\sum_{m,\boldsymbol{k}} F_m(\boldsymbol{k}, t)\, P_m(-\boldsymbol{k}) + \sum_{m,\boldsymbol{k}} \int_{-\infty}^{0} e^{\varepsilon t_1}\Big(F_m(\boldsymbol{k}, t+t_1)\, \dot{P}_m(-\boldsymbol{k}, t_1) + \frac{\partial F_m(\boldsymbol{k}, t+t_1)}{\partial t} P_m(-\boldsymbol{k}, t_1)\Big)\, dt_1\Big\}. \qquad (22.82)$$

We eliminate the time derivatives of the parameters F_m by means of the relations (22.10), which in the linear approximation in the velocities have the form

$$\begin{aligned}
\frac{\partial \beta(\boldsymbol{x}, t)}{\partial t} &= \left(\frac{\partial p}{\partial u}\right)_n \operatorname{div} \boldsymbol{u}(\boldsymbol{x}, t), \\
\frac{\partial \nu_i(\boldsymbol{x}, t)}{\partial t} &= -\left(\frac{\partial p}{\partial n_i}\right)_u \operatorname{div} \boldsymbol{u}(\boldsymbol{x}, t), \\
\frac{\partial \boldsymbol{u}(\boldsymbol{x}, t)}{\partial t} &= \frac{u+p}{\langle \rho \rangle} \nabla \beta(\boldsymbol{x}, t) - \sum_i \frac{\langle n_i \rangle}{\langle \rho \rangle} \nabla \nu_i(\boldsymbol{x}, t),
\end{aligned} \tag{22.83}$$

or, in Fourier components,

$$\begin{aligned}
\frac{\partial \beta(\boldsymbol{k}, t)}{\partial t} &= \left(\frac{\partial p}{\partial u}\right)_n i\boldsymbol{k} \cdot \boldsymbol{u}(\boldsymbol{k}, t), \\
\frac{\partial \nu_i(\boldsymbol{k}, t)}{\partial t} &= -\left(\frac{\partial p}{\partial n_i}\right)_u i\boldsymbol{k} \cdot \boldsymbol{u}(\boldsymbol{k}, t), \\
\frac{\partial \boldsymbol{u}(\boldsymbol{k}, t)}{\partial t} &= \frac{u+p}{\langle \rho \rangle} i\boldsymbol{k}\beta(\boldsymbol{k}, t) - \sum_i \frac{\langle n_i \rangle}{\langle \rho \rangle} i\boldsymbol{k}\nu_i(\boldsymbol{k}, t).
\end{aligned} \tag{22.83a}$$

Substituting (22.83a) into (22.82) and using the relation

$$\dot{\rho}(-\boldsymbol{k}) = i\boldsymbol{k} \cdot \boldsymbol{p}(-\boldsymbol{k}), \tag{22.84}$$

we obtain

$$\rho = Q^{-1} \exp\left\{-\sum_{m, \boldsymbol{k}} F_m(\boldsymbol{k}, t) P_m(-\boldsymbol{k}) + \right.$$
$$\left. + \sum_{m, \boldsymbol{k}} \int_{-\infty}^{0} e^{\varepsilon t_1} J_m(-\boldsymbol{k}, t_1) X_m(\boldsymbol{k}, t+t_1)\, dt_1 \right\}, \tag{22.85}$$

where we have introduced the fluxes of energy, momentum, and particle number

$$\begin{aligned}
J_0(-\boldsymbol{k}) &= \dot{H}(-\boldsymbol{k}) - \frac{u+p}{\langle \rho \rangle} \dot{\rho}(-\boldsymbol{k}), \\
J_1(-\boldsymbol{k}) &= \dot{\boldsymbol{p}}(-\boldsymbol{k}) - \left(\frac{\partial p}{\partial u}\right)_n i\boldsymbol{k}H(-\boldsymbol{k}) - \sum_i \left(\frac{\partial p}{\partial n_i}\right)_u i\boldsymbol{k}n_i(-\boldsymbol{k}), \\
J_{i+1}(-\boldsymbol{k}) &= \dot{n}_i(-\boldsymbol{k}) - \frac{\langle n_i \rangle}{\langle \rho \rangle} \dot{\rho}(-\boldsymbol{k})
\end{aligned} \tag{22.85a}$$

and the thermodynamic forces

$$\begin{aligned} X_0(\boldsymbol{k}, t) &= \beta(\boldsymbol{k}, t) \\ X_1(\boldsymbol{k}, t) &= -\boldsymbol{u}(\boldsymbol{k}, t), \\ X_{i+1}(\boldsymbol{k}, t) &= -\nu_i(\boldsymbol{k}, t) \qquad (i \geqslant 1). \end{aligned} \tag{22.85b}$$

Expanding (22.85) in a series in powers of the thermodynamic forces, we obtain, analogously to (22.18),

$$\rho \cong \left\{ 1 - \int_0^1 (e^{-A\tau} B e^{A\tau} - \langle B \rangle_l) \, d\tau \right\} \rho_l, \tag{22.86}$$

where

$$B = -\sum_{m,\,k} \int_{-\infty}^{0} e^{\varepsilon t_1} J_m(-\boldsymbol{k}, t_1) X_m(\boldsymbol{k}, t + t_1) \, dt_1. \tag{22.86a}$$

Using (22.86), we obtain linear relations between the thermodynamic forces and fluxes:

$$\langle J_m(\boldsymbol{k}) \rangle = \langle J_m(\boldsymbol{k}) \rangle_l + \sum_{n,\,\boldsymbol{k}_1} \int_{-\infty}^{0} e^{\varepsilon t_1} (J_m(\boldsymbol{k}), J_n(-\boldsymbol{k}_1)) X_n(\boldsymbol{k}_1, t + t_1) \, dt_1,$$

where in the quantum time correlation function in the integrand, the averaging is performed over the state of statistical equilibrium and, consequently, because of the spatial isotropy, only the terms with $\mathbf{k}_1 = \mathbf{k}$ are nonzero:

$$\langle J_m(\boldsymbol{k}) \rangle = \langle J_m(\boldsymbol{k}) \rangle_l + \sum_{n} \int_{-\infty}^{0} e^{\varepsilon t_1} (J_m(\boldsymbol{k}), J_n(-\boldsymbol{k}, t_1)) X_n(\boldsymbol{k}, t + t_1) \, dt_1. \tag{22.87}$$

The transport equations (22.87) were used by the author and Tischenko [233] to construct nonlocal hydrodynamic equations with memory.

If we neglect the retardation, then

$$\langle J_m(\boldsymbol{k}) \rangle = \langle J_m(\boldsymbol{k}) \rangle_l + \sum_n \mathcal{L}_{mn}(\boldsymbol{k}) X_n(\boldsymbol{k}), \tag{22.87a}$$

where

$$\mathcal{L}_{mn}(\boldsymbol{k}) = \int_{-\infty}^{0} e^{\varepsilon t_1} (J_m(\boldsymbol{k}), J_n(-\boldsymbol{k}, t_1)) \, dt_1 \tag{22.87b}$$

are the kinetic coefficients. Taking into account that, according to Curie's theorem, only those correlators coupling quantities of the same tensor dimensionality are nonzero, and that, according to (21.15),

$$\begin{aligned} \langle \dot{\rho}(\boldsymbol{k}) \rangle &= -i\boldsymbol{k} \cdot \langle \boldsymbol{p}(\boldsymbol{k}) \rangle = \langle \dot{\rho}(\boldsymbol{k}) \rangle_l, \\ \langle H(\boldsymbol{k}) \rangle &= \langle H(\boldsymbol{k}) \rangle_l, \\ \langle n_m(\boldsymbol{k}) \rangle &= \langle n_m(\boldsymbol{k}) \rangle_l, \end{aligned} \tag{22.88}$$

we write the linear relations (22.87a) in the form

$$\begin{aligned} \langle \dot{H}(\boldsymbol{k}) \rangle &= \langle \dot{H}(\boldsymbol{k}) \rangle_l + \mathscr{L}_{00}(\boldsymbol{k})\,\beta(\boldsymbol{k}, t) - \sum_{n \geqslant 1} \mathscr{L}_{0,n+1}(\boldsymbol{k})\,\nu_n(\boldsymbol{k}, t), \\ \langle \dot{n}_m(\boldsymbol{k}) \rangle &= \langle \dot{n}_m(\boldsymbol{k}) \rangle_l + \mathscr{L}_{m0}(\boldsymbol{k})\,\beta(\boldsymbol{k}, t) - \sum_{n \geqslant 1} \mathscr{L}_{m,n+1}(\boldsymbol{k})\,\nu_n(\boldsymbol{k}, t), \\ \langle \dot{\boldsymbol{p}}(\boldsymbol{k}) \rangle &= \langle \dot{\boldsymbol{p}}(\boldsymbol{k}) \rangle_l - \mathscr{L}_{11}(\boldsymbol{k}) \cdot \boldsymbol{u}(\boldsymbol{k}, t). \end{aligned} \tag{22.89}$$

Analogous relations without retardation were obtained by Kubo, Yokota, and Nakajima [28] by means of Onsager's hypothesis on the nature of the damping of fluctuations. These authors considered a one-component liquid and did not take into account momentum transport, which is described by the last equation of the system (22.89). The improper integrals that they obtained in (22.87b) were in the sense of Cesaro rather than in the sense of Abel, since they did not take the causality condition into account explicitly (see §§21.3 and 22.3). Moreover, in the expressions for the fluxes, they did not take into account the additional terms compensating the nondissipative motion of the center of mass, since momentum transport was not considered.

The linear relations (22.87a) have, generally speaking, nonlocal character, because of the dependence of $\mathscr{L}_{mn}(\boldsymbol{k})$ on $\mathbf{k}$. If we expand $\mathscr{L}_{mn}(\boldsymbol{k})$ in a series in $\mathbf{k}$, confining ourselves to the first nonvanishing terms of the expansion, we obtain

$$\begin{aligned} \mathscr{L}_{mn}(\boldsymbol{k}) &\cong V L_{mn} k^2 \qquad (m, n = 0, 2, 3, \ldots), \\ \mathscr{L}_{11}^{\alpha\beta}(\boldsymbol{k}) &= VT\eta\left(k^2\delta_{\alpha\beta} + \frac{1}{3}k_\alpha k_\beta\right) + VT\zeta k_\alpha k_\beta = \\ &= VT\eta(k^2\delta_{\alpha\beta} - k_\alpha k_\beta) + VT\left(\zeta + \frac{4}{3}\eta\right)k_\alpha k_\beta, \end{aligned} \tag{22.90}$$

since odd powers of k are absent, because of the spatial isotropy. If (22.90) is taken into account, the linear relations (22.89) take the local form

$$\begin{aligned} \langle \dot{H}(\boldsymbol{k})\rangle &= \langle \dot{H}(\boldsymbol{k})\rangle_l + VL_{00}k^2\beta(\boldsymbol{k},\ t) - V\sum_{n\geqslant 1} L_{0,n+1}k^2\nu_n(\boldsymbol{k},\ t),\\ \langle \dot{n}_m(\boldsymbol{k})\rangle &= \langle \dot{n}_m(\boldsymbol{k})\rangle_l + VL_{m0}k^2\beta(\boldsymbol{k},\ t) - V\sum_{n\geqslant 1} L_{m,n+1}k^2\nu_n(\boldsymbol{k},\ t), \qquad (22.91)\\ \langle \dot{\boldsymbol{p}}(\boldsymbol{k})\rangle &= \langle \dot{\boldsymbol{p}}(\boldsymbol{k})\rangle_l - VT\eta\left(k^2\boldsymbol{u}(\boldsymbol{k},\ t) + \tfrac{1}{3}\boldsymbol{k}\,\boldsymbol{k}\cdot\boldsymbol{u}(\boldsymbol{k},\ t)\right) - VT\zeta\boldsymbol{k}\,\boldsymbol{k}\cdot\boldsymbol{u}(\boldsymbol{k},\ t)\ . \end{aligned}$$

It follows from (22.90) and (22.91) that the kinetic coefficients L_{mn}, η, and ζ are equal to

$$\begin{aligned} L_{mn} &= \lim_{k\to 0}\frac{\mathscr{L}_{mn}(\boldsymbol{k})}{Vk^2} \qquad (m,\ n = 0,\ 2,\ 3,\ \ldots),\\ \eta &= \lim_{k\to 0}\frac{1}{2VTk^4}\sum_{\alpha\beta}\mathscr{L}_{11}^{\alpha\beta}(\boldsymbol{k})\left(k^2\delta_{\alpha\beta} - k_\alpha k_\beta\right), \qquad (22.92)\\ \zeta + \frac{4}{3}\eta &= \lim_{k\to 0}\frac{\boldsymbol{k}\cdot\mathscr{L}_{11}\cdot\boldsymbol{k}}{VTk^4}. \end{aligned}$$

In the expression for η, we have used the fact that

$$\sum_{\alpha\beta}(k_\alpha k_\beta - k^2\delta_{\alpha\beta})^2 = 2k^4.$$

One can easily convince oneself that the expressions (22.92) for the kinetic coefficients are equivalent to the expressions (22.53a) obtained earlier. In fact,†

$$L_{00} = \lim_{k\to 0}\frac{1}{Vk^2}\int_{-\infty}^{0} e^{\varepsilon t}\left(J_0(\boldsymbol{k}),\ J_0(-\boldsymbol{k},\ t)\right)dt, \qquad (22.93)$$

where

$$J_0(\boldsymbol{k}) = \dot{H}(\boldsymbol{k}) - \frac{u+p}{\langle\rho\rangle}\dot{\rho}(\boldsymbol{k}), \qquad (22.93a)$$

or, since

$$\dot{H}(\boldsymbol{k}) = -i\boldsymbol{k}\cdot\boldsymbol{j}_H(\boldsymbol{k}), \qquad \dot{\rho}(\boldsymbol{k}) = -i\boldsymbol{k}\cdot\boldsymbol{p}(\boldsymbol{k}),$$

we have

$$J_0(\boldsymbol{k}) = -i\boldsymbol{k}\cdot\left(\boldsymbol{j}_H(\boldsymbol{k}) - \frac{u+p}{\langle\rho\rangle}\boldsymbol{p}(\boldsymbol{k})\right) = -i\boldsymbol{k}\cdot\boldsymbol{j}_Q(\boldsymbol{k}). \qquad (22.93b)$$

Because of the spatial uniformity of the system, we can transform

---

†In such integrals, here and below, it is assumed that first $V\to\infty$ and then $\mathbf{k}\to 0$.

formula (22.93) to the form

$$L_{00} = \lim_{k\to 0} \frac{1}{Vk^2} \int_{-\infty}^{0} e^{\varepsilon t} (\boldsymbol{j}_Q(\boldsymbol{k}), \boldsymbol{j}_Q(-\boldsymbol{k}, t)) : \boldsymbol{kk}\, dt$$

$$= \lim_{k\to 0} \frac{1}{3V} \int_{-\infty}^{0} e^{\varepsilon t} (\boldsymbol{j}_Q(\boldsymbol{k}) \cdot \boldsymbol{j}_Q(-\boldsymbol{k}, t))\, dt, \qquad (22.94)$$

where the dot denotes the scalar product of the fluxes. In fact, the flux tensor can be decomposed into transverse and longitudinal parts:

$$(j_Q^\alpha(\boldsymbol{k}), j_Q^\beta(-\boldsymbol{k}, t)) = \frac{A(k, t)}{k^2}(k_\alpha k_\beta - \delta_{\alpha\beta}k^2) - B(k, t)\frac{k_\alpha k_\beta}{k^2}.$$

We obtain formula (22.94) if we take into account that

$$\lim_{k\to 0} \int_{-\infty}^{0} e^{\varepsilon t} A(k, t)\, dt = \lim_{k\to 0} \int_{-\infty}^{0} e^{\varepsilon t} B(k, t)\, dt,$$

since, as k → 0 the tensor $(j_Q^\alpha(\boldsymbol{k}), j_Q^\beta(-\boldsymbol{k}, t))$ must reduce to the unit tensor by virtue of the isotropy of space.

Formula (22.94) coincides with the first of the formulas (22.53a).

Analogously, we write the last formula of (22.92):

$$\zeta + \frac{4}{3}\eta = \lim_{k\to 0} \frac{1}{Vk^4} \int_{-\infty}^{0} e^{\varepsilon t}\, \boldsymbol{k}\cdot(J_1(\boldsymbol{k}), J_1(-\boldsymbol{k}, t))\cdot\boldsymbol{k}\, dt, \qquad (22.94a)$$

where

$$J_1(\boldsymbol{k}) = -i\boldsymbol{k}\cdot\left(T(\boldsymbol{k}) - \left(\frac{\partial p}{\partial u}\right)_n H(\boldsymbol{k})U - \sum_i \left(\frac{\partial p}{\partial n_i}\right)_u n_i(\boldsymbol{k})U\right)$$

$$= -i\boldsymbol{k}\cdot\left\{\mathring{T}(\boldsymbol{k}) + U\left(p(\boldsymbol{k}) - \left(\frac{\partial p}{\partial u}\right)_n H(\boldsymbol{k}) - \sum_i \left(\frac{\partial p}{\partial n_i}\right)_u n_i(\boldsymbol{k})\right)\right\},$$

U is the unit tensor, $p(\boldsymbol{k}) = -i\boldsymbol{k}\cdot T(\boldsymbol{k})$, and $\mathring{T}(\boldsymbol{k})$ is the divergenceless part of the tensor

$$T(\boldsymbol{k}) = \mathring{T}(\boldsymbol{k}) + p(\boldsymbol{k})U. \qquad (22.95)$$

We denote

$$\Delta p(\boldsymbol{k}) = p(\boldsymbol{k}) - \left(\frac{\partial p}{\partial u}\right)_n H(\boldsymbol{k}) - \sum_i \left(\frac{\partial p}{\partial n_i}\right)_u n_i(\boldsymbol{k}).$$

Then $J_1(\boldsymbol{k}) = -i\boldsymbol{k}\cdot\{\mathring{T}(\boldsymbol{k}) + U\,\Delta p(\boldsymbol{k})\}$. Consequently,

$$\zeta + \frac{4}{3}\eta = \lim_{k\to 0}\frac{1}{TVk^4}\int_{-\infty}^{0} e^{\varepsilon t}\,\boldsymbol{kk} : (\mathring{T}(\boldsymbol{k}),\ \mathring{T}(-\boldsymbol{k},\ t)) : \boldsymbol{kk}\,dt + \lim_{k\to 0}\frac{1}{TV}\int_{-\infty}^{0} e^{\varepsilon t}\,(\Delta p(\boldsymbol{k}),\ \Delta p(-\boldsymbol{k},\ t))\,dt$$

or

$$\zeta + \frac{4}{3}\eta = \lim_{k\to 0}\frac{1}{TV}\int_{-\infty}^{0} e^{\varepsilon t}\,(\Delta p(\boldsymbol{k}),\ \Delta p(-\boldsymbol{k},\ t))\,dt + \lim_{k\to 0}\frac{2}{15TV}\int_{-\infty}^{0} e^{\varepsilon t}\,(\mathring{T}(\boldsymbol{k}) : \mathring{T}(-\boldsymbol{k},\ t))\,dt. \tag{22.96}$$

Analogously, we also obtain expressions for ζ and η separately:

$$\zeta = \lim_{k\to 0}\frac{1}{TV}\int_{-\infty}^{0} e^{\varepsilon t}\,(\Delta p(\boldsymbol{k}),\ \Delta p(-\boldsymbol{k},\ t))\,dt,$$
$$\eta = \lim_{k\to 0}\frac{1}{10TV}\int_{-\infty}^{0} e^{\varepsilon t}\,(\mathring{T}(\boldsymbol{k}) : \mathring{T}(-\boldsymbol{k},\ t))\,dt, \tag{22.97}$$

which coincide with (22.54c) and (22.54b).

Thus, the formulas (22.92) for the kinetic coefficients are equivalent to the formulas (22.53a) obtained earlier.

In choosing the fluxes (22.85a) and the thermodynamic forces (22.85b) in the entropy balance equation

$$\frac{\partial S(\boldsymbol{x})}{\partial t} + \operatorname{div} j_S(\boldsymbol{x}) = \sigma(\boldsymbol{x}) \tag{22.98}$$

it is convenient to put

$$j_S(\boldsymbol{x}) = \sum_m F_m(\boldsymbol{x},\ t)\langle j_m(\boldsymbol{x},\ t)\rangle_l + \beta(\boldsymbol{x},\ t)\,\boldsymbol{v}(\boldsymbol{x},\ t)\,p(\boldsymbol{x}), \tag{22.99a}$$

$$\sigma(\boldsymbol{x}) = \sum_m (\langle \dot{P}_m(\boldsymbol{x})\rangle - \langle \dot{P}_m(\boldsymbol{x})\rangle_l)\,F_m(\boldsymbol{x},\ t). \tag{22.99b}$$

Then the total entropy production is positive,

$$\int \sigma(\boldsymbol{x})\,d\boldsymbol{x} = \sum_{m,\,\boldsymbol{k}} (\langle \dot{P}_m(\boldsymbol{k})\rangle - \langle \dot{P}_m(\boldsymbol{k})\rangle_l)\,F_m(-\boldsymbol{k})$$
$$= \sum_{m,\,n,\,\boldsymbol{k}} F_n(\boldsymbol{k})\,\mathscr{L}_{mn}(\boldsymbol{k})\,F^*_m(\boldsymbol{k}) \geqslant 0. \qquad (22.100)$$

The examples considered show that, to construct the hydrodynamic equations, we can use either of the expressions (22.1) and (22.80) for the nonequilibrium statistical operator.

§ 23. Relaxation Processes

23.1. General Theory

Up to this point, we have assumed that the macroscopic state of a system can be completely characterized by giving the fields of the temperature, the mass velocity, and the chemical potentials of the components. However, this is not always so. For example, in the case when the system consists of weakly interacting subsystems, between which exchange of energy is difficult, the approa to statistical equilibrium occurs in two stages: partial equilibriu is first established in the subsystems, and this then tends slowly to complete statistical equilibrium, if there are no factors impeding this. A single temperature is insufficient to describe the state of such a system, and we must introduce different temperatures for its subsystems.

Such a situation may occur both for different components, because of a large difference in their masses (for example, in an electron–ion plasma [101–103]), and for different internal degree of freedom of molecules [100, 104–107], electron spins or nuclear spins [108, 109]. The thermodynamic theory of relaxation processes in gases and liquids was developed by Kneser [110], Leontovich and Mandel'shtam [111, 112], and by many other authors (c the monograph [113]). A formal scheme for taking into account the internal degrees of freedom of molecules on the basis of the Boltzmann kinetic equation was developed by Wang Chang and Uhlenbeck [106]. Their results were refined by Snider [114], who took account of degenerate states. For further developments in this direction, see the book [107] and the thorough and original review by Dahler and Hoffmann [228], in which the method of the nonequilibrium statistical operator, as described in this chapter, is used.

Sometimes, the system cannot be characterized by a single mass velocity (for example, in supersonic flows, when the velocity field has excessively large gradients through the shock-wave front and the fundamental assumption of the linear dissipative theory, i.e., that the velocity gradients are small, is violated). In this case, in order not to go beyond the limits of linear dissipative processes, one uses the two-fluid model with two velocity fields, in front of and behind the shock-wave front [104, 105]. The equations of two-fluid hydrodynamics are derived in the papers [115-117]. One velocity field is also insufficient for the derivation of the hydrodynamics of a superfluid [118-121].

The general scheme described in §§ 21 and 22 for constructing the nonequilibrium statistical operator can also be generalized to relaxing systems. For this, we must formulate the conservation laws in more detail than we did earlier, i.e., for each weakly interacting subsystem separately. We have already considered conservation laws of this type in §19.5, where the subsystems were characterized by the quantum numbers of internal degrees of freedom.

The conservation laws for the energy, particle number and momentum for the i-th subsystem have the form

$$\begin{aligned} &\frac{\partial H_i(\boldsymbol{x})}{\partial t} + \operatorname{div} \boldsymbol{j}_{H_i}(\boldsymbol{x}) = J_{H_i}(\boldsymbol{x}), \\ &\frac{\partial n_i(\boldsymbol{x})}{\partial t} + \operatorname{div} \boldsymbol{j}_i(\boldsymbol{x}) = J_i(\boldsymbol{x}), \\ &\frac{\partial \boldsymbol{p}_i(\boldsymbol{x})}{\partial t} + \operatorname{Div} T_i(\boldsymbol{x}) = \boldsymbol{f}_i(\boldsymbol{x}), \end{aligned} \tag{23.1}$$

where $H_i(\boldsymbol{x})$, $n_i(\boldsymbol{x})$, and $\boldsymbol{p}_i(\boldsymbol{x})$ are the densities of energy, particle number and momentum of the i-th subsystem, $\boldsymbol{j}_{H_i}(\boldsymbol{x})$, $\boldsymbol{j}_i(\boldsymbol{x})$, and $T_i(\boldsymbol{x})$ are the corresponding fluxes of energy, particle number, and momentum, $J_{H_i}(\boldsymbol{x})$ is the rate of change of the energy of the i-th subsystem, $\boldsymbol{f}_i(\boldsymbol{x})$ is the density of the force of interaction of the i-th subsystem with all the other subsystems, and $J_i(\boldsymbol{x})$ is the density of particle sources.

The total densities of energy, mass, and momentum

$$H(\boldsymbol{x}) = \sum_i H_i(\boldsymbol{x}), \quad \rho(\boldsymbol{x}) = \sum_i m_i n_i(\boldsymbol{x}), \quad \boldsymbol{p}(\boldsymbol{x}) = \sum_i \boldsymbol{p}_i(\boldsymbol{x}) \tag{23.2}$$

satisfy the conservation laws

$$\frac{\partial H(\boldsymbol{x})}{\partial t} + \operatorname{div} \boldsymbol{j}_H(\boldsymbol{x}) = 0,$$
$$\frac{\partial \rho(\boldsymbol{x})}{\partial t} + \operatorname{div} \boldsymbol{p}(\boldsymbol{x}) = 0, \tag{23.3}$$
$$\frac{\partial \boldsymbol{p}(\boldsymbol{x})}{\partial t} + \operatorname{Div} T(\boldsymbol{x}) = 0.$$

The conservation laws (23.1) can be written in the form of one equation for quantities depending on two indices:

$$\frac{\partial P_{mi}(\boldsymbol{x})}{\partial t} + \nabla \cdot j_{mi}(\boldsymbol{x}) = J_{mi}(\boldsymbol{x}), \tag{23.4}$$

where we have introduced the notation

$$\begin{aligned} &P_{0i}(\boldsymbol{x}) = H_i(\boldsymbol{x}), && j_{0i}(\boldsymbol{x}) = \boldsymbol{j}_{H_i}(\boldsymbol{x}), && J_{0i}(\boldsymbol{x}) = J_{H_i}(\boldsymbol{x}), \\ &P_{1i}(\boldsymbol{x}) = \boldsymbol{p}_i(\boldsymbol{x}), && j_{1i}(\boldsymbol{x}) = T_i(\boldsymbol{x}), && J_{1i}(\boldsymbol{x}) = \boldsymbol{f}_i(\boldsymbol{x}), \\ &P_{2i}(\boldsymbol{x}) = m_i n_i(\boldsymbol{x}), && j_{2i}(\boldsymbol{x}) = m_i \boldsymbol{j}_i(\boldsymbol{x}), && J_{2i}(\boldsymbol{x}) = m_i J_i(\boldsymbol{x}). \end{aligned} \tag{23.5}$$

The operators $J_{mi}(\boldsymbol{x})$ satisfy the supplementary conditions

$$\sum_i J_{H_i}(\boldsymbol{x}) = 0, \quad \sum_i \boldsymbol{f}_i(\boldsymbol{x}) = 0, \quad \sum_i m_i J_i(\boldsymbol{x}) = 0 \tag{23.6}$$

or

$$\sum_i J_{mi}(\boldsymbol{x}) = 0, \tag{23.6a}$$

which denote the conservation of the total energy, momentum, and mass.

The operators (23.5) can describe subsystems with different internal degrees of freedom, but can also have another meaning. For example, if we consider a system in which chemical reactions occur, then the index i indicates the type of molecule (reagents and reaction products), while the index m indicates the type of conserved quantity (energy, mass, or momentum).

Following [4, 5], we apply the general scheme described in §21.1 for constructing the nonequilibrium statistical operator to a system with the conservation laws (23.4).

The subsystems are characterized by the quantities $P_{mi}(\mathbf{x})$; consequently, the nonequilibrium statistical operator is equal to

$$\rho = Q^{-1}\exp\left\{-\sum_{m,i}\varepsilon\int\int_{-\infty}^{0} e^{\varepsilon t_1} F_{im}(\boldsymbol{x}, t+t_1)\, P_{mi}(\boldsymbol{x}, t_1)\, dt_1\, d\boldsymbol{x}\right\} =$$
$$= Q^{-1}\exp\left\{-\sum_{m,i}\int F_{im}(\boldsymbol{x}, t)\, P_{mi}(\boldsymbol{x})\, d\boldsymbol{x} + \right.$$
$$+\sum_{m,i}\int\int_{-\infty}^{0} e^{\varepsilon t_1}\Big(\dot{P}_{mi}(\boldsymbol{x}, t_1)\, F_{im}(\boldsymbol{x}, t+t_1) +$$
$$\left. + P_{mi}(\boldsymbol{x}, t_1)\frac{\partial F_{im}(\boldsymbol{x}, t+t_1)}{\partial t_1}\Big)\, dt_1\, d\boldsymbol{x}\right\} \qquad (\varepsilon\to 0) \tag{23.7}$$

or, after integration by parts,

$$\rho = Q^{-1}\exp\left\{-\sum_{m,i}\int F_{im}(\boldsymbol{x}, t)\, P_{mi}(\boldsymbol{x})\, d\boldsymbol{x} + \right.$$
$$+\sum_{m,i}\int\int_{-\infty}^{0} e^{\varepsilon t_1}\Big(j_{mi}(\boldsymbol{x}, t_1)\cdot\nabla F_{im}(\boldsymbol{x}, t+t_1) + P_{mi}(\boldsymbol{x}, t_1)\frac{\partial F_{im}(\boldsymbol{x}, t+t_1)}{\partial t_1} +$$
$$\left. + J_{mi}(\boldsymbol{x}, t_1)\, F_{im}(\boldsymbol{x}, t+t_1)\Big)\, dt_1\, d\boldsymbol{x}\right\}, \tag{23.8}$$

where

$$\begin{aligned} F_{i0}(\boldsymbol{x}, t) &= \beta_i(\boldsymbol{x}, t),\\ F_{i1}(\boldsymbol{x}, t) &= -\beta_i(\boldsymbol{x}, t)\, v_i(\boldsymbol{x}, t),\\ F_{i2}(\boldsymbol{x}, t) &= -\beta_i(\boldsymbol{x}, t)\left(\frac{\mu_i(\boldsymbol{x}, t)}{m_i} - \frac{1}{2}v_i^2(\boldsymbol{x}, t)\right), \end{aligned} \tag{23.9}$$

$\beta_i(\mathbf{x}, t)$ is the inverse temperature of the i-th subsystem, $\mu_i(\mathbf{x}, t)$ is its chemical potential, and $\mathbf{v}_i(\mathbf{x}, t)$ is its mass velocity, which we have introduced for the possible generalization to two-fluid hydrodynamics.

We choose the parameters $F_{im}(\mathbf{x}, t)$ in such a way that they have the meaning of thermodynamic parameters; this is achieved if we put

$$\langle P_{mi}(\boldsymbol{x})\rangle = \langle P_{mi}(\boldsymbol{x})\rangle_l. \tag{23.10}$$

Indeed, then

$$\frac{\delta \ln Q_l}{\delta F_{im}(\boldsymbol{x}\ \ t)} = -\langle P_{mi}(\boldsymbol{x})\rangle_l = -\langle P_{mi}(\boldsymbol{x})\rangle, \tag{23.11}$$

where

$$\langle \ldots \rangle_l = \mathrm{Tr}\,(\rho_l \ldots),$$

$$\rho_l = Q_l^{-1} \exp\left\{ -\sum_{mi} \int F_{im}(\boldsymbol{x},\, t)\, P_{mi}(\boldsymbol{x})\, d\boldsymbol{x} \right\} \tag{23.12}$$

is the local-equilibrium distribution, and

$$Q_l = \mathrm{Tr}\, \exp\left\{ -\sum_{mi} \int F_{im}(\boldsymbol{x},\, t)\, P_{mi}(\boldsymbol{x})\, d\boldsymbol{x} \right\} \tag{23.13}$$

is the statistical functional corresponding to it.

The relations (23.11) are the thermodynamic equalities for relaxing systems and confirm the interpretation of $\beta_i(\mathbf{x}, t)$, $\mu_i(\mathbf{x}, t)$ and $\mathbf{v}_i(\mathbf{x}, t)$ as the inverse temperature, chemical potential, and mass velocity of the subsystems.

To elucidate the physical meaning of the concept of the temperature of a subsystem, it is convenient to express the thermodynamic equalities in terms of functional derivatives of the entropy

$$S = -\langle \ln \rho_l \rangle = \ln Q_l + \sum_{mi} \int F_{im}(\boldsymbol{x},\, t) \langle P_{mi}(\boldsymbol{x}) \rangle\, d\boldsymbol{x}. \tag{23.14}$$

Taking the variation of (23.14) and taking (23.11) into account, we obtain

$$\frac{\delta S}{\delta \langle P_{mi}(\boldsymbol{x}) \rangle} = F_{im}(\boldsymbol{x},\, t), \tag{23.15}$$

and, consequently,

$$\beta_i(\boldsymbol{x},\, t) = \frac{\delta S}{\delta \langle H_i'(\boldsymbol{x}) \rangle}. \tag{23.16}$$

where $\langle H_i'(\boldsymbol{x}) \rangle$ is the energy density of the i-th subsystem in the accompanying coordinate frame, or, for the spatially uniform case

$$\beta_i = \frac{\partial S}{\partial \langle H_i' \rangle}, \qquad \langle H_i' \rangle = \int \langle H_i'(\boldsymbol{x}) \rangle\, d\boldsymbol{x}, \tag{23.16a}$$

i.e., the inverse temperature of the i-th subsystem is equal to the derivative of the entropy with respect to the average energy of the i-th subsystem.

The temperature of a subsystem need not necessarily be positive, but there is nothing paradoxical in this, since β_i^{-1} is not the temperature of the thermostat, as it was in the equilibrium case. A more detailed discussion of the meaning of negative temperature is given in §23.2 of this chapter.

We shall calculate the change of the entropy (23.14) with time. Taking into account that

$$\frac{\partial \ln Q_l}{\partial t} = -\sum_{mi} \int \frac{\partial F_{im}(\boldsymbol{x}, t)}{\partial t} \langle P_{mi}(\boldsymbol{x}) \rangle \, d\boldsymbol{x}, \tag{23.17}$$

we obtain

$$\begin{aligned} \frac{\partial S}{\partial t} &= \sum_{mi} \int F_{im}(\boldsymbol{x}, t) \left\langle \frac{\partial P_{mi}(\boldsymbol{x})}{\partial t} \right\rangle d\boldsymbol{x} \\ &= -\sum_{mi} \int F_{im}(\boldsymbol{x}, t) \nabla \langle j_{mi}(\boldsymbol{x}) \rangle \cdot d\boldsymbol{x} + \sum_{mi} \int F_{im}(\boldsymbol{x}, t) \langle J_{mi}(\boldsymbol{x}) \rangle \, d\boldsymbol{x}, \end{aligned} \tag{23.18}$$

or, after integration by parts,

$$\begin{aligned} \frac{\partial S}{\partial t} = &-\sum_{mi} \int F_{im}(\boldsymbol{x}, t) \langle j_{mi}(\boldsymbol{x}) \rangle \cdot d\sigma + \\ &+ \sum_{mi} \int \langle j_{mi}(\boldsymbol{x}) \rangle \cdot \nabla F_{im}(\boldsymbol{x}) \, d\boldsymbol{x} + \sum_{mi} \int F_{im}(\boldsymbol{x}, t) \langle J_{mi}(\boldsymbol{x}) \rangle \, d\boldsymbol{x}. \end{aligned} \tag{23.18a}$$

Below, we shall confine ourselves to the case of one mass velocity, and obtain the balance equation for the entropy density.

We introduce the entropy density $S(\mathbf{x})$ and the density $\Phi(\mathbf{x})$ of the Massieu–Planck function by the relations

$$S = \int S(\boldsymbol{x}) \, d\boldsymbol{x}, \quad \Phi = \ln Q_l = \int \Phi(\boldsymbol{x}) \, d\boldsymbol{x}. \tag{23.19}$$

Then

$$S(\boldsymbol{x}) = \sum_{im} F_{im}(\boldsymbol{x}, t) \langle P_{mi}(\boldsymbol{x}) \rangle + \Phi(\boldsymbol{x}) \tag{23.20}$$

satisfies the balance equation

$$\frac{\partial S(\boldsymbol{x})}{\partial t} = -\operatorname{div} \boldsymbol{j}_S(\boldsymbol{x}) + \sigma(\boldsymbol{x}), \tag{23.21}$$

where

$$j_S(\boldsymbol{x}) = \sum_{mi} F_{im}(\boldsymbol{x}, t) \langle j_{mi}(\boldsymbol{x}) \rangle + \boldsymbol{v}(\boldsymbol{x}, t) \Phi(\boldsymbol{x}) \tag{23.22}$$

is the entropy flux density, and

$$\sigma(\boldsymbol{x}) = \sum_{im} (\langle j_{mi}(\boldsymbol{x}) \rangle - \langle j_{mi}(\boldsymbol{x}) \rangle_l) \cdot \nabla F_{im}(\boldsymbol{x}, t) + \sum_{im} \langle J_{mi}(\boldsymbol{x}) \rangle F_{im}(\boldsymbol{x}, t) \tag{23.23}$$

is the entropy production.

In deriving (23.21)-(23.23), we have used the relation

$$\sum_{im} \langle j_{mi}(\boldsymbol{x}) \rangle_l \cdot \nabla F_{im}(\boldsymbol{x}, t) = -\operatorname{div}(\boldsymbol{v}(\boldsymbol{x}, t) \Phi(\boldsymbol{x})), \tag{23.24}$$

which can be obtained analogously to (20.73) if we take into account the thermodynamic equalities

$$\langle n_i(\boldsymbol{x}) \rangle_l = \frac{\partial \Phi(\boldsymbol{x})}{\partial \nu_i(\boldsymbol{x})}, \qquad \langle H_i'(\boldsymbol{x}) \rangle_l = -\frac{\partial \Phi(\boldsymbol{x})}{\partial \beta_i(\boldsymbol{x})}, \qquad \nu_i = \beta_i \mu_i \tag{23.25}$$

and the relations

$$\Phi(\boldsymbol{x}) = \sum_i \Phi_i(\boldsymbol{x}) = \sum_i \beta_i p_i, \qquad p_i = \langle T_i'(\boldsymbol{x}) \rangle_l; \tag{23.25a}$$

the prime, as usual, denotes the frame moving with velocity **v**.

The sources J_{mi} **(x)** are not independent, since they are connected by the relations (23.6a).

After eliminating the l-th source, we obtain

$$\sigma(\boldsymbol{x}) = \sum_{im} (\langle j_{mi}(\boldsymbol{x}) \rangle - \langle j_{mi}(\boldsymbol{x}) \rangle_l) \cdot \nabla F_{im}(\boldsymbol{x}, t) + \\ + \sum_{im} \langle J_{mi}(\boldsymbol{x}) \rangle (F_{im}(\boldsymbol{x}, t) - F_{lm}(\boldsymbol{x}, t)). \tag{23.26}$$

Introducing in place of ∇F_{im} the thermodynamic forces

$$\begin{aligned} X_{i0} &= \nabla \beta_i(\boldsymbol{x}, t), \\ X_{i1} &= -\beta_i(\boldsymbol{x}, t) \nabla \boldsymbol{v}(\boldsymbol{x}, t), \\ X_{i2} &= -\frac{1}{m_i} \nabla \nu_i(\boldsymbol{x}, t) = -\frac{1}{m_i} \nabla \beta_i(\boldsymbol{x}, t) \mu_i(\boldsymbol{x}, t), \end{aligned} \tag{23.27}$$

we rewrite (23.26) in the explicit form

$$\sigma(x) = \sum_i \langle j'_{H_i}(x)\rangle \nabla\beta_i - \\ - \sum_i \beta_i (\langle T'_i(x)\rangle - \langle T'_i(x)\rangle_l) : \nabla v - \sum_i \langle j'_i(x)\rangle \nabla\nu_i + \\ + \sum_i \langle J'_{H_i}(x)\rangle (\beta_i - \beta_l) - \sum_i m_i \langle J_i(x)\rangle \left(\frac{\nu_i}{m_i} - \frac{\nu_l}{m_l}\right), \tag{23.26a}$$

where

$$J'_{H_i}(x) = J_{H_i}(x) - \mathfrak{f}_i(x) \cdot v.$$

Comparing (23.26a) with (22.39), we note that, in relaxational systems, new sources of entropy [the last two terms in (23.26a)], connected with the exchange of energy and particles between the subsystems, have been added.

Introducing a more compact matrix notation for the fluxes and sources

$$j^{mi}(x) = \begin{Bmatrix} j_{mi}(x) \\ J_{mi}(x) \end{Bmatrix} \tag{23.28}$$

and for the thermodynamic forces

$$X_{im}(x, t) = \{\nabla F_{im}(x, t),\ F_{im}(x, t) - F_{lm}(x, t)\}, \tag{23.28a}$$

we write (23.26) in the form

$$\sigma(x) = \sum_{i, m} (\langle j^{mi}(x)\rangle - \langle j^{mi}(x)\rangle_l) \cdot X_{im}(x, t), \tag{23.29}$$

since

$$\langle J_0(x)\rangle_l = \langle J_H(x)\rangle_l = \langle J'_H(x)\rangle_l + \langle \mathfrak{f}(x)\rangle_l \cdot v(x, t) = 0,$$
$$\langle J_1(x)\rangle = \langle \mathfrak{f}(x)\rangle_l = \langle \mathfrak{f}'(x)\rangle_l = 0,$$

where the prime denotes averaging in the accompanying frame, and we have taken into account formulas (19.42b) and (19.46c) and the fact that the velocity varies little over distances of the order of the range of action of the forces.

We obtain the linear relations between the fluxes, sources, and thermodynamic forces by averaging the conservation laws

(23.4) with the statistical operator (23.8) and confining ourselves to terms linear in the thermodynamic forces. In the stationary case, we obtain

$$\langle j^{mi}(\boldsymbol{x})\rangle = \langle j^{mi}(\boldsymbol{x})\rangle_l + \sum_{m_1 i_1} \int L_{mi}^{m_1 i_1}(\boldsymbol{x}, \boldsymbol{x}') \cdot X_{i_1 m_1}(\boldsymbol{x}')\, d\boldsymbol{x}', \tag{23.30}$$

where

$$L_{mi}^{m_1 i_1}(\boldsymbol{x}, \boldsymbol{x}') = \int_{-\infty}^{0} e^{\varepsilon t} \left(j^{mi}(\boldsymbol{x}), j^{m_1 i_1}(\boldsymbol{x}', t)\right) dt$$

are the kinetic coefficients. Substituting (23.30) into (23.29), we obtain

$$\sigma(\boldsymbol{x}) = \sum_{\substack{im \\ i_1 m_1}} \int X_{i_1 m_1}(\boldsymbol{x}) \cdot L_{mi}^{m_1 i_1}(\boldsymbol{x}, \boldsymbol{x}') \cdot X_{im}(\boldsymbol{x}')\, d\boldsymbol{x}', \tag{23.31}$$

with, as before,

$$\int \sigma(\boldsymbol{x})\, d\boldsymbol{x} > 0.$$

An important particular case of the problem under consideration is provided by irreversible processes in a spatially uniform system consisting of weakly interacting subsystems [for example, exchange of energy between components of a mixture which have different temperatures (cf. §23.4), or a chemical reaction in a homogeneous phase (cf. §23.5)]. The conservation laws for the energy and particle number in this case have the form

$$\dot{H}_i = J_{H_i} = \frac{1}{i\hbar}[H_i, H], \qquad \dot{N}_i = J_{N_i} = \frac{1}{i\hbar}[N_i, H], \tag{23.32}$$

where H_i and N_i are the energy and particle number of the i-th subsystem, with

$$\sum_i J_{H_i} = 0, \qquad \sum_i J_{N_i} = 0. \tag{23.32a}$$

Corresponding to the conservation laws (23.32) are the quasi-integrals of motion

$$\tilde{H}_i = H_i - \int_{-\infty}^{0} e^{\varepsilon t} \dot{H}_i(t)\, dt, \qquad \tilde{N} = N_i - \int_{-\infty}^{0} e^{\varepsilon t} \dot{N}_i(t)\, dt \tag{23.32b}$$

and, in the stationary case, the nonequilibrium statistical operator

$$\rho = Q^{-1}\exp\left\{-\sum_i \beta_i(\tilde{H}_i - \mu_i\tilde{N}_i)\right\} = Q^{-1}\exp\left\{-\sum_i \beta_i(H_i - \mu_i N_i) + \right.$$

$$\left. + \sum_i (\beta_i - \beta_1)\int_{-\infty}^{0} e^{\varepsilon t}\dot{H}_i(t)\,dt - \sum_i (\beta_i\mu_i - \beta_1\mu_1)\int_{-\infty}^{0} e^{\varepsilon t}\dot{N}_i(t)\,dt\right\}. \quad (23.33)$$

Averaging (23.32) with (23.33), we obtain

$$\begin{aligned}\langle \dot{H}_i\rangle &= \sum_m \{L_{\dot{H}_i\dot{H}_m}(\beta_m - \beta_1) - L_{\dot{H}_i\dot{N}_m}(\beta_m\mu_m - \beta_1\mu_1)\},\\ \langle \dot{N}_i\rangle &= \sum_m \{L_{\dot{N}_i\dot{H}_m}(\beta_m - \beta_1) - L_{\dot{N}_i\dot{N}_m}(\beta_m\mu_m - \beta_1\mu_1)\},\end{aligned} \quad (23.34)$$

where

$$\begin{aligned}L_{\dot{H}_i\dot{H}_m} &= \int_{-\infty}^{0} e^{\varepsilon t}(\dot{H}_i, \dot{H}_m(t))\,dt,\\ L_{\dot{H}_i\dot{N}_m} &= L_{\dot{N}_m\dot{H}_i} = \int_{-\infty}^{0} e^{\varepsilon t}(\dot{H}_i, \dot{N}_m(t))\,dt,\\ L_{\dot{N}_i\dot{N}_m} &= \int_{-\infty}^{0} e^{\varepsilon t}(\dot{N}_i, \dot{N}_m(t))\,dt\end{aligned} \quad (23.34a)$$

are the kinetic coefficients.

Below, in §23.2 and 23.3, we shall examine concrete examples of relaxation processes for a system of nuclear spins in solids and for conduction electrons in semiconductors. In the theory of chemical reaction rates, the assumption of a small difference of chemical potentials is frequency not fulfilled and it is important to allow for nonlinear effects; these will be considered in §23.5.

23.2. Relaxation of Nuclear Spins in a Crystal [46]

As an example of the application of the method, we shall consider, following Buishvili [46], the relaxation of nuclear spins interacting with magnetic impurities and with the lattice. We write the Hamiltonian of the system in the form

$$H = H_I + H_d + H_l + H_{Id} + H_{dl}. \quad (23.35)$$

Here,

$$H_I = -\omega_n \sum_i I_i^z \quad (23.35a)$$

is the Zeeman energy of the nuclei in the constant magnetic field, I_i^z is the z-component of the nuclear spin, and ω_n is the precessi frequency of the nuclear spin;

$$H_d = \frac{1}{2} \sum_{\substack{i, j \\ \alpha, \beta}} u_{ij}^{\alpha\beta} I_i^\alpha I_j^\beta \tag{23.35b}$$

is the dipole–dipole interaction of the nuclear spins; H_l is the Hamiltonian of the lattice, which we shall not need explicitly;

$$H_{Id} = \sum_{\substack{i, j \\ \alpha, \beta}} v_{ij}^{\alpha\beta} I_i^\alpha S_j^\beta \tag{23.35c}$$

is the interaction of the electron and nuclear subsystems, and S is the electron spin of the magnetic impurity. In (23.35c), we car retain, to a good approximation, only the terms inducing a nuclea spin flip:

$$H_{Id} = \frac{1}{2} \sum_{i, n} \left(v^{-z}(i, n) I_i^+ + v^{+z}(i, n) I_i^- \right) S_n^z,$$
$$I_i^{\pm} = I_i^x \pm i I_i^y. \tag{23.35}$$

Finally, H_{dl} is the interaction of the nuclear spins with the lattice.

We shall confine ourselves to the spatially uniform case, wh the nuclear magnetization does not depend on position (distance from the impurity); otherwise, it is necessary to take the spin di fusion of the nuclei into account [43, 44, 46].

We shall regard the nuclear Zeeman subsystem H_I, the dipole–dipole reservoir H_d and the lattice together with the othe interactions, $H_l + H_{Id} + H_{dl}$, as weakly interacting subsystems. T exchange of energy between them is described by the operator equations

$$\frac{dH_I}{dt} = \frac{1}{i\hbar} [H_I, H] = \frac{1}{i\hbar} [H_I, H_{Id}] \equiv K_I,$$
$$\frac{dH_d}{dt} = \frac{1}{i\hbar} [H_d, H] = \frac{1}{i\hbar} [H_d, H_{Id} + H_{dl}] \equiv K_d, \tag{23.3}$$

since H_I commutes with H_d. The fact that the right-hand side

(23.36) are small enables us to regard the subsystems as quasi-independent.

For the stationary case, the statistical operator corresponding to the chosen subsystems has the form

$$\rho = Q^{-1}\exp\{-\beta_I\tilde{\tilde{H}}_I - \beta_d\tilde{\tilde{H}}_d - \beta(H - \tilde{\tilde{H}}_I - \tilde{\tilde{H}}_d)\}$$
$$= Q^{-1}\exp\Big\{-\beta_I H_I - \beta_d H_d - \beta(H - H_I - H_d) + \\ + \int_{-\infty}^{0} e^{\varepsilon t}(\beta_I - \beta) K_I(t)\,dt + \int_{-\infty}^{0} e^{\varepsilon t}(\beta_d - \beta) K_d(t)\,dt\Big\}, \qquad (23.37)$$

where β_{I}, β_{d}, and β are the inverse temperatures of the nuclear Zeeman reservoir (NZR), the dipole–dipole reservoir (DDR) and the lattice respectively. The concept of the DDR temperature was first introduced by Provotorov in papers on the theory of nuclear magnetic resonance [35].

Although the nonequilibrium statistical operator (23.37) corresponds to a stationary state, it can also be applied for a non-stationary state, if we assume that β_{I} and β_{d} depend slowly on time.

The average, calculated by means of (23.37), of the operator balance equations (23.36) gives the relaxation equations

$$\frac{d}{dt}\langle H_I\rangle_l = \langle K_I\rangle = \sum_i L_{Ii}(\beta_i - \beta),$$
$$\frac{d}{dt}\langle H_d\rangle_l = \langle K_d\rangle = \sum_i L_{di}(\beta_i - \beta) \quad (i = I,\, d), \qquad (23.38)$$

where

$$L_{ij} = \int_{-\infty}^{0} e^{\varepsilon t}(K_i,\, K_j(t))\,dt \qquad (23.38a)$$

are the kinetic coefficients. In Eqs. (23.38), we have taken into account that

$$\langle H_I\rangle = \langle H_I\rangle_l, \qquad \langle H_d\rangle = \langle H_d\rangle_l,$$

where the index l denotes averaging with the quasi-equilibrium operator

$$\rho_l = Q_l^{-1}\exp\{-\beta_I H_I - \beta_d H_d - \beta(H - H_I - H_d)\}, \qquad (23.39)$$

and have used the fact that

$$\langle K_I \rangle_l = \langle K_d \rangle_l = 0.$$

We shall express derivatives of the average energies in terms of derivatives of the inverse temperatures:

$$\begin{aligned} \frac{d\langle H_I \rangle_l}{dt} &\cong \frac{d\langle H_I \rangle_l}{d\beta_I}\frac{d\beta_I}{dt} = -\langle H_I^2 \rangle_l \frac{d\beta_I}{dt}, \\ \frac{d\langle H_d \rangle_l}{dt} &\cong \frac{d\langle H_d \rangle_l}{d\beta_d}\frac{d\beta_d}{dt} = -\langle H_d^2 \rangle_l \frac{d\beta_d}{dt}, \end{aligned} \tag{23.40}$$

where we have neglected terms with $\langle H_d H_I \rangle$. Taking (23.40) into account, we can rewrite the relaxation equations in the form

$$\frac{d\beta_I}{dt} = -\frac{\beta_I - \beta}{\tau_I} + \frac{\beta_d - \beta}{\tau_{Id}}, \qquad \frac{d\beta_d}{dt} = -\frac{\beta_d - \beta}{\tau_d} + \frac{\beta_I - \beta}{\tau_{dI}}, \tag{23.41}$$

where τ_I, τ_d, τ_{Id}, and τ_{dI} are relaxation times, connected with the kinetic coefficients (23.38a) by the relations

$$\begin{aligned} \tau_I &= L_{II}^{-1} \langle H_I^2 \rangle_l, & \tau_{Id} &= -L_{Id}^{-1} \langle H_I^2 \rangle_l, \\ \tau_d &= L_{dd}^{-1} \langle H_d^2 \rangle_l, & \tau_{dI} &= -L_{dI}^{-1} \langle H_d^2 \rangle_l. \end{aligned} \tag{23.42}$$

For spin systems, we can make use of the high-temperature approximation and expand the exponential in (23.39) in all the quantities apart from βH_l. Then the formulas for the relaxation times are transformed into

$$\begin{aligned} \tau_I^{-1} &= \frac{\mathrm{Tr}(1)\beta}{\mathrm{Tr}(H_I^2)} \int_0^\beta \int_{-\infty}^0 e^{\varepsilon t} \frac{\mathrm{Tr}\left(e^{-\beta H_l} K_I K_I(t + i\hbar\tau)\right)}{\mathrm{Tr}\, e^{-\beta H_l}} \, d\tau\, dt, \\ \tau_d^{-1} &= \frac{\mathrm{Tr}(1)\beta}{\mathrm{Tr}(H_d^2)} \int_0^\beta \int_{-\infty}^0 e^{\varepsilon t} \frac{\mathrm{Tr}\left(e^{-\beta H_l} K_d K_d(t + i\hbar\tau)\right)}{\mathrm{Tr}\, e^{-\beta H_l}} \, d\tau\, dt, \\ \tau_{Id}^{-1} &= -\frac{\mathrm{Tr}(1)\beta}{\mathrm{Tr}(H_I^2)} \int_0^\beta \int_{-\infty}^0 e^{\varepsilon t} \frac{\mathrm{Tr}\left(e^{-\beta H_l} K_I K_d(t + i\hbar\tau)\right)}{\mathrm{Tr}\, e^{-\beta H_l}} \, d\tau\, dt, \\ \tau_{dI}^{-1} &= -\frac{\mathrm{Tr}(1)\beta}{\mathrm{Tr}(H_d^2)} \int_0^\beta \int_{-\infty}^0 e^{\varepsilon t} \frac{\mathrm{Tr}\left(e^{-\beta H_l} K_d K_I(t + i\hbar\tau)\right)}{\mathrm{Tr}\, e^{-\beta H_l}} \, d\tau\, dt. \end{aligned} \tag{23.42a}$$

We note that the traces in (23.42a) are taken over the eigenfunctions of the spin matrices.

For the subsequent calculation of the relaxation times from formulas (23.42a), see [46], and for a discussion of the results, see [122].

The nonequilibrium statistical operator enables us to introduce naturally the concept of a spin temperature, for example, the spin temperature β_I^{-1} of the nuclei. It may be different from the temperature β^{-1} of the lattice and can even turn out to be negative. For nonequilibrium statistical thermodynamics, there is nothing paradoxical in this, since $T_I = \beta_I^{-1}$ is not the temperature of a thermostat, but is defined by the relation

$$\beta_I = \frac{\partial S}{\partial \langle H_I \rangle}, \tag{23.43}$$

where S is the entropy.

Negative temperatures have also been formally introduced for the equilibrium case [92]; this is possible for systems of which the energy spectrum has an upper bound, such as, for example, spin systems. Otherwise, the partition function for negative temperatures would diverge. However, real systems with a spectrum bounded from above always interact with a system with no upper bound to the spectrum (for example, a lattice) and, therefore, the spectrum of the combined system does not have an upper bound; consequently, negative temperatures can be introduced consistently only for the nonequilibrium case. For experimental confirmations of the existence of negative temperatures, see the papers [123, 124].

The important case of the action of an alternating magnetic field on spin systems is a nonstationary nonequilibrium process. However, this problem can be formally reduced to the stationary case, if we first eliminate the alternating field of frequency ω by transforming to a coordinate frame rotating with frequency ω, and then, in this system, introduce the nonequilibrium statistical operator. This was done by Buishvili [46] and leads naturally to the concept of a rotating-frame temperature. Unfortunately, if there is interaction between the spin system and the lattice it is impossible to eliminate completely the time dependence of the Hami tonian by transforming to the rotating coordinate frame. Therefore, one first obtains equations disregarding the lattice, and then introduces into the resulting equations the appropriate terms describing the influence of the lattice. Another more consistent method of taking the alternating field into account, also developed by Buishvili [46a], consists in first regarding the alternating clas cal field as a quantum subsystem, and then making its temperature tend to infinity, while the quantum correlators of the field variables are replaced by classical correlators. Allowing for the alternating magnetic field enables us to use the nonequilibrium statistical operator to study the dynamic polarization of nuclei. For other applications of this method to nuclear magnetic relaxation, see [193-200, 250]. Pokrovskiĭ [157], Muller [191], and Bikl and Kalashnikov [242] have taken an alternating field into account in the method of the nonequilibrium statistical operator.

23.3. Spin-Lattice Relaxation of Conduction Electrons in Semiconductors in a Magnetic Field [53b]

We shall study one more example of the application of the no equilibrium statistical operator in the theory of relaxation processes, namely, spin-lattice relaxation of conduction electrons in a quantizing magnetic field, following the work of Kalashnikov [53 This problem was studied by a kinetic-equation method in references [125, 126].

We write the Hamiltonian of conduction electrons interacting with optical phonons in a magnetic field in the form

$$H = H_k + H_s + H_p + H_{ep} + H_{pl} + H_l, \qquad (23.44$$

where H_k is the kinetic energy of the electrons, and H_s is their Zeeman energy. The sum of H_k and H_s, i.e., the energy H_e of fre electrons in a magnetic field, is equal to

$$H_e = H_k + H_s = \sum_{\nu\sigma} \varepsilon_{\nu\sigma} a^+_{\nu\sigma} a_{\nu\sigma}, \qquad (23.44a$$

where

$$\varepsilon_{\nu\sigma} = \frac{p_z^2}{2m} + \hbar\omega_0\left(n + \frac{1}{2}\right) + \frac{1}{2}\,\sigma g \mu_0 H \qquad (\sigma = \pm 1) \qquad (23.44b$$

are the energy levels of a free electron in a magnetic field parallel to the z axis; $\nu = (n, p_x, p_z)$ are its quantum numbers, g is the spectroscopic splitting factor of the conduction electrons, μ_0 is the Bohr magneton, $\omega_0 = eH/mc$ is the Larmor frequency, and the last term in (23.44b) gives the Zeeman energy H_s of the electrons. Further,

$$H_p = \sum_{q\lambda} \hbar\Omega_{q\lambda} C^{+}_{q\lambda} C_{q\lambda}. \tag{23.44c}$$

is the energy of the optical phonons, where **q** and λ are respectively the momentum and polarization index of the optical phonon with energy $\hbar\Omega_{q\lambda}$;

$$H_{ep} = \sum_{\substack{\nu\sigma,\, \nu'\sigma' \\ q,\, \lambda}} \left(U^{q\lambda}_{\nu'\sigma',\, \nu\sigma} C_{q\lambda} + \overset{*}{U}{}^{q\lambda}_{\nu\sigma,\, \nu'\sigma'} C^{+}_{q\lambda}\right) a^{+}_{\nu'\sigma'} a_{\nu\sigma} \tag{23.44d}$$

is the electron–phonon interaction; H_{pl} is the energy of the interaction of the optical phonons with the thermostat; H_l is the energy of the thermostat. For example, if the nonelectronic relaxation of the optical phonons is connected with their decay into two acoustic phonons and with the inverse process, then

$$H_{pl} = \sum_{qq'\lambda\lambda'\lambda''} \left(\Phi^{qq'q-q'}_{\lambda\lambda'\lambda''} C_{q\lambda} b^{+}_{q'\lambda'} b^{+}_{q-q',\, \lambda''} + \overset{*}{\Phi}{}^{qq'q-q'}_{\lambda\lambda'\lambda''} C^{+}_{q\lambda} b_{q'\lambda'} b_{q-q',\, \lambda''}\right), \tag{23.44e}$$

$$H_l = \sum_{q\lambda} \hbar\omega_{q\lambda} b^{+}_{q\lambda} b_{q\lambda}, \tag{23.44f}$$

where $\omega_{q\lambda}$ is the frequency of the acoustic phonons.

We shall regard H_s, H_p and the remaining part of the Hamiltonian of the crystal as weakly interacting subsystems. The mutual exchange of energy between the subsystems is described by the operator equations

$$\begin{aligned} \frac{dH_s}{dt} &= \frac{1}{i\hbar}[H_s, H] = \frac{1}{i\hbar}[H_s, H_{ep}] \equiv \dot{H}_{s\,(p)}, \\ \frac{dH_p}{dt} &= \frac{1}{i\hbar}[H_p, H] = \frac{1}{i\hbar}[H_p, H_{ep} + H_{pl}] \equiv \dot{H}_{p\,(e)} + \dot{H}_{p\,(l)}. \end{aligned} \tag{23.45}$$

Corresponding to the chosen subsystems is the stationary nonequilibrium operator

$$\rho = Q^{-1} \exp\{-\beta_s \tilde{H}_s - \beta_p \tilde{H}_p - \beta(H - \tilde{H}_s - \tilde{H}_p - \mu N)\}, \tag{23.46}$$

where

$$\tilde{H}_s = H_s - \int_{-\infty}^{0} e^{\varepsilon t} \dot{H}_{s\,(p)}(t)\,dt,$$
$$\tilde{H}_p = H_p - \int_{-\infty}^{0} e^{\varepsilon t} (\dot{H}_{p\,(e)}(t) + \dot{H}_{p\,(l)}(t))\,dt; \tag{23.47}$$

and N is the total number of electrons. Consequently,

$$\rho = Q^{-1} \exp\Big\{- \beta_s H_s - \beta_p H_p - \beta (H - H_s - H_p - \mu N) + \int_{-\infty}^{0} e^{\varepsilon t}\, (\beta_s - \beta)\, \dot{H}_{s\,(p)}(t)\,dt + \int_{-\infty}^{0} e^{\varepsilon t} (\beta_p - \beta) (\dot{H}_{p\,(e)}(t) + \dot{H}_{p\,(l)}(t))\,dt\Big\}, \tag{23.48}$$

where β_s^{-1} is the spin temperature of the current carriers, and β_p^{-1} is the temperature of the "hot" phonons. The other degrees of freedom of the crystal are assigned a constant temperature β^{-1}.

In the linear approximation in the thermodynamic forces $\beta_s - \beta$ and $\beta_p - \beta$, we write the nonequilibrium statistical operator (23.48) in the form

$$\rho = \Big\{1 - (\beta_s - \beta)\,\beta^{-1} \int_0^{\beta} d\tau \Big[H_s(i\hbar\tau) - \langle H_s\rangle_0 - \beta \frac{\partial \mu}{\partial \beta_s} (N(i\hbar\tau) - \langle N \rangle)\Big] - (\beta_p - \beta)\,\beta^{-1} \int_0^{\beta} d\tau\, [H_p(i\hbar\tau) - \langle H_p\rangle_0] + (\beta_s - \beta)\,\beta^{-1} \int_0^{\beta} d\tau \int_{-\infty}^{0} dt\, e^{\varepsilon t} \dot{H}_{s\,(p)}(t + i\hbar\tau) + (\beta_p - \beta)\,\beta^{-1} \int_0^{\beta} d\tau \int_{-\infty}^{0} dt\, e^{\varepsilon t}\, (\dot{H}_{p\,(e)}(t + i\hbar\tau) + \dot{H}_{p\,(l)}(t + i\hbar\tau))\Big\}\,\rho_0, \tag{23.49}$$

where the nonequilibrium chemical potential has been expanded in $\beta_s - \beta$, and ρ_0 is the equilibrium statistical operator disregarding the interaction.

We find the quantity $\partial\mu/\partial\beta_s$ from the condition $\langle N\rangle_l = \langle N\rangle_0$, i.e.,

$$(N, H_s) = \beta \frac{\partial\mu}{\partial\beta_s} (N, N). \qquad (23.49a)$$

Averaging the balance equations (23.45) with the operator (23.49) we obtain the relaxation equations

$$\begin{aligned} \frac{d}{dt}\langle H_s\rangle &= (\beta_s - \beta) L_{ss} + (\beta_p - \beta) L_{sp}, \\ \frac{d}{dt}\langle H_p\rangle &= (\beta_s - \beta) L_{ps} + (\beta_p - \beta) L_{pp}, \end{aligned} \qquad (23.50)$$

where

$$\begin{aligned} L_{ss} &= \int_{-\infty}^{0} dt\, e^{\varepsilon t} (\dot{H}_{s(p)}, \dot{H}_{s(p)}(t)), \\ L_{pp} &= \int_{-\infty}^{0} dt\, e^{\varepsilon t} (\dot{H}_{p(e)} + \dot{H}_{p(l)}, \dot{H}_{p(e)}(t) + \dot{H}_{p(l)}(t)), \\ L_{sp} &= \int_{-\infty}^{0} dt\, e^{\varepsilon t} (\dot{H}_{s(p)}, \dot{H}_{p(e)}(t) + \dot{H}_{p(l)}(t)), \\ L_{ps} &= \int_{-\infty}^{0} dt\, e^{\varepsilon t} (\dot{H}_{p(e)} + \dot{H}_{p(l)}, \dot{H}_{s(p)}(t)) \end{aligned} \qquad (23.50a)$$

are the kinetic coefficients.

For the calculation of the kinetic coefficients (23.50a) in powers of the electron–phonon interaction (23.44d), see the paper [53c]. The calculational scheme described in this subsection has been applied by Kalashnikov to the theory of spin-lattice relaxation in semiconductors with magnetic impurities in a quantizing magnetic field [53c] and to the theory of hot electrons [54]. (See also [201–203, 206, 242, 244, 245, 247].)

23.4. Energy Exchange between Two Weakly Interacting Subsystems [55]

Up to this point, we have everywhere confined ourselves to the case of weakly nonequilibrium systems, when it is sufficient to take into account only terms that are linear in the thermodynamic forces in the expressions for the fluxes, i.e., we have considered linear dissipative processes. In the theory of transport processes, one often encounters cases in which the linear approximation is not valid; for example, the rates of chemical reactions are usually nonlinear in the thermodynamic forces [27], and the electrical conductivity in semiconductors in a strong electric field may be essentially nonlinear [54]. The method of the nonequilibrium statistical operator also makes it possible to study such strongly nonequilibrium processes, i.e., it is also applicable in cases in which the usual Kubo method described in Chapter III can no longer be used.

In order to take nonlinear effects into account, we shall expand the statistical operator not in the thermodynamic forces, wh are no longer small, but in other small parameters, if such exist in the problem.

Following the work of Pokrovskii [55], we shall consider the exchange of energy between two weakly interacting subsystems, when this proceeds slowly, for example, because of a large difference in the masses of the components or, in general, because of the smallness of the appropriate effective cross section. As we shall see below, such systems may be characterized by widely different temperatures and the energy-exchange process is nonlinear in the thermodynamic forces.

We take the Hamiltonian of the system in the form

$$H = H_1 + H_2, \tag{23.51}$$

where H_1 and H_2 are the Hamiltonians of the subsystems

$$\begin{gathered} H_1 = \sum_\alpha E_\alpha a_\alpha^+ a_\alpha + u, \qquad H_2 = \sum_\mu E_\mu b_\mu^+ b_\mu, \\ u = \sum_{\alpha\mu\alpha'\mu'} \langle \alpha\mu \,|\, \Phi \,|\, \alpha'\mu' \rangle \, a_\alpha^+ b_\mu^+ b_{\mu'} a_{\alpha'}; \end{gathered} \tag{23.52}$$

α and μ are the quantum numbers of particles of the first and second types, and Φ is the interaction potential between them. For simplicity, we omit the interaction between identical particles. We may assume that this is included in the renormalized values of the energies E_α and E_μ of the elementary excitations, as is done in the theory of quantum liquids [127, 128].

We note that, since the total Hamiltonian is an integral of motion, when we divide the system into subsystems only one of the latter can be chosen independently, and it is of no importance to which subsystem the small interaction energy is assigned. If the interaction energy is important in the energy balance, it can be regarded as a separate energy reservoir; for example, the dipole–dipole interaction can be taken into account in this way (cf. §23.2).

The operators of the energy fluxes between the subsystems are equal to

$$J_1 = \dot{H}_1 = \frac{1}{i\hbar}[H_1, H] = \frac{1}{i\hbar}\sum_{\alpha\mu\alpha'\mu'}(E_{\mu'} - E_{\mu})\langle\alpha\mu\,|\,\Phi\,|\,\alpha'\mu'\rangle a_\alpha^+ b_\mu^+ b_{\mu'} a_{\alpha'},$$

$$J_2 = \dot{H}_2 = -J_1. \qquad (23.53)$$

The relations (23.53) give the balance equations for the dynamic variables and enable us to construct the nonequilibrium statistical operator. Because of the slowness of the energy exchange, the temperatures of the subsystems will vary slowly with time and we can confine ourselves to the stationary variant of the theory. Allowance for the fact that the process is nonstationary leads to terms of higher order of smallness in the expression for the energy flux.

Following the general method, we construct the statistical operator

$$\rho = Q^{-1}\exp\{-\beta_1(\widetilde{\widetilde{H}}_1 - \mu_1 N_1) - \beta_2(\widetilde{\widetilde{H}}_2 - \mu_2 N_2)\}$$
$$= Q^{-1}\exp\Big\{-\beta_1(H_1 - \mu_1 N_1) - \beta_2(H_2 - \mu_2 N_2) +$$
$$+ \int_{-\infty}^{0} e^{\varepsilon t}(\beta_1 - \beta_2)\dot{H}_1(t)\,dt\Big\}, \qquad (23.54)$$

where β_1 and β_2 are the inverse temperatures, μ_1 and μ_2 are the chemical potentials, and N_1 and N_2 are the particle-number operators of the subsystems.

The operator $\dot{H}_1$ contains a small parameter, since the energy exchange is assumed to be slow. We average the flux (23.53) over the distribution (23.54), confining ourselves to terms of second order in the small parameter. We obtain

$$\langle \dot{H}_1 \rangle = (\beta_1 - \beta_2) \int_{-\infty}^{0} dt\, e^{\varepsilon t} \int_0^1 d\tau \langle \dot{H}_1 e^{-\tau B} \dot{H}_1(t) e^{\tau B} \rangle_l, \tag{23.55}$$

where $\langle \ldots \rangle_l$ denotes averaging over the quasi-equilibrium distribution

$$\rho_l = Q_l^{-1} e^{-B}, \tag{23.56}$$

$$B = \beta_1 (H_1 - \mu_1 N_1) + \beta_2 (H_2 - \mu_2 N_2), \tag{23.57}$$

where the interaction u has been omitted in H_1.

The relation (23.55) has the appearance of a linear relation between the thermodynamic force $\beta_1 - \beta_2$ and the flux $\langle \dot{H}_1 \rangle$; in fact, however, it is nonlinear in $\beta_1 - \beta_2$, since the averaging is performed not over the equilibrium distribution, but over the quasi-equilibrium distribution (23.56).

The relation (23.55) is also valid in the case when the interaction between the particles is taken into account in the operators H_1 and H_2. Then the operator (23.53) will include additional terms depending on the interaction potentials between identical particles.

We shall calculate the average energy flux (23.55). Substituting (23.53), into (23.55) and integrating over τ, we obtain

$$\langle \dot{H}_1 \rangle = - \int_{-\infty}^{0} dt\, e^{\varepsilon t} (\beta_1 - \beta_2) \sum_{\substack{\alpha\mu\alpha'\mu' \\ \alpha_1\mu_1\alpha_1'\mu_1'}} \frac{1}{i\hbar} \left(E_{\mu_1'} - E_{\mu_1} \right) (E_{\mu'} - E_{\mu}) \times$$

$$\times \langle \alpha\mu | \Phi | \alpha'\mu' \rangle \langle \alpha_1\mu_1 | \Phi | \alpha_1'\mu_1' \rangle \frac{G^{\alpha\mu\alpha'\mu'}_{\alpha_1'\mu_1'\alpha_1\mu_1}(-t)}{\beta_1 (E_\alpha - E_{\alpha'}) + \beta_2 (E_\mu - E_{\mu'})}, \tag{23.58}$$

where we have introduced the Green function

$$G^{\alpha\mu\alpha'\mu'}_{\alpha_1'\mu_1'\alpha_1\mu_1}(t - t_1) =$$

$$= (i\hbar)^{-1} \theta(t - t_1) \left\langle \left[a_\alpha^+ b_\mu^+ b_{\mu'} a_{\alpha'},\, e^{\frac{-iH(t-t_1)}{\hbar}} a_{\alpha_1}^+ b_{\mu_1}^+ b_{\mu_1'} a_{\alpha_1'} e^{\frac{iH(t-t_1)}{\hbar}} \right] \right\rangle_l \tag{23.59}$$

which is a generalization to the case of a quasi-equilibrium ensemble of the two-time Green functions (15.48) studied in §16.

In the case of weak interaction, in the Green function (23.59) we may neglect the interaction in the Heisenberg picture for the operators, since in (23.58) there is already a factor of second order in the interaction. Therefore, the Green function (23.59) can be calculated directly by pairing the operators in accordance with Wick's theorem:

$$G^{\alpha\mu\alpha'\mu'}_{\alpha_1'\mu_1'\alpha_1\mu_1}(t-t_1) = (i\hbar)^{-1}\theta(t-t_1)e^{\frac{i}{\hbar}(E_\alpha+E_\mu-E_{\alpha'}-E_{\mu'})(t-t_1)}\times$$
$$\times\{n_\alpha n_\mu(1\pm n_{\alpha'})(1\pm n_{\mu'})-n_{\alpha'}n_{\mu'}(1\pm n_\alpha)(1\pm n_\mu)\}\times$$
$$\times\,\delta_{\alpha\alpha_1'}\delta_{\mu\mu_1'}\delta_{\mu'\mu_1}\delta_{\alpha'\alpha_1}, \qquad (23.60)$$

where the upper sign is taken for Bose statistics and the lower for Fermi statistics, and n_α and n_μ are the occupation numbers,

$$n_\alpha = \left(e^{\beta_1(E_\alpha-\mu_1)}\mp 1\right)^{-1}, \qquad n_\mu = \left(e^{\beta_2(E_\mu-\mu_2)}\mp 1\right)^{-1}. \qquad (23.60a)$$

Substituting (23.60) into (23.58), we obtain

$$\langle\dot{H}_1\rangle = -(\beta_1-\beta_2)\sum_{\alpha\mu\alpha'\mu'}\frac{1}{\hbar^2}(E_\mu-E_{\mu'})^2\frac{|\langle\alpha\mu|\Phi|\alpha'\mu'\rangle|^2}{\beta_1(E_\alpha-E_{\alpha'})+\beta_2(E_\mu-E_{\mu'})}\times$$
$$\times\int_{-\infty}^{0} e^{-\frac{i}{\hbar}(E_\alpha+E_\mu-E_{\alpha'}-E_{\mu'})t+\varepsilon t}\,dt\times$$
$$\times\{n_\alpha n_\mu(1\pm n_{\alpha'})(1\pm n_{\mu'})-n_{\alpha'}n_{\mu'}(1\pm n_\alpha)(1\pm n_\mu)\}. \qquad (23.61)$$

Performing the time integration in (23.61), taking the relation (16.32) into account and noting that the principal-value integrals give no contribution, we obtain

$$\langle\dot{H}_1\rangle = \sum_{\alpha\mu\alpha'\mu'}E_\mu w^{\alpha'\mu'}_{\alpha\mu}\{n_\alpha n_\mu(1\pm n_{\alpha'})(1\pm n_{\mu'})-n_{\alpha'}n_{\mu'}(1\pm n_\alpha)(1\pm n_\mu)\}, \qquad (23.62)$$

where

$$w^{\alpha'\mu'}_{\alpha\mu} = \frac{2\pi}{\hbar}|\langle\alpha\mu|\Phi|\alpha'\mu'\rangle|^2\delta(E_\alpha+E_\mu-E_{\alpha'}-E_{\mu'}) \qquad (23.62a)$$

is the transition probability per unit time in the Born approximation.

We note that the factor $\beta_1-\beta_2$ was cancelled in going from (23.61) to (23.62), although the nonlinear effect of the dependence on β_1 and β_2 remained, since n_α and $n_{\alpha'}$ depend on β_1 and n_μ and $n_{\mu'}$ on β_2.

For a nondegenerate gas with $\mu_1 = \mu_2$, we have

$$n_\alpha n_\mu (1 \pm n_{\alpha'})(1 \pm n_{\mu'}) - n_{\alpha'} n_{\mu'} (1 \pm n_\alpha)(1 \pm n_\mu) \cong$$
$$\cong n_\alpha n_\mu \left(1 - \frac{n_{\alpha'} n_{\mu'}}{n_\alpha n_\mu}\right) = n_\alpha n_\mu \left(1 - e^{-(\beta_1 - \beta_2)(E_\mu - E_{\mu'})}\right),$$

where we have used the fact that energy is conserved in a collision:

$$E_\alpha + E_\mu = E_{\alpha'} + E_{\mu'}.$$

Consequently,

$$\langle \dot{H}_1 \rangle = \sum_{\alpha\mu\alpha'\mu'} E_\mu w_{\alpha\mu}^{\alpha'\mu'} n_\alpha n_\mu \left(1 - e^{-(\beta_1 - \beta_2)(E_\mu - E_{\mu'})}\right). \qquad (23.62b)$$

Thus, we have obtained an expression for the energy flux that is nonlinear in the thermodynamic forces $\beta_1 - \beta_2$ and agrees with the result which follows from the kinetic equation.

The connection of Eq. (23.62) with the kinetic equation is obvious, since it can be written in the form

$$\langle \dot{H}_1 \rangle = - \langle \dot{H}_2 \rangle = - \sum_\mu E_\mu \frac{\partial n_\mu}{\partial t}, \qquad (23.63)$$

where

$$\frac{\partial n_\mu}{\partial t} = - \sum_{\alpha\alpha'\mu'} w_{\alpha\mu}^{\alpha'\mu'} \{ n_\alpha n_\mu (1 \pm n_{\alpha'})(1 \pm n_{\mu'}) - n_{\alpha'} n_{\mu'} (1 \pm n_\alpha)(1 \pm n_\mu) \} \qquad (23.64)$$

is the *kinetic equation* for the occupation numbers. The right-hand side of Eq. (23.64) is the collision integral.

The kinetic equation (23.64) can be derived directly, if we average the operator

$$\dot{n}_\mu = \frac{\partial}{\partial t} (b_\mu^+ b_\mu)$$

over the statistical distribution (23.54) expanded in powers of the interaction, as was done in the calculation of $\langle \dot{H}_1 \rangle$.

It can be seen from Eq. (23.63) that, in the nonlinear theory, in place of a product of thermodynamic forces with fluxes, we have a sum of products of the energies of the subsystems with the collision integrals.

For small $\beta_1 - \beta_2$, expanding the exponential in (23.62b) in a series in $\beta_1 - \beta_2$ and retaining the linear terms, we obtain a linear relation between the thermodynamic forces and the flux:

$$\langle \dot{H}_1 \rangle = L_{\dot{H}_1 \dot{H}_1} (\beta_1 - \beta_2), \tag{23.65}$$

where

$$L_{\dot{H}_1 \dot{H}_1} = \sum_{\alpha\mu\alpha'\mu'} \frac{1}{2} (E_\mu - E_{\mu'})^2 \, w_{\alpha\mu}^{\alpha'\mu'} n_\alpha n_\mu \tag{23.65a}$$

is the kinetic coefficient for the rate of energy transfer.

The rate (23.58) of energy transfer can also be calculated easily in the case of low density of particles (or of elementary excitations). In order to write out the expressions (23.58) explicitly, we need to calculate the Green function (23.59) for low density. Differentiating (23.59) with respect to t, we obtain an equation for the Green function (23.60):

$$\begin{aligned} i\hbar \frac{\partial}{\partial t} G_{\alpha_1'\mu_1'\alpha_1\mu_1}^{\alpha\mu\alpha'\mu'}(t) + \sum_{\alpha_2\mu_2} \Big\{ \langle \alpha_1'\mu_1' \,|\, H^{(2)} \,|\, \alpha_2\mu_2 \rangle \, G_{\alpha_2\mu_2\alpha_1\mu_1}^{\alpha\mu\alpha'\mu'}(t) - \\ - G_{\alpha_1'\mu_1'\alpha_2\mu_2}^{\alpha\mu\alpha'\mu'}(t) \langle \alpha_2\mu_2 \,|\, H^{(2)} \,|\, \alpha_1\mu_1 \rangle \Big\} + F_{\alpha_1'\mu_1'\alpha_1\mu_1}^{\alpha\mu\alpha'\mu'}(t) = \\ = \delta(t)\, \delta_{\alpha\alpha_1'} \delta_{\mu\mu_1'} \delta_{\alpha'\alpha_1} \delta_{\mu'\mu_1} K_{\alpha\mu\alpha'\mu'}, \end{aligned} \tag{23.66}$$

where $H^{(2)}$ is the two-particle Hamiltonian, F(t) is a term containing Green functions of higher order, which we shall not write out explicitly, and K is the average commutator of the operators $\overset{+}{a}_\alpha \overset{+}{b}_\mu b_{\mu'} a_{\alpha'}$ and $\overset{+}{a}_{\alpha'} \overset{+}{b}_{\mu'} b_\mu a_\alpha$.

We shall consider the limiting case of a low-density gas, when we can confine ourselves to the binary-collision approximation. Then the term F(t) describing collisions of higher order can be omitted, thereby closing the chain of equations for the Green functions. The average commutator K in this approximation is equal to

$$\begin{aligned} K_{\alpha\mu\alpha'\mu'} = i\hbar G_{\alpha\mu\alpha'\mu'}^{\alpha\mu\alpha'\mu'}(+0) \cong \\ \cong n_\alpha n_\mu (1 \pm n_{\alpha'})(1 \pm n_{\mu'}) - n_{\alpha'} n_{\mu'} (1 \pm n_\alpha)(1 \pm n_\mu). \end{aligned} \tag{23.67}$$

In our approximation ($n_\alpha \ll 1$, $n_\mu \ll 1$), we can assume the distribution to be a Boltzmann distribution.

The solution of Eq. (23.66) has the form

$$G^{\alpha\mu\alpha'\mu'}_{\alpha_1'\mu_1'\alpha_1\mu_1}(t) = \frac{1}{i\hbar}\,\theta(t)\left\langle \alpha'\mu' \left| e^{-\frac{i}{\hbar}H^{(2)}t} \right| \alpha_1\mu_1 \right\rangle \left\langle \alpha_1'\mu_1' \left| e^{\frac{i}{\hbar}H^{(2)}t} \right| \alpha\mu \right\rangle K_{\alpha\mu\alpha'\mu'}. \tag{23.68}$$

Substituting (23.68) into (23.58), we obtain

$$\langle \dot{H}_1 \rangle = -\int_{-\infty}^{0} dt\, e^{\varepsilon t}\,(\beta_1 - \beta_2)\frac{1}{(i\hbar)^2} \sum_{\substack{\alpha\mu\alpha'\mu' \\ \alpha_1\mu_1\alpha_1'\mu_1'}} \left\langle \alpha\mu \left| \Phi e^{\frac{i}{\hbar}H^{(2)}t} \right| \alpha'\mu' \right\rangle \times$$

$$\times \langle \alpha'\mu' \,|\, (h_2(-t) - E_\mu)\, f_{\alpha\mu}(h_1(-t), h_2(-t)) \,|\, \alpha_1\mu_1 \rangle \times$$

$$\times \left\langle \alpha_1\mu_1 \left| \Phi e^{-\frac{i}{\hbar}H^{(2)}t} \right| \alpha_1'\mu_1' \right\rangle \times \langle \alpha_1'\mu_1' \,|\, h_2(t) - E_{\mu_1} \,|\, \alpha\mu \rangle, \tag{23.69}$$

where

$$H^{(2)} = h_1 + h_2 + \Phi, \tag{23.70}$$

and h_1 and h_2 are the single-particle Hamiltonians of particles of the first and second type; their time argument indicates the Heisenberg picture;

$$f_{\alpha\mu}(E_{\alpha'}, E_{\mu'}) = K_{\alpha\mu\alpha'\mu'}\,[\beta_1(E_\alpha - E_{\alpha'}) + \beta_2(E_\mu - E_{\mu'})]^{-1}$$

$$f_{\alpha\mu}(h_1(-t), h_2(-t)) = e^{-(i/\hbar)H^{(2)}t} f_{\alpha\mu}(h_1, h_2)\, e^{(i/\hbar)H^{(2)}t} \tag{23.71}$$

We note that, in the matrix elements

$$\langle \alpha'\mu' \,|\, (h_2(-t) - E_\mu)\, f_{\alpha\mu}(h_1(-t), h_2(-t)) \,|\, \alpha_1\mu_1 \rangle$$

and

$$\langle \alpha_1'\mu_1' \,|\, h_2(t) - E_{\mu_1} \,|\, \alpha\mu \rangle$$

we can omit the dependence on t, which is effectively equivalent to neglecting terms of order $v^3t^3/V = (vt/L)^3$ in comparison with unity, where v is the relative velocity of the colliding particles and $V = L^3$ is the volume of the system (cf. [214]).

In fact, because of the factor $e^{\varepsilon t}$, the integrand in (23.69) is appreciably different from zero only at times $t \sim \varepsilon^{-1}$; therefore, $v^3t^3/V \sim v^3/V\varepsilon^3$, whence it follows that, with the correct order of the limits, when first $V \to \infty$ and then $\varepsilon \to 0$, the time dependence of the matrix elements can be neglected in our case. In this case, waves reflected from the boundaries of the volume are automatically excluded [214]. A similar procedure is discussed in the formal theory of scattering of Gell-Mann and Goldberger [84] (see Appendix I).

It follows from the formal theory of scattering that, for large times considerably greater than the collision time τ_c, the matrix elements of the pair-interaction operator with the two-particle evolution operator, $\Phi e^{\pm (i/\hbar) H^{(2)} t}$ can be expressed in terms of the scattering matrix [85], i.e., for $|t| \gg \tau_c$,

$$\left\langle \alpha\mu \left| \Phi e^{\pm \frac{i}{\hbar} H^{(2)} t} \right| \alpha'\mu' \right\rangle = \\ = e^{\pm \frac{i}{\hbar} (E_{\alpha'} + E_{\mu'}) t} \begin{cases} \langle \alpha\mu | T | \alpha'\mu' \rangle & \text{for} \quad +, \\ \langle \alpha\mu | T^{+} | \alpha'\mu' \rangle & \text{for} \quad -, \end{cases} \tag{23.72}$$

where $\langle \alpha\mu | T | \alpha'\mu' \rangle$ are the matrix elements of the scattering T-matrix. Substituting the asymptotic expressions for the matrix elements into (23.69) and integrating over the time, we obtain the previous equation (23.62), but with the transition probability expressed in terms of the T-matrix:

$$w_{\alpha\mu}^{\alpha'\mu'} = \frac{2\pi}{\hbar} | \langle \alpha\mu | T | \alpha'\mu' \rangle |^2 \delta (E_\alpha + E_\mu - E_{\alpha'} - E_{\mu'}). \tag{23.73}$$

Therefore, all the previous conclusions are conserved for the low-density case.

In particular, n_μ also satisfies the kinetic equation (23.64), but the transition probability (23.73) in (23.64) corresponds not to a small interaction, as it did before, but to a low density.

That the entropy production is positive was proved earlier only for linear dissipative processes. We shall consider the entropy production for a nonlinear process of energy exchange between subsystems.

According to the general definition (20.13), the entropy in our case is equal to

$$S = - \langle \ln \rho_l \rangle_l = - \langle \ln \rho_l \rangle, \tag{23.74}$$

i.e.,

$$S = \sum_i \beta_i \langle H_i \rangle + \ln Q_l. \tag{23.75}$$

Differentiating with respect to time and using the fact that

$$\frac{d \ln Q_l}{dt} = -\sum_i \dot{\beta}_i \langle H_i \rangle, \tag{23.76}$$

we obtain

$$\frac{dS}{dt} = \sum_i \beta_i \langle \dot{H}_i \rangle = (\beta_1 - \beta_2) \langle \dot{H}_1 \rangle. \tag{23.77}$$

We shall show that

$$\frac{dS}{dt} = (\beta_1 - \beta_2)^2 \int_{-\infty}^{0} e^{\varepsilon t}\, dt \int_0^1 d\tau \, \langle \dot{H}_1 e^{-\tau B} \dot{H}_1(t)\, e^{\tau B} \rangle_l > 0. \tag{23.78}$$

The inequality (23.78) has already been used earlier for the case when the averaging was performed over the equilibrium distribution.

Taking (23.62) into account, we write the entropy production (23.77) in the form

$$\frac{dS}{dt} = \frac{1}{2} \sum_{\alpha\mu\alpha'\mu'} (\beta_1 - \beta_2)(E_\mu - E_{\mu'})\, w_{\alpha\mu}^{\alpha'\mu'} \times$$
$$\times \{ n_\alpha n_\mu (1 \pm n_{\alpha'})(1 \pm n_{\mu'}) - n_{\alpha'} n_{\mu'} (1 \pm n_\alpha)(1 \pm n_\mu) \}; \tag{23.79}$$

consequently,

$$\frac{dS}{dt} = \frac{1}{2} \sum_{\alpha\mu\alpha'\mu'} w_{\alpha\mu}^{\alpha'\mu'} n_\alpha n_\mu (1 \pm n_{\alpha'})(1 \pm n_{\mu'}) \times$$
$$\times (\beta_1 - \beta_2)(E_\mu - E_{\mu'}) \left(1 - e^{-(\beta_1 - \beta_2)(E_\mu - E_{\mu'})}\right) \geqslant 0, \tag{23.80}$$

since, for any x,

$$x(1 - e^{-x}) \geqslant 0 \tag{23.81}$$

and all the other factors in (23.80) are positive.

Thus, we have proved that, in the nonlinear process considered, the entropy production is positive.

23.5. Rates of Chemical Reactions

Chemical reactions in a homogeneous phase are an example of nonlinear irreversible processes, similar to relaxation processes to which the method of the nonequilibrium statistical operator can be very simply applied.

We shall assume that the chemical reactions occur sufficiently slowly, so that there is time for a spatially homogeneous state, with equal temperatures of the reagents and reaction products, to be established in the volume in which the reaction is occurring. We shall consider the simplest reactions in such a system.

Suppose that only one binary reaction, between the molecules A and B with formation of the molecules C and D, occurs in the system, i.e.,

$$A+B\rightleftarrows C+D. \tag{23.82}$$

We shall assume for simplicity that the reaction occurs in the gas phase.

We take the Hamiltonian of the system in the form

$$H=\sum_{i=1}^{4}H_i+u, \tag{23.83}$$

where H_1 and H_2 are the Hamiltonians of the reagents, and H_3 and H_4 are the Hamiltonians of the reaction products; u is the interaction leading to the reaction. For it, we take the model form

$$u=\sum_{\alpha_1\alpha_2\alpha_1'\alpha_2'}\left\{\Phi_{\alpha_1\alpha_2}^{\alpha_1'\alpha_2'}a_{\alpha_1}^{+}b_{\alpha_2}^{+}c_{\alpha_1'}d_{\alpha_2'}+\overset{*}{\Phi}{}_{\alpha_1\alpha_2}^{\alpha_1'\alpha_2'}d_{\alpha_2'}^{+}c_{\alpha_1'}^{+}b_{\alpha_2}a_{\alpha_1}\right\}, \tag{23.84}$$

where $a_{\alpha_1}^{+}$, $b_{\alpha_2}^{+}$, $c_{\alpha_1'}^{+}$, and $d_{\alpha_2'}^{+}$ are creation operators for the molecules A, B, C, and D in the states α_1, α_2, α_1', and α_2', and a_{α_1}, b_{α_2}, $c_{\alpha_1'}$, and $d_{\alpha_2'}$ are annihilation operators for the molecules in the corresponding states. Thus, the second term in (23.84) describes the forward reaction of (23.82) and the first term describes the back reaction. The operator (23.84) is analogous to the corresponding interaction operator (23.52) considered in §23.4 of this chapter.

The model nature of the Hamiltonian (23.83) consists in the fact that we are assuming the matrix elements $\Phi_{\alpha_1\alpha_2}^{\alpha_1'\alpha_2'}$ to be known from quantum mechanical calculations. Such a Hamiltonian is suitable for the theory of chemical reactions in the gas phase, since in a liquid, in collisions of the reagent molecules, part of the energy will be transferred to the liquid in the form of elementary excitations and this process must be taken into account in the interaction operator.

The operator of the total particle number

$$N = \sum_{i=1}^{4} N_i, \tag{23.85}$$

where

$$\begin{aligned} N_1 &= \sum_{\alpha} a_{\alpha}^{+} a_{\alpha}, \quad N_2 = \sum_{\alpha} b_{\alpha}^{+} b_{\alpha}, \\ N_3 &= \sum_{\alpha} c_{\alpha}^{+} c_{\alpha}, \quad N_4 = \sum_{\alpha} d_{\alpha}^{+} d_{\alpha}, \end{aligned} \tag{23.85a}$$

is conserved in time, since

$$\dot{N} = \frac{1}{i\hbar} [N, H] = 0, \tag{23.86}$$

although N_i varies in time as a result of the reaction,

$$\dot{N}_i = \frac{1}{i\hbar} [N_i, H] \qquad (i = 1, 2, 3, 4). \tag{23.87}$$

All these fluxes are expressed in terms of one reaction-rate operator

$$J = \dot{N}_1 = \sum_{\alpha_1 \alpha_2 \alpha_1' \alpha_2'} \frac{1}{i\hbar} \left\{ \Phi_{\alpha_1 \alpha_2}^{\alpha_1' \alpha_2'} a_{\alpha_1}^{+} b_{\alpha_2}^{+} c_{\alpha_1'} d_{\alpha_2'} - \overset{*}{\Phi}{}_{\alpha_1 \alpha_2}^{\alpha_1' \alpha_2'} d_{\alpha_2'}^{+} c_{\alpha_1'}^{+} b_{\alpha_2} a_{\alpha_1} \right\}, \tag{23.88}$$

$$\dot{N}_1 = J, \quad \dot{N}_2 = J, \quad \dot{N}_3 = -J, \quad \dot{N}_4 = -J \tag{23.89}$$

or

$$\dot{N}_i = \nu_i J, \tag{23.89a}$$

where ν_i are the stoichiometry numbers in the equation for the reaction, i.e., $\nu_1 = \nu_2 = 1$ and $\nu_3 = \nu_4 = -1$.

Corresponding to the conservation laws (23.89a), we have the statistical operator

$$\begin{aligned} \rho &= Q^{-1} \exp\left\{ -\beta \left(H - \sum_i \mu_i \tilde{\tilde{N}}_i \right) \right\} \\ &= Q^{-1} \exp\left\{ -\beta \left(H - \sum_i \mu_i N_i + \sum_i \int_{-\infty}^{0} e^{\varepsilon t} \mu_i \dot{N}_i(t)\, dt \right) \right\}, \end{aligned} \tag{23.90}$$

where μ_1 and μ_2 are the chemical potentials of the reagents, and μ_3 and μ_4 are those of the reaction products. Taking (23.89a) into account, we write (23.90) in the form

$$\rho = Q^{-1}\exp\left\{-\beta\left(H - \sum_i \mu_i N_i + \sum_i \mu_i \nu_i \int_{-\infty}^{0} e^{\varepsilon t} J(t)\,dt\right)\right\}$$
$$= Q^{-1}\exp\left\{-\beta\left(H - \sum_i \mu_i N_i - A \int_{-\infty}^{0} e^{\varepsilon t} J(t)\,dt\right)\right\}, \tag{23.91}$$

where

$$A = -\sum_i \mu_i \nu_i \tag{23.92}$$

is the chemical affinity in the terminology of de Donder, which, in the linear approximation, plays the role of the thermodynamic force.

The operator J contains a small parameter, namely, the matrix elements of the transition accompanied by the chemical reaction.

We shall average the reaction rate (23.88) over the distribution (23.91) and expand the statistical operator in the small parameter contained in (23.88). We obtain

$$\langle J\rangle = \langle \dot{N}_1\rangle = A\beta \int_{-\infty}^{0} dt\, e^{\varepsilon t} \int_{0}^{1} d\tau \langle \dot{N}_1 e^{-\tau B} \dot{N}_1(t) e^{\tau B}\rangle_l, \tag{23.93}$$

where $\langle\ldots\rangle_l$ denotes averaging over the quasi-equilibrium distribution

$$\rho_l = Q_l^{-1} e^{-B}, \qquad B = \beta\left(H - \sum_i \mu_i N_i\right). \tag{23.94}$$

If in (23.93) we replace the averaging over the quasi-equilibrium state by averaging over the equilibrium state, we obtain a linear relation between the reaction rate and the affinity:

$$\langle N_1\rangle = \beta A L_{\dot{N}_1\dot{N}_1}, \tag{23.93a}$$

where

$$L_{\dot{N}_1\dot{N}_1} = \beta^{-1} \int_{-\infty}^{0}\int_{0}^{\beta} e^{\varepsilon t} \langle \dot{N}_1 \dot{N}_1(t + i\hbar\tau)\rangle_0\, d\tau\, dt, \tag{23.93b}$$

i.e., the relation obtained by Yamamoto [129, 130].

We expand the formula (23.93) in powers of the interaction, but for any A.

Taking into account that

$$e^{-B\tau} a_\alpha e^{B\tau} = e^{\beta(E_\alpha - \mu_1)\tau} a_\alpha,$$
$$e^{-B\tau} a_\alpha^+ e^{B\tau} = e^{-\beta(E_\alpha - \mu_1)\tau} a_\alpha^+$$

we obtain

$$e^{-B\tau} a_{\alpha_1}^+ b_{\alpha_2}^+ c_{\alpha_1'} d_{\alpha_2'} e^{B\tau} =$$
$$= \exp\left\{-\beta\left[\left(E_{\alpha_1} + E_{\alpha_2} - E_{\alpha_1'} - E_{\alpha_2'}\right) + A\right]\tau\right\} a_{\alpha_1}^+ b_{\alpha_2}^+ c_{\alpha_1'} d_{\alpha_2'}. \tag{23.95}$$

In the Heisenberg picture for the operator $\dot{N}_1(t)$, we can neglect the interaction, since in (23.93) there is already a factor of second order in the interaction; the averages, therefore are easily calculated by Wick's theorem using (23.95). Performing the integration over τ and t in (23.93), as we did in expanding formula (23.55), we obtain for the chemical reaction rate the expression

$$\langle J \rangle = \langle \dot{N}_1 \rangle = \sum_{\alpha_1\alpha_2\alpha_1'\alpha_2'} w_{\alpha_1\alpha_2}^{\alpha_1'\alpha_2'} \left\{ n_{\alpha_1} n_{\alpha_2} \left(1 \pm n_{\alpha_1'}\right)\left(1 \pm n_{\alpha_2'}\right) - \right.$$
$$\left. - n_{\alpha_1'} n_{\alpha_2'} \left(1 \pm n_{\alpha_1}\right)\left(1 \pm n_{\alpha_2}\right)\right\}, \tag{23.96}$$

where

$$w_{\alpha_1\alpha_2}^{\alpha_1'\alpha_2'} = \frac{2\pi}{\hbar} \left| \Phi_{\alpha_1\alpha_2}^{\alpha_1'\alpha_2'} \right|^2 \delta\left(E_{\alpha_1} + E_{\alpha_2} - E_{\alpha_1'} - E_{\alpha_2'}\right) \tag{23.97}$$

is the reaction probability in unit time.

In the right-hand side of Eq. (23.96) we have the summed collision integral of the kinetic equation (23.64), with the transition probabilities (23.97). It is easy to convince oneself, as in the preceding subsection, that the occupation numbers satisfy the kinetic equation.

For a nondegenerate gas, $n_\alpha \ll 1$, and the reaction rate is equa[l] to

$$\langle \dot{N}_1 \rangle = (1 - e^{-\beta A}) \sum_{\alpha_1\alpha_2\alpha_1'\alpha_2'} w_{\alpha_1\alpha_2}^{\alpha_1'\alpha_2'} n_{\alpha_1} n_{\alpha_2}, \tag{23.98}$$

where A is the chemical affinity (23.92).

If the system is close to the state of statistical equilibrium, i.e.,

$$\beta|A| \ll 1, \tag{23.99}$$

then the reaction rate is equal to

$$\langle \dot{N}_1 \rangle = L_{\dot{N}_1\dot{N}_1}\beta A, \tag{23.100}$$

where

$$L_{\dot{N}_1\dot{N}_1} = \sum_{\alpha_1\alpha_2\alpha_1'\alpha_2'} w_{\alpha_1\alpha_2}^{\alpha_1'\alpha_2'} n_{\alpha_1} n_{\alpha_2} \tag{23.101}$$

is the kinetic coefficient having the meaning of the forward reaction rate. Thus under the condition (23.99), the total chemical reaction rate is proportional to the chemical affinity, and A is the thermodynamic force.

The condition for chemical equilibrium is that the rate of the chemical reaction vanishes, i.e., that the chemical affinity is equal to zero:

$$A = -\sum_i \mu_i \nu_i = 0. \tag{23.102}$$

For a mixture of ideal gases, the chemical potential is equal to [92]

$$\mu_i = \beta^{-1} \ln p_i + \chi_i, \tag{23.103}$$

where

$$p_i = pc_i \tag{23.103a}$$

is the partial pressure of the i-th component, p is the pressure, $c_i = N_i/N$ is the concentration of particles of the i-th component, and χ_i is a function of temperature, easily calculated from the partition function of an ideal gas. If we take (23.103) into account, the chemical equilibrium condition takes the form of the law of mass action:

$$\prod_i p_i^{\nu_i} = e^{-\beta \sum_i \nu_i \chi_i} = K, \tag{23.104}$$

where K is the chemical equilibrium constant, which depends only on the temperature.

Taking (23.104) into account, we can write the kinetic coefficient (23.101) in the form

$$L_{\dot{N}_1\dot{N}_1} = \varkappa c_1 c_2, \tag{23.105}$$

where

$$\varkappa = e^{\beta(\chi_1+\chi_2)} p^2 \sum_{\alpha_1\alpha_2\alpha_1'\alpha_2'} w_{\alpha_1\alpha_2}^{\alpha_1'\alpha_2'} e^{-\beta(E_{\alpha_1}+E_{\alpha_2})} \tag{23.106}$$

is the rate constant of the forward reaction. The relation (23.105) expresses the kinetic law of mass action.

Taking (23.89a) into account, we can write the linear relation (23.100) between the relation rate and the chemical affinity in the form

$$\frac{\langle \dot{N}_i \rangle}{\nu_i} = \frac{d\xi}{dt} = L_{\dot{N}_1\dot{N}_1} \beta A, \tag{23.107}$$

where we have introduced the parameter

$$\xi = \frac{\langle N_i \rangle - \langle N_i \rangle_0}{\nu_i} \tag{23.108}$$

which is the progress variable, or degree of advancement, of the chemical reaction, and $\langle N_i \rangle_0$ is the equilibrium concentration of the product i.

The chemical affinity A can be expressed in terms of the derivative of the entropy with respect to the progress variable ξ of the reaction:

$$S = \ln Q_l + \beta \langle H \rangle - \beta \sum_i \mu_i \langle N_i \rangle. \tag{23.109}$$

In fact,

$$\frac{\partial S}{\partial t} = -\beta \sum_i \mu_i \langle \dot{N}_i \rangle = \beta A \frac{d\xi}{dt}, \tag{23.110}$$

since the derivative of ln Q_l and the terms with derivatives of β and μ_i cancel, and $\langle H \rangle$ is assumed to be constant. Consequently,

$$A = T \left(\frac{\partial S}{\partial \xi} \right)_{\langle H \rangle}. \tag{23.111}$$

In a state of statistical equilibrium, A = 0, since the reaction rate goes to zero. Close to equilibrium, we expand A in powers of the deviation of ξ from the equilibrium value ξ_0:

$$A = T\left(\frac{\partial^2 S}{\partial \xi^2}\right)_{\xi=\xi_0} (\xi - \xi_0). \tag{23.112}$$

Substituting (23.112) into (23.107), we obtain the relaxation equation for the progress variable of the reaction

$$\frac{d\xi}{dt} = -\frac{\xi - \xi_0}{\tau}, \tag{23.113}$$

where

$$\tau = -\frac{1}{L_{\dot{N}_1\dot{N}_1}\left(\frac{\partial^2 S}{\partial \xi^2}\right)_{\xi=0}} > 0 \tag{23.114}$$

is the relaxation time of the reaction. Integrating (23.113), we obtain the law by which the progress variable of the reaction changes with time:

$$\xi - \xi_0 = (\xi(0) - \xi_0)\, e^{-t/\tau}, \tag{23.115}$$

i.e., the progress variable of the reaction tends exponentially to the equilibrium value ξ_0.

The relaxation time introduced in (23.114) corresponds to constant energy and constant particle number. Analogous relations can also be obtained under other thermodynamic conditions [27,131].

In the general case, the rate of a chemical reaction is nonlinear in the affinity, i.e.,

$$\langle \dot{N}_1 \rangle = L_{\dot{N}_1\dot{N}_1}(1 - e^{-\beta A}). \tag{23.116}$$

The treatment given above is also valid for an arbitrary number of components when different chemical reaction between them are possible. Then,

$$\langle \dot{N}_i \rangle = \sum_m L_{\dot{N}_i\dot{N}_m}(1 - e^{-\beta A_m}), \tag{23.117}$$

where

$$L_{\dot{N}_i\dot{N}_m} = \sum_{\alpha_1\alpha_2\ldots\alpha_1'\alpha_2'\ldots} w_{\alpha_1\alpha_2\ldots}^{\alpha_1'\alpha_2'\ldots}(i, m)\, n_{\alpha_1} n_{\alpha_2} \ldots \tag{23.118}$$

is the kinetic coefficient,

$$A_m = -\sum_i \mu_i \nu_{im} \tag{23.119}$$

is the chemical affinity, and ν_{im} is the stoichiometric coefficient with which the substance i takes part in the m-th reaction.

If (23.103) is taken into account, the chemical equilibrium condition

$$A_m = -\sum_i \mu_i \nu_{im} = 0 \tag{23.120}$$

takes the form

$$\prod_i p_i^{\nu_{im}} = e^{-\beta \sum_i \nu_{im} \chi_i} = K_m, \tag{23.121}$$

where K_m is the chemical equilibrium constant of the m-th reaction.

The kinetic mass-action law in this case takes the form

$$L_{\dot{N}_i \dot{N}_m} = \varkappa_i \prod_m c_m^{\nu_{im}}, \tag{23.122}$$

where

$$\varkappa_i = e^{\beta \sum_m \nu_{im} \chi_i} p^{\sum_m \nu_{im}} \sum_{\alpha_1 \alpha_2 \ldots \alpha_1' \alpha_2' \ldots} w_{\alpha_1 \alpha_2 \ldots}^{\alpha_1' \alpha_2' \ldots} e^{-\beta \sum_m \nu_{im} E_{\alpha_m}} \tag{23.122a}$$

and $\nu_{im} > 0$ everywhere, i.e., only the positive stoichiometric coefficients are considered.

We shall consider the entropy production in chemical reactions in a homogeneous system for the nonlinear case.

The entropy for such a system is equal to (23.109), and, if we use (23.89a), its production is

$$\frac{dS}{dt} = -\beta \sum_{i=1}^{4} \mu_i \nu_i \langle J \rangle = \beta A \langle J \rangle. \tag{23.123}$$

Substituting into this the expression (23.116) for the average reaction rate, we obtain for the entropy production the expression

$$\frac{dS}{dt} = L_{\dot{N}_1\dot{N}_1}\beta A\,(1 - e^{-\beta A}) \geqslant 0, \tag{23.124}$$

which follows from the inequality (23.81). Consequently, the entropy production in a chemical reaction is positive.

Up to this point, we have considered only the balance equations for particles in chemical reactions. Analogously, one could also consider the energy balance. Although this is important for chemical reactions (see [132]), we shall not make this generalization here. The aim of this section was to show that the method of the nonequilibrium statistical operator can also describe nonlinear processes in chemical kinetics.

§ 24. The Statistical Operator for Relativistic Systems and Relativistic Hydrodynamics [5]

24.1. The Relativistic Statistical Operator

A phenomenological nonequilibrium thermodynamics for the relativistic case was developed by Eckart [133] for a one-component liquid (or gas) and was generalized for a mixture by Kluitenberg, de Groot and Mazur [134]. It is described, for example, in [82].

Following [5], we shall apply the method of constructing nonequilibrium statistical operators, described in §21, to a relativistic system. Taking relativity into account only simplifies the problem, since it makes it possible to construct invariant quantities easily, and the statistical operator and the entropy which it determines must, in the relativistic case, be invariant under Lorentz transformations [135].

The energy-momentum conservation law in the relativistic case has the form

$$\sum_{\mu=1}^{4} \frac{\partial T_{\mu\nu}(\boldsymbol{x}, t)}{\partial x_\mu} = 0, \tag{24.1}$$

where $\mathbf{x}$ is the set of coordinates x_1, x_2, x_3, and $x_4 = ict$; $T_{\mu\nu}(\mathbf{x}, t)$ is the relativistic expression, assumed known, for the symmetric tensor operator of the energy-momentum density. We shall consider a system which is characterized by only these conservation laws; in particular, the system has no electric charge or spin.

Corresponding to the conservation law (24.1) is the local integral of motion

$$\widetilde{T_{4\nu}(\boldsymbol{x})} = T_{4\nu}(\boldsymbol{x}) - \int_{-\infty}^{0} e^{\varepsilon t}\, \dot{T}_{4\nu}(\boldsymbol{x}, t)\, dt =$$

$$= T_{4\nu}(\boldsymbol{x}) + \sum_{\mu=1}^{3} \int_{-\infty}^{0} e^{\varepsilon t}\, \frac{\partial T_{\mu\nu}(\boldsymbol{x}, t)}{\partial x_\mu}\, ic\, dt. \qquad (24.2)$$

In (24.2), we have discarded the retarded solution, i.e., we have used the causality condition.

It is possible to use (24.2) to study stationary processes, although we shall proceed immediately to nonstationary processes, since these are treated even more simply in the relativistic case, because of the symmetry between the space and time coordinates.

By the usual method, but taking the conservation law (24.1) into account, we construct operators depending on time only through certain parameters $F_\nu(\mathbf{x}, t)$ ($\nu = 1, \ldots, 4$), which define the macroscopic state of the system:

$$B_{4\nu}(\boldsymbol{x}, t) = -\widetilde{F_\nu(\boldsymbol{x}, t)\, T_{4\nu}(\boldsymbol{x})} = -F_\nu(\boldsymbol{x}, t)\, T_{4\nu}(\boldsymbol{x}) +$$

$$+ \int_{-\infty}^{0} e^{\varepsilon t_1} \left\{ F_\nu(\boldsymbol{x}, t + t_1)\, \dot{T}_{4\nu}(\boldsymbol{x}, t_1) + \frac{\partial F_\nu(\boldsymbol{x}, t + t_1)}{\partial t_1}\, T_{4\nu}(\boldsymbol{x}, t_1) \right\} dt_1 =$$

$$= -F_\nu(\boldsymbol{x}, t)\, T_{4\nu}(\boldsymbol{x}) + \int_{-\infty}^{0} e^{\varepsilon t_1} \left\{ -F_\nu(\boldsymbol{x}, t + t_1)\, ic \sum_{\mu=1}^{3} \frac{\partial T_{\mu\nu}(\boldsymbol{x}, t_1)}{\partial x_\mu} + \right.$$

$$\left. + \frac{\partial F_\nu(\boldsymbol{x}, t + t_1)}{\partial t_1}\, T_{4\nu}(\boldsymbol{x}, t_1) \right\} dt_1. \qquad (24.3)$$

The parameters $F_\nu(\mathbf{x}, t)$, the physical meaning of which we shall elucidate later, are selected in such a way that the expression

$$\sum_\nu \int B_{4\nu}(\boldsymbol{x})\, d\boldsymbol{x}$$

is Lorentz-invariant. Discarding the surface integrals, we obtain for this quantity the expression

$$\sum_{\nu}\int B_{4\nu}(\boldsymbol{x}, t)\,d\boldsymbol{x} = -\sum_{\nu}\int F_{\nu}(\boldsymbol{x}, t)\,T_{4\nu}(\boldsymbol{x})\,d\boldsymbol{x} + \\ + \sum_{\mu,\,\nu=1}^{4}\int\int_{-\infty}^{0} e^{\varepsilon t_1}\frac{\partial F_{\nu}(\boldsymbol{x}, t+t_1)}{\partial x_{\mu}}\,T_{\mu\nu}(\boldsymbol{x}, t_1)\,d\boldsymbol{x}\,ic\,dt_1. \tag{24.4}$$

Using the usual technique, we construct the statistical operator [5]

$$\rho = Q^{-1}\exp\left\{-\sum_{\nu}\int B_{4\nu}(\boldsymbol{x}, t)\,d\boldsymbol{x}\right\} = Q^{-1}\exp\left\{\sum_{\nu}\int F_{\nu}(\boldsymbol{x}, t)\,T_{4\nu}(\boldsymbol{x})\,d\boldsymbol{x} - \right. \\ \left. -\sum_{\mu,\,\nu}\int\int_{-\infty}^{0} e^{\varepsilon t_1}\frac{\partial F_{\nu}(\boldsymbol{x}, t+t_1)}{\partial x_{\mu}}\,T_{\mu\nu}(\boldsymbol{x}, t_1)\,d\boldsymbol{x}\,ic\,dt_1\right\}, \tag{24.5}$$

the parameters $F_{\nu}(\mathbf{x}, t)$ being determined from the conditions

$$\langle T_{4\nu}(\boldsymbol{x})\rangle = \langle T_{4\nu}(\boldsymbol{x})\rangle_{l}, \tag{24.6}$$

where

$$\rho_{l} = Q_{l}^{-1}\exp\left\{\sum_{\nu}\int F_{\nu}(\boldsymbol{x}, t)\,T_{4\nu}(\boldsymbol{x})\,d\boldsymbol{x}\right\} \tag{24.7}$$

is the relativistic local-equilibrium statistical operator.

24.2. Thermodynamic Equalities

We shall now elucidate the physical meaning of the parameters $F_{\nu}(\mathbf{x}, t)$. We put

$$F_{\nu}(\boldsymbol{x}, t) = -\beta(\boldsymbol{x}, t)\,iu_{\nu}(\boldsymbol{x}, t), \tag{24.8}$$

where u_{ν} is a four-dimensional velocity, i.e.,

$$\sum_{\nu} u_{\nu}^{2}(\boldsymbol{x}, t) = -1. \tag{24.8a}$$

This ensures the Lorentz invariance of the statistical operator (24.5), since $\int T_{4\nu}(\boldsymbol{x})\,d\boldsymbol{x}$ transforms as a four-vector, if the field is nonzero only in a finite region of space [136].

We shall study the local-equilibrium state (24.7)

$$\rho_l = Q_l^{-1} \exp\left\{ - \sum_\nu \int \beta(\boldsymbol{x}, t)\, i u_\nu(\boldsymbol{x}, t)\, T_{4\nu}(\boldsymbol{x})\, d\boldsymbol{x} \right\}, \tag{24.9}$$

where

$$Q_l = \operatorname{Tr} \exp\left\{ - \sum_\nu \int \beta(\boldsymbol{x}, t)\, i u_\nu(\boldsymbol{x}, t)\, T_{4\nu}(\boldsymbol{x})\, d\boldsymbol{x} \right\}. \tag{24.9a}$$

We transform to the moving frame in which the space components of the four-vector u_ν are equal to zero, $u_1' = u_2' = u_3' = 0$, $u_4' = i$, i.e.,

$$u_\nu' = i\delta_{\nu 4}, \tag{24.10}$$

which satisfies the condition (24.8a). Here and below, we use primes to denote components of vectors and tensors in this moving frame.

In this accompanying frame, the statistical operator (24.9) has the usual nonrelativistic form

$$\rho_l = Q_l^{-1} \exp\left\{ - \int \beta(\boldsymbol{x}, t)\, H'(\boldsymbol{x})\, d\boldsymbol{x} \right\}, \tag{24.11}$$

where

$$H'(\boldsymbol{x}) = - T_{44}'(\boldsymbol{x}) \tag{24.11a}$$

is the Hamiltonian density in the moving frame. Formulas (24.11) and (24.11a) confirm the correctness of the definition (24.9).

We choose u_ν in such a way that the variations of $\ln Q_l$ with respect to u_1, u_2, and u_3 vanish, i.e.,

$$\frac{\delta \ln Q_l}{\delta u_\nu(\boldsymbol{x}, t)} = 0 \qquad (\nu = 1, 2, 3). \tag{24.12}$$

We have already applied a similar condition in §20, when we considered an ordinary, rather than a four-dimensional, velocity. This condition ensured that the parameter $\mathbf{v}(\mathbf{x}, t)$ was chosen to be the average mass velocity, as required to satisfy the thermodynamic equalities. The condition (24.12) means that we are defining the densities of the mechanical quantities on which the statistical operator depends in a local frame moving with an element of the liquid.

We shall calculate the variation of (24.9a) taking (24.8a) into account:

$$\delta \ln Q_l = -i \sum_{\nu} \int \beta(\boldsymbol{x}, t) \left\{ \langle T_{4\nu}(\boldsymbol{x}) \rangle - \langle T_{44}(\boldsymbol{x}) \rangle \frac{u_\nu(\boldsymbol{x}, t)}{u_4(\boldsymbol{x}, t)} \right\} \delta u_\nu(\boldsymbol{x}, t)\, d\boldsymbol{x},$$

since

$$\sum_{\nu=1}^{3} u_\nu \delta u_\nu + u_4 \delta u_4 = 0 \qquad (\nu = 1, 2, 3).$$

From the condition (24.12), we obtain

$$\frac{u_\nu(\boldsymbol{x}, t)}{u_4(\boldsymbol{x}, t)} = \frac{\langle T_{4\nu}(\boldsymbol{x}) \rangle}{\langle T_{44}(\boldsymbol{x}) \rangle} = -ic \frac{\langle G_\nu(\boldsymbol{x}) \rangle}{\langle H(\boldsymbol{x}) \rangle}, \tag{24.13}$$

where

$$G_\nu(\boldsymbol{x}) = \frac{1}{ic} T_{4\nu}(\boldsymbol{x}), \qquad H(\boldsymbol{x}) = -T_{44}(\boldsymbol{x}) \tag{24.14}$$

are the densities of the momentum and energy respectively.

If we introduce the usual three-dimensional local velocity

$$v_\nu(\boldsymbol{x}, t) = c^2 \frac{\langle G_\nu(\boldsymbol{x}) \rangle}{\langle H(\boldsymbol{x}) \rangle} = ic \frac{\langle T_{4\nu}(\boldsymbol{x}) \rangle}{\langle T_{44}(\boldsymbol{x}) \rangle} \qquad (\nu = 1, 2, 3), \tag{24.15}$$

we can write the relation (24.13) in the form

$$\frac{u_\nu(\boldsymbol{x}, t)}{u_4(\boldsymbol{x}, t)} = -\frac{iv_\nu(\boldsymbol{x}, t)}{c}, \tag{24.16}$$

or, taking (24.8a) into account,

$$u_\nu(\boldsymbol{x}, t) = \frac{v_\nu(\boldsymbol{x}, t)}{c\sqrt{1 - \frac{v^2(\boldsymbol{x}, t)}{c^2}}}, \qquad u_4(\boldsymbol{x}, t) = \frac{i}{\sqrt{1 - \frac{v^2(\boldsymbol{x}, t)}{c^2}}}. \tag{24.16a}$$

Thus, for the local velocities $v_\nu(\mathbf{x}, t)$ and $u_\nu(\mathbf{x}, t)$, we obtain the well-known relativistic relations.

It remains to elucidate the meaning of the parameter $\beta(\mathbf{x}, t)$. For this, we shall calculate the variation of the partition function (24.9a) with respect to $\beta(\mathbf{x}, t)$ at constant $u_\nu(\mathbf{x}, t)$:

$$\frac{\delta \ln Q_l}{\delta \beta(\boldsymbol{x}, t)} = -\sum_{\nu=1}^{4} i u_\nu(\boldsymbol{x}, t) \langle T_{4\nu}(\boldsymbol{x}) \rangle, \tag{24.17}$$

whence, taking (24.13), (24.8a), (24.16a), and (24.14) into account, we obtain the thermodynamic equality

$$\frac{\delta \ln Q_l}{\delta \beta(\boldsymbol{x}, t)} = \langle T_{44}(\boldsymbol{x})\rangle \sqrt{1 - \frac{v^2(\boldsymbol{x}, t)}{c^2}} = -\langle H(\boldsymbol{x})\rangle \sqrt{1 - \frac{v^2(\boldsymbol{x}, t_1)}{c^2}}, \qquad (24.17a)$$

which is analogous to the first equation of the system (20.21a) and goes over into it in the nonrelativistic limit $v(\mathbf{x}, t) \ll c$ and for zero chemical potential μ.

It follows from (24.17) that $\beta^{-1}(\mathbf{x}, t)$ plays the role of the invariant "proper" temperature, and the quantity

$$\beta^{-1}(\boldsymbol{x}, t)\left(1 - \frac{v^2(\boldsymbol{x}, t)}{c^2}\right)^{-1/2}$$

plays the role of the ordinary, noninvariant temperature.

24.3. The Equations of Relativistic Hydrodynamics

The statistical operator (24.5) enables us to obtain the equations of relativistic hydrodynamics. For this, we obtain linear relations between the average energy-momentum tensor and the thermodynamic forces $\partial F_\nu/\partial x_\mu$, assuming that the lattice are small:

$$\langle T_{\mu\nu}(\boldsymbol{x})\rangle = \langle T_{\mu\nu}(\boldsymbol{x})\rangle_l - \sum_{\mu_1\nu_1} \int \int_{-\infty}^{0} e^{\varepsilon t_1} (T_{\mu\nu}(\boldsymbol{x}), T_{\mu_1\nu_1}(\boldsymbol{x}', t_1)) \frac{\partial F_{\nu_1}(\boldsymbol{x}', t+t_1)}{\partial x'_{\mu_1}} d\boldsymbol{x}' \, ic\, dt_1, \qquad (24.18)$$

where $(T_{\mu\nu}, T_{\mu_1\nu_1})$ is a quantum correlation function, and $\langle\ldots\rangle_l$ denotes averaging over the local-equilibrium distribution (24.9).

The expression (24.18) contains within itself all the irreversible processes that can occur in a system with one energy-momentum conservation law, i.e., thermal conduction and shear and bulk viscosity, but is inconvenient for an isotropic medium, since processes of different tensor dimensionality are not separated in it. Below, we shall give another expression, not so general but more convenient, for the irreversible fluxes in an isotropic medium.

In order to construct operators describing the irreversible fluxes, in the tensor $T_{\mu\nu}$, we must separate out the part describing the convective motion with hydrodynamic velocity u_μ; we have already defined the latter by the condition (24.12).

We note that any vector F_μ can be decomposed into a sum of vectors, of which one is parallel to u_μ and the other is perpendicular to it, i.e.,

$$F_\mu = f u_\mu + f_\mu. \tag{24.19}$$

From the condition that f_μ and u_μ are orthogonal

$$\sum_\mu f_\mu u_\mu = 0, \tag{24.19a}$$

taking (24.8a) into account, we find

$$f = \left(\sum_\mu u_\mu F_\mu\right)\left(\sum_\mu u_\mu^2\right)^{-1} = -\sum_\mu u_\mu F_\mu. \tag{24.19b}$$

We shall express f_μ in terms of the tensor

$$\Delta_{\mu\nu} = \delta_{\mu\nu} + u_\mu u_\nu, \tag{24.20}$$

where $\delta_{\mu\nu}$ is the Kronecker symbol. The tensor $\Delta_{\mu\nu}$ is orthogonal to u_μ:

$$\sum_\mu u_\mu \Delta_{\mu\nu} = 0. \tag{24.21}$$

Putting

$$f_\mu = \sum_\nu \Delta_{\mu\nu} F_\nu, \tag{24.22}$$

we see that this vector is indeed the component of F_μ perpendicular to u_μ. In fact, multiplying (24.19) by $\Delta_{\mu\nu}$ and summing over μ, using (24.21) we obtain the expression (24.22).

The tensor $\Delta_{\mu\nu}$ plays the same role in the relativistic theory as the Kronecker symbol in the nonrelativistic theory. To elucidate its meaning, we transform to the coordinate frame rotating with the hydrodynamic velocity (24.10). In this frame, the tensor $\Delta_{\mu\nu}$ has the very simple form:

$$\Delta'_{\mu\nu} = \delta_{\mu\nu} - \delta_{\mu 4}\delta_{\nu 4}, \tag{24.22a}$$

or, in matrix form,

$$\Delta' = \begin{pmatrix} 1 & 0 & 0 & 0 \\ 0 & 1 & 0 & 0 \\ 0 & 0 & 1 & 0 \\ 0 & 0 & 0 & 0 \end{pmatrix}. \tag{24.22b}$$

A similar division into principal components of the relative hydrodynamic motion can also be performed for any tensor, and, in particular, for $T_{\mu\nu}$:

$$T_{\mu\nu} = \varepsilon u_\mu u_\nu + P_\mu u_\nu + P_\nu u_\mu + P_{\mu\nu}, \tag{24.23}$$

where

$$\sum_\mu P_\mu u_\mu = 0, \qquad \sum_\nu P_{\mu\nu} u_\nu = 0. \tag{24.23a}$$

The coefficients of the expansion (24.23) are equal to

$$\begin{aligned} \varepsilon &= \sum_{\mu, \lambda} u_\mu T_{\mu\lambda} u_\lambda, \\ P_\mu &= -\sum_{\nu\lambda} \Delta_{\mu\nu} T_{\nu\lambda} u_\lambda = c^{-1} q_\mu, \\ P_{\mu\nu} &= \sum_{\mu_1\nu_1} \Delta_{\mu\mu_1} T_{\mu_1\nu_1} \Delta_{\nu_1\nu}, \end{aligned} \tag{24.24}$$

as can be seen by direct inspection. We shall omit the arguments **x** and t wherever this can lead to no confusion.

The quantities (24.24) have a simple physical meaning: ε is the density of the internal energy unconnected with the convective motion; $P_\mu c = q_\mu$ is the heat flux; $P_{\mu\nu}$ is the stress tensor. All these quantities are operators or dynamic variables.

In order to elucidate the physical meaning of the expressions (24.24), we write them in the frame moving with the hydrodynamic velocity. Using (24.10) and (24.22), we obtain

$$\begin{aligned} \varepsilon &= -T'_{44}, \\ P'_\mu &= -iT'_{\mu 4} + \delta_{\mu 4} i T'_{44}, \\ P'_{\mu\nu} &= T'_{\mu\nu} - \delta_{\mu 4} T'_{4\nu} - \delta_{\nu 4} T'_{\mu 4} + \delta_{\mu 4}\delta_{\nu 4} T'_{44}. \end{aligned} \tag{24.24a}$$

Taking (24.14) into account, we can write these relations in the form

$$\begin{aligned} \varepsilon &= H', \\ P'_\mu &= cG'_\mu \quad (\mu = 1, 2, 3), \qquad P'_4 = 0, \\ P'_{\mu\nu} &= T'_{\mu\nu} \quad (\mu, \nu = 1, 2, 3), \qquad P'_{\nu 4} = 0. \end{aligned} \tag{24.24b}$$

In the frame moving with the hydrodynamic velocity, all the quantities must coincide with their nonrelativistic expressions.

Thus, ε does indeed have the meaning of the energy density, P_μ is the energy flux in the accompanying frame, i.e., the heat flux divided by c, and $P_{\mu\nu}$ is the stress tensor.

We introduce the viscous-shear stress tensor $\pi_{\mu\nu}$, which, like $P_{\mu\nu}$, is orthogonal to u_μ, but has zero trace:

$$P_{\mu\nu} = \pi_{\mu\nu} + p\Delta_{\mu\nu}, \tag{24.25}$$

where

$$p = \frac{1}{3}\sum_\mu P_{\mu\mu}. \tag{24.25a}$$

Then,

$$\sum_\mu \pi_{\mu\nu} u_\nu = 0, \tag{24.26}$$

and, in addition,

$$\sum_\mu \pi_{\mu\mu} = 0, \tag{24.26a}$$

since

$$\sum_\mu \Delta_{\mu\mu} = \sum_\mu (1 + u_\mu^2) = 3.$$

Taking (24.25) into account, we write the expansion (24.23) of the tensor $T_{\mu\nu}$ in its principal components,

$$T_{\mu\nu} = \varepsilon u_\mu u_\nu + p\Delta_{\mu\nu} + P_\mu u_\nu + P_\nu u_\mu + \pi_{\mu\nu}. \tag{24.27}$$

Thus, we have decomposed the tensor $T_{\mu\nu}$ into three parts, having scalar, tensor, and vector character with respect to the operators ε, p, P_μ, and $\pi_{\mu\nu}$. This separation, which is applied in phenomenological relativistic hydrodynamics [133, 134], enables us to separate the scalar, vector, and tensor processes.

After local-equilibrium averaging, the first two terms in (24.27) have the meaning of the energy-momentum tensor of an ideal liquid and describe nondissipative processes. The next two terms give the heat flux and the last term gives the viscous flux of momentum. These parts describe irreversible processes.

For a local-equilibrium state, the average values of only the first two terms are nonzero:

$$\langle T_{\mu\nu}\rangle_l = (\langle\varepsilon\rangle_l + \langle p\rangle_l)\, u_\mu u_\nu + \langle p\rangle_l\, \delta_{\mu\nu}, \tag{24.28}$$

since in the accompanying frame the average values of the vectors and of the off-diagonal elements of the tensors are equal to zero (see §20.5),

$$\langle P_\mu \rangle_l = 0, \qquad \langle \pi_{\mu\nu} \rangle_l = 0. \tag{24.29}$$

We shall represent the statistical operator (24.5) in a form such that the scalar, vector, and tensor processes are separated in it. Taking into account (24.27) and the orthogonality of $\partial u_\mu/\partial x_\nu$ to u_μ,

$$\sum_\mu \frac{\partial u_\mu}{\partial x_\nu} u_\mu = 0, \tag{24.30}$$

which follows from (24.8a), we obtain

$$\sum_{\mu\nu} T_{\mu\nu} \frac{\partial \beta u_\mu}{\partial x_\nu} = \beta \sum_{\mu\nu} \pi_{\mu\nu} \frac{\partial u_\mu}{\partial x_\nu} - \sum_\mu \beta P_\mu \left(\beta^{-1} \frac{\partial \beta}{\partial x_\mu} - \frac{1}{c} D u_\mu \right) - \frac{1}{c} \varepsilon D\beta + \beta p \operatorname{div} \boldsymbol{u}, \tag{24.31}$$

$$\operatorname{div} \boldsymbol{u} = \sum_\mu \frac{\partial u_\mu}{\partial x_\mu},$$

where

$$D = c \sum_\mu u_\mu \frac{\partial}{\partial x_\mu} \tag{24.32}$$

is a scalar operator, having the meaning of a total (or barycentric derivative in relativistic theory. In fact, in the moving frame (24.10), the operator D coincides with the time derivative in this frame:

$$D = \frac{\partial}{\partial t'} . \tag{24.32a}$$

In the expression (24.31), which appears in the exponent of the statistical operator, terms containing operators of different tensor dimensionality have been separated. Taking (24.31) into account, we write the statistical operators in the form

$$\begin{aligned} \rho = Q^{-1} \exp \Bigg\{ & - \sum_\nu \int \beta(\boldsymbol{x}, t)\, i u_\nu (\boldsymbol{x}, t)\, T_{4\nu}(\boldsymbol{x})\, d\boldsymbol{x} - \\ & - \int \int_{-\infty}^{0} e^{\varepsilon t_1} c\beta(\boldsymbol{x}, t + t_1) \Bigg[\sum_{\mu\nu} \pi_{\mu\nu}(\boldsymbol{x}, t_1) \frac{\partial u_\mu (\boldsymbol{x}, t + t_1)}{\partial x_\nu} - \sum_\mu P_\mu (\boldsymbol{x}, t_1) \times \\ & \times \left(\beta^{-1}(\boldsymbol{x}, t + t_1) \frac{\partial \beta (\boldsymbol{x}, t + t_1)}{\partial x_\mu} - \frac{1}{c} D u_\mu (\boldsymbol{x}, t + t_1) \right) + p(\boldsymbol{x}, t_1) \operatorname{div} \boldsymbol{u}(\boldsymbol{x}, t + t_1) - \\ & - \frac{1}{c} \varepsilon(\boldsymbol{x}, t_1) \beta^{-1}(\boldsymbol{x}, t + t_1) D\beta(\boldsymbol{x}, t + t_1) \Bigg] dt_1\, d\boldsymbol{x} \Bigg\}. \end{aligned} \tag{24.33}$$

The quantities

$$\frac{\partial u_\mu}{\partial x_\nu}, \quad \beta^{-1}\frac{\partial \beta}{\partial x_\mu} - \frac{1}{c} D u_\mu, \qquad \operatorname{div} \boldsymbol{u}$$

play the role of the thermodynamic forces conjugate to the fluxes

$$\langle \pi_{\mu\nu}(\boldsymbol{x})\rangle, \qquad \langle P_\mu(\boldsymbol{x})\rangle, \qquad \langle p(\boldsymbol{x})\rangle - \langle p(\boldsymbol{x})\rangle_l,$$

and the quantity $D\beta$ can be expressed in terms of div **u** by means of the equations of ideal hydrodynamics. In fact, for a local-equilibrium state, for the energy-momentum tensor we have, according to (24.28),

$$\langle T_{\mu\nu}(\boldsymbol{x})\rangle_l = h u_\mu u_\nu + p\delta_{\mu\nu}, \tag{24.34}$$

where

$$h = \langle \varepsilon \rangle_l + \langle p \rangle_l \tag{24.34a}$$

is the heat function or enthalpy per unit volume, and

$$p = \langle p \rangle_l \tag{24.34b}$$

is the pressure. Equating the four-dimensional divergence of (24.34) to zero,

$$\frac{\partial}{\partial x_\nu}\langle T_{\mu\nu}(\boldsymbol{x})\rangle_l = 0, \tag{24.35}$$

we obtain the hydrodynamic equation for a relativistic system with neglect of dissipative processes:

$$u_\mu \frac{\partial h u_\nu}{\partial x_\nu} + h u_\nu \frac{\partial u_\mu}{\partial x_\nu} + \frac{\partial p}{\partial x_\mu} = 0, \tag{24.36}$$

where a summation over indices appearing twice is implied. Taking the scalar product of this equation with u_μ and taking (24.8a) into account, we find

$$-\frac{\partial h u_\nu}{\partial x_\nu} + u_\mu \frac{\partial p}{\partial x_\mu} = 0, \tag{24.37}$$

or, using the notation (24.32),

$$-ch \operatorname{div} \boldsymbol{u} + D(p - h) = 0. \tag{24.38}$$

In our case, p and h are functions of β and, consequently,

$$D\beta = \left(\frac{\partial (p-h)}{\partial \beta}\right)^{-1} hc \operatorname{div} \boldsymbol{u} = \beta \frac{\partial p}{\partial \varepsilon} c \operatorname{div} \boldsymbol{u}, \tag{24.39}$$

since

$$p-h=-\varepsilon, \qquad h=-\beta\frac{\partial p}{\partial \beta}. \tag{24.39a}$$

Consequently, we can eliminate the quantity $D\beta$ in the exponent of the statistical operator (24.33):

$$\rho=Q^{-1}\exp\Big\{-\sum_{\nu}\int\beta(\boldsymbol{x},t)\,iu_{\nu}(\boldsymbol{x},t)\,T_{4\nu}(\boldsymbol{x})\,d\boldsymbol{x}-$$
$$-\int\int_{-\infty}^{0}e^{\varepsilon t_1}c\beta(\boldsymbol{x},t+t_1)\Big[\sum_{\mu\nu}\pi_{\mu\nu}(\boldsymbol{x},t_1)\frac{\partial u_{\mu}(\boldsymbol{x},t+t_1)}{\partial x_{\nu}}-$$
$$-\sum_{\mu}P_{\mu}(\boldsymbol{x},t_1)\Big(\beta^{-1}(\boldsymbol{x},t+t_1)\frac{\partial\beta(\boldsymbol{x},t+t_1)}{\partial x_{\mu}}-\frac{1}{c}Du_{\mu}(\boldsymbol{x},t+t_1)\Big)+$$
$$+p'(\boldsymbol{x},t_1)\operatorname{div}\boldsymbol{u}(\boldsymbol{x},t+t_1)\Big]dt_1d\boldsymbol{x}\Big\}, \tag{24.40}$$

where

$$p'(\boldsymbol{x},t)=p(\boldsymbol{x},t)-\frac{\partial p}{\partial\varepsilon}\varepsilon(\boldsymbol{x},t).$$

We now obtain linear relations between the average fluxes and the thermodynamic forces by assuming that the latter are small, confining ourselves to the linear terms and using the Curie theorem [27], according to which only fluxes of the same tensor dimensionality can be coupled:

$$\langle\pi_{\mu\nu}(\boldsymbol{x})\rangle=-\sum_{\mu_1\nu_1}\int\int_{-\infty}^{0}e^{\varepsilon t_1}(\pi_{\mu\nu}(\boldsymbol{x}),\pi_{\mu_1\nu_1}(\boldsymbol{x}',t_1))\,\beta(\boldsymbol{x}',t+t_1)\times$$
$$\times\frac{\partial u_{\mu_1}(\boldsymbol{x}',t+t_1)}{\partial x'_{\nu_1}}c\,dt_1d\boldsymbol{x}', \tag{24.41a}$$

$$\langle P_{\mu}(\boldsymbol{x})\rangle=c^{-1}\langle q_{\mu}(\boldsymbol{x},t)\rangle=$$
$$=\sum_{\nu}\int\int_{-\infty}^{0}e^{\varepsilon t_1}(P_{\mu}(\boldsymbol{x}),P_{\nu}(\boldsymbol{x}',t_1))\,\beta(\boldsymbol{x}',t+t_1)\times$$
$$\times\Big(\beta^{-1}(\boldsymbol{x}',t+t_1)\frac{\partial\beta(\boldsymbol{x}',t+t_1)}{\partial x'_{\nu}}-\frac{1}{c}Du_{\nu}(\boldsymbol{x}',t+t_1)\Big)c\,dt_1d\boldsymbol{x}', \tag{24.41b}$$

$$\langle p(\boldsymbol{x})\rangle-\langle p(\boldsymbol{x})\rangle_l=$$
$$=-\int\int_{-\infty}^{0}e^{\varepsilon t_1}(p'(\boldsymbol{x}),p'(\boldsymbol{x}',t_1))\,\beta(\boldsymbol{x}',t+t_1)\operatorname{div}\boldsymbol{u}(\boldsymbol{x}',t+t_1)\,c\,dt_1d\boldsymbol{x}'. \tag{24.41c}$$

We shall use the spatial isotropy of the system and simplify the expressions for the correlation functions occurring in the relations (24.41a) and (24.41b).

In the moving frame, the tensors (P', P') and (π', π') have the usual form:

$$(P'_\mu, P'_\nu) = L_P \delta_{\mu\nu},$$
$$(\pi'_{\mu\nu}, \pi'_{\mu_1\nu_1}) = L_\pi \frac{1}{2}\left\{\delta_{\mu\mu_1}\delta_{\nu\nu_1} + \delta_{\mu\nu_1}\delta_{\nu\mu_1} - \frac{2}{3}\delta_{\mu\nu}\delta_{\mu_1\nu_1}\right\}, \tag{24.42}$$

where L_p and L_π are scalar constants, μ, ν, μ_1, $\nu_1 = 1, 2, 3$, and the time components are equal to zero; then $\delta_{\mu\nu} = \Delta_{\mu\nu}$, where the prime denotes the function (24.20) in the moving frame.

Returning to the original frame, we obtain

$$(P_\mu, P_\nu) = L_P \Delta_{\mu\nu},$$
$$(\pi_{\mu\nu}, \pi_{\mu_1\nu_1}) = L_\pi \frac{1}{2}\left\{\Delta_{\mu\mu_1}\Delta_{\nu\nu_1} + \Delta_{\mu\nu_1}\Delta_{\nu\mu_1} - \frac{2}{3}\Delta_{\mu\nu}\Delta_{\mu_1\nu_1}\right\}. \tag{24.43}$$

The tensors (24.43) satisfy the conditions, following from (24.23a), that they be orthogonal to the four-velocity, and also satisfy the condition (24.26a) and spatial isotropy.

We find the scalars L_p and L_π by calculating the complete contraction of the left- and right-hand sides of (24.43):

$$L_P(\boldsymbol{x}, \boldsymbol{x}', t) = \frac{1}{3}(\boldsymbol{P}(\boldsymbol{x}) \cdot \boldsymbol{P}(\boldsymbol{x}', t)),$$
$$L_\pi(\boldsymbol{x}, \boldsymbol{x}', t) = \frac{1}{5}(\pi(\boldsymbol{x}) : \pi(\boldsymbol{x}', t)). \tag{24.44}$$

In deriving the second relation of (24.44), we have made use of the fact that

$$\operatorname{Tr} \Delta^2 = \operatorname{Tr} \Delta = 3,$$

since

$$\sum_\lambda \Delta_{\mu\lambda}\Delta_{\lambda\nu} = \Delta_{\mu\nu}.$$

Taking (24.43a) into account, we write the relations (24.41a)-(24.41c) in the form

$$\langle \pi_{\mu\nu}(\boldsymbol{x})\rangle = -\sum_{\mu_1\nu_1} \int \int_{-\infty}^{0} e^{\varepsilon t_1} L_\pi(\boldsymbol{x}, \boldsymbol{x}', t_1)\beta(\boldsymbol{x}', t + t_1) \times$$
$$\times \frac{1}{2}\left\{\Delta_{\mu\mu_1}\Delta_{\nu\nu_1}\left(\frac{\partial u_{\mu_1}(\boldsymbol{x}', t+t_1)}{\partial x'_{\nu_1}} + \frac{\partial u_{\nu_1}(\boldsymbol{x}', t+t_1)}{\partial x'_{\mu_1}}\right) - \right.$$
$$\left. - \frac{2}{3}\Delta_{\mu\nu}\Delta_{\mu_1\nu_1}\frac{\partial u_{\mu_1}(\boldsymbol{x}', t+t_1)}{\partial x'_{\nu_1}}\right\} c\, dt_1 d\boldsymbol{x}',$$

$$\langle P_\mu(\boldsymbol{x})\rangle = \sum_\nu \int \int_{-\infty}^{0} e^{\varepsilon t_1} L_P(\boldsymbol{x}, \boldsymbol{x}', t_1)\, \beta(\boldsymbol{x}', t+t_1) \times$$
$$\times \Delta_{\mu\nu}\left(\beta^{-1}(\boldsymbol{x}', t+t_1)\frac{\partial \beta(\boldsymbol{x}', t+t_1)}{\partial x'_\nu} - \frac{1}{c} Du_\nu(\boldsymbol{x}', t+t_1)\right) c\, dt_1 d\boldsymbol{x}', \tag{24.45}$$
$$\langle p(\boldsymbol{x})\rangle - \langle p(\boldsymbol{x})\rangle_l =$$
$$= -\int \int_{-\infty}^{0} e^{\varepsilon t_1} L_p(\boldsymbol{x}, \boldsymbol{x}', t_1)\, \beta(\boldsymbol{x}', t+t_1)\, \mathrm{div}\, \boldsymbol{u}(\boldsymbol{x}', t+t_1)\, c\, dt_1 d\boldsymbol{x}',$$

where

$$L_p(\boldsymbol{x}, \boldsymbol{x}', t_1) = (p'(\boldsymbol{x}), p'(\boldsymbol{x}', t_1)). \tag{24.45a}$$

If we neglect the retardation and the spatial dispersion in (24.45), these relations go over into the linear relations of relativistic hydrodynamics:

$$\langle \pi_{\mu\nu}\rangle = -\eta c \sum_{\mu_1\nu_1} \frac{1}{2}\left\{ \Delta_{\mu\mu_1}\Delta_{\nu\nu_1}\left(\frac{\partial u_{\mu_1}}{\partial x_{\nu_1}} + \frac{\partial u_{\nu_1}}{\partial x_{\mu_1}}\right) - \frac{2}{3}\Delta_{\mu\nu}\Delta_{\mu_1\nu_1}\frac{\partial u_{\mu_1}}{\partial x_{\nu_1}}\right\},$$
$$\langle P_\mu\rangle = c^{-1}\langle q_\mu\rangle = \lambda c \sum_\nu \Delta_{\mu\nu}\left(\beta^{-1}\frac{\partial\beta}{\partial x_\nu} - \frac{1}{c} Du_\nu\right), \tag{24.46}$$
$$\langle p\rangle - \langle p\rangle_l = -\zeta c\, \mathrm{div}\, \boldsymbol{u},$$

obtained, except for the last relation, by Eckart [133] (see also [82]) by the method of phenomenological nonequilibrium thermodynamics. In statistical nonequilibrium thermodynamics, in addition to these relations we obtain explicit expressions for the kinetic coefficients the shear viscosity η, bulk viscosity ζ, and thermal conductivity:

$$\eta = \frac{\beta}{5}\int\int_{-\infty}^{0} e^{\varepsilon t}(\pi(\boldsymbol{x}) : \pi(\boldsymbol{x}', t))\, dt\, d\boldsymbol{x}',$$
$$\lambda = \frac{\beta}{3}\int\int_{-\infty}^{0} e^{\varepsilon t}(\boldsymbol{P}(\boldsymbol{x})\cdot \boldsymbol{P}(\boldsymbol{x}', t))\, dt\, d\boldsymbol{x}', \tag{24.47}$$
$$\zeta = \beta\int\int_{-\infty}^{0} e^{\varepsilon t}(p'(\boldsymbol{x}), p'(\boldsymbol{x}', t))\, dt\, d\boldsymbol{x}'.$$

The expression (24.46) for the heat flux contains the relativistic term $(Du_\nu)/c$, which shows that a heat flux in a one-component system is induced not only by a temperature gradient, but also by acceleration.

24.4. Charge Transport Processes

Up to this point, we have considered only energy-momentum conservation. In the relativistic regime the number of particles is not conserved and if we treat particle-number transport we must take into account the appearance of particles as a result of various reactions, i.e., in the balance equations for the number of particles we must add sources, as in the theory of chemical reactions (see §23.5).

In relativistic theory, apart from energy-momentum, the different types of charge (electric, baryon, lepton, etc.) are conserved. We shall study the conservation law for charge of any one type (the generalization to different types of charge presents no difficulties):

$$\sum_{\nu=1}^{4} \frac{\partial j_\nu(\boldsymbol{x}, t)}{\partial x_\nu} = 0, \tag{24.48}$$

where $j_\nu(x)$ is the density of the four-vector of a current with space components j_1, j_2, and j_3 (three-dimensional current) and with time component $j_4 = -ic\rho$, where ρ is the charge density.

Corresponding to the conservation law (24.48) is the local integral of motion

$$\widetilde{j_4(\boldsymbol{x})} = j_4(\boldsymbol{x}) + ic \sum_{\nu=1}^{3} \int_{-\infty}^{0} e^{\varepsilon t} \frac{\partial j_\nu(\boldsymbol{x}, t)}{\partial x_\nu}\, dt, \tag{24.49}$$

which transforms like the fourth component of a vector.

We construct the quantity $B_4(\boldsymbol{x}, t)$ transforming in the same way as $\widetilde{j_4(\boldsymbol{x})}$, but containing the parameters $\varphi(\boldsymbol{x}, t)$ (some auxiliary scalar field) and $\beta(\boldsymbol{x}, t)$ (the inverse temperature):

$$\begin{aligned} B_4(\boldsymbol{x}, t) &= -\widetilde{i\beta(\boldsymbol{x}, t)\varphi(\boldsymbol{x}, t) j_4(\boldsymbol{x})} = -i\beta(\boldsymbol{x}, t)\varphi(\boldsymbol{x}, t) j_4(\boldsymbol{x}) + \\ &\quad + i\int_{-\infty}^{0} e^{\varepsilon t_1} \frac{d}{dt_1} \beta(\boldsymbol{x}, t+t_1)\varphi(\boldsymbol{x}, t+t_1) j_4(\boldsymbol{x}, t_1)\, dt_1 \\ &= -i\beta(\boldsymbol{x}, t)\varphi(\boldsymbol{x}, t) j_4(\boldsymbol{x}) + i\int_{-\infty}^{0} e^{\varepsilon t_1} \Big\{ j_4(\boldsymbol{x}, t_1) \frac{\partial}{\partial t_1} \beta(\boldsymbol{x}, t+t_1)\varphi(\boldsymbol{x}, t+t_1) - \\ &\quad - ic\beta(\boldsymbol{x}, t+t_1)\varphi(\boldsymbol{x}, t+t_1) \sum_{\nu=1}^{3} \frac{\partial j_\nu(\boldsymbol{x}, t_1)}{\partial x_\nu} \Big\}\, dt_1. \end{aligned} \tag{24.50}$$

The operator $B_4(\mathbf{x}, t)$ is an integral of Liouville's equation as $\varepsilon \to 0$. Its corresponding invariant is

$$\int B_4(\boldsymbol{x}, t)\, d\boldsymbol{x} = -\int \beta(\boldsymbol{x}, t)\, \varphi(\boldsymbol{x}, t)\, i j_4(\boldsymbol{x})\, d\boldsymbol{x} + \\ + i \sum_{\nu} \int \int_{-\infty}^{0} e^{\varepsilon t_1}\, j_\nu(\boldsymbol{x}, t_1) \frac{\partial}{\partial x_\nu} (\beta(\boldsymbol{x}, t + t_1)\, \varphi(\boldsymbol{x}, t + t_1))\, ic\, dt_1\, d\boldsymbol{x}, \quad (24.51)$$

where we have discarded the surface integrals.

In the statistical operator (24.5), we must also take into account the invariant (24.51):

$$\rho = Q^{-1} \exp\left\{ -\int \left(\sum_{\nu} B_{4\nu}(\boldsymbol{x}, t) + \frac{1}{c} B_4(\boldsymbol{x}, t) \right) d\boldsymbol{x} \right\}. \quad (24.52)$$

The nonequilibrium statistical operator (24.52) can be applied in the study of energy-momentum and charge transport processes in a spatially nonuniform system. We write it in explicit form:

$$\rho = Q^{-1} \exp\left\{ \int \beta(\boldsymbol{x}, t) \left(-\sum_{\nu} u_\nu(\boldsymbol{x}, t)\, i T_{4\nu}(\boldsymbol{x}) + \frac{i}{c} \varphi(\boldsymbol{x}, t)\, j_4(\boldsymbol{x}) \right) d\boldsymbol{x} - \right. \\ - \int \int_{-\infty}^{0} e^{\varepsilon t_1} \left(\sum_{\mu\nu} T_{\mu\nu}(\boldsymbol{x}, t_1) \frac{\partial F_\nu(\boldsymbol{x}, t + t_1)}{\partial x_\mu} + \right. \\ \left. \left. + j_\nu(\boldsymbol{x}, t_1) \frac{\partial}{\partial x_\nu} \beta(\boldsymbol{x}, t + t_1) \frac{i}{c} \varphi(\boldsymbol{x}, t + t_1) \right) ic\, dt_1\, d\boldsymbol{x} \right\}. \quad (24.53)$$

The quantities $\partial F_\nu / \partial x_\mu$ and $\partial \beta\varphi / \partial x_\mu$ play the role of the thermodynamic forces. If they are small, we obtain linear relations of the type (24.41a)-(24.41c) for the average fluxes of energy, momentum and charge. Here it is convenient, as previously, to separate out the convective motion in $T_{\mu\nu}$ by means of (24.23), and similarly in j_ν. If the particles possess spin, then, in addition to the conservation of energy-momentum (24.1) and of charge (24.48), we must take into account conservation of angular momentum, and this can be done by the same method.

For ordinary gases, quantum-hydrodynamic effects are very small. Quantum hydrodynamics finds application in another field, namely, in the theory of multiple production of particles in collisions of fast nucleons with nuclei [137, 138].

§ 25. Kinetic Equations

25.1. Generalized Kinetic Equations [56]

Up to this point, we have studied transport equations for the hydrodynamic regime, when the nonequilibrium state can be described macroscopically by a set of a small number of hydrodynamic or thermodynamic) parameters: the temperatures and chemical potentials of the components, the mass velocity, and so on. In our study of strongly nonequilibrium processes for systems with weak interaction in the spatially uniform case (§§23.4 and 23.5), it was found that the average occupation numbers satisfy a kinetic equation [see Eq. (23.64)]. We shall show that this is not fortuitous and that the method of the nonequilibrium statistical operator can also be applied, following the work of Pokrovskiĭ [156, 157], to the kinetic stage, if we choose the parameters describing the state of the system in the appropriate way. (For a discussion of the meaning of the kinetic and hydrodynamic stages, see pp. 301-302.)

We shall consider a quantum-mechanical system with Hamiltonian

$$H = H_0 + H_1, \tag{25.1}$$

where H_0 is the Hamiltonian of the free particles or quasi-particles, and H_1 is the Hamiltonian of the interaction, which we shall assume to be small.

We shall assume that, to describe a nonequilibrium state for time scales that are not too short, a set of quantities $\langle P_k \rangle$, where the brackets denote nonequilibrium averaging, is sufficient. For example, for a spatially uniform state of a gas, we can choose

$$P_k = a_k^+ a_k, \tag{25.2}$$

so that

$$\langle P_k \rangle = \langle a_k^+ a_k \rangle \tag{25.3}$$

is the distribution function over the states k. For the spatially nonuniform case, we can choose

$$P_k = a_{k+q}^+ a_k, \tag{25.4}$$

so that

$$\langle a_{k+q}^+ a_k \rangle \tag{25.5}$$

is the distribution function characterizing a spatially nonuniform state of the gas. Thus, the distribution function $\langle P_k \rangle$ can be regarded as a thermodynamic parameter, and this enables us to extend the general scheme of nonequilibrium thermodynamics to kinetic processes too. Since a kinetic equation, generally speaking, is nonlinear, we must consider a nonlinear variant of the theory.

We note that the operators P_k often satisfy simple commutation relations with the free-particle Hamiltonian

$$[H_0,\ P_k] = \sum_l \alpha_{kl} P_l, \tag{25.6}$$

where α_{kl} are certain numerical coefficients determining the free evolution of the operators P_k. Peletminskiĭ and Yatsenko [36] constructed a generalized kinetic equation for the average values of such operators.

In the particular case when the P_k are chosen as in (25.4), and

$$H_0 = \sum_k E_k a_k^+ a_k, \tag{25.7}$$

we have

$$[H_0,\ a_{k+q}^+ a_k] = (E_{k+q} - E_k)\, a_{k+q}^+ a_k. \tag{25.8}$$

With the choice (25.2) we have $\alpha_{kl} = 0$.

In most problems, it is sufficient to consider only operators satisfying the condition (25.6), although it is sometimes convenient to include also operators which do not satisfy this condition; we shall not consider such cases.

The operators P_k obey the equations of motion

$$\frac{\partial P_k}{\partial t} = \frac{1}{i\hbar}[P_k,\ H] = -\frac{1}{i\hbar}\sum_l \alpha_{kl} P_l + \frac{1}{i\hbar}[P_k,\ H_1]. \tag{25.9}$$

Corresponding to the equations of motion (25.9) is the nonequilibrium statistical operator, constructed in accordance with the usual rules (see §21) and satisfying Liouville's equations in the limit $\varepsilon \to 0$

$$\begin{aligned}\rho &= Q^{-1} \exp\left\{-\sum_k \widetilde{F_k(t) P_k}\right\} \\ &= Q^{-1} \exp\left\{-\sum_k \varepsilon \int_{-\infty}^{0} e^{\varepsilon t_1} F_k(t + t_1)\, P_k(t_1)\, dt_1\right\},\end{aligned} \tag{25.10}$$

or

$$\rho = Q^{-1} \exp \left\{ -\sum_k F_k(t)\, P_k + \right.$$

$$\left. + \int_{-\infty}^{0} dt_1\, e^{\varepsilon t_1} \sum_k \left(F_k(t+t_1)\, \dot{P}_k(t_1) + \frac{dF_k(t+t_1)}{dt} P_k(t_1) \right) \right\}, \qquad (25.11)$$

where the $F_k(t)$ are certain parameters associated with the $\langle P_k \rangle$. This dependence is determined from the supplementary conditions

$$\langle P_k \rangle = \langle P_k \rangle_q, \qquad (25.12)$$

where the notation

$$\langle \ldots \rangle = \mathrm{Tr}\,(\rho \ldots) \qquad (25.13)$$

indicates averaging with the nonequilibrium statistical operator (25.11), and

$$\langle \ldots \rangle_q = \mathrm{Tr}\,(\rho_q \ldots) \qquad (25.14)$$

denotes averaging with the quasi-equilibrium statistical operator

$$\rho_q = Q_q^{-1} \exp \left\{ -\sum_k F_k(t)\, P_k \right\}, \qquad (25.14a)$$

$$Q_q = \mathrm{Tr}\, \exp \left\{ -\sum_k F_k(t)\, P_k \right\}. \qquad (25.14b)$$

The quasi-equilibrium statistical operator (25.14a) is constructed analogously to the local-equilibrium statistical operator (20.10), which was considered in §20, although the statistical operator (25.14a) can describe strongly nonequilibrium states and is not connected with the concept of a local temperature.

For the nonequilibrium statistical operator (25.11), the relations

$$\frac{\partial \ln Q_q}{\partial F_k} = \frac{\partial \Phi}{\partial F_k} = -\langle P_k \rangle_q = -\langle P_k \rangle, \qquad (25.15)$$

serve as an analog of the thermodynamic equalities, where

$$\Phi = \Phi(\ldots F_k \ldots) = \ln Q_q \qquad (25.16)$$

is the analog of the Massieu–Planck function. If k takes a continuous series of values, the sums in (25.14a) go over into integrals, and the functions into functionals.

The thermodynamic equalities can also be represented in the form

$$\frac{\partial S}{\partial \langle P_k \rangle} = F_k(t), \tag{25.17}$$

where

$$S = \Phi + \sum_k \langle P_k \rangle F_k(t) \tag{25.18}$$

is the entropy of the nonequilibrium state.

The entropy production is equal to

$$\dot{S} = \sum_k \langle \dot{P}_k \rangle F_k(t), \tag{25.19}$$

since

$$\dot{\Phi} = -\sum_k \langle P_k \rangle \dot{F}_k(t). \tag{25.20}$$

The fact that the relations (25.15)-(25.18) have the same form as the thermodynamic equalities does not mean that the nonequilibrium state described by the statistical operator (25.11) is close to a state of statistical equilibrium.

The average value of (25.9), calculated with the operator (25.11),

$$\frac{\partial \langle P_k \rangle}{\partial t} = -\frac{1}{i\hbar} \sum_l \alpha_{kl} \langle P_l \rangle + \frac{1}{i\hbar} \langle [P_k, H_1] \rangle, \tag{25.21}$$

is the generalized kinetic equation for $\langle P_k \rangle$, since the average commutator of P_k with the interaction operator can be expressed in terms of $\langle P_k \rangle$ by means of (25.11) and (25.12). The first term in the right-hand side of (25.21) expresses the free collisionless evolution of the distribution function $\langle P_k \rangle$, and the second term is the collision integral.

In the exponent of the statistical operator (25.11) we eliminate the time derivatives of the parameters $F_k(t + t_1)$. We have

$$\begin{aligned} \frac{dF_k(t)}{dt} &= \sum_l \frac{\partial F_k}{\partial \langle P_l \rangle} \langle \dot{P}_l \rangle \\ &= -\frac{1}{i\hbar} \sum_{lm} \frac{\partial F_k}{\partial \langle P_l \rangle} \alpha_{lm} \langle P_m \rangle + \sum_l \frac{\partial F_k}{\partial \langle P_l \rangle} \frac{1}{i\hbar} \langle [P_l, H_1] \rangle. \end{aligned} \tag{25.22}$$

Further, we note that

$$\sum_{k,\,l} F_k \alpha_{kl} \langle P_l \rangle = \left\langle \left[H_0, \sum_k F_k P_k \right] \right\rangle_q \equiv 0, \tag{25.23}$$

since $\sum_k F_k P_k$ commutes with ρ_q. Differentiating the identity (25.23) with respect to F_k, we obtain

$$\sum_m \alpha_{km} \langle P_m \rangle + \sum_{ml} F_m \alpha_{ml} \frac{\partial \langle P_l \rangle}{\partial F_k} = 0. \tag{25.24}$$

Taking into account that

$$\frac{\partial \langle P_l \rangle}{\partial F_k} = - \frac{\partial^2 \ln Q_q}{\partial F_k \, \partial F_l} = \frac{\partial \langle P_k \rangle}{\partial F_l}, \tag{25.25}$$

we multiply (25.24) by $\partial F_i / \partial \langle P_k \rangle$ and sum over k. As a result, we obtain

$$\sum_{k,\,l} \frac{\partial F_i}{\partial \langle P_k \rangle} \alpha_{kl} \langle P_l \rangle + \sum_m F_m \alpha_{mi} = 0. \tag{25.26}$$

By means of (25.26), we bring (25.22) to the form

$$\frac{dF_k(t)}{dt} = \frac{1}{i\hbar} \sum_l \alpha_{lk} F_l(t) + \sum_l \frac{\partial F_k}{\partial \langle P_l \rangle} \frac{1}{i\hbar} \langle [P_l, \; H_1] \rangle. \tag{25.27}$$

This equation can be regarded as a kinetic equation in the variables F_k.

Substituting (25.9) and (25.27) into (25.11), we write the non-equilibrium statistical operator in the form

$$\rho = Q^{-1} \exp \Bigg\{ - \sum_k F_k(t) \, P_k + \int_{-\infty}^{0} dt_1 \, e^{\varepsilon t_1} \sum_k \frac{1}{i\hbar} \Bigg([P_k(t_1), \; H_1(t_1)] \, F_k(t + t_1) + \sum_l \frac{\partial F_k(t + t_1)}{\partial \langle P_l \rangle} \langle [P_l, \; H_1] \rangle^{t + t_1} \, P_k(t_1) \Bigg) \Bigg\}, \tag{25.28}$$

where the superscript $t + t_1$ on the average means that the average is taken with the statistical operator (25.11) taken at time $t + t_1$.

It can be seen from (25.28) that the integral term in the exponent is of at least first order of smallness in the interaction.

We shall seek an expansion of the collision integral

$$S_k = \frac{1}{i\hbar}\langle [P_k, H_1]\rangle \tag{25.29}$$

as a series in powers of the interaction. Since the operator being averaged in (25.29) is itself of first order in the interaction, in the expansion of the exponential in (25.28) (see §22.2) we can confine ourselves to the first-order terms, so that we obtain an expansion exact up to and including terms of second order. The expansion of (25.29) starts from the first-order terms and has the form

$$S_k(\ldots\langle P_i\rangle\ldots) = S_k^{(1)} + S_k^{(2)} + \ldots, \tag{25.30}$$

where

$$S_k^{(1)} = \frac{1}{i\hbar}\langle [P_k, H_1]\rangle_q, \tag{25.31}$$

$$S_k^{(2)} = S_k'^{(2)} + S_k''^{(2)}, \tag{25.32}$$

$$S_k'^{(2)} = -\sum_l \frac{1}{\hbar^2}\int_{-\infty}^{0} dt_1\, e^{\varepsilon t_1}([H_1, P_k], [H_1(t_1), P_l(t_1)])\, F_l(t+t_1), \tag{25.33}$$

$$S_k''^{(2)} = -\sum_{m,l} \frac{1}{\hbar^2}\int_{-\infty}^{0} dt_1\, e^{\varepsilon t_1}([H_1, P_k], P_l(t_1))\, \frac{\partial F_l(t+t_1)}{\partial\langle P_m\rangle}\langle [H_1(t_1), P_m(t_1)]\rangle_q \tag{25.34}$$

and we have introduced the usual notation for the quantum correlation functions:

$$(B, C) = \int_0^1 d\tau\, \langle B(e^{-\tau A} C e^{\tau A} - \langle C\rangle_q)\rangle_q, \tag{25.35}$$

$$A = \sum_k F_k(t) P_k. \tag{25.36}$$

The formulas (25.31)-(25.34) already give the collision integral of the generalized kinetic equation exact up to and including terms of second order in the interaction, although they can be further simplified. In the collision integrals (25.33) and (25.34), we can omit the interaction in the Heisenberg picture for the operators, since in these formulas there is already a factor of second order in the interaction. Moreover, in (25.33) we can put

$$\sum_l F_l(t+t_1)\, P_l(t_1) = \sum_l F_l(t)\, P_l, \tag{25.37}$$

since, if the interaction is neglected, this sum is an integral of motion. In fact, (25.28) is an integral of motion when $\varepsilon \to 0$, and the integral terms in its exponent are proportional to the interaction. Noting also that

$$\sum_l e^{-\tau A}[H_1(t_1),\ P_l F_l(t)]\, e^{\tau A} = \frac{d}{d\tau}\, e^{-\tau A} H_1(t_1)\, e^{\tau A}, \tag{25.38}$$

and performing the integration over τ, we obtain

$$S_k'^{(2)} = -\frac{1}{\hbar^2} \int_{-\infty}^{0} dt\, e^{\varepsilon t} \langle [H_1(t),\ [H_1,\ P_k]\,]\rangle_q, \tag{25.39}$$

i.e., this part of the collision integral is proportional to the Fourier component of the retarded Green function at $\omega = i\varepsilon$, or to the spectral intensity of the time correlation function at zero frequency (see §16).

We now transform $S_k''^{(2)}$ by writing it in explicit form:

$$S_k''^{(2)} = -\sum_{l,m} \frac{1}{\hbar^2} \int_{-\infty}^{0} dt_1\, e^{\varepsilon t_1} \int_0^1 d\tau\, \langle [H_1,\ P_k]\, e^{-\tau A} \frac{\partial F_l(t+t_1)}{\partial \langle P_m\rangle} \times$$
$$\times \langle [H_1(t_1),\ P_m(t_1)]\rangle_q (P_l(t_1) - \langle P_l(t_1)\rangle_q)\, e^{\tau A}\rangle_q. \tag{25.40}$$

We shall prove that

$$\frac{d}{dt_1} \sum_{l,m} \left[\frac{\partial F_l(t+t_1)}{\partial \langle P_m\rangle} \langle [H_1(t_1'),\ P_m(t_1)]\rangle_q (P_l(t_1) - \langle P_l(t_1)\rangle_q)\right] = 0, \tag{25.41}$$

whence it follows that we can put $t_1 = 0$ everywhere in (25.40), except in $H_1(t_1)$.

With neglect of the interaction, we shall calculate the time derivative of the matrix

$$\frac{\partial \langle P_k\rangle}{\partial F_n} = -\frac{\partial^2 \ln Q_q}{\partial F_n\, \partial F_k} = -\int_0^1 d\tau\, \langle P_k\, e^{-\tau A} (P_n - \langle P_n\rangle_q)\, e^{\tau A}\rangle_q. \tag{25.42}$$

We have

$$\frac{d}{dt}\frac{\partial\langle P_k\rangle}{\partial F_n}=-\sum_m \frac{\partial^3 \ln Q_q}{\partial F_n\,\partial F_k\,\partial F_m}\frac{dF_m}{dt}$$
$$=-\sum_{m,\,j}\frac{1}{i\hbar}\frac{\partial^3 \ln Q_q}{\partial F_n\,\partial F_k\,\partial F_m}\alpha_{jm}F_j. \tag{25.43}$$

Differentiating the identity (25.23) with respect to F_n and F_k, we find

$$\sum_{j,\,m}\frac{\partial^3 \ln Q_q}{\partial F_n\,\partial F_k\,\partial F_m}\alpha_{jm}F_j+\sum_m\frac{\partial^2 \ln Q_q}{\partial F_n\,\partial F_m}\alpha_{km}+\sum_m\frac{\partial^2 \ln Q_q}{\partial F_k\,\partial F_m}\alpha_{nm}=0. \tag{25.44}$$

By means of (25.44), (25.43), and (25.42), we obtain

$$\frac{d}{dt}\frac{\partial\langle P_k\rangle}{\partial F_n}=-\frac{1}{i\hbar}\sum_m\left(\frac{\partial\langle P_m\rangle}{\partial F_n}\alpha_{kn}+\frac{\partial\langle P_k\rangle}{\partial F_m}\alpha_{nm}\right). \tag{25.45}$$

Differentiating the identity

$$\sum_n\frac{\partial F_n}{\partial\langle P_m\rangle}\frac{\partial\langle P_m\rangle}{\partial F_k}=\delta_{nk}, \tag{25.46}$$

with respect to t, multiplying the result by $\partial F_k/\partial\langle P_i\rangle$ and summing over k, we find

$$\frac{d}{dt}\frac{\partial F_n}{\partial\langle P_i\rangle}=\frac{1}{i\hbar}\sum_k \alpha_{kn}\frac{\partial F_k}{\partial\langle P_i\rangle}+\frac{1}{i\hbar}\sum_m \alpha_{mi}\frac{\partial F_n}{\partial\langle P_m\rangle}, \tag{25.47}$$

whence, using (25.9), we obtain (25.41). Therefore, in (25.40) we can omit t_1 everywhere except in $H_1(t_1)$.

Noting also that

$$\int_0^1 d\tau\,\langle[H_1,\ P_k]\,e^{-\tau A}(P_n-\langle P_n\rangle_q)\,e^{\tau A}\rangle_q=i\hbar\frac{\partial S_k^{(1)}}{\partial F_n}, \tag{25.48}$$

we write (25.40) in the form

$$S_k''^{(2)}=-\frac{i}{\hbar}\int_{-\infty}^{0}dt\,e^{\varepsilon t}\left\langle\left[H_1(t),\ \sum_m P_m\frac{\partial S_k^{(1)}}{\partial\langle P_m\rangle}\right]\right\rangle_q, \tag{25.49}$$

or, combining (25.39) and (25.49), we obtain

$$S_k^{(2)} = -\frac{1}{\hbar^2}\int_{-\infty}^{0} dt\, e^{\varepsilon t}\left\langle\left[H_1(t),\ [H_1,\ P_k] + i\hbar\sum_m P_m \frac{\partial S_k^{(1)}}{\partial\langle P_m\rangle}\right]\right\rangle_q . \qquad (25.50)$$

The formulas (25.31) and (25.50) give the expansion of the collision integral in powers of the interaction to terms of second order. In this form, they were obtained by Peletminskii and Yatsenko [36] by another method.

25.2. Nonideal Quantum Gases

As a simple example, we shall consider the construction of kinetic equations for nonideal quantum gases. In this case, the free-particle Hamiltonian is

$$H_0 = \sum_{\boldsymbol{k}} E_{\boldsymbol{k}} a^+_{\boldsymbol{k}} a_{\boldsymbol{k}}, \qquad E_{\boldsymbol{k}} = \frac{\hbar^2 k^2}{2m}, \qquad (25.51)$$

and the interaction Hamiltonian is

$$H_1 = \frac{1}{2V}\sum_{\substack{\boldsymbol{k}_1\boldsymbol{k}_2\boldsymbol{k}'_1\boldsymbol{k}'_2 \\ \boldsymbol{k}_1+\boldsymbol{k}_2=\boldsymbol{k}'_1+\boldsymbol{k}'_2}} \Phi(\boldsymbol{k}_1\boldsymbol{k}_2 \,|\, \boldsymbol{k}'_1\boldsymbol{k}'_2)\, a^+_{\boldsymbol{k}_1} a^+_{\boldsymbol{k}_2} a_{\boldsymbol{k}'_1} a_{\boldsymbol{k}'_2}, \qquad (25.52)$$

where

$$\Phi(\boldsymbol{k}_1\boldsymbol{k}_2 \,|\, \boldsymbol{k}'_1\boldsymbol{k}'_2) = \frac{1}{2}\left(\nu(\boldsymbol{k}_1 - \boldsymbol{k}'_1) \pm \nu(\boldsymbol{k}_1 - \boldsymbol{k}'_2)\right) \qquad (25.53)$$

is the matrix element of the interaction, and

$$\nu(\boldsymbol{k}) = \int \Phi(\boldsymbol{x})\, e^{i\,\boldsymbol{k}\cdot\boldsymbol{x}}\, d\boldsymbol{x} \qquad (25.54)$$

is the Fourier component of the interaction potential.

The matrix element (25.53) is symmetrized for Bose statistics and antisymmetrized for Fermi statistics. This can be done conveniently, since for Bose statistics a permutation of the operators $a_{\boldsymbol{k}'_1}$ and $a_{\boldsymbol{k}'_2}$, equivalent to a permutation of the indices $\mathbf{k}_1$ and $\mathbf{k}_2$, does not change the sign of the product, while for Fermi statistics, the sign is reversed. If we do not make this symmetrization (or antisymmetrization), similar combinations of Fourier components will appear nevertheless in the final results.

As our basic operators, we choose

$$P_k = a_k^+ a_k = n_k; \tag{25.55}$$

then

$$[H_0, n_k] = 0 \tag{25.56}$$

and we have $\alpha_{kl} = 0$ in the relations (25.6).

The commutator of the interaction operator with n_k is equal to

$$[H_1, n_k] = \frac{1}{V} \sum_{\substack{k_1 k_2 k_1' \\ k_1 + k_2 = k_1' + k}} \Phi(k_1 k_2 | k_1' k) \left(a_{k_1}^+ a_{k_2}^+ a_{k_1'} a_k - a_k^+ a_{k_1'}^+ a_{k_2} a_{k_1} \right), \tag{25.57}$$

where we have taken into account the symmetry (or antisymmetry) properties of the matrix elements (25.53).

It is now easy to calculate the collision integral (25.39), using Wick's theorem and the formula (16.32),

$$S_k^{\prime(2)} = -\frac{1}{\hbar^2} \int_{-\infty}^{0} dt\, e^{\varepsilon t} \langle [H_1(t), [H_1, n_k]] \rangle_q =$$

$$= \sum_{\substack{k_1 k_2 k_1' \\ k_1 + k_2 = k_1' + k}} w(k_1 k_2 | k_1' k) \left\{ \bar{n}_{k_1} \bar{n}_{k_2} \left(1 \mp \bar{n}_{k_1'}\right) (1 \mp \bar{n}_k) - \right.$$

$$\left. - \left(1 \mp \bar{n}_{k_1}\right)\left(1 \mp \bar{n}_{k_2}\right) \bar{n}_{k_1'} \bar{n}_k \right\}, \tag{25.58}$$

where

$$w(k_1 k_2 | k_1' k) = \frac{4\pi}{\hbar} |\Phi(k_1 k_2 | k_1' k)|^2 \delta\left(E_{k_1} + E_{k_2} - E_{k_1'} - E_k\right) \tag{25.59}$$

is the transition probability in unit time in the Born approximation, and

$$\bar{n}_k = \langle n_k \rangle_q \tag{25.60}$$

are the average occupation numbers of the state **k**. The other collision operators (25.31) and (25.49) are equal to zero,

$$S_k^{(1)} = S_k^{\prime\prime(2)} = 0.$$

Finally, we obtain the kinetic equation for a quantum Bose or Fermi gas in the form

$$\frac{\partial n_k}{\partial t} = - \sum_{k_1 k' k_1'} w(kk_1 | k'k_1') \left\{ \bar{n}_k \bar{n}_{k_1} \left(1 \mp \bar{n}_{k'}\right)\left(1 \mp \bar{n}_{k_1'}\right) - \right.$$
$$\left. - \left(1 \mp \bar{n}_k\right)\left(1 \mp \bar{n}_{k_1}\right) \bar{n}_{k'} \bar{n}_{k_1'} \right\}, \qquad (25.61)$$

where the plus sign is for a Bose gas, and the minus for a Fermi gas.

The third approximation for the collision operator of quantum gases was obtained by Bogolyubov and Gurov [139] (see the monograph [140]) and the fourth by Bar'yakhtar, Peletminskii, and Yatsenko [141].

The kinetic equation for quantum gases with the collision operator (25.61) was first obtained by Uehling and Uhlenbeck [142]. This problem was treated later by many authors [143-145], who started from the Wigner mixed coordinate-momentum representation (see §14.2). For more details on the derivation of the quantum kinetic equation, see the monograph by Fujita [146] and also [147-149].

Kinetic equations of the type (25.61) are applicable to nondegenerate quantum gases. For degenerate gases, the kinetic equations must be constructed not for the particle distribution functions, but for the distribution functions of the elementary excitations. For example, a kinetic equation for the elementary excitations in a nonideal Bose gas was obtained by Bogolyubov [150].

For a nonideal degenerate Bose gas, the distribution functions (25.3) are insufficient, and one must also consider the functions $\langle a_k a_{-k} \rangle$, i.e.,

$$P_k = \left\{ a_k^+ a_k,\ a_k a_{-k} \right\}.$$

This scheme is carried out in paper [151], by the method of Peletminskii and Yatsenko [36]. An analogous situation also holds in the theory of superconductivity.

25.3. The Kinetic Equation for Electrons in a Metal

We shall consider one more example of a quantum kinetic equation, the Bloch equation for electrons in a metal. In this case,

$$H = H_0 + H_1, \qquad (25.62)$$

where

$$H_0 = \sum_{k,\sigma} E_{k\sigma} a^+_{k\sigma} a_{k\sigma} + \sum_q \hbar\omega_q b^+_q b_q \tag{25.62a}$$

is the Hamiltonian of the free electrons and phonons, and

$$H_1 = \frac{1}{\sqrt{V}} \sum_{\substack{k_1 k_2 q,\ \sigma \\ k_1 - k_2 = q}} v_q \left(\frac{\hbar}{2\omega_q}\right)^{1/2} (b_q + b^+_{-q})\, a^+_{k_1\sigma} a_{k_2\sigma} \tag{25.62b}$$

is the Hamiltonian of the interaction of the electrons with the lattice phonons (see the footnote to p. 168).

To derive the kinetic equation for the electrons in the spatially uniform case, we choose

$$P_k = a^+_{k\sigma} a_{k\sigma} = n_{k\sigma}. \tag{25.63}$$

The kinetic equations for $\langle n_{k\sigma} \rangle$ has the form

$$\frac{\partial \langle n_{k\sigma} \rangle}{\partial t} = \frac{1}{i\hbar} \langle [n_{k\sigma}, H_1] \rangle = S_k^{(1)} + S_k^{(2)} + \cdots \tag{25.64}$$

Noting that

$$S_k^{(1)} = \frac{1}{i\hbar} \langle [n_{k\sigma}, H_1] \rangle_q = 0,$$

we write the kinetic equation (25.64) in the form

$$\begin{aligned} \frac{\partial \bar{n}_k}{\partial t} &= -\frac{1}{\hbar^2} \int_{-\infty}^{0} dt\, e^{\varepsilon t} \langle [H_1(t),\ [H_1, n_{k\sigma}]] \rangle_q \\ &= \frac{1}{V} \sum_{\substack{k_1, q \\ k_1 - k = q}} \frac{\pi}{\omega_q} |v_q|^2 \{(\overline{N}_q + 1)\bar{n}_{k_1}(1 - \bar{n}_k) - \overline{N}_q (1 - \bar{n}_{k_1}) \bar{n}_k\} \times \\ &\quad \times \delta(E_{k_1} - E_k - \hbar\omega_q) - \frac{1}{V} \sum_{\substack{k_1 q \\ k_1 - k = -q}} \frac{\pi}{\omega_q} |v_q|^2 \times \\ &\quad \times \{(\overline{N}_q + 1)\bar{n}_k (1 - \bar{n}_{k_1}) - \overline{N}_q \bar{n}_{k_1} (1 - \bar{n}_k)\} \delta(E_{k_1} - E_k + \hbar\omega_q), \end{aligned} \tag{25.65}$$

where

$$\bar{n}_k = \langle n_{k\sigma} \rangle_q, \qquad \overline{N}_q = \langle N_q \rangle_q = \langle b^+_q b_q \rangle_q \tag{25.66}$$

are the electron and phonon distribution functions.

The equation (25.65) is the well-known Bloch equation on which is constructed the theory of the electrical conductivity and thermal conductivity of metals and semiconductors [152–155].

In the same way, we can derive the kinetic equation for the phonon distribution function:

$$\frac{\partial\langle N_q\rangle}{\partial t}=\frac{1}{i\hbar}\langle[N_q,\ H_1]\rangle, \qquad (25.67)$$

where the right-hand side can be expanded easily by perturbation theory. This equation is considered in the book [153].

Equally simply, one can also obtain other kinetic equations, for example, the Peierls equation for phonons in a lattice [156], when the collisions are due to the effects of anharmonicity.

A special case of the kinetic equations is the equation for a small subsystem interacting with a large one in an equilibrium state (i.e., with a thermostat). For such a subsystem, the probability of a forward transition is not equal to the probability of the inverse transition, since exchange of energy with the thermostat is possible in the transition. Therefore, the transition probability is not simply the square of the matrix element of the perturbation, as in (25.59), but must depend on the temperature. The method of the nonequilibrium statistical operator is convenient for obtaining equations of such a type, as was shown by Pokrovskii [157]. Also considered in this paper was a particular case of the derivation of equations for spin systems with allowance for off-diagonal terms (the Redfield equations [158]) and of equations for average spin operators, i.e., the Bloch equations [158] (see also [211]). Using the method of the nonequilibrium statistical operator, it is also very simple to derive master equations [231].

An important field of application of kinetic equations is that of completely or partially ionized plasmas and plasma-like media. To construct a theory of transport processes in such systems, it is necessary to go outside the framework of ordinary perturbation theory of powers of the interaction and consider the effects of polarization of the medium; otherwise, divergences appear in the kinetic equation. A kinetic equation for a plasma with allowance for the polarization was first obtained by Balescu [176, 177] by Prigogine's method [178], and by Lenard [179] by the method of Bogolyubov [1]. A fairly simple method of deriving this equation was given by Klimontovich [163]. By the same method, he developed a statistical theory of inelastic processes in a plasma [180].

The kinetic theory of an electron liquid in metals was developed by Silin [181, 182], who predicted spin waves in nonferromagnetic metals.

In this book, we shall not concern ourselves with the theory of transport processes in a plasma, since this is a large field in its own right. Besides, the method of the nonequilibrium statistical operator [2-5] has not yet been applied to a plasma, although it is possible to do this.

§ 26. The Kramers – Fokker – Planck Equations [162]

In many problems of nonequilibrium statistical mechanics (the Brownian motion of a particle in a liquid, relaxation in a system of oscillators, the theory of homogeneous nucleation, etc.), one considers the evolution of a small subsystem in contact with a large subsystem, to be called the thermostat, in thermodynamic equilibrium. In the case of weak interaction between them, this evolution is described by the Kramers–Fokker–Planck equation, which was first derived by Kramers [161] by means of the theory of Markovian processes, starting from the Langevin equation with a phenomenological friction constant. Later, this equation was obtained by Kirkwood [13] for the particular case of Brownian motion in a liquid. Kirkwood succeeded in deriving an expression for the friction coefficient in terms of the autocorrelation function of the forces acting on the Brownian particle. The first to obtain the Fokker–Planck equation from the equations of mechanics (classical and quantum), and also expressions for the coefficients in this equation in terms of correlation functions of the perturbing forces, were Krylov and Bogolyubov [169] as long ago as 1939, long before the work of Kirkwood [13] (see also the paper by Bogolyubov [169a]). Unfortunately, these important papers were not easily available, and because of this were not sufficiently widely known in their time.

In this section, following the work of Bashkirov and the author [162], we shall give a derivation of the Kramers–Fokker–Planck equation for the case of classical statistical mechanics, using the method described in §21.

26.1. General Method

Suppose that we have N identical subsystems, in contact with a thermostat and not interacting with each other. The total Hamiltonian of such a system has the form

$$H = \sum_i H_1(p_i, q_i) + H_2(P, Q) + \sum_i U(p_i, q_i, P, Q), \tag{26.1}$$

where $H_1(p_i, q_i)$ is the Hamiltonian of the i-th small subsystem with dynamic variables p_i and q_i, $H_2(P, Q)$ is the Hamiltonian of the thermostat, P, Q are the set of its dynamic variables, and $U(p_i, q_i, P, Q)$ is the interaction potential between the i-th subsystem and the thermostat.

The macroscopic state of the complete system is characterized by (in addition to the thermodynamic variables of the thermostat) the distribution function $f(p, q, t)$ of the subsystems in phase space. Corresponding to this quantity is the dynamic variable

$$n(p, q) = \sum_i \delta(p - p_i)\,\delta(q - q_i), \tag{26.2}$$

which is the mixed density of the subsystems in phase space, so that

$$f(p, q, t) = \langle n(p, q) \rangle, \tag{26.3}$$

where $\langle \ldots \rangle$ denotes averaging with some nonequilibrium distribution function, to be considered below. We note that the integral of (26.2) over the phase variables p and q is equal to the total number of small subsystems $\int n(p, q)\,dp\,dq = N$. In the particular case when the small subsystems are, for example, spherically symmetric Brownian particles, p_i and q_i are the ordinary momentum $\mathbf{p}_i$ and coordinate $\mathbf{q}_i$ of the i-th Brownian particle, and the expression (26.2) defines the density in the six-dimensional phase space

$$n(\mathbf{p}, \mathbf{q}) = \sum_i \delta(\mathbf{p} - \mathbf{p}_i)\,\delta(\mathbf{q} - \mathbf{q}_i), \tag{26.2a}$$

which was widely used by Klimontovich to construct kinetic equations in the theory of a nonequilibrium plasma [163].

We shall first consider the general case, when (26.2) is the density in a multidimensional phase space the dimensions of which are determined by the number of canonically conjugate dynamic variables p_i, q_i of one small subsystem.

The mixed density (26.2) satisfies the equation of motion

$$\dot{n}(p, q) = \{n(p, q), H\} = -\frac{\partial}{\partial q} j_1(p, q) - \frac{\partial}{\partial p} j_2(p, q), \tag{26.4}$$

where $\{\ldots\}$ is the classical Poisson bracket (2.10),

$$j_1(p, q) = \left(\frac{\partial H_1(p, q)}{\partial p} + \frac{\partial U(p, q, P, Q)}{\partial p}\right) n(p, q), \tag{26.5a}$$

$$j_2(p, q) = -\left(\frac{\partial H_1(p, q)}{\partial q} + \frac{\partial U(p, q, P, Q)}{\partial q}\right) n(p, q). \tag{26.5b}$$

According to the general method of constructing a nonequilibrium distribution function, to the conservation law (26.4) there corresponds the nonequilibrium distribution function

$$\begin{aligned} \rho &= Q^{-1} \exp\left\{-\beta H + \beta \int dp\, dq\, \varepsilon \int_{-\infty}^{0} dt_1\, e^{\varepsilon t_1} \varphi(p, q, t+t_1)\, n(p, q, t_1)\right\} \\ &= Q^{-1} \exp\Big\{-\beta H + \beta \int dp\, dq\, \varphi(p, q, t)\, n(p, q) - \\ &\quad - \beta \int dp\, dq \int_{-\infty}^{0} dt_1\, e^{\varepsilon t_1} [\varphi(p, q, t+t_1)\, \dot{n}(p, q, t_1) + \\ &\quad + \dot{\varphi}(p, q, t+t_1)\, n(p, q, t_1)]\Big\} \end{aligned} \tag{26.6a}$$

or, if we take (26.4) into account,

$$\begin{aligned} \rho &= Q^{-1} \exp\Big\{-\beta H + \beta \int dp\, dq\, \varphi(p, q, t)\, n(p, q) - \\ &\quad - \beta \int dp\, dq \int_{-\infty}^{0} dt_1\, e^{\varepsilon t_1} \Big[\frac{\partial \varphi(p, q, t+t_1)}{\partial q} j_1(p, q, t_1) + \\ &\quad + \frac{\partial \varphi(p, q, t+t_1)}{\partial p} j_2(p, q, t_1) + \dot{\varphi}(p, q, t+t_1)\, n(p, q, t_1)\Big]\Big\}, \end{aligned} \tag{26.6b}$$

where $\varphi(p, q, t)$ is a function of $\langle n(p, q)\rangle$; it will be eliminated below and will not appear in the final results. This function is determined from the condition

$$\langle n(p, q)\rangle = \langle n(p, q)\rangle_l, \tag{26.7}$$

where $\langle\ldots\rangle$ denotes averaging with the total distribution function (26.6b), and $\langle\ldots\rangle_l$ denotes averaging with the local-equilibrium or quasi-equilibrium distribution function†

$$\rho_l = Q_l^{-1}\exp\left\{-\beta H + \beta\int dp\,dq\,\varphi(p, q, t)\,n(p, q)\right\}. \tag{26.8}$$

To eliminate the derivative $\dot{\varphi}(p, q, t)$ from (26.6b), we differentiate both parts of the equality (26.7) with respect to the time. Then, for the left-hand side, we obtain

$$\begin{aligned}\frac{d}{dt}\langle n(p, q)\rangle &= -\frac{\partial}{\partial q}\langle j_1(p, q)\rangle - \frac{\partial}{\partial p}\langle j_2(p, q)\rangle \cong \\ &\cong -\frac{\partial H_1(p, q)}{\partial p}\frac{\partial}{\partial q}\langle n(p, q)\rangle + \frac{\partial H_1(p, q)}{\partial q}\frac{\partial}{\partial p}\langle n(p, q)\rangle \cong \\ &\cong \left(-\frac{\partial H_1(p, q)}{\partial p}\frac{\partial \varphi}{\partial q} + \frac{\partial H_1(p, q)}{\partial q}\frac{\partial \varphi}{\partial p}\right)\beta\langle n(p, q)\rangle_l,\end{aligned} \tag{26.9a}$$

where we have neglected terms of the type

$$\left\langle\frac{\partial U(p, q, P, Q)}{\partial p}\right\rangle \text{ and } \left\langle\frac{\partial U(p, q, P, Q)}{\partial q}\right\rangle$$

and have made use of the equality (26.4) [cf. (26.24)]. We now differentiate the right-hand side of (26.8):

$$\begin{aligned}\frac{d}{dt}\langle n(p, q)\rangle_l &= \beta\left\langle\int dp'\,dq'\,\dot{\varphi}(p', q', t)\,[n(p, q)(n(p', q') - \langle n(p', q')\rangle_l)]\right\rangle_l \\ &= \beta(\dot{\varphi}(p, q) - \bar{\dot{\varphi}})\langle n(p, q)\rangle_l \cong \beta\dot{\varphi}(p, q, t)\langle n(p, q)\rangle_l,\end{aligned} \tag{26.9b}$$

where we have neglected the term

$$\bar{\dot{\varphi}} = \int dp\,dq\,\dot{\varphi}(p, q, t)\langle n(p, q)\rangle_l, \tag{26.10}$$

which is of the order of magnitude of the average force in the local equilibrium state.

†It is better to call the distribution function (26.8) a quasi-equilibrium rather than a local-equilibrium distribution function, since it can describe strongly nonequilibrium states. In fact

$$\rho_l = Q_l^{-1}\exp\left\{-\beta H + \beta\sum_i \varphi(p_i, t)\right\} = Q_l^{-1}\,e^{-\beta H}\prod_i e^{\beta\varphi(p_i, q_i, t)},$$

and the single-particle distribution function $\exp[\beta\,\varphi(p_i, q_i, t)]$ can differ greatly from the equilibrium function.

Equating (26.9a) and (26.9b), we obtain the collisionless kinetic equation

$$\dot{\varphi}(p, q, t) = -\frac{\partial H_1(p, q)}{\partial p}\frac{\partial \varphi(p, q, t)}{\partial q} + \frac{\partial H_1(p, q)}{\partial q}\frac{\partial \varphi(p, q, t)}{\partial p}, \tag{26.11}$$

which we could have written down immediately, if we had assumed that φ(p, q, t) is a function of the single-particle distribution function ⟨n(p, q)⟩; for this such an equation is obvious.

Substituting (26.11) into (26.7b), we obtain

$$\rho = Q^{-1}\exp\Big\{-\beta H + \beta\int dp\, dq\, \varphi(p, q, t)\, n(p, q) - \\ -\beta\int dp\, dq \int_{-\infty}^{0} dt_1\, e^{\varepsilon t_1}\Big[\frac{\partial \varphi(p, q, t+t_1)}{\partial q} j_1'(p, q, t_1) + \\ + \frac{\partial \varphi(p, q, t+t_1)}{\partial q} j_2'(p, q, t_1)\Big]\Big\} \tag{26.12}$$

or, in the linear approximation in the interaction between the subsystem and the thermostat,

$$\rho = \rho_l - \rho_l\beta\int dp\, dq \int_{-\infty}^{0} dt_1\, e^{\varepsilon t_1}\Big[\frac{\partial \varphi(p, q, t+t_1)}{\partial q}\left(j_1'(p, q, t_1) - \langle j_1'(p, q, t_1)\rangle_l\right) + \\ + \frac{\partial \varphi(p, q, t+t_1)}{\partial p}\left(j_2'(p, q, t_1) - \langle j_2'(p, q, t_1)\rangle_l\right)\Big], \tag{26.13}$$

where

$$j_1'(p, q) = j_2(p, q) - \frac{\partial H_1(p, q)}{\partial p} n(p, q) = \frac{\partial U(p, q, P, Q)}{\partial p} n(p, q), \tag{26.14a}$$

$$j_2'(p, q) = j_2(p, q) + \frac{\partial H_1(p, q)}{\partial q} n(p, q) = -\frac{\partial U(p, q, P, Q)}{\partial q} n(p, q). \tag{26.14b}$$

We proceed now to the derivation of the equation for the distribution function f(p, q, t) = ⟨n(p, q)⟩. For this, we average the exact mechanical equation of motion (26.4) over all the dynamic variables of the system; this gives

$$\frac{\partial f}{\partial t} + \frac{\partial H_1(p, q)}{\partial p}\frac{\partial f}{\partial q} - \frac{\partial H_1(p, q)}{\partial q}\frac{\partial f}{\partial p} = -\frac{\partial}{\partial q}\langle j_1'(p, q)\rangle - \frac{\partial}{\partial p}\langle j_2'(p, q)\rangle. \tag{26.15}$$

This equation is none other than an equation of Bogolyubov's chain [1] for the distribution functions. In fact, in the left-hand side of (26.15), we have a distribution function in the phase space of the subsystem, and in the right-hand side, a distribution function of higher order in the phase space of the dynamic variables of both the subsystem and the thermostat. To decouple this equation, in its right-hand side we use the distribution function (26.13) obtained above. Then,

$$\langle j_1'(p, q)\rangle = \langle j_1'(p, q)\rangle_l - \beta \int dp'\, dq' \int_{-\infty}^{0} dt_1\, e^{\varepsilon t_1} \Big[\frac{\partial \varphi(p', q', t+t_1)}{\partial q'} \times$$

$$\times \langle j_1'(p, q)\big(j_1'(p', q', t_1) - \langle j_1'(p', q', t_1)\rangle_l\big)\rangle_l +$$

$$+ \frac{\partial \varphi(p', q', t+t_1)}{\partial p'} \langle j_1'(p, q)\big(j_2'(p', q', t_1) - \langle j_2'(p', q', t_1)\rangle_l\big)\rangle_l \Big]. \qquad (26.16)$$

We shall study the second term in the right-hand side of (26.16):

$$-\beta \int dp'\, dq' \int_{-\infty}^{0} dt_1\, e^{\varepsilon t_1} \frac{\partial \varphi(p', q', t+t_1)}{\partial q'} \Big\langle \frac{\partial U(p, q, P, Q)}{\partial p} n(p, q) \times$$

$$\times \Big(\frac{\partial U(p', q', P, Q, t_1)}{\partial p'} n(p', q', t_1) - \langle j_1'(p', q', t_1)\rangle_l \Big) \Big\rangle_l =$$

$$= -\beta \int_{-\infty}^{0} dt_1\, e^{\varepsilon t_1} \Big[\frac{\partial \varphi(p, q, t+t_1)}{\partial q} f(p, q, t) \Big\langle \frac{\partial U(p, q, P, Q)}{\partial p} \frac{\partial U(p, q, P, Q, t_1)}{\partial p} \Big\rangle_0 -$$

$$- f(p, q, t) \Big\langle \frac{\partial U(p, q, P, Q)}{\partial p} \Big\rangle_0 \overline{\Big\langle \frac{\partial \varphi(t+t_1)}{\partial q} \frac{\partial U(p, q, P, Q, t_1)}{\partial p} \Big\rangle_0} \Big] \simeq$$

$$\simeq -\beta \int_{-\infty}^{0} dt_1\, e^{\varepsilon t_1} \frac{\partial \varphi(p, q, t+t_1)}{\partial q} f(p, q, t) \times$$

$$\times \Big\langle \frac{\partial U(p, q, P, Q)}{\partial p} \frac{\partial U(p, q, P, Q, t_1)}{\partial p} \Big\rangle_0, \qquad (26.17a)$$

where we have neglected the correlation between the different small subsystems, which corresponds to a higher order of smallness in the interaction U.

We have replaced the local-equilibrium averaging $\langle \ldots \rangle_l$ of quantities small in U by averaging over the conditional equilibrium distribution:

$$\langle \ldots n(p, q)\rangle_l \simeq f(p, q, t) \langle \ldots \rangle_0, \qquad (26.18)$$

$$\langle \ldots \rangle_0 = \int dP\, dQ \ldots \frac{e^{-\beta H_0}}{\int dP\, dQ\, e^{-\beta H_0}}, \qquad (26.19)$$

where

$$H_0 = H_1(p, q) + H_2(P, Q) + U(p, q, P, Q); \qquad (26.19a)$$

the line over the average in (26.17a) denotes the averaging (26.10). In addition, we have neglected terms of second order in the "average forces"

$$\left\langle \frac{\partial U(p, q, P, Q)}{\partial q} \right\rangle_0 \text{ and } \left\langle \frac{\partial U(p, q, P, Q)}{\partial p} \right\rangle_0,$$

which are very small.

In particular, if, for example, the potential depends on q only through the difference Q − q, and the Hamiltonian (26.1) is an even function of Q, then the average force

$$\left\langle \frac{\partial U(p, q, P, Q)}{\partial p} \right\rangle_0$$

is exactly equal to zero, as is the case, for example, for a Brownia particle in a liquid.

These arguments can also be used to calculate the local-equilibrium flux

$$\langle j_1'(p, q)\rangle_l = \left\langle \frac{\partial U(p, q, P, Q)}{\partial p} n(p, q)\right\rangle_l = \left\langle \frac{\partial U(p, q, P, Q)}{\partial p} \right\rangle_0 f(p, q, t), \tag{26.20}$$

which can also be equal to zero, but is, in the general case, a smal quantity; as will be shown below, taking it into account leads to a renormalization $H_1 \rightarrow H_1 + \langle U\rangle_0$ of the energy of the subsystem.

We can investigate the third term in the right-hand side of (26.16) in an analogous way. It is equal to

$$\beta f(p, q, t) \int_{-\infty}^{0} dt_1\, e^{\varepsilon t_1} \frac{\partial \varphi(p, q, t+t_1)}{\partial p} \left\langle \frac{\partial U(p, q, P, Q)}{\partial p} \frac{\partial U(p, q, P, Q, t_1)}{\partial q} \right\rangle_0 \tag{26.17b}$$

in the same approximation as (26.17a).

Thus,

$$\begin{aligned}\langle j_1'(p, q)\rangle = \left\langle \frac{\partial U(p, q, P, Q)}{\partial p} \right\rangle_0 f(p, q, t) - \\ - \beta f(p, q, t) \int_{-\infty}^{0} dt_1\, e^{\varepsilon t_1} \frac{\partial \varphi(p, q, t+t_1)}{\partial q} \\ \times \left\langle \frac{\partial U(p, q, P, Q)}{\partial p} \frac{\partial U(p, q, P, Q, t_1)}{\partial p} \right\rangle_0 + \beta f(p, q, t) \int_{-\infty}^{0} dt_1\, e^{\varepsilon t_1} \frac{\partial \varphi(p, q, t+t_1)}{\partial p} \\ \times \left\langle \frac{\partial U(p, q, P, Q)}{\partial p} \frac{\partial U(p, q, P, Q, t_1)}{\partial q} \right\rangle_0\end{aligned} \tag{26.21}$$

and, analogously,

$$\langle j_2'(p, q)\rangle = -\left\langle \frac{\partial U(p, q, P, Q)}{\partial q}\right\rangle_0 f(p, q, t) + \beta f(p, q, t) \times$$
$$\times \int_{-\infty}^{0} dt_1\, e^{\varepsilon t_1} \frac{\partial \varphi(p, q, t+t_1)}{\partial q} \left\langle \frac{\partial U(p, q, P, Q)}{\partial q} \frac{\partial U(p, q, P, Q, t_1)}{\partial p}\right\rangle_0 -$$
$$- \beta f(p, q, t) \int_{-\infty}^{0} dt_1\, e^{\varepsilon t_1} \frac{\partial \varphi(p, q, t+t_1)}{\partial p} \left\langle \frac{\partial U(p, q, P, Q)}{\partial q} \frac{\partial U(p, q, P, Q, t_1)}{\partial q}\right\rangle_0 . \tag{26.22}$$

We now eliminate the derivatives $\partial\varphi/\partial q$ and $\partial\varphi/\partial p$ from (26.21) and (26.22). Here we can confine ourselves to the zeroth order in the interaction potential, by putting

$$f(p, q, t) = f_l(p, q, t) \cong Q_l^{-1} \exp\{-\beta(H_1(p, q) - \varphi(p, q, t))\}, \tag{26.23}$$

whence

$$\begin{aligned} \frac{\partial \varphi(p, q, t+t_1)}{\partial p} &= \frac{\partial H_1(p, q)}{\partial p} + kT \frac{\partial \ln f(p, q, t+t_1)}{\partial p}, \\ \frac{\partial \varphi(p, q, t+t_1)}{\partial q} &= \frac{\partial H_1(p, q)}{\partial q} + kT \frac{\partial \ln f(p, q, t+t_1)}{\partial q}. \end{aligned} \tag{26.24}$$

Substituting the expressions (26.21) and (26.22) into the right-hand side of (26.15) and taking (26.24) into account, we obtain

$$\frac{\partial f}{\partial t} + \frac{\partial (H_1(p, q) + \langle U(p, q, P, Q)\rangle_0)}{\partial p}\frac{\partial f}{\partial q} - \frac{\partial (H_1(p, q) + \langle U(p, q, P, Q)\rangle_0)}{\partial q}\frac{\partial f}{\partial p} =$$
$$= -\frac{\partial}{\partial q}\Bigg[-\beta f(p, q, t) \int_{-\infty}^{0} dt_1\, e^{\varepsilon t_1} \left(\frac{\partial H_1(p, q)}{\partial q} + kT \frac{\partial \ln f(p, q, t+t_1)}{\partial q}\right) \times$$
$$\times \left\langle \frac{\partial U(p, q, P, Q)}{\partial p} \frac{\partial U(p, q, P, Q, t_1)}{\partial p}\right\rangle_0 +$$
$$+ \beta f(p, q, t) \int_{-\infty}^{0} dt_1\, e^{\varepsilon t_1} \left(\frac{\partial H_1(p, q)}{\partial p} + kT \frac{\partial \ln f(p, q, t+t_1)}{\partial p}\right) \times$$
$$\times \left\langle \frac{\partial U(p, q, P, Q)}{\partial p} \frac{\partial U(p, q, P, Q, t_1)}{\partial q}\right\rangle_0 \Bigg] -$$
$$- \frac{\partial}{\partial p}\Bigg[\beta f(p, q, t) \int_{-\infty}^{0} dt_1\, e^{\varepsilon t_1} \left(\frac{\partial H_1(p, q)}{\partial q} + kT \frac{\partial \ln f(p, q, t+t_1)}{\partial q}\right) \times$$
$$\times \left\langle \frac{\partial U(p, q, P, Q)}{\partial q} \frac{\partial U(p, q, P, Q, t_1)}{\partial p}\right\rangle_0 -$$
$$- \beta f(p, q, t) \int_{-\infty}^{0} dt_1\, e^{\varepsilon t_1} \left(\frac{\partial H_1(p, q)}{\partial p} + kT \frac{\partial \ln f(p, q, t+t_1)}{\partial p}\right) \times$$
$$\times \left\langle \frac{\partial U(p, q, P, Q)}{\partial q} \frac{\partial U(p, q, P, Q, t_1)}{\partial q}\right\rangle_0 \Bigg]. \tag{26.25}$$

This equation describes the evolution of the distribution function of a subsystem with coordinates p and q in contact with a thermostat; it can be regarded as the Liouville equation for an open system. The right-hand side (the collision integral) shows that the evolution of the distribution function f(p, q, t) at time t depends on the state of the system at the previous times $-\infty < t + t_1 \leq 0$. Equations of such a type are usually called non-Markovian. A Markovian equation is obtained from (26.25) in the particular case when the attenuation of the time correlation functions

$$\left\langle \frac{\partial U}{\partial p} \frac{\partial U(t_1)}{\partial p} \right\rangle_0, \quad \left\langle \frac{\partial U}{\partial p} \frac{\partial U(t_1)}{\partial q} \right\rangle_0, \text{ etc.}$$

is so rapid that the factors multiplying them, of the type

$$\frac{\partial H_1}{\partial q} + kT \frac{\partial \ln f(t+t_1)}{\partial q},$$

do not have time to change significantly and can be taken outside the integrals over time. In the Markovian approximation, (26.25) has the form

$$\frac{\partial f}{\partial t} + \frac{\partial (H_1(p, q) + \langle U(p, q, P, Q)\rangle_0)}{\partial p} \frac{\partial f}{\partial q} - \frac{\partial (H_1(p, q) + \langle U(p, q, P, Q)\rangle_0)}{\partial q} \frac{\partial f}{\partial p} =$$

$$= \frac{\partial}{\partial q}\left[L_{11}(p, q)\left(\frac{\partial H_1(p, q)}{\partial q} f(p, q, t) + kT \frac{\partial f(p, q, t)}{\partial q}\right) - L_{12}(p, q)\left(\frac{\partial H_1(p, q)}{\partial p} f(p, q, t) + kT \frac{\partial f(p, q, t)}{\partial p}\right)\right] -$$

$$- \frac{\partial}{\partial p}\left[L_{21}(p, q)\left(\frac{\partial H_1(p, q)}{\partial q} f(p, q, t) + kT \frac{\partial f(p, q, t)}{\partial q}\right) - L_{22}(p, q)\left(\frac{\partial H_1(p, q)}{\partial p} f(p, q, t) + kT \frac{\partial f(p, q, t)}{\partial p}\right)\right], \qquad (26.26)$$

where we have introduced the following kinetic coefficients:

$$
\begin{aligned}
L_{11}(p,\ q) &= \beta \int_{-\infty}^{0} dt\, e^{\varepsilon t} \left\langle \frac{\partial U(p,\ q,\ P,\ Q)}{\partial p}\, \frac{\partial U(p,\ q,\ P,\ Q,\ t)}{\partial p} \right\rangle_0 , \\
L_{12}(p,\ q) &= \beta \int_{-\infty}^{0} dt\, e^{\varepsilon t} \left\langle \frac{\partial U(p,\ q,\ P,\ Q)}{\partial p}\, \frac{\partial U(p,\ q,\ P,\ Q,\ t)}{\partial q} \right\rangle_0 , \\
L_{21}(p,\ q) &= \beta \int_{-\infty}^{0} dt\, e^{\varepsilon t} \left\langle \frac{\partial U(p,\ q,\ P,\ Q)}{\partial q}\, \frac{\partial U(p,\ q,\ P,\ Q,\ t)}{\partial p} \right\rangle_0 , \\
L_{22}(p,\ q) &= \beta \int_{-\infty}^{0} dt\, e^{\varepsilon t} \left\langle \frac{\partial U(p,\ q,\ P,\ Q)}{\partial q}\, \frac{\partial U(p,\ q,\ P,\ Q,\ t)}{\partial q} \right\rangle_0 .
\end{aligned}
\tag{26.27}
$$

The equation (26.26) obtained is the Kramers–Fokker–Planck equation describing the behavior of a small subsystem in a thermostat. It can be regarded as the generalization of Liouville's equation to the case of a nonisolated system. This equation has been obtained by many authors [161, 164-168] by means of the theory of stochastic processes, the kinetic coefficients being expressed in terms of the transition probability, which was regarded as a given characteristic of the random process.

The application to this problem of the method of the nonequilibrium statistical operator has made it possible not only to derive the Kramers–Fokker–Planck equation, but also to obtain expressions (26.27) for the kinetic coefficients occurring in it in terms of correlation functions of the forces acting on the subsystem.

26.2. Particular Cases

The simplest example of the problem under consideration is the Brownian motion of a heavy particle in a gas or liquid, when the small interaction between the particle and the fluid is due to the large difference between the mass M of the Brownian particle and the mass m of a particle of the fluid.

In this case, the Hamiltonian has the form

$$H = \sum_i \frac{p_i^2}{2M} + H_2(P,\ Q) + \sum_{i,\ j} U(|\mathbf{q}_i - \mathbf{Q}_j|). \tag{26.28}$$

For the density (26.2a), we obtain an equation of motion of the type (26.4) with the fluxes

$$j_1(p, q) = \frac{p}{M} n(p, q),$$
$$j_2(p, q) = -\sum_j \frac{\partial U(q - Q_j)}{\partial q} n(p, q). \tag{26.29}$$

The nonequilibrium statistical operator (26.13) in our case takes the form

$$\rho = \rho_l - \rho_l \beta \int dp\, dq \int_{-\infty}^{0} dt_1\, e^{\varepsilon t_1} \frac{\partial \varphi(p, q, t + t_1)}{\partial p} j_2(p, q, t_1); \tag{26.30}$$

here, the average local flux $\langle j_2 \rangle_l$ is equal to zero, since the fluid is assumed to be in equilibrium.

Because of the large difference in the masses ($M \gg m$), the Markovian approximation (26.26) is completely adequate for the motion of the Brownian particles:

$$\frac{\partial f}{\partial t} + \frac{p}{M} \cdot \frac{\partial f}{\partial q} = \frac{\partial}{\partial p} \cdot \zeta(q) \left(\frac{p}{M} f + kT \frac{\partial f}{\partial p} \right), \tag{26.31}$$

where

$$\zeta(q) = \beta \int_{-\infty}^{0} dt\, e^{\varepsilon t} \sum_{ij} \left\langle \frac{\partial U(q - Q_i)}{\partial q} \frac{\partial U(q - Q_j, t)}{\partial q} \right\rangle_0 \tag{26.32}$$

is the friction coefficient, expressed in terms of the correlator of the forces acting on the Brownian particle.

By the same method, we can obtain the Kramers–Fokker–Planck equation for Brownian motion in a liquid that is nonuniform in the temperature [57].

The relaxation of a harmonic oscillator weakly interacting with an equilibrium system of similar oscillators serves as another interesting example.

In angle-action variables (α, J), the Hamiltonian of the total system has the form

$$H = \omega \sum_i J_i + U(\alpha_1, \ldots, \alpha_n, J_1, \ldots, J_n). \tag{26.33}$$

A feature of this system is the dependence of the interaction potential U both on the generalized coordinates α_i and on the generalized momenta J_i.

The density (26.2) in angle-action space has the form

$$n(\alpha, J) = \sum_i \delta(\alpha - \alpha_i)\,\delta(J - J_i). \tag{26.34}$$

By applying the general scheme to the system (26.33), we obtain the corresponding Kramers–Fokker–Planck equation [161] (for more detail, see [162]; a generalization of the method is given in [204]). For the application of this scheme to the theory of homogeneous nucleation, see the paper [205].

§ 27. Extremal Properties of the Nonequilibrium Statistical Operator [170, 189]

The equilibrium distribution functions and statistical operators for all the Gibbsian ensembles correspond to the maximum of the information entropy for the different given external conditions, as was shown in §§4 and 10. The local-equilibrium distribution also corresponds to the maximum information entropy for given distributions of energy, momentum and particle number as functions of space and time. In §§21-26, we constructed nonequilibrium statistical operators from quasi-integrals of motion, without connecting these distributions with an extremum of the information entropy.

Attempts have repeatedly been made to construct the nonequilibrium statistical operator from the extremum of the information entropy [71, 72]; usually, however, only quasi-equilibrium distributions, which do not describe irreversible processes, have been obtained. In this section, we shall show, following the work of the author and Kalashnikov [170, 189], that one can obtain a statistical operator describing irreversible processes from the extremum of the information entropy, if we require the extremum of the information entropy for fixed thermodynamic coordinates, not only for the given moment of time, but also for all past times. It turns out that this statistical operator coincides with the nonequilibrium statistical operator obtained using the quasi-integrals of motion and considered in §§21-26.†

27.1. Extremal Properties of the Quasi-Equilibrium Distribution [170]

Our subsequent account will be concerned with both the hydrodynamic and the kinetic stages of a nonequilibrium process; therefore, we shall consider first the extremal properties of the quasi-equilibrium distribution describing such states.

Let the nonequilibrium state be defined by a set of average values of certain operators P_m, where m is an index which can take continuous and discrete values. To describe the hydrodynamic stage of a nonequilibrium process for the P_m, we must choose the operators of the densities (21.3a) of the energy, momentum, and particle number, or their Fourier components. To describe the kinetic stage, for the P_m we can choose the occupation numbers (25.4) of the single-particle states.

The quasi-equilibrium (or local-equilibrium) operator is determined from the extremum of the information entropy (10.1)

$$S_i = - \operatorname{Tr}(\rho \ln \rho) \tag{27.1}$$

under the supplementary conditions that

$$\operatorname{Tr}(\rho P_m) = \langle P_m \rangle_q^t \tag{27.2}$$

be constant, and that the normalization be conserved

$$\operatorname{Tr} \rho = 1. \tag{27.3}$$

In fact, the conditional extremum of the functional (27.1) corresponds to the unconditional extremum of the functional

$$L(\rho) = - \operatorname{Tr}(\rho \ln \rho) - \sum_m F_m \operatorname{Tr}(\rho P_m) - (\Phi - 1) \operatorname{Tr} \rho, \tag{27.4}$$

where F_m and $\Phi - 1$ are Lagrange multipliers. From the condition

$$\delta L(\rho) = - \operatorname{Tr} \left\{ \left(\ln \rho + \Phi + \sum_m P_m F_m(t) \right) \delta\rho \right\} = 0 \tag{27.5}$$

†As shown in the paper [243], it is not necessary to require that the information entropy be an extremum at all times in the past. It is sufficient to require only that it be extremal in the infinitely remote past.

it follows that, corresponding to the extremum, we have the quasi-equilibrium statistical operator

$$\rho_q = \exp\left\{-\Phi - \sum_m P_m F_m(t)\right\} \equiv \exp\{-S(t, 0)\},$$
$$\Phi = \ln \operatorname{Tr} \exp\left\{-\sum_m P_m F_m(t)\right\}. \tag{27.6}$$

In the particular case of the hydrodynamic regime, the parameters F_m have the meaning of the thermodynamic parameters (21.6a), which depend on the time and space coordinates. In this case, the summation over m in formula (27.6) also implies integration over **x**. In the case of the kinetic regime, the P_m can be chosen as in (25.3) or (25.4), and then the summation over m goes over into an integration over the momenta.

The quasi-equilibrium statistical operator (27.6) is not an integral of Liouville's equation and cannot give a correct description of irreversible processes; nevertheless, the properties of the nonequilibrium statistical operator are closely connected with the properties of the quasi-equilibrium operator (27.6) (see §27.3 and Appendix III).

In the case of statistical equilibrium, the quasi-equilibrium distribution (27.6) goes over into the Gibbsian distribution

$$\rho_0 = \exp\left\{-\Phi_0 - \sum_m F_m^0 P_m\right\}, \tag{27.6a}$$

where

$$P_0 = \int H(\boldsymbol{x})\, d\boldsymbol{x} = H, \qquad F_0^0 = \beta,$$
$$P_1 = \int \boldsymbol{p}(\boldsymbol{x})\, d\boldsymbol{x} = \boldsymbol{P}, \qquad F_1^0 = -\beta \boldsymbol{v},$$
$$P_2 = \int n(\boldsymbol{x})\, d\boldsymbol{x} = N, \qquad F_2^0 = -\beta\left(\mu - \frac{m}{2} v^2\right),$$

and this distribution corresponds not only to an extremum of the information entropy but is also an integral of Liouville's equation.

The thermodynamic entropy and the logarithm of the partition function (the Massieu–Planck functional) for the distribution (27.6) are connected by the relation

$$S = \Phi + \sum_m \langle P_m \rangle_q^t F_m(t), \qquad \langle \ldots \rangle_q^t = \operatorname{Tr}(\rho_q \ldots), \tag{27.7}$$

which can be regarded as the generalization to the quasi-equilibrium case of the Legendre transformation of equilibrium thermodynamics. Taking the variation of the normalization condition for the operator (27.6) and using the relation (27.7), we obtain

$$\delta\Phi = -\sum_m \langle P_m\rangle_q^t \,\delta F_m(t), \quad \delta S = \sum_m F_m(t)\,\delta\langle P_m\rangle_q^t, \tag{27.8}$$

whence follow the thermodynamic equalities

$$\langle P_m\rangle_q^t = -\frac{\delta\Phi}{\delta F_m(t)}, \tag{27.9}$$

$$F_m(t) = \frac{\delta S}{\delta\langle P_m\rangle_q^t} \tag{27.10}$$

and the Gibbs–Helmholtz relations

$$S = \Phi - \sum_m F_m(t)\frac{\delta\Phi}{\delta F_m(t)}, \quad \Phi = S - \sum_m \langle P_m\rangle_q^t \frac{\delta S}{\delta\langle P_m\rangle_q^t}. \tag{27.11}$$

The relations (27.7)-(27.11) differ from the equilibrium thermodynamic equalities only in the replacement of the partial derivatives by functional derivatives, if the m are continuous indices.

We shall find the relation between the second functional derivatives of S and Φ and the quantum correlation functions in a quasi-equilibrium state, by differentiating the equalities (27.9) and (27.10)

$$\frac{\delta\langle P_m\rangle_q^t}{\delta F_n(t)} = -\frac{\delta^2\Phi}{\delta F_m(t)\,\delta F_n(t)} = \frac{\delta\langle P_n\rangle_q^t}{\delta F_m(t)} = -(P_n, P_m)^t, \tag{27.12}$$

$$\frac{\delta F_m(t)}{\delta\langle P_n\rangle_q^t} = \frac{\delta^2 S}{\delta\langle P_n\rangle_q^t\,\delta\langle P_m\rangle_q^t} = \frac{\delta F_n(t)}{\delta\langle P_m\rangle_q^t}, \tag{27.13}$$

$$\sum_{m'} \frac{\delta^2\Phi}{\delta F_m(t)\,\delta F_{m'}(t)}\,\frac{\delta^2 S}{\delta\langle P_{m'}\rangle_q^t\,\delta\langle P_n\rangle_q^t} = -\delta_{mn}, \tag{27.14}$$

where the correlation functions $(P_n, P_m)^t$ have the form

$$(P_n, P_m)^t = \int_0^1 d\tau\,\langle P_n(e^{-\tau S(t,0)}P_m e^{\tau S(t,0)} - \langle P_m\rangle_q^t)\rangle_q^t. \tag{27.15}$$

The relation (27.14) has been used already in §6 [see (6.24)].

27.2. Derivation of the Nonequilibrium Statistical Operator from the Extremum of the Information Entropy [170, 189]

The quasi-equilibrium statistical operator (27.6) corresponds to the extremum of the information entropy (27.1) for given $\langle P_m \rangle^t$ for a fixed moment of time t. Consequently, it is a function of $F_m(t)$ (if the m are discrete indices) for a given moment of time t, and no account is taken of "memory" effects, i.e., the possible functional dependence of ρ on $F_m(t + t')$ at past times $-\infty < t \leq 0$. Irreversible processes are frequently characterized by such retardation, which leads to dispersion of the kinetic coefficients.

We shall show that the nonequilibrium statistical operator which we applied in §§21-26 can be determined from the extremum of the information entropy (27.1) under the supplementary conditions that

$$\operatorname{Tr}(\rho P_m(t')) = \langle P_m \rangle^{t+t'} \tag{27.16}$$

are given in the interval $-\infty < t' \leq 0$, i.e., not only for the given time t, but also for all past times, and that the normalization is conserved

$$\operatorname{Tr}\rho = 1. \tag{27.17}$$

In (27.16), $P_m(t')$ indicates the Heisenberg picture, i.e., the evolution of the system in time in accordance with Liouville's equation. Thus, the supplementary condition (27.16) has a dynamic character and includes information about the evolution of the system, whereas the supplementary condition (27.2) has a static character and includes information only on the state of the system at the given moment of time.

This conditional extremum with the "memory" effect corresponds to the unconditional extremum of the functional

$$L(\rho) = -\operatorname{Tr}(\rho \ln \rho) - (\tilde{\Phi} - 1)\operatorname{Tr}\rho - \int_{-\infty}^{0} dt' \sum_m G_m(t, t') \operatorname{Tr}(\rho P_m(t')), \tag{27.18}$$

where $\tilde{\Phi} - 1$ and $G_m(t, t')$ are Lagrange multipliers. It follows from the condition that the functional (27.18) be an extremum that

$$\delta L(\rho) = -\operatorname{Tr}\left\{\left[\ln \rho + \tilde{\Phi} + \sum_m \int_{-\infty}^{0} dt' \, G_m(t, t') P_m(t')\right] \delta\rho\right\} = 0, \tag{27.19}$$

whence we find

$$\rho = \exp\left\{-\tilde{\Phi} - \int_{-\infty}^{0} dt' \sum_m G_m(t, t') P_m(t')\right\}. \tag{27.20}$$

The Lagrange multipliers are determined from the condition (27.16) and the normalization (27.17). Taking the variation of the normalization condition with respect to $G_m(t, t')$, we obtain, taking (27.16) into account,

$$\frac{\delta\tilde{\Phi}}{\delta G_m(t, t')} = -\langle P_m(t')\rangle^t = -\langle P_m\rangle^{t+t'}. \tag{27.21}$$

If the P_m are integrals of motion, then $P_m(t') = P_m$ and the statistical operator (27.20) must go over into the Gibbsian distribution (27.6a), i.e., the integral

$$\int_{-\infty}^{0} G_m(t, t')\, dt'$$

must converge to a constant quantity F_m^0. This can be achieved by putting

$$G_m(t, t') = F_m^0 \frac{d}{dt'} e^{\varepsilon t'} = \varepsilon e^{\varepsilon t'} F_m^0 \qquad (\varepsilon > 0).$$

Taking this property and the relation (27.21) into account, we can conveniently choose the Lagrange multipliers in the form†

$$G_m(t, t') = \varepsilon e^{\varepsilon t'} F_m(t + t'), \tag{27.22}$$

where $F_m(t + t')$ are parameters conjugate to the $\langle P_m\rangle^{t+t'}$. We then obtain the statistical operator

$$\rho = \exp\left\{-\tilde{\Phi} - \varepsilon \int_{-\infty}^{0} dt'\, e^{\varepsilon t'} \sum_m F_m(t + t') P_m(t')\right\}, \tag{27.23}$$

which coincides with the nonequilibrium statistical operator (21.10; (25.10) obtained earlier.‡ In the calculation of the averages, the parameter $\varepsilon > 0$ tends to zero after the thermodynamic limit has been taken.

---

†One can show that this choice is unique, if we impose only the requirement that the information entropy be an extremum in the remote past [243].

‡The normalization factor $Q = \exp\tilde{\Phi}$ of the distribution (27.28) was denoted previously, in §§21 and 22, by $Q = \exp\Phi$.

Thus, we have shown that the nonequilibrium statistical operator (27.23) corresponds to an extremum (the maximum) of the information entropy when the averages $\langle P_m \rangle^{t_1}$ are given at all past times t_1 in the interval $-\infty < t_1 \leq t$.

The nonequilibrium statistical operator can be written in a more compact form

$$\rho = \exp\left\{-\widetilde{\Phi} - \sum_m \widetilde{P_m F_m(t)}\right\},$$
$$\widetilde{\Phi} = \ln \operatorname{Tr} \exp\left\{-\sum_m \widetilde{P_m F_m(t)}\right\}, \tag{27.24}$$

where we have introduced the operation of taking the invariant (or quasi-invariant) part of an operator with respect to motion with the Hamiltonian H; as always, we denote this operation by a wavy line over the operators:

$$\widetilde{P_m F_m(t)} \equiv \varepsilon \int_{-\infty}^{0} dt' \, e^{\varepsilon t'} P_m(t') F_m(t+t') =$$
$$= P_m F_m(t) - \int_{-\infty}^{0} dt' \, e^{\varepsilon t'} \{\dot{P}_m(t') F_m(t+t') + P_m(t') \dot{F}_m(t+t')\}, \tag{27.25}$$
$$P_m(t) = e^{\frac{itH}{\hbar}} P_m e^{-\frac{itH}{\hbar}},$$
$$\dot{P}_m = \frac{1}{i\hbar}[P_m, H], \quad \dot{F}_m(t) = \frac{dF_m(t)}{dt}.$$

In the final results, $\varepsilon \to +0$ after the volume of the system tends to infinity. The operators $\widetilde{P_m F_m(t)}$ satisfy Liouville's equation

$$\frac{\partial}{\partial t}(\widetilde{P_m F_m(t)}) + \frac{1}{i\hbar}[(\widetilde{P_m F_m(t)}), H] =$$
$$= \varepsilon \int_{-\infty}^{0} dt' \, e^{\varepsilon t'} \{\dot{P}_m(t') F_m(t+t') + P_m(t') \dot{F}_m(t+t')\} \tag{27.26}$$

when $\varepsilon \to +0$. This is why we call (27.25) the invariant (or quasi-invariant) part of the products $P_m F_m(t)$ with respect to evolution with the Hamiltonian H. Thus, they are integrals of motion when $\varepsilon \to +0$. It is clear that, in this limit, the statistical operator (27.24) constructed from the operators (27.25) will also be an integral of Liouville's equation. The operation of taking the invariant part, which smooths the oscillating terms, is used in the formal theory of scattering to impose the boundary conditions excluding the advanced solutions of the Schrödinger equation [84] (see Appendix I); we too shall use this operation to derive Lagrange multipliers (27.22) such that the nonequilibrium statistical operator (27.23) is a retarded solution of Liouville's equations.

The parameters $F_m(t)$ of the nonequilibrium statistical operator are chosen so that $F_m(t)$ and $\langle P_m \rangle^t$ be thermodynamically conjugate parameters; this is achieved if on the $F_m(t)$ we impose the conditions

$$\langle P_m \rangle^t = \langle P_m \rangle^t_q, \tag{27.27}$$

where

$$\langle P_m \rangle^t_q = \mathrm{Tr}\,(\rho_q P_m) = \mathrm{Tr}\,(e^{-S(t,\,0)}\,P_m).$$

Indeed, we then have

$$\frac{\delta \Phi}{\delta F_m(t)} = -\,\langle P_m \rangle^t_q = -\,\langle P_m \rangle^t, \tag{27.28}$$

and consequently,

$$\delta \Phi = -\sum_m \langle P_m \rangle^t\,\delta F_m(t). \tag{27.29}$$

By definition, the entropy is equal to

$$S = -\,\mathrm{Tr}\,(\rho_q \ln \rho_q) = \Phi + \sum_m \langle P_m \rangle^t F_m(t), \tag{27.30}$$

whence follow, if we take (27.29) into account, the thermodynamic equalities

$$\delta S = \sum_m F_m(t)\,\delta \langle P_m \rangle^t, \tag{27.31}$$

$$F_m(t) = \frac{\delta S}{\delta \langle P_m \rangle^t}, \tag{27.31a}$$

which are the same as for the quasi-equilibrium distribution (27.10)

27.3. Connection between the Nonequilibrium and Quasi-Equilibrium Statistical Operators [184]

The nonequilibrium statistical operator (27.24) is closely related to the quasi-equilibrium statistical operator (27.6). It can be constructed from the quasi-equilibrium operator, if we take the quasi-invariant part of the logarithm of the latter:

$$\rho = \exp\{\widetilde{\ln \rho_q}\} = \exp\left\{\varepsilon \int_{-\infty}^{0} dt'\, e^{\varepsilon t'} e^{\frac{it'H}{\hbar}} \ln \rho_q(t+t')\, e^{-\frac{it'H}{\hbar}}\right\} =$$

$$= \exp\{-\widetilde{S(t,\, 0)}\}, \qquad (27.32)$$

or

$$\rho = \exp\left\{-\varepsilon \int_{-\infty}^{0} dt'\, e^{\varepsilon t'}\, S(t+t',\, t')\right\} =$$

$$= \exp\left\{-S(t,\, 0) + \int_{-\infty}^{0} dt'\, e^{\varepsilon t'}\, \dot{S}(t+t',\, t')\right\}, \qquad (27.32a)$$

where

$$S(t,\, 0) = \Phi + \sum_m P_m F_m(t) \qquad (27.33)$$

is the **entropy operator**, and

$$\dot{S}(t,\, 0) = \frac{\partial S(t,\, 0)}{\partial t} + \frac{1}{i\hbar}[S(t,\, 0),\, H],$$
$$\dot{S}(t,\, t') = e^{it'H/\hbar}\, \dot{S}(t,\, 0)\, e^{-it'H/\hbar} \qquad (27.34)$$

is the **entropy production operator.**

The argument t + t' of $\rho_q(t + t')$ denotes the time dependence through the parameters $F_m(t + t')$; the first argument of $S(t + t', t')$ denotes the time dependence through the parameters, and the second argument denotes the time dependence through the Heisenberg picture of the operators $P_m(t')$.

We shall demand that it follow from the normalization (27.6) that (27.32) is also normalized. The relation (27.32) determines the following connection between the logarithms of the normalization factors of the quasi-equilibrium (27.6) and nonequilibrium (27.24) operators:

$$\widetilde{\Phi} = \varepsilon \int_{-\infty}^{0} dt'\, e^{\varepsilon t'}\, \Phi(t+t') = \Phi(t) - \int_{-\infty}^{0} dt'\, e^{\varepsilon t'}\, \dot{\Phi}(t+t'), \qquad (27.35)$$

if the conditions (27.27) are imposed on the functions $F_m(t)$.

In fact, the variations of the left- and right-hand sides of formula (27.35) with respect to the functions $F_m(t + t')$ are, respectively, equal to

$$\delta\tilde{\Phi} = -\varepsilon \int_{-\infty}^{0} dt'\, e^{\varepsilon t'} \sum_m \langle P_m(t')\rangle^t\, \delta F_m(t+t'), \tag{27.36}$$

$$\varepsilon\delta \int_{-\infty}^{0} dt'\, e^{\varepsilon t'}\, \Phi(t+t') = -\varepsilon \int_{-\infty}^{0} dt'\, e^{\varepsilon t'} \sum_m \langle P_m\rangle_q^{t+t'}\, \delta F_m(t+t'). \tag{27.36a}$$

By virtue of the equalities (27.27), these variations are the same. Moreover, for the concrete choice of the functions $F_m(t)$, corresponding to the statistical-equilibrium distribution (27.6a), namely $F_m(t) = F^0_m$, we have

$$\rho = \rho_q = \rho_0, \qquad \tilde{\Phi} = \Phi = \Phi_0.$$

This proves the relation (27.35).

The entropy production operator (27.24) can be written in the form

$$\dot{S}(t,\, 0) = \sum_m \{\dot{P}_m F_m(t) + P_m \dot{F}_m(t)\} + \dot{\Phi} =$$
$$= \sum_m \{\dot{P}_m F_m(t) + (P_m - \langle P_m\rangle^t)\, \dot{F}_m(t)\}. \tag{27.37}$$

In this case, the average entropy production $\dot{S}$ can be written down as follows:

$$\dot{S} = \langle \dot{S}(t,\, 0)\rangle^t = \sum_m \langle \dot{P}_m\rangle^t F_m(t). \tag{27.38}$$

It follows from (27.38) that the quantities $F_m(t)$ play the role of the thermodynamic forces, and $\langle \dot{P}_m\rangle^t$ that of the conjugate fluxes As we have seen in §22.3, the fact that the entropy production is positive is connected with the choice of the retarded form of the integrals of motion (27.25).

In a quasi-equilibrium state, we have

$$\langle \dot{S}(t,\, 0)\rangle_q = \sum_m \langle \dot{P}_m\rangle_q F_m(t) = \mathrm{Tr}\left\{\frac{1}{i\hbar}\,[S(t,\, 0),\, H]\, e^{-S(t,\, 0)}\right\} = 0, \tag{27.38a}$$

where we have used a cyclic permutation of the operators in the trace. Consequently, the entropy production operator (27.37) can be written in the form

$$\dot{S}(t, 0) = \Delta \sum_m \{\dot{P}_m F_m(t) + P_m \dot{F}_m(t)\}, \tag{27.37a}$$

where

$$\Delta\dot{P}_m = \dot{P}_m - \langle \dot{P}_m \rangle_q, \qquad \Delta P_m = P_m - \langle P_m \rangle_q.$$

It is easily verified that the nonequilibrium statistical operator (27.32) is an integral of Liouville's equation when $\varepsilon \to 0$. Indeed, we have

$$\frac{\partial \rho}{\partial t} + \frac{1}{i\hbar}[\rho, H] =$$

$$= -\varepsilon \int_{-\infty}^{0} dt' \, e^{\varepsilon t'} \int_0^1 d\tau \, e^{-\tau \widetilde{S}(t, 0)} \sum_m \{(\dot{P}_m(t') - \langle \dot{P}_m(t') \rangle^t) F_m(t+t') +$$

$$+ (P_m(t') - \langle P_m(t') \rangle^t) \dot{F}_m(t+t')\} e^{\tau \widetilde{S}(t, 0)} \rho, \tag{27.39}$$

where $\varepsilon \to +0$ after the volume V of the system goes to infinity. With this order of calculating the limits, the following relations hold [186]:

$$\begin{aligned} \lim_{\varepsilon \to 0,\ V \to \infty} \langle P_m(t') \rangle^t &= \langle P_m \rangle^{t+t'}, \\ \lim_{\varepsilon \to 0,\ V \to \infty} \langle \dot{P}_m(t') \rangle^t &= \langle \dot{P}_m \rangle^{t+t'} \end{aligned} \tag{27.40}$$

In essence, we do not require that the operator satisfy Liouville's equation exactly; it is sufficient that the properties (27.40) be fulfilled for every operator.

The introduction of the quasi-integrals of motion (27.25) can be regarded, in connection with the ideas of Bogolyubov on "quasi-averages" [171, 172], as the introduction of infinitesimally small forces into the Liouville equation, which then tend to zero in the calculation of the averages, after the volume of the system tends to infinity (see Appendix III and the papers [186, 187]).

27.4. Generalized Transport Equations [56, 170]

In nonequilibrium statistical mechanics, to describe the evolution in time of a nonequilibrium state, along with the thermodynamic equalities we need to know the equations of motion of the average values of the dynamic quantities, i.e., the generalized transport equations (or generalized kinetic equations). We have already considered such equations, for the kinetic regime in §25, and for the hydrodynamic regime in §§21-24.

The generalized transport equations describing the time evolution of the averages $\langle P_m\rangle^t$ or of the functions $F_m(t)$ associated with them can be obtained by averaging the equations of motion for the operators P_m over the nonequilibrium distribution (27.24); this, together with the conditions (27.27), gives

$$\frac{d}{dt}\langle P_m\rangle^t = \frac{d}{dt}\langle P_m\rangle_q^t = \sum_{m'} \frac{\delta\langle P_m\rangle_q^t}{\delta F_{m'}(t)} \dot{F}_{m'}(t) = \langle \dot{P}_m\rangle^t = \frac{1}{i\hbar}\langle [P_m, H]\rangle^t. \tag{27.41}$$

We call these equations generalized transport equations, including in this concept all possible balance equations of the theory of irreversible processes, for example, kinetic equations for different particles or quasi-particles (§§25 and 26), balance equations for the energy, particle number and momentum (§22), relaxation equations (§23), etc. If there exists a small parameter, the right-hand side of Eq. (27.41) can be expanded in a series in powers of this parameter, and this leads, generally speaking, to integral equations for $F_m(t)$ or $\langle P_m\rangle^t$. To obtain hydrodynamic equations, gradients of the thermodynamic parameters are taken as the small parameters (§22), to obtain relaxation equations, the differences of these gradients are taken as the small parameters (§23), and to obtain kinetic equations we take the small parameter to be the interaction between the particles or quasi-particles (§25).

To conclude this subsection, we find explicit generalized transport equations for the simple specific case when the Hamiltonian has the form

$$H = H_0 + V, \tag{27.42}$$

where V is a small perturbation. We have already considered this case in §25, following the paper [56]. We return to this example in order to demonstrate the convenience of using the entropy production operator (27.34) and to write the generalized transport equation (27.41) in explicit form.

We suppose, as in §25, that the equations of motion of the operators P_m have the form

$$\dot{P}_m = \frac{1}{i\hbar}[P_m, H_0 + V] = -\frac{1}{i\hbar}\sum_n \alpha_{mn} P_n + \dot{P}_{m\,(V)}, \tag{27.43}$$

where α_{mn} is a c-number matrix defined by the commutation relations (25.6), and

$$\dot{P}_{m\,(V)} = \frac{1}{i\hbar}[P_m, V]. \tag{27.44}$$

In this case, taking (25.22) into account, we can write the entropy production operator (27.37) in the form

$$\dot{S}(t, 0) = \Delta \sum_m \left\{ \dot{P}_{m\,(V)} F_m(t) + \sum_l P_m \frac{\delta F_m(t)}{\delta \langle P_l \rangle^t} \langle \dot{P}_{l\,(V)} \rangle^t + \right.$$
$$\left. + \left(\frac{i}{\hbar} \sum_l \alpha_{ml} F_m(t) P_l + \frac{i}{\hbar} \sum_{ll'} P_m \frac{\delta F_l(t)}{\delta \langle P_m \rangle^t} \alpha_{ll'} \langle P_{l'} \rangle^t \right) \right\}. \tag{27.45}$$

The expansion of the operator $\dot{S}(t, 0)$ in powers of V starts from terms of first order in V, since the sum of terms in the round brackets in (27.45) is identically equal to zero because of the identity (25.23). Consequently,

$$\dot{S}(t, 0) = \Delta \sum_m \left\{ \dot{P}_{m\,(V)} F_m(t) + \sum_l P_m \frac{\delta F_m(t)}{\delta \langle P_l \rangle^t} \langle \dot{P}_{l\,(V)} \rangle^t \right\}. \tag{27.46}$$

Thus, the integral term in the exponent in the expression (27.32a) for the nonequilibrium statistical operator is small and ρ can be expanded in powers of it:

$$\rho = \exp\{-\widetilde{S}(t, 0)\}$$
$$= \left\{ 1 + \int_{-\infty}^{0} dt'\, e^{\varepsilon t'} \int_0^1 d\tau\, e^{-\tau S(t,0)} \dot{S}(t+t', t') e^{\tau S(t,0)} + \ldots \right\} e^{-S(t,0)}. \tag{27.47}$$

Now we can obtain explicit expressions for the right-hand side of the generalized transport equation (27.41) that are exact up to and including terms of second order in V:

$$\langle \dot{P}_m \rangle^t = \frac{i}{\hbar} \sum_n \alpha_{mn} \langle P_n \rangle^t + \langle \dot{P}_{m\,(V)} \rangle_q^t +$$
$$+ \int_{-\infty}^{0} dt'\, e^{\varepsilon t'} \sum_n (\dot{P}_{m\,(V)}, \dot{P}_{n\,(V)}(t'))^t F_n(t+t') +$$
$$+ \int_{-\infty}^{0} dt'\, e^{\varepsilon t'} \sum_{nl} (\dot{P}_{m\,(V)}, P_n(t'))^t \frac{\delta F_n(t+t')}{\delta \langle P_l \rangle^{t+t'}} \langle \dot{P}_{l\,(V)} \rangle_q^{t+t'} + \ldots, \tag{27.48}$$

where the brackets $(\ldots, \ldots)^t$ denote the correlation functions (27.15).

In the important particular case when $\langle \dot{P}_{n(V)} \rangle_q^t = 0$, the generalized transport equation (27.48) takes the form

$$\frac{d}{dt} \langle P_m \rangle = - \sum_n (P_m, P_n)^t \dot{F}_n(t)$$
$$= \sum_n \left\{ \frac{i}{\hbar} \alpha_{mn} \langle P_n \rangle^t + \int_{-\infty}^{0} dt' \, e^{\varepsilon t'} (\dot{P}_{m\,(V)}, \dot{P}_{n\,(V)}(t'))^t F_n(t+t') \right\}. \tag{27.49}$$

In the right-hand side of Eq. (27.49), time correlation functions, calculated over the quasi-equilibrium state, of the fluxes appear. They determine either the collision operator or the kinetic equation or the kinetic coefficients.

We remark that expansions of the type (27.47) make it possible to write the entropy production approximately in the following simple form:

$$\dot{S} = \int_{-\infty}^{0} dt' \, e^{\varepsilon t'} (\dot{S}(t, 0), \dot{S}(t+t', t'))^t. \tag{27.50}$$

Thus, the entropy production is determined by the correlation functions of the entropy production operators.†

27.5. Generalized Transport Equations and Prigogine's and Glansdorff's Criteria for the Evolution of Macroscopic Systems [170]

We shall consider the conditions to which we are led by the requirement that the entropy in a quasi-equilibrium state be a maximum, and shall show, following the paper [170], that these conditions give the criteria established by Prigogine and Glansdorff [173, 174], for the evolution of macroscopic systems and lead, in the particular case when the kinetic coefficients are constant, to Prigogine's theorem of minimum entropy production [175, 27] (see also [227]).

†For a generalization of the formula (27.50), see the paper [271], where it is shown that the exact expression for the entropy production can be represented in the same form, but with a modified definition of the correlation functions.

We shall consider the time derivatives of the functionals $\dot{S}$ [cf. (27.38)] and $\dot{\Phi}$:

$$\dot{\Phi} = -\sum_m \langle P_m \rangle^t \dot{F}_m(t). \tag{27.51}$$

We obtain

$$\ddot{S} = \sum_m \{\langle \dot{P}_m \rangle^t \dot{F}_m(t) + \langle \ddot{P}_m \rangle^t F_m(t)\}, \tag{27.52}$$

$$\ddot{\Phi} = -\sum_m \{\langle \dot{P}_m \rangle^t \dot{F}_m(t) + \langle P_m \rangle^t \ddot{F}_m(t)\}. \tag{27.53}$$

Using (27.49), (27.9), (27.10), and (27.12), we can bring the first terms in the right-hand sides of formulas (27.52) and (27.53) to the form

$$\frac{d_F \dot{S}}{dt} = \sum_m \langle \dot{P}_m \rangle^t \dot{F}_m(t) = -\sum_{m,\, n} \frac{\delta^2 \Phi}{\delta F_m(t)\, \delta F_n(t)} \dot{F}_m(t) \dot{F}_n(t) =$$

$$= -\sum_{m,\, n} (P_m,\ P_n)^t \dot{F}_m(t) \dot{F}_n(t), \tag{27.54}$$

$$\frac{d_P \dot{\Phi}}{dt} = -\sum_{m,\, n} \frac{\delta^2 S}{\delta \langle P_m \rangle^t \delta \langle P_n \rangle^t} \langle \dot{P}_m \rangle^t \langle \dot{P}_n \rangle^t = -\frac{d_F \dot{S}}{dt}. \tag{27.55}$$

They have the meaning of the rate of change of the entropy production due to change of the thermodynamic forces, and the rate of change of $\dot{\Phi}$ due to change of the $\langle P_m \rangle^t$.

The condition that the entropy be a maximum in a quasi-equilibrium state means that the quadratic form $\delta^2 S$ is negative-definite, i.e.,

$$\sum_{m,\, n} \frac{\delta^2 S}{\delta \langle P_m \rangle^t \delta \langle P_n \rangle^t} \delta \langle P_m \rangle^t \delta \langle P_n \rangle^t < 0. \tag{27.56}$$

Hence, if we take (27.55) and (27.54) into account, it follows that

$$\frac{d_F \dot{S}}{dt} < 0, \qquad \frac{d_P \dot{\Phi}}{dt} > 0. \tag{27.57}$$

The first of these relations forms the content of the general criterion of the evolution of macroscopic systems [173, 174] (the Glansdorff–Prigogine theorem), which states that, in a real irreversible process, a diminution $d_F\dot{S}/dt$ of part of the entropy production occurs.

The second of the relations (27.57) is another formulation of the general evolution criterion according to which, in a real irreversible process, an increase $d_P\dot{\Phi}/dt$ occurs. This theorem was established in the paper [170].

We shall show now that, in the approximation linear in the thermodynamic forces, the generalized transport equations satisfy the relations

$$\ddot{S}\equiv\frac{d_F\dot{S}}{dt}+\frac{d_P\dot{S}}{dt}=2\frac{d_F\dot{S}}{dt}<0, \tag{27.58}$$

$$\ddot{\Phi}\equiv\frac{d_P\dot{\Phi}}{dt}+\frac{d_F\dot{\Phi}}{dt}=2\frac{d_P\dot{\Phi}}{dt}>0, \tag{27.59}$$

i.e., in other words, that these processes are accompanied by a diminution of the entropy production and by an increase of the rate of change in time of the functional Φ.

The relation (27.58) is Prigogine's theorem of minimum entropy production [175, 27]. The theorem (27.59) concerning the maximum of the functional was proved in the paper [170].

P r o o f . Let the deviations of the thermodynamic forces $F_m(t)$ from their equilibrium values F_m^0

$$\Delta F_m=F_m(t)-F_m^0 \tag{27.60}$$

be small and assume that the relation between ΔF_m and ΔP_m

$$\Delta P_m=\langle P_m\rangle^t-\langle P_m\rangle_0 \tag{27.61}$$

is linear, i.e.,

$$\Delta\langle P_m\rangle=-\sum_n\frac{\delta^2\Phi_0}{\delta F_m(t)\,\delta F_n(t)}\Delta F_n, \tag{27.62}$$

$$\Delta F_m=\sum_n\frac{\delta^2 S_0}{\delta\langle P_m\rangle^t\,\delta\langle P_n\rangle^t}\Delta\langle P_n\rangle. \tag{27.63}$$

The subscript 0 on the quantities S_0 and Φ_0 denotes that, after taking the functional derivatives, we must put $F_m(t)=F_m^0$ in the resulting correlation functions; $\langle P_m \rangle_0$ denote averages over the equilibrium state.

Further, it is obvious that

$$\sum_m \dot{P}_m F_m^0 = 0 \text{ and } \sum_m \langle \dot{P}_m \rangle^t F_m^0 = 0, \tag{27.64}$$

where we put $H = \sum_m P_m F_m^0$, and for weakly equilibrium states there exists a linear relation between the fluxes and the thermodynamic forces:

$$\langle \dot{P}_m \rangle^t = \sum_n L_{mn}^F \Delta F_n, \tag{27.65}$$

where L_{mn}^F are kinetic coefficients; it follows from this that the entropy production (27.38) is equal to:

$$\dot{S} = \sum_m F_m \langle \dot{P}_m \rangle^t = \sum_{m,\, n} L_{mn}^F \Delta F_m \Delta F_n. \tag{27.66}$$

Linear relations also hold for the rates of change $\dot{F}_m$ of the thermodynamic forces and the increments $\langle P_n \rangle$:

$$\dot{F}_m = \sum_n L_{mn}^P \Delta \langle P_n \rangle, \tag{27.67}$$

where L_{mn}^P are kinetic coefficients connected with the L_{mn}^F by the relations

$$\begin{aligned} L_{mn}^P &= \sum_{m',\, n'} \frac{\delta^2 S_0}{\delta \langle P_m \rangle^t \delta \langle P_{m'} \rangle^t} L_{m'n'}^F \frac{\delta^2 S_0}{\delta \langle P_{n'} \rangle^t \delta \langle P_n \rangle^t}, \\ L_{mn}^F &= \sum_{m',\, n'} \frac{\delta^2 \Phi_0}{\delta F_m(t)\, \delta F_{m'}(t)} L_{m'n'}^P \frac{\delta^2 \Phi_0}{\delta F_{n'}(t)\, \delta F_n(t)} \end{aligned} \tag{27.68}$$

and satisfying the Onsager reciprocity relations

$$L_{mn}^F = L_{nm}^F, \qquad L_{mn}^P = L_{nm}^P. \tag{27.69}$$

Using the relations (27.65) and (27.69), we obtain

$$\begin{aligned} \frac{d_F \dot{S}}{dt} &= \sum_m \dot{F}_m(t) \langle \dot{P}_m \rangle^t = \sum_{mn} L_{mn}^F \Delta \dot{F}_m \Delta F_n, \\ \frac{d_P \dot{S}}{dt} &= \sum_m F_m(t) \langle \ddot{P}_m \rangle^t = \sum_{mn} L_{mn}^F \Delta F_m \Delta \dot{F}_n = \frac{d_F \dot{S}}{dt}, \end{aligned} \tag{27.70}$$

which, together with the first of formulas (27.57), proves the first of the inequalities (27.58). Analogously, using (27.63) and (27.69), we obtain

$$\frac{d_P\dot{\Phi}}{dt} = -\sum_m \langle \dot{P}_m \rangle^t \dot{F}_m(t) = -\sum_{mn} L^P_{mn} \Delta\langle \dot{P}_m \rangle \Delta \langle P_n \rangle,$$

$$\frac{d_F\dot{\Phi}}{dt} = -\sum_m \langle P_m \rangle^t \ddot{F}_m(t) = -\sum_{mn} L^P_{mn} \Delta\langle P_m \rangle \Delta \langle \dot{P}_n \rangle = -\frac{d_P\dot{\Phi}}{dt}, \qquad (27.71)$$

which, together with (27.57), proves the second of the inequalities (27.58).

Thus, the generalized transport equations satisfy the evolution criteria of nonequilibrium phenomenological thermodynamics.

Appendix I

Formal Scattering Theory in Quantum Mechanics

The formal theory of scattering is described in many textbooks and monographs [1-3], but it is far from always the case that sufficient elucidation is given of the very important question of how the limiting processes (making the dimensions L of the system go to infinity and making the parameter ε characterizing the switching on of the interaction go to zero) are performed. What is very important is that the result depends on the order in which these limits are taken. This question is elucidated with complete clarity in the paper by Gell-Mann and Goldberger [4], of which we shall give a short account, since the question of the order of the limits is also fundamental in nonequilibrium statistical mechanics (see §21).

In the quantum-mechanical description of scattering, the total Hamiltonian H of the colliding particles is divided into two parts K and V, where K is the Hamiltonian of the noninteracting particles and V is the interaction between them. It is assumed that V tends sufficiently rapidly to zero as the particles move apart. The transition probability per unit time from one free state to another is sought.

The complete system is described by the Schrödinger equation

$$i\hbar \frac{\partial \Psi(t)}{\partial t} = (K + V)\Psi(t). \tag{I.1}$$

An important feature of the problem is that the interaction V exists at every moment of time, although the scattering process occurs between states without interaction.

In the absence of the interaction, the Schrödinger equation has the form

$$i\hbar \frac{\partial \Phi(t)}{\partial t} = K\Phi(t), \tag{I.2}$$

and its stationary solutions are

$$\Phi_i(t) = \Phi_i e^{-\frac{iE_i}{\hbar}t}. \tag{I.3}$$

We need to calculate the differential effective cross section of scattering from the state Φ_j to the state Φ_i under the influence of the interaction V. The initial state Φ_j is used for the characteristics of the true state Ψ_j of the real system. Knowing $\Psi_j(t)$, we can find the probability that the system undergoes a transition to one of the final states Φ_i by the time t.

We now discuss the question of how to formulate correctly the scattering boundary conditions to the Schrödinger equation (I.1). Suppose that we observe the scattering process at time t = 0. We must formulate mathematically a physical procedure for preparing the quantum-mechanical state Ψ_j up to the time t = 0 that the transition occurs, i.e., for t < 0 (i.e., for fixing the energy and direction of the beam).

If we simply assume that at some remote time t = T prior to the collision the wave function Ψ_j was equal to the wave function of the free state

$$\Psi_j(t) = e^{-\frac{iH}{\hbar}(t-T)}\Phi_j(T), \qquad H = K + V,$$

then such a boundary condition contains the nonphysical element of an "instantaneous" switching on of the interaction V at t = T. In reality, the interaction is switched on gradually, and, therefore, such boundary conditions are inconvenient.

We can impose boundary conditions in another way, by representing the incident wave train as an average over some time interval τ in the past

$$\frac{1}{\tau}\int_{-\tau}^{0} e^{-\frac{i}{\hbar}H(t-T)}\Phi_j(T)\,dT$$

and making τ go to infinity at the end of the calculations, i.e., performing a "time-smoothing" operation. Such a boundary condition is also inconvenient, since it leads to insufficiently well-defined expressions, which require additional procedures to make their meaning precise.

The most convenient boundary condition is that the wave function Ψ_j for t < 0 be put equal to

$$\Psi_j^{(\varepsilon)}(t) = \varepsilon\int_{-\infty}^{0} e^{\varepsilon T} e^{-\frac{i}{\hbar}H(t-T)}\Phi_j(T)\,dT, \tag{I.4}$$

where $\varepsilon \to +0$ at the end of the calculations. Here, we are also performing a "time-smoothing," since

$$\varepsilon\int_{-\infty}^{0} e^{\varepsilon T}\,dT = 1,$$

but the factor $e^{\varepsilon T}$ distinguishes the "past," and so the averaging (I.4) has a "causal" character.

We must, however, exercise care, since in addition to the limit $\varepsilon \to 0$, we must also perform another limiting process $L \to \infty$ (the functions Φ_i are normalized to unity in the large volume L^3). The time τ of the switching-on of the interaction is ε^{-1} in order of magnitude and cannot be greater than the time of propagation of the wave packet over a distance L, i.e., than the quantity L/v, where v is the group velocity,

$$\varepsilon^{-1} \ll L/v;$$

consequently, as $L^{-3} \to 0$ and $\varepsilon^{-1} \to \infty$, the quantity $\varepsilon^{-1}L^{-3}$ must tend to zero. This means that we must first take the limit $L^3 \to \infty$, and then $\varepsilon \to 0$.

Together with this rule for the limits $L \to \infty$ and $\varepsilon \to 0$, the condition (I.4) ensures the selection of the correct retarded causal solutions of the Schrödinger equation. In fact, if $\varepsilon^{-1} < L/v$, then waves reflected from the boundaries of the system, i.e., incoming waves, are excluded, since the extent of the wave train in time, ε^{-1}, is shorter than the time necessary for it to propagate over the distance L. The great convenience of the boundary condition (I.4) compared with the Sommerfeld condition lies in the fact that the causality condition is imposed more automatically, without a detailed analysis of the outgoing waves. The boundary condition (I.4) can be justified by the method of wave packets [5]. A boundary condition analogous to (I.4) is applied to Liouville's equation in §21 of this book. It is clear that its meaning also consists in the selection of the retarded solutions (see Appendix III).

We now calculate the probability of quantum transitions between states as a function of time. The probability that a system which is described by the wave function $\Psi_j(t)$ is found in the state Φ_i by time t is, according to the basic rules of quantum mechanics,

$$w_{ij}(t) = | f_{ij}(t) |^2 N_j^{-1}, \tag{I.5}$$

where

$$f_{ij}(t) = \left(\Phi_i^*(t)\, \Psi_j(t)\right) \tag{I.5a}$$

is the transition probability amplitude, and

$$N_j = \left(\Psi_j^*(t)\, \Psi_j(t)\right) \tag{I.5b}$$

is a normalization constant, which, since the Hamiltonian is Hermitian, does not depend on time.

Taking (I.3) into account, we can write Eq. (I.4) in the form

$$\Psi_j(t) = e^{-\frac{iHt}{\hbar}}\, \varepsilon \int_{-\infty}^{0} e^{\varepsilon T} e^{\frac{i}{\hbar}(H-E_j)T}\, \Phi_j\, dT \tag{I.6}$$

or, after performing the integration over T,

$$\Psi_j(t) = e^{-\frac{i}{\hbar}Ht} \frac{\varepsilon}{\varepsilon + \frac{i}{\hbar}(H - E_j)}\, \Phi_j. \tag{I.7}$$

The function Φ_j satisfies the equation

$$(H - E_j)\, \Phi_j = V \Phi_j, \tag{I.8}$$

and, therefore, Eq. (I.7) for t = 0 can be written in the form

$$\Psi_j(0) = \Phi_j + \frac{1}{(E_j - H) + i\varepsilon\hbar} V\Phi_j. \tag{I.9}$$

In place of the explicit expression (I.9) for $\Psi_j(0)$, we can write the equivalent equation

$$\Psi_j(0) = \Phi_j + \frac{1}{(E_j - K) + i\varepsilon\hbar} V\Psi_j(0), \tag{I.10}$$

which is called the Lippmann–Schwinger equation. Iteration of Eq. (I.10) gives a series in powers of V. The factor

$$G^+(E_j) = \lim_{\varepsilon \to +0} \frac{1}{(E_j - K) + i\varepsilon\hbar}$$

has the meaning of a retarded Green function.

By means of (I.10) we obtain for the transition amplitude the expression

$$f_{ij}(0) = \delta_{ij} + \frac{1}{(E_j - E_i) + i\varepsilon\hbar} R_{ij}(\varepsilon), \tag{I.11}$$

where

$$R_{ij}(\varepsilon) = \left(\Phi_i^* V \Psi_j(0)\right) \tag{I.12}$$

is the reaction matrix. The equation (I.11) is convenient in that it shows explicitly the existence of a singularity of f_{ij} when $E_i = E_j$ and $\varepsilon \to 0$.

The operator $R_{ij}(\varepsilon)$ is a smooth function of the energy after the passage to the limit, the entire singularity being contained in the factor

$$\frac{1}{(E_j - E_i) + i\varepsilon\hbar}.$$

However, the limit operation $L^3 \to \infty$ still cannot be applied to $R_{ij}(\varepsilon)$, since, because of the normalization of Φ_i to unity in the volume L^3, $R_{ij}(\varepsilon)$ is proportional to L^{-3}. Therefore, it is convenient to introduce the operator

$$\lim_{\substack{L \to \infty \\ \varepsilon \to +0}} R_{ij}(\varepsilon) L^3 = \Re_{ij}, \tag{I.13}$$

which no longer has singularities at $E_i = E_j$.

To calculate the derivative of f_{ij} at t = 0, we write (I.5a) in the form

$$f_{ij}(t) = \left(\Phi_i^* e^{\frac{i}{\hbar}(E_i - H)t} \Psi_j(0)\right), \tag{I.14}$$

whence it follows that

$$\dot{f}_{ij}(0) = \frac{i}{\hbar}\left(\Phi_i^* (E_i - H) \Psi_j(0)\right), \tag{I.15}$$

or, if we use (I.8),

$$\dot{f}_{ij}(0) = -\frac{i}{\hbar}\left(\Phi_i^* V \Psi_j(0)\right) = -\frac{i}{\hbar} R_{ij}(\varepsilon). \tag{I.15a}$$

This relation justifies us in calling R_j the reaction matrix, since it is proportional to the rate of change of the transition amplitude f_{ij}.

An expression for the rate of change of the modulus of the transition amplitude follows from (I.11) and (I.15a):

$$\left[\frac{d}{dt} | f_{ij}(t) |^2\right]_{t=0} = \frac{1}{\hbar} 2\delta_{ij} \operatorname{Im} R_{jj} + \frac{2\varepsilon}{(E_j - E_i)^2 + \varepsilon^2\hbar^2} | R_{ij}(\varepsilon) |^2. \tag{I.16}$$

It remains to calculate the normalization constant N_j. It follows from the completeness condition for the set of functions Φ_j, if we use (I.5a), that

$$\sum_i | f_{ij} |^2 = N_j. \tag{I.17}$$

It follows from (I.16) and (I.17), and from the fact that the normalization constant N_j does not depend on time, that

$$\frac{2}{\hbar} \operatorname{Im} R_{jj}(\varepsilon) + \sum_i \frac{2\varepsilon}{(E_j - E_i)^2 + \varepsilon^2\hbar^2} | R_{ij}(\varepsilon) |^2 = 0. \tag{I.18}$$

From (I.17) and (I.11), we obtain for N_j the expression

$$N_j = 1 + \frac{2}{\varepsilon\hbar} \operatorname{Im} R_{jj}(\varepsilon) + \sum_i \frac{1}{(E_j - E_i)^2 + \varepsilon^2\hbar^2} | R_{ij}(\varepsilon) |^2 \tag{I.19}$$

or, taking (I.18) into account,

$$N_j = 1 + \frac{1}{\varepsilon\hbar} \operatorname{Im} R_{jj}(\varepsilon). \tag{I.19a}$$

Noting that R_{jj} is of order L^{-3}, we obtain that N_j tends to unity in our double limiting process.

The differential effective cross section of the transition $j \to i$ $(i \neq j)$ is equal to the transition probability per unit time (I.16) divided by the flux vL^{-3}, where v is the relative velocity of the colliding systems. Therefore, it follows from (I.16) that

$$\sigma_{ij} = \lim_{\varepsilon \to +0} \lim_{L \to \infty} \frac{2\varepsilon}{(E_j - E_i)^2 + \varepsilon^2\hbar^2} | R_{ij}(\varepsilon) |^2 L^3 v^{-1}. \tag{I.20}$$

In the limiting process $\varepsilon \to 0$, the factor

$$\frac{2\varepsilon}{(E_j - E_i)^2 + \varepsilon^2\hbar^2}$$

tends to $(2\pi/\hbar)\delta(E_j - E_i)$. The final states j lie in the continuous spectrum, so that one observes transitions not to a given state i, but to a small interval of final states; therefore, we must average (I.20) over a small interval of final states. This operation corresponds to the "coarse-graining" in statistical mechanics. In this averaging, $\delta(E_j - E_i)$ is removed and in its place appears $\rho(E_j)L^3$ – the density of states in momentum space in a volume V per unit energy interval at energy E_j. Finally,

$$\sigma_{ij} = 2\pi \,|\, \mathfrak{R}_{ij} \,|^2 \rho(E_j)\, v^{-1}, \tag{I.21}$$

where σ_{ij} is already calculated over a small range of final states (usually over an element of solid angle).

Up to this point, we have assumed that $j \neq i$, i.e., that the initial state is not the same as the final state. It is obvious that a single state alone cannot influence the transition probability that we have calculated, but the change of w_{jj} in time is important for calculating the change of occupation of the initial level.

It follows from Eq. (I.16) for i = j that

$$N_j \frac{d}{dt} w_{jj} \Big|_{t=0} = \frac{2}{\hbar} \operatorname{Im} R_{jj}(\varepsilon) + \frac{2}{\varepsilon\hbar} \,|\, R_{jj}(\varepsilon) \,|^2. \tag{I.22}$$

Now, on the contrary, in the limit $L \to \infty$, $\varepsilon \to +0$ the second term in (I.22) is vanishingly small compared with the first.

Taking the limit $\varepsilon \to +0$ in (I.18), we obtain

$$\frac{-2 \operatorname{Im} \mathfrak{R}_{jj}}{L^3} = \frac{1}{L^3} \sum_{i \neq j} \frac{2\pi}{L^3} \,|\, \mathfrak{R}_{jj} \,|^2 \delta(E_j - E_i), \tag{I.23}$$

i.e., the relation

$$\sum_{i \neq j} \sigma_{ij} = -\frac{2}{v} \operatorname{Im} \mathfrak{R}_{jj}, \tag{I.23a}$$

which gives the connection between the imaginary part of the scattering matrix and the total effective cross section. This relation, which stems from the conservation of the normalization, is known as the optical theorem.

The boundary conditions for the quantum-mechanical collision problem can be formulated by means of the introduction of infinitesimally small sources selecting the retarded solutions of the Schrödinger equation [6].

We note that the Schrödinger equation (I.1) is invariant under the time-reversal transformation, i.e., under the replacements $t \to -t$ and $i \to -i$ and reversal of the magnetic field. In addition, the solution of Eq. (I.1) is sensitive to the introduction of an infinitesimally small source violating this symmetry.

The boundary conditions selecting the retarded solutions of the Schrödinger equation in formal scattering theory, in the variant of Gell-Mann and Goldberger [4], can be obtained if, following [6], we introduce into (I.1) for $t \leq 0$ and infinitesimally small source violating the symmetry of the Schrödinger equation with respect to time reversal:

$$\frac{\partial \Psi_\varepsilon(t)}{\partial t} - \frac{1}{i\hbar} H\Psi_\varepsilon(t) = -\varepsilon(\Psi_\varepsilon(t) - \Phi(t)), \tag{I.24}$$

where $\varepsilon \to +0$ after the volume of the system tends to infinity, and $\Phi(t)$ is the wave function of the free motion of the particles, with Hamiltonian K. The infinitesimally small source (I.24) has been introduced in such a way that it is equal to zero when $\Psi(t) = \Phi(t)$, i.e., in the absence of the interaction. It does indeed violate the symmetry of the Schrödinger equation with respect to time reversal, since in this transformation the left-hand side of Eq. (I.24) changes sign, while the right-hand side remains unchanged. The sign of ε is chosen so that we obtain the retarded rather than the advanced solutions.

We write Eq. (I.24) in the form

$$\frac{d}{dt}\left(e^{\varepsilon t}\Psi_\varepsilon(t, t)\right) = \varepsilon e^{\varepsilon t}\Phi(t, t), \tag{I.25}$$

where

$$\Psi_\varepsilon(t, t) = e^{-Ht/i\hbar}\Psi_\varepsilon(t), \quad \Phi(t, t) = e^{-Ht/i\hbar}\Phi(t). \tag{I.25a}$$

Integrating (I.25) from $-\infty$ to t, we have

$$\Psi_\varepsilon(t) = \varepsilon \int_{-\infty}^{t} e^{\varepsilon(t_1 - t)} e^{-H(t_1 - t)/i\hbar}\Phi(t_1)\,dt_1 = \varepsilon \int_{-\infty}^{0} e^{\varepsilon t'} e^{-Ht'/i\hbar}\Phi(t + t')\,dt'. \tag{I.26}$$

Putting t = 0 in (I.26), we obtain the scattering-theory boundary condition in the Gell-Mann–Goldberger form

$$\Psi_\varepsilon(0) = \varepsilon \int_{-\infty}^{0} e^{\varepsilon t} e^{-Ht/i\hbar}\Phi(t)\,dt, \tag{I.27}$$

which we have already studied above.

Appendix II

MacLennan's Statistical Theory of Transport Processes

MacLennan's derivation of a statistical theory of transport processes [1, 2] is based on the introduction of external forces of a nonconservative nature, which describe the influence of the surroundings or of a thermostat on the given system, i.e., the influence of energy- and particle-reservoirs and movable pistons in contact with the system. We shall describe this method briefly, since it is similar to the method of the non-equilibrium statistical operator [3-6], described in §21, and leads to the same results. A comparison of these two methods enables us to treat the same problems from different points of view and to gain a better understanding of their physical meaning.

We shall consider classical systems with total Hamiltonian H_u – the Hamiltonian of the universe in MacLennan's terminology. One must not attach any great significance to this term; as before, we shall be interested in the evolution of a small subsystem accessible to our measurements, and questions of the structure of universe will not be touched on. We have

$$H_u = H + H_s + U, \tag{II.1}$$

where H is the Hamiltonian of the system under consideration, H_s is the Hamiltonian of the surroundings, and U is the Hamiltonian of the interaction of the system with the surroundings.

The distribution function f_u of the total system obeys Liouville's equation (2.11)

$$\frac{\partial f_u}{\partial t} + \{f_u, H_u\} = 0, \tag{II.2}$$

where $\{ \ldots \}$ is the classical Poisson bracket (2.10). The distribution functions of the system under consideration and of the surroundings are, respectively, equal to

$$f = \int f_u \, d\Gamma_s, \quad g = \int f_u \, d\Gamma, \tag{II.3}$$

where $d\Gamma_s$ and $d\Gamma$ are phase-volume elements of the surroundings and of the given system. If f_u is normalized, then f and g are also normalized:

$$\int f\,d\Gamma = 1, \quad \int g\,d\Gamma_s = 1,$$
$$\int f_u\,d\Gamma\,d\Gamma_s = 1. \tag{II.4}$$

We shall obtain an equation for f by integrating (II.2) over the phase space $d\Gamma_s$ of the surroundings:

$$\frac{\partial f}{\partial t} + \{f, H\} + \int \{f_u, H_s\}\,d\Gamma_s + \int \{f_u, U\}\,d\Gamma_s = 0, \tag{II.5}$$

since H depends only on the variables of the system under consideration. The third term in this equation goes to zero, since the integrand is equal to a divergence in the phase space of the surroundings. The last term can be simplified if we introduce a function X describing the correlation of the system with the surroundings:

$$f_u = fgX. \tag{II.6}$$

Then,

$$\frac{\partial f}{\partial t} + \{f, H\} + \frac{\partial (fF_\alpha)}{\partial p_\alpha} = 0, \tag{II.7}$$

where

$$F_\alpha = -\int gX \frac{\partial U}{\partial q_\alpha}\,d\Gamma_s, \tag{II.8}$$

q_α and p_α are the coordinates and momenta of the system, and a summation over α is implied in (II.7). The quantity F_α has the meaning of a "force" representing the action of the surroundings on the system. If there is no correlation between the system and the surroundings, i.e., X = 1, then

$$F_\alpha = -\frac{\partial}{\partial q_\alpha}\int gU\,d\Gamma_s,$$

i.e., the force F_α is conservative and its effect can be represented by an additional term in the Hamiltonian. But in the general case, the force F_α is nonconservative.

We note that the derivation of Eq. (II.7) from Eq. (II.2) is analogous to the derivation of the chain of coupled equations in the Bogolyubov–Born–Green–Kirkwood–Yvonne (BBGKY) method [7].

We introduce minus the logarithm of the distribution function, $\eta = -\ln f$:

$$f = e^{-\eta}. \tag{II.9}$$

Then Eq. (II.7) can be written in the form

$$\frac{\partial \eta}{\partial t} + \{\eta, H\} + F_\alpha \frac{\partial \eta}{\partial p_\alpha} = \frac{\partial F_\alpha}{\partial p_\alpha}, \tag{II.10}$$

or, if we introduce the total derivative of the dynamic variable η

$$\frac{d\eta}{dt} = \frac{\partial \eta}{\partial t} + \{\eta, H\} + F_\alpha \frac{\partial \eta}{\partial p_\alpha}, \tag{II.11}$$

then Eq. (II.10) can be rewritten in the form

$$\frac{d\eta}{dt} = \frac{\partial F_\alpha}{\partial p_\alpha}. \tag{II.12}$$

It can be seen from this equation that if the forces F_α depend on the momenta, the total derivative of η is nonzero.

For what follows, we must establish the form of the external forces occurring in the right-hand side of Eq. (II.12). We shall assume that they are determined by the thermodynamic variables (the temperature, chemical potential, and velocity) characterizing the surroundings, and not by the details of the microscopic state of the latter. Then the term $\partial F_\alpha / \partial t$ in Eq. (II.10) is connected with the flux of entropy into the system. For it we can take the expression

$$\frac{\partial F_\alpha}{\partial p_\alpha} = - \int j_S(x)\, ds, \tag{II.13}$$

where $\mathbf{j}_S(\mathbf{x})$ is the entropy flux density (including the work performed on the system), and $\mathbf{ds}$ is an element of the surface bounding the system;

$$j_S(x) = \beta(x, t)\left[j_H(x) - \left(\mu(x, t) - \frac{1}{2} mv^2(x)\right) j(x) - v(x, t) \cdot T(x)\right], \tag{II.14}$$

$\mathbf{j}_H(\mathbf{x})$, $\mathbf{j}(\mathbf{x})$, and $T(\mathbf{x})$ are the dynamic variables of the fluxes of energy, particle number, and momentum (see §19).

Some instructive arguments for such a choice of entropy sources can be given by considering a system of discrete sources [8]; however, this is, in essence, the basic assumption of MacLennan's theory, namely, the assumption that the influence of the surroundings can be characterized by the functions $\beta(\mathbf{x}, t)$, $\mu(\mathbf{x}, t)$, and $\mathbf{v}(\mathbf{x}, t)$. In the expression (II.14), the first term corresponds to the contribution of the energy flux, the second to the contribution of the particle flux, and the third arises from the work done.

Below, for simplicity, we shall first consider the case when only energy exchange with thermostats occurs, i.e., when the equation for η has the form

$$\frac{d\eta}{dt} = - \int \beta(x, t)\, j_H(x) \cdot ds. \tag{II.15}$$

We transform the surface integral in (II.15) into a volume integral, taking the energy conservation law (19.16) into account:

$$\frac{\partial H(\boldsymbol{x})}{\partial t} + \nabla \cdot \boldsymbol{j}_H(\boldsymbol{x}) = 0. \tag{II.16}$$

After integrating by parts, we obtain

$$\frac{d\eta}{dt} = \frac{d}{dt} \int \beta(\boldsymbol{x}, t) H(\boldsymbol{x})\, d\boldsymbol{x} - \int \left\{ \boldsymbol{j}_H(\boldsymbol{x}) \cdot \nabla\beta(\boldsymbol{x}) + H(\boldsymbol{x}) \frac{\partial \beta(\boldsymbol{x}, t)}{\partial t} \right\} d\boldsymbol{x}. \tag{II.17}$$

This equation for η has many solutions, since we can always add a solution of the homogeneous equation

$$\frac{d\eta}{dt} = 0. \tag{II.18}$$

In order to obtain the particular solution of interest to us, we must specify a further initial condition. We suppose that at $t = -\infty$ the system was in statistical equilibrium and was described by the Gibbsian canonical ensemble:

$$\eta\,|_{t=-\infty} = \alpha + \beta H. \tag{II.19}$$

The initial condition (II.19) corresponds to the usual situation in which one starts from the equilibrium state of the system and brings it into a nonequilibrium state by means of an external perturbation, as in the indirect methods of linear-response theory (see the beginning of Chapter IV). It can be verified that the solution of Eq. (II.17) with the initial condition (II.19) has the form

$$\eta(t) = \alpha + \int \beta(\boldsymbol{x}, t) H(\boldsymbol{x})\, d\boldsymbol{x} -$$

$$- \int \int_{-\infty}^{t} \left[\boldsymbol{j}_H(\boldsymbol{x}, t' - t) \cdot \nabla\beta(\boldsymbol{x}, t') + H(\boldsymbol{x}, t' - t) \frac{\partial \beta(\boldsymbol{x}, t')}{\partial t'} \right] d\boldsymbol{x}\, dt'. \tag{II.20}$$

It is assumed that $\beta(\mathbf{x}, t) \to \beta$ as $t \to -\infty$ rapidly enough for the integrals to converge. The function

$$f = e^{-\eta} \tag{II.21}$$

is the required distribution function.

In the more general case when exchange of particles and momentum with the surroundings is also taken into account, we obtain the formula (21.10e) of Chapter IV.

Thus, MacLennan's method [1, 2] and the method of the nonequilibrium statistical operator [3-6] lead to the same expressions for the distribution function.

Appendix III

Boundary Conditions for the Statistical Operators

In the theory of nonequilibrium processes, we study solutions of the quantum Liouville equation for the statistical operator ρ

$$\frac{\partial \rho}{\partial t} + \frac{1}{i\hbar}[\rho, H] = 0, \tag{III.1}$$

or of the classical Liouville equation for the distribution function f

$$\frac{\partial f}{\partial t} + \{f, H\} = 0, \tag{III.2}$$

where $\{\ldots, \ldots\}$ is the classical Poisson bracket. We shall show, following [1], how to formulate boundary conditions to the Liouville equation by means of infinitesimally small sources. Below, we shall discuss the boundary conditions for the quantum case only, since the classical case is analogous.

The formal solution of Liouville's equation (III.1) is found very simply:

$$\rho(t) = U(t, t_0)\, \rho(t_0)\, U^{+}(t, t_0), \tag{III.3}$$

where $\rho(t_0)$ is an arbitrary statistical operator at the initial time t_0, and $U(t, t_0)$ is the evolution operator. However, the formal solution (III.3) can be useful only if the statistical operator $\rho(t_0)$ at the initial time is well chosen. For example, if the state is close to statistical equilibrium, then in the Kubo theory one chooses $t_0 = -\infty$ and $\rho(-\infty)$ to be in the form of the equilibrium statistical operator. But in the general case, the formal solution (III.3) is of little use for describing a nonequilibrium process. Consequently, the fundamental problem of nonequilibrium statistical mechanics is not the determination of exact formal solutions of Liouville's equation, but the choice of the correct boundary conditions for it and the construction of solutions in the sense of quasi-averages, as in the quantum theory of collisions.

A state with the given average values $\langle P_m \rangle$ can be described by the quasi-equilibrium statistical operator

$$\rho_q = \exp\left\{-\Phi - \sum_m P_m F_m(t)\right\} \equiv \exp\{-S(t, 0)\}, \tag{III.4}$$

where

$$\Phi = \ln \operatorname{Tr} \exp\left\{-\sum_m P_m F_m(t)\right\} \tag{III.5}$$

is the Massieu–Planck function, and $F_m(t)$ are parameters conjugate to the average values

$$\langle P_m \rangle_q^t = \operatorname{Tr}(\rho_q P_m). \tag{III.6}$$

The quasi-equilibrium operator (III.4) ensures that the thermodynamic equalities for its parameters Φ, $F_m(t)$, and $S = -\langle \ln \rho_q \rangle_q$, are satisfied, i.e., that

$$\frac{\delta \Phi}{\delta F_m(t)} = -\langle P_m \rangle_q^t, \quad \frac{\delta S}{\delta \langle P_m \rangle_q^t} = F_m(t), \tag{III.7}$$

i.e., the parameters $F_m(t)$ and $\langle P_m \rangle_q^t$ are indeed thermodynamically conjugate.† However, the statistical operator (III.4) does not satisfy Liouville's equation and does not describe irreversible processes. As we shall see below, it can be used to formulate boundary conditions to Liouville's equation (III.1), in the same way as the free-particle wave function is used to formulate boundary conditions to the Schrödinger equation in the quantum theory of scattering.

The quantum Liouville equation (III.1), like the classical equation (III.2), is symmetric under the time-reversal transformation (in the classical case, this means the replacement $t \to -t$ and the reversal of the momenta of all the particles and of the direction of the magnetic field). However, the solution of Liouville's equation is unstable with respect to small perturbations violating this symmetry of the equation.

Following [1], we introduce into Liouville's equation an infinitesimally small source satisfying the following requirements:

The quasi-equilibrium statistical operator (III.4) ensures that the thermodynamic equalities for its parameters Φ, $F_m(t)$, and $\langle P_m \rangle_q^t$ are satisfied, i.e., that

1. The source violates the time-reversal symmetry of Liouville's equation, or in other words, violates the complete isolation of the system. In addition, it tends to zero as $\varepsilon \to 0$, this limit being taken after the thermodynamic limiting process.

†If the indices m take on a discrete series of values, the functional derivatives in (III.7) go over into ordinary partial derivatives.

2. The source selects the retarded solutions of Liouville's equation. This condition determines the sign of ε, i.e., if we introduce a source as in (I.24), then $\varepsilon > 0$ and $\varepsilon \to +0$. The advanced solutions would give not an increase but a decrease of entropy [2].
3. The source vanishes for ρ equal to the quasi-equilibrium statistical operator ρ_q (III.4). In the particular case of statistical equilibrium, the source must be absent.

Two methods can be suggested for introducing an infinitesimally small source, satisfying these requirements, into Liouville's equation.

The first method consists in introducing an infinitesimally small source into the right-hand side of Liouville's equation (III.1)

$$\frac{\partial \rho_\varepsilon}{\partial t} + \frac{1}{i\hbar}[\rho_\varepsilon, H] = -\varepsilon(\rho_\varepsilon - \rho_q),$$

where $\varepsilon \to +0$ after the thermodynamic limiting process in the calculation of the averages. The equation (III.8) is analogous to Eq. (1.24) of the quantum theory of collisions. This is the only form which satisfies, in addition to the conditions 1-3, the requirement that the source be linear in ρ_ε. The infinitesimally small source in (III.8) does indeed violate the time-reversal symmetry of Eq. (III.1) since under time reversal the left-hand side of Eq. (III.8) changes sign while the right-hand side remains unchanged†.

We write Eq. (III.8) in the form

$$\frac{d}{dt}\left(e^{\varepsilon t}\rho_\varepsilon(t, t)\right) = \varepsilon e^{\varepsilon t}\rho_q(t, t), \tag{III.9}$$

where

$$\begin{aligned} \rho_\varepsilon(t, t) &= U^+(t, 0)\,\rho_\varepsilon(t, 0)\,U(t, 0), \\ \rho_q(t, t) &= U^+(t, 0)\,\rho_q(t, 0)\,U(t, 0), \\ U(t, 0) &= \exp\left\{\frac{Ht}{i\hbar}\right\} \end{aligned} \tag{III.10}$$

(H does not depend on time) and we have introduced the notation

$$\rho_\varepsilon = \rho_\varepsilon(t, 0), \qquad \rho_q = \rho_q(t, 0). \tag{III.11}$$

Integrating Eq. (III.9) from $-\infty$ to t and assuming that

$$\lim_{t \to -\infty} e^{\varepsilon t}\rho(t, t) = 0,$$

†It can be shown [9, 10, 11] that the introduction of infinitesimally small sources into Liouville's equation is equivalent to the boundary condition

$$e^{it_1 H/\hbar}\left(\rho(t + t_1) - \rho_q(t + t_1)\right)e^{-it_1 H/\hbar} \to 0$$

as $t_1 \to -\infty$ after the thermodynamic limiting process.

we obtain

$$\rho_\varepsilon (t, t) = \varepsilon \int_{-\infty}^{t} e^{\varepsilon(t_1 - t)} \rho_q(t_1, t_1)\, dt_1 = \varepsilon \int_{-\infty}^{0} e^{\varepsilon t'} \rho_q(t + t', t + t')\, dt'. \tag{III.12}$$

Consequently, the required nonequilibrium statistical operator has the form

$$\rho_\varepsilon = \rho_\varepsilon (t, 0) = \widetilde{\rho_q (t, 0)} = \varepsilon \int_{-\infty}^{0} e^{\varepsilon t'} \rho_q (t + t', t')\, dt', \tag{III.13}$$

where the wavy line denotes the operation of taking the quasi-invariant part. The statistical operator (III.13) was obtained earlier from different considerations in a paper by Kalashnikov and the author [3]. By means of integration by parts, the nonequilibrium statistical operator (III.13) can be written conveniently in the form [3]

$$\rho_\varepsilon = \rho_q + \int_{-\infty}^{0} dt'\, e^{\varepsilon t'} \int_{0}^{1} d\tau\, e^{-\tau S(t+t', t')}\, \dot{S}(t + t', t')\, e^{(\tau - 1) S(t+t', t')}, \tag{III.14}$$

where

$$\begin{gathered} \dot{S}(t, 0) = \frac{\partial S(t, 0)}{\partial t} + \frac{1}{i\hbar} [S(t, 0), H], \\ \dot{S}(t, t') = U^+(t', 0)\, \dot{S}(t, 0)\, U(t', 0) \end{gathered} \tag{III.15}$$

is the entropy production operator.

We choose the parameters $F_m(t)$ occurring in the expression for the entropy operator from the condition that the averages of the P_m, calculated with the nonequilibrium statistical operator (III.13), coincide with their averages over the quasi-equilibrium statistical operator (III.4):

$$\langle P_m \rangle^t = \langle P_m \rangle^t_q, \tag{III.16}$$

where

$$\langle \ldots \rangle^t = \lim_{\varepsilon \to +0} \mathrm{Tr}\, (\rho_\varepsilon \ldots). \tag{III.17}$$

Then $\langle P_m \rangle^t$ and $F_m(t)$ become conjugate parameters, since

$$\frac{\delta \Phi}{\delta F_m(t)} = - \langle P_m \rangle^t_q = - \langle P_m \rangle^t. \tag{III.18}$$

By means of the nonequilibrium statistical operator (III.13), we can calculate the average value of any operator A:

$$\langle A \rangle = \lim_{\varepsilon \to +0} \mathrm{Tr}\, (\rho_\varepsilon A) = \langle\!\langle A \rangle\!\rangle. \tag{III.19}$$

In Bogolyubov's terminology [4, 5], such averages are called quasi-averages. If we apply the averaging operation (III.19) to the operators $\dot{P}_m$, then, taking (III.16) into account, we obtain the transport equations

$$\frac{\partial}{\partial t} \langle P_m \rangle_q^t = \langle \dot{P}_m \rangle^t = \lim_{\varepsilon \to +0} \mathrm{Tr}\, (\rho_\varepsilon \dot{P}_m) = \prec \dot{P}_m \succ. \tag{III.20}$$

Consequently, the transport equations are equations for quasi-averages.

The second method of introducing infinitesimally small sources is based on the fact that the logarithm of a statistical operator satisfying Liouville's equation also satisfies Liouville's equation

$$\frac{\partial \ln \rho}{\partial t} + \frac{1}{i\hbar} [\ln \rho, H] = 0; \tag{III.21}$$

this is connected with the properties of Poisson brackets (both quantum and classical). Consequently, infinitesimally small sources can be introduced not only into (III.1), but also into (III.21).

If we demand that the infinitesimally small source satisfy the conditions 1-3 and, in addition, be linear in ln ρ, we obtain

$$\frac{\partial \ln \rho_\varepsilon}{\partial t} + \frac{1}{i\hbar} [\ln \rho_\varepsilon, H] = -\varepsilon (\ln \rho_\varepsilon - \ln \rho_q), \tag{III.22}$$

where $\varepsilon \to +0$ after the thermodynamic limiting process. In fact, the source in (III.22) violates the time-reversal symmetry of Eq. (III.21) and accords with the other conditions 1-3. We write Eq. (III.22) in the form

$$\frac{d}{dt} \left(e^{\varepsilon t} \ln \rho_\varepsilon (t, t)\right) = \varepsilon e^{\varepsilon t} \ln \rho_q (t, t). \tag{III.23}$$

Integrating (III.23) from $-\infty$ to t, we obtain

$$\ln \rho_\varepsilon (t, t) = \varepsilon \int_{-\infty}^{t} e^{\varepsilon (t_1 - t)} \ln \rho_q (t_1, t_1)\, dt_1 = \varepsilon \int_{-\infty}^{0} e^{\varepsilon t'} \ln \rho_q (t + t', t + t')\, dt', \tag{III.24}$$

and, consequently, the nonequilibrium statistical operator that we seek has the form

$$\rho_\varepsilon = \rho_\varepsilon (t, 0) = \exp \{\widetilde{\ln \rho_q (t, 0)}\} = \exp \left\{ -\varepsilon \int_{-\infty}^{0} dt'\, e^{\varepsilon t'} \ln \rho_q (t + t', t') \right\}, \tag{III.25}$$

where $\varepsilon \to +0$ after the thermodynamic limiting process in the calculation of averages.

The nonequilibrium statistical operator (III.25) was obtained earlier from other paper [2-6]. After integration by parts, it can be written conveniently in the form

$$\rho_\varepsilon = \exp \{-\widetilde{S (t, 0)}\} = \exp \left\{ -S (t, 0) + \int_{-\infty}^{0} dt'\, e^{\varepsilon t'}\, \dot{S} (t + t', t') \right\}. \tag{III.26}$$

The parameters $F_m(t)$ occurring in the expressions for the entropy operator $S(t, 0)$ and the entropy production $\dot{S}(t, 0)$ are determined, as before, from the conditions (III.16).

The statistical operator (III.25) has some similarity with MacLennan's statistical operator [8]. The latter can be obtained from (III.25) after integration by parts, if the conservation laws are taken into account and the surface integrals discarded. The difference between this method and MacLennan's method lies in the fact that we consider infinitesimally small perturbations of Liouville's equation, and not, as MacLennan does, finite real perturbations induced by a thermostat. One could say that the introduction of the boundary conditions to Liouville's equation is an idealized convention for taking into account the influence of the thermostat.

References

Introduction

1. S. R. de Groot and P. Mazur, Nonequilibrium Thermodynamics, North-Holland, Amsterdam, 1962.
2. N. N. Bogolyubov, Problems of a Dynamical Theory in Statistical Physics, p. 11 of Vol. 1 of "Studies in Statistical Mechanics," ed. J. de Boer and G. E. Uhlenbeck, North-Holland, Amsterdam, 1962.
3. J. G. Kirwood, J. Chem. Phys. 14:180 (1946); 15:72 (1947).
4. H. S. Green, The Molecular Theory of Fluids, North-Holland, Amsterdam, 1952.
5. L. van Hove, Physica 21:517 (1955).
6. I. Prigogine, Nonequilibrium Statistical Mechanics, Interscience, New York, 1962.
7. Yu. L. Klimontovich, The Statistical Theory of Nonequilibrium Processes in a Plasma, Pergamon, Oxford, 1967.
8. J. W. Gibbs, Elementary Principles in Statistical Mechanics, in Vol. 2 of "The Collected Works of J. Willard Gibbs," Longmans, New York, 1931.
9. G. V. Chester, Rep. Progr. Phys. 26:411 (1963).
10. R. Kubo, Lectures in Theoretical Physics, 1:120 (1959) [Boulder Summer Institute for Theoretical Physics, 1958; Interscience, New York].
11. R. Kubo, p. 81 of "Statistical Mechanics of Equilibrium and Nonequilibrium," ed. J. Meixner, Proc. Int. Symp. on Statistical Mechanics and Thermodynamics, Aachen, 1964, North-Holland, Amsterdam, 1965.
12. W. Bernard and H. B. Callen, Rev. Mod. Phys. 31:1017 (1959).
13. R. Zwanzig, Ann. Rev. Phys. Chem. 16:67 (1965).
14. H. B. Callen and T. A. Welton, Phys. Rev. 83:34 (1951).
15. T. L. Hill, Statistical Mechanics, McGraw-Hill, New York, 1956.
16. L. D. Landau and E. M. Lifshitz, Statistical Physics, Pergamon, Oxford, 1969.
17. K. Huang, Statistical Mechanics, Wiley, New York, 1963.
18. S. A. Rice and P. Gray, The Statistical Mechanics of Simple Liquids, Interscience, New York, 1965.
19. A. Münster, Statistical Thermodynamics, Vol. 1, Academic, New York, 1969.

20. D. N. Zubarev, Dokl. Akad. Nauk SSSR 140:92 (1961); 162:532, 794 (1965); 164:537 (1965) [Sov. Phys. – Doklady 6:776 (1962); 10:452 (1965); 10:526 (1966); 10:850 (1966)]; Fortsch. Phys. 18:125 (1970); Teor. Mat. Fiz. 3:276 (1970) [Theor. Math. Phys. 3:505 (1970)].
21. D. N. Zubarev and V. P. Kalashnikov, Teor. Mat. Fiz. 1:137 (1969); 3:126 (1970) [Theor. Math. Phys. 1:108 (1969); 3:395 (1970)]; Physica 46:550 (1970).
22. L. L. Buishvili and D. N. Zubarev, Fiz. Tverd. Tela 7:722 (1965) [Sov. Phys. – Solid State 7:580 (1965)].
23. A. G. Bashkirov and D. N. Zubarev, Teor. Mat. Fiz. 1:407 (1969) [Theor. Math. Phys. 1: 311 (1969)].
24. L. A. Pokrovskii, Dokl. Akad. Nauk SSSR 182:317 (1968); 183:806 (1968) [Sov. Phys. – Doklady 13:911 (1969); 13:1216 (1969)].

Chapter I

1. J. W. Gibbs, Elementary Principles in Statistical Mechanics, in Vol. 2 of "The Collected Works of J. Willard Gibbs," Longmans, New York, 1931.
2. M. A. Leontovich, Statistical Physics, Moscow, Leningrad, 1944.
3. I. Prigogine, Nonequilibrium Statistical Mechanics, Interscience, New York, 1962.
4. S. Chapman and T. G. Cowling, The Mathematical Theory of Nonuniform Gases (3rd ed.), Cambridge, 1970.
5. P. Ehrenfest and T. Ehrenfest, Enzyklopädie der Mathematischen Wissenschaften, Vol. IV, Art. 32 (1907-1914). [Reprinted in "Paul Ehrenfest, Collected Scientific Papers" (M. J. Klein, ed.), p. 213, North-Holland, Amsterdam, 1959.]
6. D. Ter Haar, Rev. Mod. Phys. 27:289 (1955).
7. J. G. Kirkwood, J. Chem. Phys. 14:180 (1946); 15:72 (1947).
8. N. N. Bogolyubov and Yu. A. Mitropol'skii, Asymptotic Methods in the Theory of Nonlinear Oscillations, Moscow, 1955.
9. N. N. Bogolyubov, Problems of Dynamic Theory in Statistical Physics, p. 11 of Vol. 1 of "Studies in Statistical Mechanics," ed. J. de Boer and G. E. Uhlenbeck, North-Holland, Amsterdam, 1962.
10. M. Gell-Mann and M. L. Goldberger, Phys. Rev. 91:398 (1953).
11. "Ergodic Theories," Rend. Scuola Int. Fisica "Enrico Fermi," 14 corso, Varenna, 1960, Academic, New York, 1961.
12. R. Kurth, Axiomatics of Statistical Mechanics, Pergamon, Oxford, 1960.
13. L. van Hove, Physica 15:951 (1949).
14. C. N. Yang and T. D. Lee, Phys. Rev. 87:404 (1952).
15. D. Ruelle, Helv. Phys. Acta 36:183 (1963); Statistical Mechanics: Rigorous Results, Benjamin, Amsterdam, 1969.
16. R. L. Dobryshin, Teor. Veroyat. ee Primen. 9:626 (1964) [Theory Probab. Its Appl. (USSR) 9:566 (1964)].
17a. M. E. Fisher, Arch. Ration. Mech. Anal. 17:377 (1964).
17b. J. Ginibre, J. Math. Phys. 6:238 (1965).
17c. J. van der Linden, Physica 32:642 (1966).
17d. J. van der Linden and P. Mazur, Physica 36:491 (1967).

17e. N. N. Bogolyubov, D. Ya. Petrina and B. I. Khatset, Teor. Mat. Fiz. 1:251 (1969) [Theor. Math. Phys. 1:194 (1969)].
18. Yu. A. Krutkov, Dokl. Akad. Nauk SSSR 1:3 (1933).
19. P. Courant and D. Hilbert, Methods of Mathematical Physics, Vol. 1, Interscience, New York, 1970.
20. A. Ya. Khinchin, Statistical Mechanics, Dover, New York, 1949.
21. A. Ya. Khinchin, Tr. Mat. Inst. Akad. Nauk SSSR 33:3 (1950).
22. E. T. Jaynes, Phys. Rev. 106:620 (1957); 108:171 (1957).
23. E. T. Jaynes, Information Theory and Statistical Mechanics, in "Statistical Physics," Brandeis Lectures, Vol. 3, p. 181, Bejamin, New York, 1963.
24. S. Shubin, Dokl. Akad. Nauk SSSR 1:301 (1935).
25. R. H. Fowler and E. A. Guggenheim, Statistical Thermodynamics, Cambridge, 1939.
26. T. L. Hill, Statistical Mechanics, McGraw-Hill, New York, 1956.
27. D. N. Zubarev, Nauch. Dokl. Vyssh. Shkoly, Ser. Fiz.-Mat. Nauk, No. 6, p. 169 (1958).
28. C. E. Shannon, Papers on Information Theory and Cybernetics, Moscow, 1963.
29. A. Ya. Khinchin, Usp. Mat. Nauk 8(3):3 (1953).
30. R. Greene and H. B. Callen, Phys. Rev. 83:1231 (1951).
31. A. Einstein, Ann. Phys. (Leipzig) 33:1275 (1910).
32. A. Münster, Rend. Scuola Int. Fisica "Enrico Fermi," 10 corso Varenna, 1959, Soc. Ital. di Fisica, 1960.
33. L. D. Landau and E. M. Lifshitz, Statistical Physics, Pergamon, Oxford, 1969.
34. N. N. Bogolyubov and B. I. Khatset, Dokl. Akad. Nauk SSSR 66:321 (1949).

Chapter II

1. N. N. Bogolyubov, Lectures on Quantum Statistics, Gordon and Breach, New York, 1967.
2. J. von Neumann, Mathematical Foundations of Quantum Mechanics, Princeton, N. J., 1955.
3. J. von Neumann, Nachr. Ges. Wiss. Goettingen, Math. Phys. Kl. (1927), p. 273.
4. R. C. Tolman, The Principles of Statistical Mechanics, Oxford, 1938.
5. L. I. Mandel'shtam, Complete Collected Works, Vol. 5, Moscow, 1950.
6. F. London and E. Bauer, La théorie de l'observation en méchanique quantique, Hermann, Paris, 1939.
7. G. Ludwig, Die Grundlagen der Quantenmechanik, Springer-Verlag, Berlin, 1954.
8. D. ter Haar, Rep. Progr. Phys. 24:304 (1961).
9. W. Elsasser, Phys. Rev. 52:987 (1937).
10. U. Fano, Rev. Mod. Phys. 29:74 (1957).
11. L. D. Landau, Z. Phys. 45:430 (1927).
12. E. Wigner, Phys. Rev. 40:749 (1932).
13. "Ergodic Theories," Rend. Scuola Int. Fisica "Enrico Fermi," 14 corso, Varenna, 1960, Academic, New York, 1961.
14. D. ter Haar, Rev. Mod. Phys. 27:289 (1955).
15. S. Shubin, Dokl. Akad. Nauk SSSR 1:301 (1935).

16. R. H. Fowler and E. A. Guggenheim, Statistical Thermodynamics, Cambridge, 1939.
17. T. L. Hill, Statistical Mechanics, McGraw-Hill, New York, 1956.
18. E. T. Jaynes, Phys. Rev. 106:620 (1957); 108:171 (1957).
19. E. T. Jaynes, Information Theory and Statistical Mechanics, in "Statistical Physics," Brandeis Lectures, Vol. 3, p. 181, Benjamin, New York, 1963.
20. M. J. Klein, p. 198 of "Thermodynamics of Irreversible Processes," Rend. Scuola Int. Fisica "Enrico Fermi," 10 corso, Varenna, 1959, Soc. Ital. di Fisica, 1960.
21. F. E. Simon, p. 9 of Vol. 2 of "Temperature, Its Measurement and Control in Science and Industry," Reinhold, New York, 1955.
22. J. Wilks, The Third Law of Thermodynamics, Oxford, 1961.
23. D. N. Zubarev, Nauch. Dokl. Vyssh. Shkoly, Ser. Fiz.-Mat. Nauk, No. 6, p. 169 (1958).
24. J. W. Gibbs, Elementary Principles in Statistical Mechanics, in Vol. 2 of "The Collected Works of J. Willard Gibbs," Longmans, New York, 1931.
25. S. Ono and S. Kondo, Molecular Theory of Surface Tension in Liquids, Handbuch der Physik, Vol. 10, p. 134 (1960), Springer-Verlag, Berlin.
25a. A. I. Rusanov, Phase Equilibria and Surface Phenomena, "Khimiya," Moscow, 1967.
26. G. E. Uhlenbeck and L. Gropper, Phys. Rev. 41:79 (1932).
27. J. G. Kirkwood, Phys. Rev. 44:31 (1933).
27a. J. G. Kirkwood, Phys. Rev. 45:116 (1934).
28. D. I. Blokhintsev, J. Phys. USSR 2:71 (1940).

Chapter III

1. R. Kubo, J. Phys. Soc. Japan 12:570 (1957).
2. R. Kubo, M. Yokota, and S. Nakajima, J. Phys. Soc. Japan 12:1203 (1957).
3. R. Kubo, p. 81 of "Statistical Mechanics of Equilibrium and Nonequilibrium," ed. J. Meixner, Proc. Int. Symp. on Statistical Mechanics and Thermodynamics, Aachen, 1964, North-Holland, Amsterdam, 1965.
4. R. Kubo, Lectures in Theoretical Physics 1:120 (1959) [Boulder Summer Institute for Theoretical Physics, 1958, Interscience, New York].
5. E. W. Montroll, p. 217 of "Thermodynamics of Irreversible Processes," Rend. Scuola Int. Fisica "Enrico Fermi," 10 corso, Varenna, 1959, Soc. Ital. di Fisica, 1960.
6. L. I. Komarov, Zh. Eksp. Teor. Fiz. 48:145 (1965) [Sov. Phys. – JETP, 21:99 (1965)].
7. J. M. Luttinger, Phys. Rev. 135A:1505 (1964).
8. L. P. Kadanoff and P. C. Martin, Ann. Phys. 24:419 (1963).
9. P. C. Martin, p. 100 of "Statistical Mechanics of Equilibrium and Nonequilibrium," ed. J. Meixner, Proc. Int. Symp. on Statistical Mechanics and Thermodynamics, Aachen, 1964, North-Holland, Amsterdam, 1965.
10. G. V. Chester, Rep. Progr. Phys. 26:411 (1963).
11. W. Bernard and H. B. Callen, Rev. Mod. Phys. 31:1017 (1959).
12. D. N. Zubarev, Usp. Fiz. Nauk 71:71 (1960) [Sov. Phys. – Uspekhi 3:320 (1960)].
13. V. L. Bonch-Bruevich and S. V. Tyablikov, The Green Function Method in Statistical Mechanics, North-Holland, Amsterdam, 1962.

14. S. V. Tyablikov, Methods in the Quantum Theory of Magnetism, Plenum, New York, 1967.
15. N. N. Bogolyubov and S. V. Tyablikov, Dokl. Akad. Nauk SSSR 126:53 (1959) [Sov. Phys. – Doklady 4:589 (1959)].
16. A. Ya. Khinchin, Statistical Mechanics, Dover, New York, 1949.
17. J. L. Doob, Stochastic Processes, Wiley, New York, 1953.
18. N. N. Bogolyubov, Jr., and B. I. Sadovnikov, Zh. Eksp. Teor. Fiz. 43:677 (1962) [Sov. Phys. – JETP 16:482 (1963)].
19. B. I. Sadovnikov, Dokl. Akad. Nauk SSSR 164:785 (1965) [Sov. Phys. – Doklady 10:934 (1966)]; Physica 32:858 (1969).
20. J. C. Herzel, J. Math. Phys. 8:1650 (1967); 11:741 (1970).
21. J. G. Kirkwood, J. Chem. Phys. 14:180 (1946).
22. H. B. Callen and T. A. Welton, Phys. Rev. 83:34 (1951).
23. N. N. Bogolyubov and D. V. Shirkov, Introduction to the Theory of Quantized Fields, Interscience, New York, 1959.
24. S. S. Schweber, H. A. Bethe, and F. de Hoffmann, Mesons and Fields, Vol. 1, Row and Peterson, Evanston, Ill., 1955.
25. S. S. Schweber, An Introduction to Relativistic Quantum Field Theory, Harper and Row, New York, 1961.
26. A. I. Akhiezer and V. B. Berestetskii, Quantum Electrodynamics, Interscience, New York, 1965.
27. A. Abragam, The Principles of Nuclear Magnetism, Oxford, 1961.
28. A. A. Samokhin, Physica 32:823 (1966); Zh. Eksp. Teor. Fiz. 51:928 (1966) [Sov. Phys. – JETP 24:617 (1967)]; Fiz. Tverd. Tela 9:1597 (1967) [Sov. Phys. – Solid State 9:1257 (1967)].
29. A. A. Pervozvanskii, Random Processes in Nonlinear Control Systems, Academic, New York, 1965.
30. A. K. Zarembo and V. A. Krasil'nikov, Introduction to Nonlinear Acoustics, Moscow, 1966.
31. V. M. Fain and Ya. I. Khanin, Quantum Electronics, Pergamon, Oxford, 1969.
32. "Quantum Optics and Electronics," ed. C. De Witt, Lectures at the Summer School of Theoretical Physics, Les Houches, 1964, Gordon and Breach, New York, 1965.
33. D. A. Frank-Kamenetskii, Diffusion and Heat Exchange in Chemical Kinetics, Princeton Univ. Press, N. J., 1955.
34. N. N. Bogolyubov and Yu. A. Mitropol'skii, Asymptotic Methods in the Theory of Nonlinear Oscillations, Moscow, 1955.
35. Yu. A. Mitropol'skii, Nonstationary Processes in Nonlinear Oscillatory Systems, Kiev, 1955.
36. A. A. Andronov, A. A. Vitt, and S. É. Khaikin, Theory of Oscillators, Pergamon, Oxford, 1966.
37. I. V. Aleksandrov, The Theory of Nuclear Magnetic Resonance, Academic, New York, 1966.
38. S. A. Akhmanov, and R. V. Khokhlov, Problems of Nonlinear Optics, Moscow, 1964.
39. E. W. Montroll and J. C. Ward, Physica 25:423 (1959).
40. T. Izuyama, Progr. Theor. Phys. 25:964 (1961).

41. N. M. Plakida, Fiz. Tverd. Tela 6:3444 (1964) [Sov. Phys. – Solid State 6:2754 (1965)]; Fiz. Metal. Metalloved. 21:657 (1966) [Phys. Metals Metallogr. (USSR) 21(5):16 (1966)]; Zh. Eksp. Teor. Fiz. 53:2041 (1967) [Sov. Phys. – JETP 26:1155 (1968)].
42. J. M. Ziman, Principles of the Theory of Solids, Cambridge, 1965.
43. F. Seitz, Modern Theory of Solids, McGraw-Hill, New York, 1940.
44. D. Pines, Elementary Excitations in Solids, Benjamin, New York, 1963.
45. J. M. Ziman, Electrons and Phonons, Oxford, 1960.
46. J. Bardeen and D. Pines, Phys. Rev. 99:1140 (1955).
47. M. Gell-Mann and M. L. Goldberger, Phys. Rev. 91:398 (1953).
48. H. Nakano, Progr. Theor. Phys. 17:145 (1957).
49. S. Fujita and R. Abe, J. Math. Phys. 3:350 (1962).
50. G. V. Chester and A. Thellung, Proc. Phys. Soc. 73:745 (1959).
51. S. Fujita, Introduction to Nonequilibrium Quantum Statistical Mechanics, Saunders, Philadelphia, 1966.
52. K. Yamada, Progr. Theor. Phys. 28:299 (1962).
53. K. H. Michel and J. M. J. van Leeuwen, Physica 30:410 (1964).
54. T. Holstein, Ann. Phys. (New York) 29:410 (1964).
55. J. S. Langer, Phys. Rev. 127:5 (1962).
56. E. Verboven, Physica 26:1091 (1962).
57. R. Kubo and K. Tomita, J. Phys. Soc. Japan 9:888 (1954).
58. S. V. Tyablikov, Fiz. Tverd. Tela 2:361 (1960) [Sov. Phys. – Solid State 2:332 (1960)].
59. S. V. Tyablikov, Fiz. Tverd. Tela 2:2009 (1960) [Sov. Phys. – Solid State 2:1805 (1961)].
60. A. A. Abrikosov, L. P. Gor'kov, and I. E. Dzyaloshinskii, Quantum Field Theoretical Methods in Statistical Physics, Pergamon, Oxford, 1965.
61. D. A. Kirzhnits, Field Theoretical Methods in Many-Body Systems, Pergamon, Oxford, 1967.
62. D. Thouless, The Quantum Mechanics of Many-Body Systems, Academic, New York, 1961.
63. L. P. Kandanoff and G. Baym, Quantum Statistical Mechanics, Benjamin, New York, 1962.
64. P. C. Martin and J. Schwinger, Phys. Rev. 115:1342 (1959).
65. T. Matsubara, Progr. Theor. Phys. 14:351 (1955).
66. V. N. Kashcheev, Thermodynamics of a Ferromagnet near the Curie Point, Riga, 1966.
67. N. N. Bogolyubov, B. V. Medvedev, and M. K. Polivanov, Problems of Dispersion Relations, Moscow, 1958.
68. Sh. M. Kogan, Dokl. Akad. Nauk SSSR 126:546 (1959) [Sov. Phys. – Doklady 4:604 (1959)].
69. P. Mazur, p. 167 of "Thermodynamics of Irreversible Processes," Rend. Scuola Int. Fisica "Enrico Fermi," 10 corso, Varenna, 1959, Soc. Ital. di Fisica, 1960.
70. H. B. Callen, R. H. Swendsen, and R. A. Tahir-Kheli, Phys. Lett. 25A:505 (1967).
71. P. Bocchieri and A. Loinger, Phys. Rev. 107:337 (1957).

71a. I. C. Percival, J. Math. Phys. 2:235 (1961).
72. M. Kac, Probability and Related Topics in Physical Sciences (Vol. 1 of "Lectures in Applied Mathematics"), Interscience, London, 1959.
73. N. N. Bogolyubov and O. S. Parasyuk, Dokl. Akad. Nauk SSSR 109:717 (1956).
74. M. A. Lavrent'ev and B. V. Shabat, Methods in the Theory of Functions of a Complex Variable, Moscow, 1958.
75. E. S. Fradkin, Zh. Eksp. Teor. Fiz. 36:1286 (1959) [Sov. Phys. – JETP 9:912 (1959)]; Dokl. Akad. Nauk SSSR, 125:66 (1959) [Sov. Phys. – Doklady 4:327 (1959)]; Nucl. Phys. 12:465 (1959).
76. L. D. Landau, Zh. Eksp. Teor. Fiz. 34:262 (1958) [Sov. Phys. – JETP 7:182 (1958)].
77. N. N. Bogolyubov, Quasi-Averages in Problems of Statistical Mechanics (p. 1 of Vol. 2 of "Lectures on Quantum Statistics," Gordon and Breach, New York, 1970).
78. H. Nyquist, Phys. Rev. 32:110 (1928).
79. H. B. Callen and R. F. Greene, Phys. Rev. 86:702 (1952).
80. R. F. Greene and H. B. Callen, Phys. Rev. 88:1387 (1952).
81. H. B. Callen, M. L. Barasch, and J. L. Jackson, Phys. Rev. 88:1382 (1952).
82. S. R. de Groot and P. Mazur, Nonequilibrium Thermodynamics, North-Holland, Amsterdam, 1962.
83. H. A. Kramers, Atti. Congr. Intern. Fisici Como 2:545 (1927).
84. R. de L. Kronig, J. Opt. Soc. Amer. 12:547 (1926).
85. L. Onsager, Phys. Rev. 37:405 (1931); 38:2265 (1931).
86. L. D. Landau and E. M. Lifshitz, Mechanics of Continuous Media, Moscow, 1953.
87. K. G. Denbigh, The Thermodynamics of the Steady State, Wiley, New York, 1951.
88. I. Prigogine, Introduction to the Thermodynamics of Irreversible Processes, 3rd ed. Interscience, New York, 1967.
89. R. Haase, Thermodynamics of Irreversible Processes, Addison-Wesley, Reading, Mass., 1969.
90. S. R. de Groot, Thermodynamics of Irreversible Processes, North-Holland, Amsterdam, 1951.
91. S. M. Rytov, Theory of Electrical Fluctuations and Thermal Radiation, Moscow, 1953.
92. S. M. Rytov, Usp. Fiz. Nauk 63:657 (1957) [Adv. Phys. Sci., Vol. 63, Part II, 891 (1957) (Israel Program for Scientific Translations, Jerusalem)].
93. M. L. Levin and S. M. Rytov, Theory of Equilibrium Thermal Fluctuations in Electrodynamics, Moscow, 1967.
94. F. V. Bunkin, Zh. Eksp. Teor. Fiz. 32:338, 811 (1957) [Sov. Phys. – JETP 5:277, 665 (1957)].
95. F. V. Bunkin, Zh. Eksp. Teor. Fiz. 44:1567 (1963) [Sov. Phys. – JETP 17:1054 (1963)].
96. M. Lax, Rev. Mod. Phys. 32:25 (1960).
97. M. Lax and P. Mengert, Phys. Chem. Sol. 14:248 (1960).
98. M. Lax, Rev. Mod. Phys. 38:359 (1966).

99. M. Lax, Phys. Rev. 145:110 (1966).
100. A. S. Davydov, Quantum Mechanics, Pergamon, Oxford, 1965.
101. L. D. Landau and E. M. Lifshitz, Statistical Physics, Pergamon, Oxford, 1969.
102. L. P. Kadanoff and P. C. Martin, Phys. Rev. 124:670 (1961).
103. J. M. Blatt and T. Matsubara, Progr. Theor. Phys. 21:696 (1959).
104. O. V. Konstantinov and V. I. Perel', Zh. Eksp. Teor. Fiz. 37:786 (1959) [Sov. Phys. – JETP 10:560 (1960)].
105. N. N. Bogolyubov, Lectures on Quantum Statistics, Gordon and Breach, New York, 1967.
106. L. D. Landau and E. M. Lifshitz, Quantum Mechanics, Pergamon, Oxford, 1965.
107. D. I. Blokhintsev, Quantum Mechanics, Reidel, Dordrecht, 1964.
108. R. Becker, Electromagnetic Fields and Interactions, Blackie, London, 1964.
109. P. P. Ewald, Ann. Phys. (Leipzig) 54:519, 557 (1917); 64:253 (1921); Nachr. Ges. Wiss. Goettingen Math. Phys. Kl. Fachgruppe 2, 3:55 (1938).
110. M. Born and K. Huang, Dynamical Theory of Crystal Lattices, Oxford, 1954.
111. V. M. Agranovich and V. L. Ginzburg, Spatial Dispersion in Crystal Optics and the Theory of Excitons, Interscience, London, 1966.
112. V. P. Silin and A. A. Rukhadze, Electromagnetic Properties of Plasma and Plasm-like Media, Moscow, 1961.
113. Yu. L. Klimontovich, The Statistical Theory of Nonequilibrium Processes in a Plasma, Pergamon, Oxford, 1967.
114. R. E. Peierls, Quantum Theory of Solids, Oxford, 1955.
115. J. Bardeen and J. R. Schrieffer, Recent Developments in Superconductivity, Progr. Low Temp. Phys. (ed. C. J. Gorter) Vol. 3, p. 170, North-Holland, Amsterdam, 1961.
116. N. N. Bogolyubov, Zh. Eksp. Teor. Fiz. 34:58 (1958) [Sov. Phys. – JETP 7:41 (1958)].
117. M. A. Leontovich, Zh. Eksp. Teor. Fiz. 40:907 (1961) [Sov. Phys. – JETP 13:634 (1961)].
118. M. Lax, Phys. Rev. 172:350 (1968).
119. S. J. Miyake and R. Kubo, Phys. Rev. Lett. 9:62 (1962).
120. L. V. Keldysh, Zh. Eksp. Teor. Fiz. 47:1515 (1964) [Sov. Phys. – JETP 20:1018 (1965)].
121. G. Murayama, Phys. Lett. 31A:537 (1970).
122. K. Tani, Progr. Theor. Phys. 32:167 (1964).
123. V. P. Kalashnikov, Teor. Mat. Fiz. 9:94, 406 (1971); 11:117 (1972) [Theor. Math. Phys. 9:1003 (1971); the others not yet translated].
124. Yu. A. Tserkovnikov, Dokl. Akad. Nauk SSSR 143:832 (1962) [Sov. Phys. – Doklady 7:322 (1962)]; Teor. Mat. Fiz. 4:119 (1970) [Theor. Math. Phys. 4:720 (1970)].
125. L. M. Roth, Phys. Rev. 186:428 (1969).
126. B. V. Thompson, Phys. Rev. 131:1420 (1964).
127. R. J. Elliott and D. W. Taylor, Proc. Phys. Soc. 83:189 (1964).
128. K. Nishikawa and R. Barrie, Can. J. Phys. 41:1135 (1963).
129. K. N. Pathak, Phys. Rev. 139:A1569 (1965).
130. K. D. Schotte and U. Schotte, Phys. Rev. 185:509 (1969).

131. J. Hubbard, Proc. Roy. Soc. A276:238 (1963); A281:401 (1964); Proc. Phys. Soc. 84:455 (1964).
132. J. F. Cornwell, Proc. Roy. Soc. A284:423 (1964).
133. D.-J. Kim, Phys. Rev. 146:455 (1966).
134. H. Ueyama and T. Matsubara, Progr. Theor. Phys. 38:784 (1967).
135. T. Morita, J. Phys. Soc. Japan 28:1128 (1970); T. Morita, T. Horiguchi, and S. Katsura, J. Phys. Soc. Japan 29:84 (1970); T. Morita, 29:850 (1970).
136. F. Leoni and C. R. Natoli, J. Phys. C 3:1462 (1970).
137. N. M. Plakida, Teor. Mat. Fiz. 5:147 (1970) [Theor. Math. Phys. 5:1047 (1970)].
138. K. W. H. Stevens and G. A. Toombs, Proc. Phys. Soc. 85:1307 (1965).
139. J. F. Fernandez and H. A. Gersch, Proc. Phys. Soc. 91:505 (1967).
140. H. B. Callen, R. H. Swendsen, and R. A. Tahir-Kheli, Phys. Lett. 25A:505 (1967).
141. G. L. Lucas and G. Horwitz, J. Phys. A2:503 (1969).
142. P. C. Kwok and T. D. Schultz, J. Phys. C 2:1196 (1969).
143. F. A. Kaempffer, Concepts in Quantum Mechanics, Academic, New York, 1965.
144. G. Källén, Elementary Particle Physics, Addison-Wesley, Reading, Mass., 1964.
145. W. E. Parry and R. E. Turner, Rep. Progr. Phys. 27:23 (1964).
146. M. Ichiyanagi, J. Phys. Soc. Japan 32:604 (1972).
147. M. Porsch, Phys. Stat. Sol. 39:477 (1970).
148. L. J. Sham, Phys. Rev. 139:A1189 (1965).
149. O. Litzman, Czech. J. Phys. 15:465 (1965).
150. R. A. Tahir-Kheli and D. ter Haar, Phys. Rev. 127:88, 95 (1962).
151. R. Kishore and S. K. Joshi, Phys. Rev. 186:484 (1969).
152. A. Oguchi, Progr. Theor. Phys. 43:257 (1970).
153. H. Mamada and F. Takano, Progr. Theor. Phys. 43:1458 (1970).
154. G. E. Uhlenbeck and L. S. Ornstein, Phys. Rev. 36:823 (1930).
155. N. M. Krylov and N. N. Bogolyubov, p. 5 of Vol. II of the "Selected Works in Three Volumes," Kiev, 1970.

Chapter IV

1. N. N. Bogolyubov, Problems of Dynamic Theory in Statistical Physics, p. 11 of Vol. 1 of "Studies in Statistical Mechanics," ed. J. de Boer and G. E. Uhlenbeck, North-Holland, Amsterdam, 1962.
2. D. N. Zubarev, Dokl. Akad. Nauk SSSR 140:92 (1961) [Sov. Phys. – Doklady 6:776 (1962)].
3. D. N. Zubarev, Dokl. Akad. Nauk SSSR 162:532 (1965) [Sov. Phys. – Doklady 10:452 (1965)].
4. D. N. Zubarev, Dokl. Akad. Nauk SSSR 162:794 (1965) [Sov. Phys. – Doklady 10:526 (1966)].
5. D. N. Zubarev, Dokl. Akad. Nauk SSSR 164:537 (1965) [Sov. Phys. – Doklady 10:850 (1966)].
6. H. B. Callen and T. A. Welton, Phys. Rev. 83:34 (1951).
7. R. Zwanzig, Ann. Rev. Phys. Chem. 16:67 (1965).
8. E. W. Montroll, p. 217 of "Thermodynamics of Irreversible Processes," Rend. Scuola Int. Fisica "Enrico Fermi," 10 Corso, Varenna, 1959, Soc. Ital. di Fisica, 1960.

9. J. M. Luttinger, Phys. Rev. 135:A1505 (1964).
10. L. P. Kadanoff and P. C. Martin, Ann. Phys. (New York) 24:419 (1963); P. C. Martin, p. 100 of "Statistical Mechanics of Equilibrium and Nonequilibrium," ed. J. Meixner, Proc. Int. Symp. on Statistical Mechanics and Thermodynamics, Aachen, 1964, North-Holland, Amsterdam, 1965.
11. J. L. Jackson and P. Mazur, Physica 30:2295 (1964).
12. B. U. Felderhof and I. Oppenheim, Physica 31:1441 (1965).
13. J. G. Kirkwood, J. Chem. Phys. 14:180 (1946); 15:72 (1947).
14. M. S. Green, J. Chem. Phys. 20:1281 (1952); 22:398 (1954).
15. N. Hashitsume, Progr. Theor. Phys. 8:461 (1952); 15:369 (1956).
16. M. Lax, Rev. Mod. Phys. 32:25 (1960); 38:359 (1966); Phys. Rev. 145:110 (1966); 172:350 (1968).
17. L. Onsager and S. Machlup, Phys. Rev. 91:1505 (1953); S. Machlup and L. Onsager, Phys. Rev. 91:1512 (1953).
18. U. Uhlhorn, Ark. Fysik 17:193, 233, 257, 273, 343, 361 (1960).
19. N. G. van Kampen, Physica 23:707, 816 (1957).
20. T. Yamamoto, Progr. Theor. Phys. 10:11 (1953).
21. R. Zwanzig, Phys. Rev. 124:983 (1961); Physica 30:1109 (1964).
22. H. Mori, Progr. Theor. Phys. 33:423 (1965).
23. R. Kubo, p. 3 of Tokyo Summer Lectures in Theoretical Physics, 1965, part 1, Many-Body Theory, Syokabo, Tokyo, and Benjamin, New York, 1966.
24. R. Kubo, Rep. Progr. Phys. 29:255 (1966).
25. E. Helfand, Phys. Rev. 119:1 (1960).
26. L. Onsager, Phys. Rev. 37:405 (1931); 38:2265 (1931).
27. S. R. de Groot and P. Mazur, Nonequilibrium Thermodynamics, North-Holland, Amsterdam, 1962.
28. R. Kubo, M. Yokota, and S. Nakajima, J. Phys. Soc. Japan 12:1203 (1957); S. Nakajima, Progr. Theor. Phys. 20:948 (1958).
29. H. Mori, J. Phys. Soc. Japan 11:1029 (1956).
30. H. Mori, Phys. Rev. 112:1829 (1958).
31. H. Mori, Phys. Rev. 115:298 (1959).
32. H. S. Green, J. Math. Phys. 2:344 (1961).
33. M. H. J. Ernst, Transport Coefficients from Time Correlation Functions, Doctoral Thesis, University of Amsterdam, 1964.
34. M. I. Klinger, Fiz. Tverd. Tela 1:1225 (1959) [Sov. Phys. – Solid State 1:1122 (1960)].
35. B. N. Provotorov, Zh. Eksp. Teor. Fiz. 41:1582 (1961); 42:882 (1962) [Sov. Phys. – JETP 14:1126 (1962); 15:611 (1962)].
36. S. V. Peletminskii and A. A. Yatsenko, Zh. Eksp. Teor. Fiz. 53:1327 (1967) [Sov. Phys. – JETP 26:773 (1968)]; Teor. Mat. Fiz. 3:287 (1970) [Theor. Math. Phys. 3:513 (1970)]; S. V. Peletminskii, Teor. Mat. Fiz. 6:123 (1971) [Theor. Math. Phys. 6:88 (1971)].
37. J. A. McLennan, Jr., Phys. Fluids 4:1319 (1961).
38. J. A. McLennan, Jr., Advan. Chem. Phys., 5:261 (1963).
39. J. A. McLennan, Jr., Phys. Rev. 115:1405 (1959).
40. J. A. McLennan, Jr., Phys. Fluids 3:493 (1960).

41. L. A. Pokrovskii, Dokl. Akad. Nauk SSSR 177:1054 (1967) [Sov. Phys. – Doklady 12:1135 (1968)].
42. T. N. Khazanovich and V. A. Savchenko, Phys. Lett. 27A:615 (1968).
43. L. L. Buishvili and D. N. Zubarev, Fiz. Tverd. Tela 7:722 (1965) [Sov. Phys. – Solid State 7:580 (1965)].
44. L. L. Buishvili, Fiz. Tverd. Tela 7:1871 (1965) [Sov. Phys. – Solid State 7:1505 (1965)].
45. G. R. Khutsishvili, Zh. Eksp. Teor. Fiz. 52:1579 (1967) [Sov. Phys. – JETP 25:1050 (1967)]; Progr. Low Temp. Phys. (ed. C. J. Gorter), Vol. 6, p. 375, North-Holland, Amsterdam, 1970.
46. L. L. Buishvili, Zh. Eksp. Teor. Fiz. 49:1868 (1968) [Sov. Phys. – JETP 22:1277 (1966)].
46a. L. L. Buishvili, Fiz. Tverd. Tela 9:2157 (1967) [Sov. Phys. – Solid State 9:1695 (1968)].
47. L. L. Buishvili and M. D. Zviadadze, ZhETF Pis. Red. 6:665 (1967) [JETP Lett. 6:153 (1967)].
47a. L. L. Buishvili and M. D. Zviadadze, Phys. Lett. 25A:86 (1967).
47b. L. L. Buishvili and M. D. Zviadadze, Phys. Lett. 24A:634 (1967).
47c. L. L. Buishvili and M. D. Zviadadze, Phys. Lett. 24A:661 (1967).
47d. L. L. Buishvili and M. D. Zviadadze, Fiz. Tverd. Tela 10:2397 (1968) [Sov. Phys. – Solid State 10:1885 (1968)].
47e. L. L. Buishvili and M. D. Zviadadze, Fiz. Tverd. Tela 10:2553 (1968) [Sov. Phys. – Solid State 10:2015 (1969)].
48. L. L. Buishvili, M. D. Zviadadze, and G. R. Khutsishvili, Zh. Eksp. Teor. Fiz. 54:876 (1968) [Sov. Phys. – JETP 27:469 (1968)].
49. N. S. Bendiashvili, L. L. Buishvili, and M. D. Zviadadze, Fiz. Tverd. Tela 8:2919 (1966) [Sov. Phys. – Solid State 8:2333 (1967)].
49a. N. S. Bendiashvili, L. L. Buishvili, and M. D. Zviadadze, Fiz. Tverd. Tela 10:1224 (1968) [Sov. Phys. – Solid State 10:971 (1968)].
50. L. L. Buishvili and N. P. Giorgadze, Fiz. Tverd. Tela 10:1181 (1968) [Sov. Phys. – Solid State 10:937 (1968)].
51. L. L. Buishvili and M. D. Zviadadze, Fiz. Tverd. Tela 9:1969 (1967) [Sov. Phys. – Solid State 9:1549 (1968)].
52. V. G. Grachev, Ukr. Fiz. Zh. 13:633 (1968) [Ukr. Phys. J. 13:445 (1968)].
53. V. P. Kalashnikov, Fiz. Metal Metalloved. 22:786 (1966) [Phys. Metals Metallogr. (USSR) 22(5):132 (1966)].
53a. V. P. Kalashnikov, Fiz. Tverd. Tela 9:634 (1967) [Sov. Phys. – Solid State 9:488 (1967)].
53b. V. P. Kalashnikov, Phys. Stat. Sol. 21:775 (1967).
53c. V. P. Kalashnikov, Fiz. Tekh. Poluprov. 1:1281 (1967) [Sov. Phys. – Semicond. 1:(1069) (1967)].
54. V. P. Kalashnikov, Phys. Lett. 26A:433 (1968).
55. L. A. Pokrovskii, Dokl. Akad. Nauk SSSR 182:317 (1968) [Sov. Phys. – Doklady 13:911 (1969)].
56. L. A. Pokrovskii, Dokl. Akad. Nauk SSSR 183:806 (1968) [Sov. Phys. – Doklady 13:1216 (1969)].

57. D. N. Zubarev and A. G. Bashkirov, Phys. Lett. 25A:202 (1967); Physica 39:334 (1968).
58. G. V. Chester, Rep. Progr. Phys. 26:411 (1963).
59. S. A. Rice and P. Gray, The Statistical Mechanics of Simple Liquids, Interscience, New York, 1965.
60. I. Prigogine and G. Severne, Phys. Lett. 6:177 (1963).
61. E. G. D. Cohen and M. H. J. J. Ernst, Phys. Lett. 5:192 (1963).
62. J. A. McLennan, Jr., Progr. Theor. Phys. 30:408 (1963).
63. I. Prigogine, P. Resibois, and G. Severne, Phys. Lett. 9:317 (1964).
64. E. G. D. Cohen, J. R. Dorfman, and M. H. J. J. Ernst, Phys. Lett. 12:319 (1964).
65. D. Bohm and D. Pines, Phys. Rev. 92:609 (1953).
66. N. N. Bogolyubov and D. N. Zubarev, Zh. Eksp. Teor. Fiz. 28:129 (1955) [Sov. Phys. – JETP 1:83 (1955)].
67. D. N. Zubarev, Zh. Eksp. Teor. Fiz. 25:548 (1953).
68. D. Bohm, General Theory of Collective Coordinates, p. 401 of "The Many-Body Problem," Summer School of Theoretical Physics, Les Houches, 1958, Wiley, New York, 1959.
69. J. A. McLennan, Physica 32:689 (1966).
70. M. A. Leontovich, Statistical Physics, Moscow, 1944.
71. E. T. Jaynes, Phys. Rev. 106:620 (1957); 108:171 (1957).
72. E. T. Jaynes, Information Theory and Statistical Mechanics, in "Statistical Physics," Brandeis Lectures Vol. 3, p. 181, Benjamin, New York, 1963.
73. H. Grad, Commun. Pure Appl. Math. 5:455 (1952).
74. C. F. Curtiss, J. Chem. Phys. 24:225 (1956).
75. A. Einstein, Ann. Phys.(Leipzig) 33:1275 (1910).
76. R. F. Greene and H. B. Callen, Phys. Rev. 83:1231 (1951).
77. L. Ornstein and F. Zernike, Proc. Kon. Ned. Akad. Wetensch. 17:793 (1914); F. Zernike, ibid. 18:1520 (1916); L. Ornstein, ibid. 19:1321 (1917).
78. M. J. Klein and L. Tisza, Phys. Rev. 76:1861 (1949).
79. A. Münster, p. 23 of "Thermodynamics of Irreversible Processes," Rend. Scuola Int. Fisica "Enrico Fermi," 10 corso, Varenna, 1959, Soc. Ital. di Fisica, 1960.
80. J. L. Lebowitz and J. K. Percus, J. Math. Phys. 4:116 (1963).
81. J. L. Lebowitz and J. K. Percus, J. Math. Phys. 4:248 (1963).
82. L. D. Landau and E. M. Lifshitz, Mechanics of Continuous Media, Moscow, 1953.
83. M. Kac, Probability and Related Topics in Physical Sciences (Vol. 1 of "Lectures in Applied Mathematics"), Interscience, London, 1959.
84. M. Gell-Mann and M. L. Goldberger, Phys. Rev. 91:398 (1953).
85. M. L. Goldberger and K. M. Watson, Collision Theory, Wiley, New York, 1964.
86. L. van Hove, Physica 21:517 (1955).
87. J. Lewins, Importance: The Adjoint Function, Pergamon, Oxford, 1965.
88. V. A. Ditkin and A. P. Prudnikov, Integral Transforms and Operational Calculus, Pergamon, Oxford, 1965.
89. R. Zwanzig and R. D. Mountain, J. Chem. Phys. 43:4464 (1965); R. D. Mountain and R. Zwanzig, J. Chem. Phys. 44:2777 (1966).
90. G. P. de Vault and J. A. McLennan, Phys. Rev. 137:A724 (1965).

91. L. Boltzmann, Lectures on Gas Theory, University of California, Berkeley, 1964.
92. L. D. Landau and E. M. Lifshitz, Statistical Physics, Pergamon, Oxford, 1969.
93. R. C. Tolman, Relativity, Thermodynamics and Cosmology, Oxford, 1934.
94. Ya. B. Zel'dovich and I. D. Novikov, Relativistic Astrophysics, Moscow, 1967.
95. H. Mori, Progr. Theor. Phys. 28:763 (1962).
96. S. Chapman and T. G. Cowling, Mathematical Theory of Nonuniform Gases (3rd ed.), Cambridge, 1970.
97. H. Mori, Phys. Rev. 111:694 (1958).
98. S. Fujita, J. Math. Phys. 3:359 (1962).
99. J. A. McLennan, Jr., and R. J. Swenson, J. Math. Phys. 4:1527 (1964).
100. J. O. Hirschfelder, C. F. Curtiss, and R. B. Bird, Molecular Theory of Gases and Liquids, Wiley, New York, 1954.
101. L. D. Landau, Zh. Eksp. Teor. Fiz. 7:203 (1937) [Phys. Z. Sowjetunion 10:154 (1936)].
102. V. I. Kogan, Plasma Physics and the Problem of Controlled Thermonuclear Reactions, Vol. 1, Moscow, 1958.
103. A. A. Dougal and L. Goldstein, Phys. Rev., 109:615 (1958).
104. Ya. B. Zel'dovich and Yu. P. Raizer, Physics of Shock Waves and High-Temperature Hydrodynamic Phenomena, Vols. 1 and 2, Academic, New York, 1966, 1967.
105. E. V. Stupochenko, S. A. Losev, and A. I. Osipov, Relaxation Processes in Shock Waves, Springer-Verlag, Berlin, 1967.
106. C. S. Wang-Chang and G. E. Uhlenbeck, Transport Phenomena in Polyatomic Molecules, Univ. Michigan, CM-681 (1951); C. S. Wang-Chang, G. E. Uhlenbeck, and J. de Boer, p. 241 in Vol. 2 of 'Studies in Statistical Mechanics," ed. J. de Boer and G. E. Uhlenbeck, North-Holland, Amsterdam, 1964.
107. M. N. Kogan, Rarefied Gas Dynamics, Plenum, New York, 1969.
108. A. Abragam, The Principles of Nuclear Magnetism, Oxford, 1961.
109. C. D. Jeffries, Dynamic Nuclear Orientation, Interscience, New York, 1963.
110. H. O. Kneser, Ann. Phys. (Leipzig) 16:337 (1933); Handbuch der Physik, Vol. XI/1, p. 129, Springer-Verlag, Berlin-Göttingen-Heidelberg, 1961; Nuovo Cim. 7(Suppl.2):231 (1950).
111. M. A. Leontovich, Zh. Eksp. Teor. Fiz. 6:561 (1936).
112. L. I. Mandel'shtam and M. A. Leontovich, Zh. Eksp. Teor. Fiz. 7:438 (1937).
113. I. G. Mikhailov, V. A. Solov'ev, and Yu. P. Syrnikov, Principles of Molecular Acoustics, Moscow, 1964.
114. R. F. Snider, J. Chem. Phys. 32:1051 (1960).
115. E. Goldman and L. Sirovich, Phys. Fluids 10:1928 (1967).
116. L. Sirovich, Phys. Fluids 5:908 (1962).
117. P. Glansdorff, Phys. Fluids 5:371 (1962).
118. L. D. Landau and E. M. Lifshitz, Mechanics of Continuous Media, Moscow, 1953.
119. N. N. Bogolyubov, Hydrodynamics of a Superfluid Liquid (p. 148 of Vol. 2 of "Lectures on Quantum Statistics," Gordon and Breach, New York, 1970).
120. I. M. Khalatnikov, An Introduction to the Theory of Superfluidity, Benjamin, New York, 1965.
121. P. C. Hohenberg and P. C. Martin, Ann. Phys. (New York) 34:291 (1965).

122. G. R. Khutsishvili, Usp. Fiz. Nauk 96:441 (1968) [Sov. Phys. – Uspekhi 11:802 (1969)].
123. E. M. Purcell and R. V. Pound, Phys. Rev. 81:279 (1951).
124. A. Abragam and W. G. Proctor, Phys. Rev. 106:160 (1957); 109:1441 (1958).
125. Y. Yafet, Solid State Physics (ed. F. Seitz and D. Turnbull), Vol. 14, p. 1, Academic, New York, 1963.
126. S. T. Pavlov, Fiz. Tverd. Tela 8:900 (1966) [Sov. Phys. – Solid State 8:719 (1966)].
127. L. D. Landau, Zh. Eksp. Teor. Fiz. 30:1058 (1956) [Sov. Phys. – JETP 3:920 (1957)].
128. D. Pines and P. Nozières, Theory of Quantum Liquids, Benjamin, New York, 1966.
129. T. Yamamoto, J. Chem. Phys. 33:281 (1960).
130. H. Aroeste, Adv. Chem. Phys. 6:1 (1964).
131. R. Haase, Thermodynamics of Irreversible Processes, Addison-Wesley, Reading, Mass., 1969.
132. D. A. Frank-Kamenetskii, Diffusion and Heat Exchange in Chemical Kinetics, Princeton University Press, Princeton, N. J., 1955.
133. C. Eckart, Phys. Rev. 58:919 (1940).
134. G. A. Kluitenberg, S. R. de Groot, and P. Mazur, Physica 19:689, 1079 (1953); G. A. Kluitenberg and S. R. de Groot, ibid. 20:199 (1954).
135. W. Pauli, Theory of Relativity, Pergamon, Oxford, 1958.
136. W. Heitler, The Quantum Theory of Radiation (3rd ed.), Oxford, 1954.
137. G. A. Milekhin, Tr. Fiz. Inst. Akad. Nauk SSSR 16:50 (1961).
138. M. Namiki and C. Iso, Progr. Theor. Phys. 18:591 (1957).
139. N. N. Bogolyubov and K. P. Gurov, Zh. Eksp. Teor. Fiz. 17:614 (1947).
140. K. P. Gurov, Fundamentals of Kinetic Theory, Moscow, 1966.
141. V. G. Bar'yakhtar, S. V. Peletminskii, and A. A. Yatsenko, ITPK (Institute for Theoretical Physics, Kiev) Preprint 68-4, 1968.
142. E. A. Uehling and G. E. Uhlenbeck, Phys. Rev. 43:552 (1933).
143. H. Mori and S. Ono, Progr. Theor. Phys. 8:327 (1952).
144. J. Ross and J. G. Kirkwood, J. Chem. Phys. 22:1094 (1954).
145. D. K. Hoffman, J. J. Mueller, and C. F. Curtiss, J. Chem. Phys. 43:2878 (1965).
146. S. Fujita, Introduction to Nonequilibrium Quantum Statistical Mechanics, Saunders, Philadelphia, 1966.
147. L. P. Kadanoff and G. Baym, Quantum Statistical Mechanics, Benjamin, New York, 1962.
148. L. V. Keldysh, Zh. Eksp. Teor. Fiz. 47:1515 (1964) [Sov. Phys. – JETP 20:1018 (1965)].
149. B. Bezzerides and D. F. du Bois, Phys. Rev. 168:233 (1968).
150. N. N. Bogolyubov, Zh. Eksp. Teor. Fiz. 18:622 (1948).
151. S. V. Peletminskii, V. D. Tsukanov, and A. A. Yatsenko, ITPK Preprint 68-51, 1968.
152. J. M. Ziman, Principles of the Theory of Solids, Cambridge, 1965.
153. J. M. Ziman, Electrons and Phonons, Oxford, 1960.
154. F. J. Blatt, Physics of Electronic Conduction in Solids, McGraw-Hill, New York, 1968.

155. J. R. Drabble and H. J. Goldsmid, Thermal Conduction in Semiconductors, Pergamon, Oxford, 1961.
156. R. E. Peierls, Quantum Theory of Solids, Oxford, 1955.
157. L. A. Pokrovskii, ITPK Preprint 68-78, 1968.
158. C. P. Slichter, Principles of the Theory of Magnetic Resonance, Harper and Row, New York, 1963.
159. S. Chandresekhar, Rev. Mod. Phys. 15:1 (1943).
160. J. L. Doob, Stochastic Processes, Wiley, New York, 1953.
161. H. Kramers, Physica 7:284 (1940).
162. A. G. Bashkirov and D. N. Zubarev, Teor. Mat. Fiz. 1:407 (1969) [Theor. Math. Phys. 1:311 (1969)].
163. Yu. L. Klimontovich, The Statistical Theory of Nonequilibrium Processes in Plasma, Pergamon, Oxford, 1967.
164. J. Meixner, Z. Phys. 149:624 (1957).
165. P. G. Bergmann and J. L. Lebowitz, Phys. Rev. 99:578 (1955).
166. J. L. Lebowitz and P. G. Bergmann, Ann. Phys. (New York) 1:1 (1957).
167. M. C. Wang and G. E. Uhlenbeck, Rev. Mod. Phys. 17:323 (1945).
168. P. Mazur, Physica 25:149 (1959).
169. N. M. Krylov and N. N. Bogolyubov, p. 5 of Vol. II of the "Selected Works in Three Volumes," Kiev, 1970.
169a. N. N. Bogolyubov, On Some Statistical Methods in Mathematical Physics, Kiev, 1945.
170. D. N. Zubarev and V. P. Kalashnikov, Theor. Mat. Fiz. 1:137 (1969) [Theor. Math. Phys. 1:108 (1969)].
171. N. N. Bogolyubov, Physica 26:S1 (1960).
172. N. N. Bogolyubov, Quasi-Averages in Problems of Statistical Mechanics (p. 1 of Vol. 2 of "Lectures on Quantum Statistics," Gordon and Breach, New York, 1970).
173. P. Glansdorff and I. Prigogine, Physica 20:773 (1954).
174. "Nonequilibrium Thermodynamics, Variational Techniques, and Stability," ed. R. J. Donnelly, R. Herman, and I. Prigogine, Proc. Symp. Univ. Chicago, 1965, University of Chicago Press, 1966.
175. I. Prigogine, Introduction to the Thermodynamics of Irreversible Processes, Interscience, New York, 1967.
176. R. Balescu, Phys. Fluids 3:52 (1960); 4:94 (1961); Physica 27:693 (1961).
177. R. Balescu, Statistical Mechanics of Charged Particles, Interscience, New York, 1963.
178. I. Prigogine, Nonequilibrium Statistical Mechanics, Interscience, New York, 1962.
179. A. Lenard, Ann. Phys. (New York) 10:390 (1960).
180. Yu. L. Klimontovich, Zh. Eksp. Teor. Fiz. 52:1233 (1967); 54:136 (1968) [Sov. Phys. – JETP 25:820 (1967); 27:75 (1968)].
181. V. P. Silin, Zh. Eksp. Teor. Fiz. 33:495 (1957); 35:1243 (1958); 34:707 (1958) [Sov. Phys. – JETP 6:387 (1958); 8:870 (1959); 7:486 (1958)].
182. V. P. Silin and A. A. Rukhadze, Electromagnetic Properties of Plasma and Plasma-like Media, Moscow, 1961.
183. T. N. Khazanovich, Mol. Phys. 17:281 (1969); Phys. Lett. 29A:601 (1969).

184. D. N. Zubarev, Fortsch. Phys. 18:125 (1970).
185. D. N. Zubarev, The Method of the Nonequilibrium Statistical Operator, p. 329 of "Problems of Theoretical Physics," Moscow, 1969.
186. D. N. Zubarev, Fortsch. Phys. (in press).
187. D. N. Zubarev, Teor. Mat. Fiz. 3:276 (1970) [Theor. Math. Phys. 3:505 (1970)].
188. D. N. Zubarev, p. 1 of Lectures, VIth Annual Winter School of Theoretical Physics, Karpacz, 1969, Wroclaw, 1969.
189. D. N. Zubarev and V. P. Kalashnikov, Physica 46:550 (1970).
190. A. G. Bashkirov and D. N. Zubarev, Physica 48:137 (1970).
191. K.-H. Müller, Teor. Mat. Fiz. 5:453 (1970) [Theor. Math. Phys. 5:1276 (1970)]. Phys. Lett. 32A:179 (1970).
192. G. Röpke, Teor. Mat. Fiz. 6:294 (1971) [Theor. Math. Phys. 6:216 (1971)]; Phys. Lett. 32A:252 (1970).
193. L. L. Buishvili, M. D. Zviadadze, and G. R. Khutsishvili, Zh. Eksp. Teor. Fiz. 56:290 (1969) [Sov. Phys. – JETP 29:159 (1969)].
193a. L. L. Buishvili and M. D. Zviadadze, Dokl. Akad. Nauk SSSR 191:58 (1970) [Sov. Phys.–Doklady 15:241 (1970)]; Fiz. Metal. Metalloved. 30:680 (1970) [Phys. Metals Metallogr. (USSR) 30(4):8 (1970)].
194. L. L. Buishvili and N. S. Bendiashvili, in "Relaxation Phenomena in Solids," Moscow, 1968.
195. N. S. Bendiashvili, L. L. Buishvili, and G. R. Khutsishvili, Zh. Eksp. Teor. Fiz. 57:1231 (1969) [Sov. Phys. – JETP 30:671 (1970)].
196. L. L. Buishvili and N. P. Giorgadze, Dokl. Akad. Nauk SSSR 189:508 (1969) [Sov. Phys. – Doklady 14:1104 (1970)]; Fiz. Tverd. Tela 12:1817 (1970) [Sov. Phys. – Solid State 12:1439 (1970)]; L. L. Buishvili, Izv. Vyssh. Ucheb. Zaved. Radiofiz. 14:1364 (1971) [to be translated in Radiophys. Quant. Electron.].
197. L. L. Buishvili and G. A. Volgina, Vyssh. Ucheb. Zaved. Radiofiz. 12:1805 (1969) [to be translated in Radiophys. Quant. Electron.].
198. L. L. Buishvili and G. A. Volgina, Radiospektroskopiya (Perm') No. 6, p. 243 (1969).
199. L. L. Buishvili and M. D. Zviadadze, Fiz. Tverd. Tela 12:900 (1970) [Sov. Phys. – Solid State 12:694 (1970)].
200. N. S. Bendiashvili, L. L. Buishvili, and M. D. Zviadadze, Fiz. Tverd. Tela 11:726 (1969) [Sov. Phys. – Solid State 11:579 (1969)].
201. V. P. Kalashnikov, Theory of Spin-Lattice Relaxation of Conduction Electrons in a Strong Magnetic Field, in "Problems in the Magnetism and Strength of Solids," No. 27, IFM AN SSSR, Sverdlovsk, 1968.
202. V. P. Kalashnikov, Dokl. Akad. Nauk SSSR 186:803 (1969) [Sov. Phys. – Doklady 14:542 (1969)].
203. V. P. Kalashnikov, Theory of Nuclear-Spin Polarization by a Constant Current in Semiconductors (Feher Effect), I. Effective-Parameters Approximation, JINRD (Joint Institute for Nuclear Research, Dubna) Report P4-810 (1969); II. The Fokker–Planck Approximation, JINRD Report P4-811 (1969).
204. A. G. Bashkirov and D. N. Zubarev, Physica 48:137 (1970).
205. A. G. Bashkirov, Teor. Mat. Fiz. 3:265 (1970) [Theor. Math. Phys. 3:497 (1970)].
206. V. P. Kalashnikov, Teor. Mat. Fiz. 5:293 (1970); 6:279 (1971) [Theor. Math. Phys. 5:1159 (1970); 6:206 (1971)]; Physica 48:93 (1970).

207. A. D. Khon'kin, Dokl. Akad. Nauk SSSR, 183:1285 (1968) [Sov. Phys. – Doklady 13:1236 (1969)].
208. T. N. Khazanovich, Mekh. Polim. 6:980 (1969) [to be translated in Polym. Mech. (USSR)].
209. L. A. Pokrovskii, Teor. Mat. Fiz. 2:103 (1970); 3:143 (1970) [Theor. Math. Phys. 2:78 (1970); 3:408 (1970)].
210. K. Walasek, D. N. Zubarev, and A. L. Kuzemskii, Teor. Mat. Fiz. 5:281 (1970) [Theor. Math. Phys. 5:1150 (1970)].
211. K. Walasek and A. L. Kuzemskii, Teor. Mat. Fiz. 4:267 (1970) [Theor. Math. Phys. 4:826 (1970)].
212. L. L. Buishvili, Theoretical Investigation of Nuclear Magnetic Resonance and Dynamic Polarization of Nuclei, Doctoral Thesis, Tbilisi, 1966.
213. M. D. Zviadadze, Problems of the Quantum-Statistical Theory of Magnetic Resonance in Solids, Candidate's Thesis, Tbilisi, 1969.
214. L. A. Pokrovskii, Application of the Method of the Nonequilibrium Statistical Operator to Systems with Internal Degrees of Freedom, Candidate's Thesis, Moscow, 1969.
215. E. M. Iolin, Fiz. Tverd. Tela 12:1159 (1970) [Sov. Phys. – Solid State 12:905 (1970)].
216. M. J. Ernst, L. K. Haines, and J. R. Dorfman, Rev. Mod. Phys. 41:296 (1969).
217. A. Kugler, Z. Phys. 198:236 (1967).
218. R. A. Piccirelli, Phys. Rev. 175:77 (1968).
219. B. Robertson, Phys. Rev. 144:151 (1966); 160:175 (1967); J. Math. Phys. 11:2482 (1970).
220. B. L. Clarke and S. A. Rice, Phys. Fluids 13:271 (1970).
221. S. Grossmann, Z. Phys. 233:74 (1970).
221a. V. A. Savchenko and T. N. Khazanovich, Teor. Mat. Fiz. 4:246 (1970) [Theor. Math. Phys. 4:812 (1970)].
222. Chang-Hyun Chung and S. Yip, Phys. Rev. 182:323 (1969).
223. J. W. Dufty and J. A. MacLennan, Phys. Rev. 172:176 (1968).
224. J. W. Dufty, Phys. Rev. 176:398 (1968).
225. R. D. Puff and N. S. Gillis, Ann. Phys. (New York) 46:364 (1968).
226. P. C. Martin, Measurements and Correlation Functions, p. 37 of "Many-Body Physics," ed. C. de Witt and R. Balian, Summer School in Theoretical Physics, Les Houches, 1967, Gordon and Breach, New York, 1968.
227. P. Glansdorff and I. Prigogine, Physica 46:344 (1970).
228. J. S. Dahler and D. K. Hoffmann, Theory of Transport and Relaxation Processes in Polyatomic Fluids, p. 1 in Vol. 3 of "Transfer and Storage of Energy by Molecules," ed. G. M. Burnett and A. M. North, Interscience, New York, 1970.
229. D. N. Zubarev and V. P. Kalashnikov, Teor. Mat. Fiz. 5:406 (1970) [Theor. Math. Phys. 5:1242 (1970)].
230. D. N. Zubarev and V. P. Kalashnikov, Teor. Mat. Fiz. 7:372 (1971).[Theor. Math. Phys. 7:600 (1971)].
231. D. N. Zubarev and V. P. Kalashnikov, Physica 56:345 (1970); Phys. Lett. 34A:311 (1971).
232. V. P. Kalashnikov, Teor. Mat. Fiz. 9:94, 406 (1971); 11:117 (1972) [Theor. Math. Phys. 9:1003 (1971); others not yet translated].

233. D. N. Zubarev and S. V. Tishchenko, Phys. Lett. 33A:444 (1970); Physica 59:285 (1972); Chisl. Metod. Mekh. Splosh. Sred. 2:114 (1971).
234. D. N. Zubarev and M. Yu. Novikov, Phys. Lett. 36A:343 (1971); Teor. Mat. Fiz. (in press).
235. L. A. Pokrovskii and M. V. Sergeev, Physica (in press).
236. F. M. Kuni and B. A. Storonkin, Teor. Mat. Fiz. 9:124 (1971) [Theor. Math. Phys. 9:1024 (1971)].
237. M. E. Fisher, J. Math. Phys. 5:944 (1964).
238. M. Fixman, Advan. Chem. Phys. 6:175 (1965).
239. N. Wiener, The Fourier Integral and Certain of Its Applications, Cambridge, 1933.
240. A. D. Khon'kin, Dokl. Akad. Nauk SSSR 190:563 (1970) [Sov. Phys. – Doklady 15:55 (1970)]; Teor. Mat. Fiz. 4:253 (1970) [Theor. Math. Phys. 4:817 (1970)].
241. D. N. Zubarev and A. D. Khon'kin, Teor. Mat. Fiz. 11:No. 3 (1972) to be translated).
242. Kh. M. Bikkin and V. P. Kalashnikov, Teor. Mat. Fiz. 7:79 (1971) [Theor. Math. Fiz. 7:380 (1971)].
243. V. P. Kalashnikov and D. N. Zubarev, Physica 59:314 (1972).
244. L. D. Abuladze and L. L. Buishvili, Fiz. Tekh. Poluprov. 5:1087 (1971) [Sov. Phys. – Semicond. 5:959 (1971)].
245. L. D. Abuladze, L. L. Buishvili, and V. P. Kalashnikov, Fiz. Tverd. Tela 13:1981 (1971) [Sov. Phys. – Solid State 13:1662 (1972)].
246. R. D. Puff and N. S. Gillis, Ann. Phys. (New York) 46:364 (1968).
247. H. Böttger, Phys. Stat. Sol. 35:653 (1969); 40:309 (1970).
248. S. V. Peletminskii and V. D. Tsukanov, Teor. Mat. Fiz. 7:395 (1971) [Theor. Math. Phys. 7:617 (1971)].
249. F. C. Andrews, J. Chem. Phys., 47:3161 (1967).
250. E. D. Shaw, Phys. Rev. B2:2746 (1970).
251. M. H. Ernst, L. K. Haines, and J. R. Dorfman, Rev. Mod. Phys. 41:296 (1969).
252. G. E. Uhlenbeck and L. S. Ornstein, Phys. Rev. 36:823 (1930).

Appendix I

1. S. S. Schweber, H. A. Bethe, and F. de Hoffman, Mesons and Fields, Vol. 1, Row and Peterson, Evanston, Ill., 1955.
2. S. S. Schweber, An Introduction to Relativistic Quantum Field Theory, Harper and Row, New York, 1961.
3. A. S. Davydov, Quantum Mechanics, Pergamon, Oxford, 1965.
4. M. Gell-Mann and M. L. Goldberger, Phys. Rev. 91:398 (1953).
5. M. L. Goldberger and K. M. Watson, Collision Theory, Wiley, New York, 1964.
6. D. N. Zubarev, Teor. Mat. Fiz. 3:276 (1970) [Theor. Math. Phys. 3:505 (1970)].

Appendix II

1. J. A. MacLennan, Jr., Phys. Fluids 4:1319 (1961).
2. J. A. MacLennan, Jr., Advan. Chem. Phys. 5:261 (1963).

3. D. N. Zubarev, Dokl. Akad. Nauk SSSR 140:92 (1961) [Sov. Phys. – Doklady 6:776 (1962)].
4. D. N. Zubarev, Dokl. Akad. Nauk SSSR 162:532 (1965) [Sov. Phys. – Doklady 10:452 (1965)].
5. D. N. Zubarev, Dokl. Akad. Nauk SSSR 162:794 (1965) [Sov. Phys. – Doklady 10:526 (1966)].
6. D. N. Zubarev, Dokl. Akad. Nauk SSSR 164:537 (1965) [Sov. Phys. – Doklady 10:850 (1966)].
7. N. N. Bogolyubov, Problems of Dynamic Theory in Statistical Physics, p. 11 of Vol. 1 of "Studies in Statistical Mechanics," ed. J. de Boer and G. E. Uhlenbeck, North-Holland, Amsterdam, 1962.
8. J. A. MacLennan, Jr., Phys. Rev. 115:1405 (1959); Phys. Fluids 3:493 (1960).

Appendix III

1. D. N. Zubarev, Teor. Mat. Fiz. 3:276 (1970) [Theor. Math. Phys. 3:505 (1970)].
2. D. N. Zubarev, Dokl. Akad. Nauk SSSR 140:92 (1961) [Sov. Phys. – Doklady 6:776 (1962)].
3. D. N. Zubarev and V. P. Kalashnikov, Teor. Mat. Fiz. 3:126 (1970) [Theor. Math. Phys. 3:395 (1970)].
4. N. N. Bogolyubov, Quasi-Averages in Problems of Statistical Mechanics (p. 1 of Vol. 2 of "Lectures on Quantum Statistics," Gordon and Breach, New York, 1970).
5. N. N. Bogolyubov, Physica 26:51 (1960).
6. D. N. Zubarev, Dokl. Akad. Nauk SSSR 162:532 (1965); 164:537 (1965) [Sov. Phys. – Doklady 10:452 (1965); 10:850 (1966)].
7. D. N. Zubarev and V. P. Kalashnikov, Teor. Mat. Fiz. 1:137 (1969) [Theor. Math. Phys. 1:108 (1969)].
8. J. A. MacLennan, Phys. Fluids 4:1319 (1961); Advan. Chem. Phys. 5:261 (1963).
9. D. N. Zubarev, Fortsch. Phys. (in press).
10. D. N. Zubarev and V. P. Kalashnikov, Teor. Mat. Fiz. 7:372 (1971) [Theor. Math. Phys. 7:600 (1971)].
11. D. N. Zubarev and V. P. Kalashnikov, Physica 56:345 (1970); Phys. Lett. 34A:311 (1971).